CO_2 SEQUESTRATION AND VALORIZATION

Edited by **Cláudia do Rosario Vaz Morgado and Victor Paulo Peçanha Esteves**

CO2 Sequestration and Valorization
http://dx.doi.org/10.5772/57034
Edited by Cláudia do Rosario Vaz Morgado and Victor Paulo Peçanha Esteves

Contributors

Margarita J. Ramírez-Moreno, Issis C. Romero-Ibarra, José Ortiz-Landeros, Heriberto Pfeiffer, Bernardo Llamas, Mazadiego Luis Felipe, Robert Jandl, Huilong Lin, Valny Giacomelli Sobrinho, Aaron Chow, Aaron C Chow, Ademir De Oliveira Ferreira, Telmo Jorge Carneiro Amado, Jill M. M. Motschenbacher, Kristofor Brye, Merle Anders, Edward Gbur, Nathan A. Slaton, Michelle A. Evans-White, Romain Privat, Jean-Noel Jaubert, Jesus Jaime Guerra Santos, Julia Griselda Ceron Breton, Rosa Maria Ceron, Atl Victor Cordova Quiroz, Ramesh Agarwal, Claudia Ravazzoli, Jim Kiniry, Norman Meki, Rita Maria De Brito Alves, Jose Luiz De Medeiros, Ofelia Araujo

Published by InTech

Janeza Trdine 9, 51000 Rijeka, Croatia

Publishing Process Manager Iva Lipovic

Technical Editor InTech DTP team

Cover InTech Design team

First published March, 2014

Printed in Croatia

Additional hard copies can be obtained from orders@intechopen.com

CO2 Sequestration and Valorization, Edited by Cláudia do Rosario Vaz Morgado and Victor Paulo Peçanha Esteves

p. cm.

ISBN 978-953-51-1225-9

For EU product safety concerns:
IN TECH d.o.o., Prolaz Marije Krucifikse Kozulić 3, 51000 Rijeka, Croatia,
info@intechopen.com or visit our website at intechopen.com.

Meet the editors

Cláudia do Rosario Vaz Morgado has been associate professor at the Polytechnic School of Rio de Janeiro Federal University (UFRJ) since 1997. She received a D.Sc. in Industrial Engineering (1994) from COPPE/UFRJ. She has been the head of the human resources program in Environmental Engineering of the National Petroleum Agency since 2010. She was the founder of the Program in Environmental Engineering of the Polytechnic School and Chemistry School at UFRJ (2007-2011). She was the former head of the Environmental Committee of the Regional Council of Architecture, Engineering and Agronomy of Rio de Janeiro/CREA-RJ. Prof. Morgado also founded GESTORE, a leading group that acts in training and research in life cycle assessment, governance and sustainability.

Victor Paulo Peçanha Esteves is assistant professor of Information Technology and Environment with the Professional Master's Program of UFRJ. He has been a lecturer and senior researcher in Department of Electronic Engineering at UFRJ since 1987. He holds a B.Sc. in Electronic Engineering (1987), Executive MBA in Environmental Management (2008) and M.Sc. in Environmental Engineering (2011), all from UFRJ. He is also a civilian speaker and instructor for the Brazilian Navy. He has carried out research projects for Petrobras and other companies and public sector. His recent research interests include a georeferenced and ecological approach for carbon sequestration and biofuels.

Contents

Preface

It is well established that the principal causes of the increasing concentration of CO2 in the atmosphere are human activities, mainly the burning of fossil fuels for transportation and electricity generation.

Notwithstanding the huge efforts to obtain a cleaner energy mix, human society has far to go to achieve the oft-expressed goal of sustainable development. The reconciliation of economic development, social justice and reduction of greenhouse gas emissions is one of the main political challenges of the moment.

Strategies for large-scale mitigation of CO2 emissions using technologies for sequestration, storage and utilization of carbon, along with increasing the carbon stock in plant cover (natural or not), are priorities on the agenda of research centers and governments around the world.

Readers will find in this book fifteen chapters that reflect on, propose, study and contribute to new technologies, strategies and policies for sustainability of the planet.

The four classes of carbon sequestration presented are: natural sequestration, which uses mechanisms to maintain carbon stocks in natural plant cover; biological sequestration, which employs management techniques in agriculture and silviculture; sequestration by storage in the ocean or in geological sites by capture and injection in rock formations; and sequestration by reuse of carbon on a large scale – chemical and/or biochemical sequestration – where the effort is to achieve a balance between inputs and products in the various productive chains that use carbon as a raw material.

The first three chapters cover new strategies and technological solutions proposed for sustainability in relation to atmospheric emissions. Reducing emissions from deforestation and forest degradation (REDD) combined with payment for environmental services, use of CO2 as a raw material for chemical products and fuels within a perspective for regional synergies, and the feasibility and public acceptance of carbon sequestration in the oceans are the themes addressed.

The next six chapters present studies, simulations and proposals for management models to maintain and monitor the carbon stocks in the soil, forests and cultivated areas.

The last six chapters examine the systems and subsystems for carbon capture and geological storage (CCGS) in the phases of site characterization and selection, capture, separation and injection of CO2 and various models for thermodynamic, multicriteria and seismic simulation.

The editors would like to extend special thanks to Professor Ofélia Araújo of Rio de Janeiro Federal University and the coordinators of the PROCLIMA and PRO-CO2 programs at the Petrobras Research Center (CENPES), especially the engineers Paulo Cunha, Paulo Negrais and Wilson Grava, for their encouragement, stimulus and support for our research efforts into carbon sequestration.

The editors kindly acknowledge financial support from the Civil Construction Department of the Federal University Rio de Janeiro.

We also thank InTech Publisher for the invitation to edit this book on such an important theme to the welfare of humanity, today and in the future.

Finally, we acknowledge the contribution of the chapters' authors by providing a better understanding of the latest research in this important theme.

Research into carbon sequestration is a key to solving one of the greatest economic, environmental and social problems of this century – a complex question that requires a multidisciplinary approach involving science and technology along with collaborative policies among nations. This challenge makes this book an important source of information for researchers, policymakers and anyone with an inquiring mind on this subject.

Cláudia do R. Vaz Morgado and Victor P. P. Esteves
Polytechnic School
Rio de Janeiro Federal University

Chapter 1

REDD Roses for a Green Lady – Target Setting for Deforestation in the Brazilian Amazon

Valny Giacomelli Sobrinho

Additional information is available at the end of the chapter

http://dx.doi.org/10.5772/57288

1. Introduction

In the lyrics written by Sid Tepper and Roy C. Bennett of a popular 1948 song, a broken-hearted guy, who had the day before argued with his girlfriend, rushed to the florist to buy some "red roses for a blue lady". His hope was that those pretty flowers could chase her blues away. In short, he wished some red flowers could compensate for the damage he had caused to his lover's heart.

Likewise, climate policy has recently[1] come up with the REDD mechanism (Reduced Emissions from Deforestation and forest Degradation) to protect natural standing forests before they are damaged by deforestation or degradation. Within the REDD framework, carbon credits can be earned for deforestation avoidance, rather than as in forestry CDM (Clean Development Mechanism), for afforestation or reforestation.

Since deforestation gives off nearly one-fifth (1.6 Gt) of global carbon emissions [2], avoiding it is claimed to be the most effective and cheapest way of control. However, until recently deforestation avoidance had been kept out of international climate accords, mainly because of

1 The first time REDD came into the UNFCCC (United Nations Framework Convention on Climate Change) agenda was in 2005, when a mechanism for reducing deforestation and forest degradation was proposed by Papua New Guinea, Costa Rica and eight other countries. Two years earlier, Brazilian researchers had already come up with a similar tool backed by international payments for reduced deforestation. Between the Conference of the Parties (COP) in Bali, Indonesia, in 2007 (COP-13), and in Copenhagen, Denmark, in 2009 (COP-15), different REDD proposals emerged. As the scope of the mechanism was getting wider, its abbreviation was getting longer with the addition of "D's" and plus (+) signs. Chronologically, it started with RED — short for Reducing Emissions from Deforestation; next, it became REDD — when forest Degradation was added; then, it turned out to be REDD+ — so as to encompass conservation and enhancement of forest stocks through Sustainable Forest Management (SFM) techniques; at present, it has often been labelled REDD++, including reforestation and afforestation — that is, reforestation of non-forested areas [1]. Throughout this chapter, simply REDD will be employed, regardless of its several chronological meanings. However, afforestation and reforestation will be considered to fall into forestry-CDM — the branch of the Clean Development Mechanism suited to unnatural forests.

the fear that credits supplied in return could flood carbon markets, thereby lowering the price of carbon credits [2]. But, with the closing, in 2012, of the first commitment period of the Kyoto Protocol (2008-2012), avoiding deforestation has been taken up as a post-2012 proposal for an international agreement that might bring in developing (non-Annex I) countries also to comply with emission targets.

Although the forest conservation feature of REDD means a step forward regarding carbon offsets from either energy or forestry-CDM [3], the credits arising from avoiding deforestation sound like roses offered as rewards for *not* causing damages to any "green lady" (forest). However, public incentive systems seldom pay people for not doing something — in the case of REDD, for not deforesting or not degrading. Hence, since a reward means a payment due for not doing something, it might encourage people to do what is already forbidden. In addition, it could make room for compensation seekers, whose anti-social and imprudent behaviour would otherwise have been inhibited by moral or legal censure [4].

Anyway, such a reward largely rests upon the creation of a market for REDD credits, in which Payments for Ecosystem Services (PES) are made to landowners and users, to compensate them for keeping forests — which are worth more alive than dead. PES schemes are defined as voluntary, conditional transactions — the payment is only made if the service is delivered — with at least one seller, one buyer, and a well-defined environmental service [2]. However, none of these conditions are yet in place in most REDD countries: the commodity is hard to quantify, the sellers are not well defined, big buyers do not exist and the rules of the game are not well established [5].

Besides these practical hindrances, REDD architecture is not without ideological critiques. Polanyi's [6] followers have charged not only REDD with having a privatising and marketing nature, but also PES with allowing elites and large-scale commercial actors to profit from what is called "forest capitalism". The gist of the criticism is that the largest share of deforestation is carried out by larger and wealthier players than by the poor, who REDD notwithstanding claims to be the greatest beneficiaries of the payments for avoiding deforestation [5]. Thus, of course, the bulk of these reward payments is expected to go to the players whose monetary foregone benefits are the greatest, when deforestation is halted.

Another pitfall of REDD and PES is that they rely too much on financial support. Although forests are assigned a high value in public debate, the relatively low returns accruing to forest management are due to the many and diverse non-financial benefits that forests can provide [7]. Both global REDD regimes and national REDD strategies call for the provision of financial resources — in the form of compensation or incentive payments — to alter the political economy of commercial access to forest resources [2]. Yet, REDD finance hinges on political will, institutional settings and long bureaucratic chains to work properly [2], [5].

At the national level, REDD policies demand the removal of support and subsidies for companies or activities that increase the pressure towards forest conversion — agricultural expansion, destructive logging, settlement schemes, plantation and road development in forest areas [2]. At the international level, a large proportion of finance will need to come from developed countries. Therefore, the scale of such finance will depend on the political will within these countries [5]. Nevertheless, developed countries are not likely to transfer payments to fragile states, where long-term efforts would be required to create or reform institutions, strengthen governance processes and build capacity to deal with the new models of forest management underlying REDD policies [2].

In any event, disturbances of this sort rest not only upon the promise to serve different actors and interests, but also upon that to bridge the environment and development agendas. While this proposal sounds so appealing and distinct from past efforts in the forestry sector, it not only has turned REDD into a successful idea, but also made it move from single (carbon) to multiple objectives. Such a move, though, is now threatening and overshadowing the main characteristic of REDD, which comes down to large-scale funding and performance-based support. So far, the vast majority of both developed and developing countries lack concrete strategies on how to implement REDD. Therefore, REDD finance remains unresolved, because the cost of reducing emissions from deforestation also depends on the strategy chosen [5].

Arguably, new strategies must emphasise carbon-stock protection [2]. Many studies have indicated that, after wood production, carbon sequestration is the most valuable output from forests [7]. However, a conservation market (REDD credits) for forest protection should draw on a form of compensation for producing something additional (new carbon stored), rather than on a reward for "not doing something" (not deforesting or not degrading) [4]. To begin with, this is supposed to help the political economy of REDD build the argument for its public support [4], thus allaying the mistrust that lurks over it of rewarding the wealthiest forest users.

A further step towards that shift lies in emphasising a *stock maintenance* rather than an *emissions avoidance* approach. This turns carbon conservation strategies from output (performance) into input-driven ones. After all, forest sequestration of carbon emissions is primarily a matter of forestland availability rather than of emissions avoided. If forest stocks are maintained or even increased, they do not emit carbon, which is kept there. Although this might sound like a "two-sides-of-the-same-coin" problem, the stock maintenance approach highlights the positive side of conservation — the stewardship of carbon stocks — whereas the emissions avoidance approach stresses its negative side — the discouragement or closure of activities causing emissions to rise in spite of delivering economic benefits. Quite often, the latter is seen as an unproductive strategy while the former might well evoke the production of some useful commodity.

The commodity at stake is carbon storage rather than money. Yet, REDD finance claims that economic and monetary incentives can, through price signals, alter the decisions of individual land users and compensate them for foregone benefits from not converting or degrading the forest [2]. Since carbon storage is increasingly needed, the demand for carbon credits is expected to go up, thereby generating finance for forest conservation [5]. At present, however, there is deep uncertainty as to whether and how a future international climate agreement would value carbon sequestration provided by forests. In addition, when standing forests compete with high-value agricultural and mining commodities, no one can ensure anymore whether and how REDD funding — particularly output, results-based finance — would be available in the future. So far, from current REDD finance, it is widely recognised that more money alone cannot solve the deforestation problem and that the expectations of more money can even make it worse [8].

As argued elsewhere [3], [9], [10], since the ascent of money in modern societies, a community's wealth now has two components: real goods, accumulated through real investments, and fiat or paper "goods", made of nothing. Unlike the former, the latter kind of wealth is not subject to the natural law of decrease, the entropy law. Whereas the accumulation of real goods, which hold use value, does meet physical limits, there is no limit at all to the accumulation of "virtual wealth" [10], used for exchange only [9]. So, in this world, where the substance is exchanged for the shadow, what determines the value of money is the amount of wealth people prefer to do without, and that is the same as the amount of credit they retain as money [10]. In monetary economies, the greater this "virtual wealth" is, the higher the price of the real goods it can afford; conversely, the smaller the former, the lower the latter[2]. It is puzzling how the protection of environmental goods and services should be commanded by such a virtual wealth rationale [3], [9].

In this regard, a non-monetary Bio-Economic model for carbon Sequestration by Forests (BESF) [3] is applied to a deforestation scenario taking place in the Brazilian "Legal Amazon" region (*Amazônia Legal*, Figure 1), to show how forest stocks can be prevented from being degraded. After all, if these stocks wither away, they will no longer be able to store carbon emissions. The model assumes that the growth of natural stocks follows an upper limited path. Therefore, if emissions from economic growth (k) have to be taken in by standing natural forests, their stocks could not fall below a certain limit. Those forest storage and tree growth constraints cap emissions.

Building upon fishery and forest bio-economics, in the BESF model emitters play the catchers, whereas natural forests supply the catch — namely, the environmental service of emissions storage. But unlike in fishery and forest models, the "catch", in this case, is an input rather than an output. As emissions increase, so does the demand for their storage, and the supply of forest stocks goes down. Thus, forest stocks turn out to be priced biophysically rather than monetarily.

Such a biophysically set price is called the *bio-economic exchange rate* (ε). Found by dynamic optimisation methods, it works as a *shadow price* measuring the shortage of the environmental input (carbon removal) on demand. At any given time, the more (less) this service is demanded, the less (more) of it is carried over into the future. However, the supply of current removal stocks varies across the carbon sinks. Such spatial differences, measured by the *bio-diversity ratio* λ, are called "exports" (Z) of carbon removal to elsewhere and correspond to *conservation*

2 In monetary economics, this relationship is known as the "quantity theory of money". "The quantity theory is a mechanistic proposition strangely alien to the assumptions of rational maximising behaviour on which classical and neoclassical theories generally rely ... It ignores the effects of the returns to holding money on the amounts economic agents choose to hold ... Money holdings depend ... on the opportunity costs, the expected changes in the value of money and the real yields of other assets into which the same funds could be placed" [11]. Of course, these remarks follow a Keynesian theoretical tradition, within which money is thought to be an asset rather than a token of bank liabilities to current account holders. Taking money as liabilities dates back to the days when bank-notes were "promises to pay", handed over at once as a receipt to depositors who had voluntarily given up gold to the bank, which, in turn, promised to repay them on demand [10].

loans (REDD credits) between carbon sinks. On the other hand, the provision of such an environmental service in the future is "imported" (borrowed) from the savings of removal stocks that happened in the past. Therefore, these transfers of carbon removal over time are called "imports" (*M*) and are referred to as *compensation loans* (CDM offsets) between the sinks.

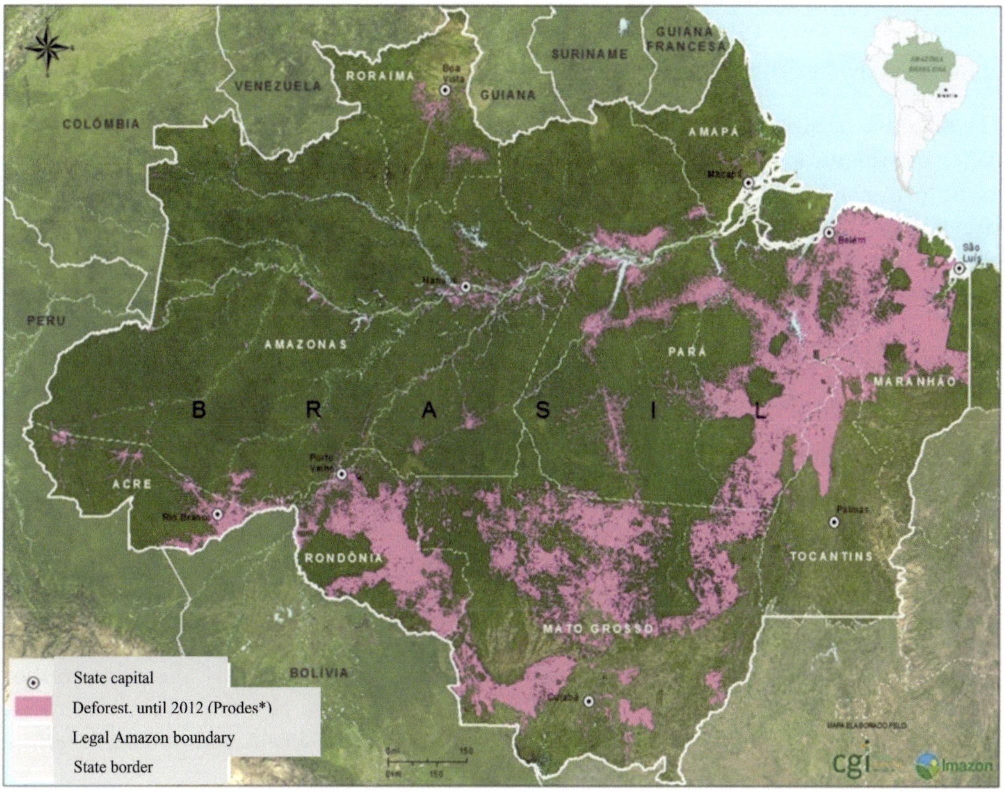

(*) Prodes Project — Satellite-borne monitoring of the Brazilian Amazon Rainforest [23]; Source: [12]

Figure 1. Deforestation accumulated until 2012 in the Brazilian Legal Amazon

Whereas *Z* means that the ecological burden of removing carbon emissions is "exported" to elsewhere, *M* implies carrying the ecological burden within an economy's boundaries over time [3]. The former translates into an ecological credit ($0 < \varepsilon < 1$) and the latter into an ecological debt ($\varepsilon > 1$).

When conservation is low (high), compensation is supposed to decrease (increase), unless ε depreciates (appreciates). Since ε exchanges future carbon removal stocks (*M*) for current ones (*Z*), its rise (depreciation) means that *Z* is relatively deteriorating, whereas its fall (appreciation) means that *Z* is relatively increasing. Otherwise, for a given ε, *Z* and *M* vary positively with it (Figure 2.b).

Just like bond markets are grounded in existing loan supply (savings) and demand (investment), removal loans for either carbon conservation or compensation are backed up by the biological growth of actual stocks (removal supply) of forest sinks set aside for curbing emissions from economic growth (removal demand). Forest-wide, Z grows with λ, whose growth, in turn, causes ε to fall (appreciate). On the other hand, the faster (slower) the speed k of economic activity, the greater (smaller) the demand for M and the depreciation (rise) of ε should be. Hence, at this forestland level of aggregation, ε is a negative function of the spatial distribution of carbon sinks (λ) and a positive function of emissions given off by the growth of the economy over time (k) (Figure 2.a).

The objective of probing these relationships is to demonstrate that the macro-scale determination of ε (Figure 2.a) can help find: a) the optimal supply of conservation and compensation in the loan market for carbon removal (Figure 2.b); b) biophysically attainable deforestation targets at the micro-scale (Figure 2.c), according to the rates at which emissions from economic growth ($\ln k$) are given off and the ratio of forest to deforested land— the bio-diversity ratio — varies ($\ln \lambda$).This analysis will be carried out for three scenarios:

i. BAU (Business-As-Usual), in which $\ln k > \ln \lambda$;

ii. REDD, in which $\ln k < \ln \lambda$;

iii. CDM, in which $\ln k = \ln \lambda$.

To start with, the methodology sections will describe the BESF model, its geometry, basic assumptions, parameters, functions and variables. After that, empirical data on deforestation in Brazilian Amazonia will be used to account for the model equilibrium points — both at the micro and macro-scale. Finally, the aforementioned three scenarios will be assessed to determine, for either of them, how much forestland would have to be used for conservation (REDD) and for offsetting carbon emissions (CDM). According to this allocation, carbon sequestration provided by forests could be "paid" at its real, biophysical value, rather than according to the virtual monetary benefits such an environmental service is supposed to deliver.

2. Research question and analytical framework

The creation of a market for carbon is based on the assumption that monetary payments for carbon storage might make economic agents opt for forest conservation rather than forest conversion [4]. Although such payments might sound like a working solution, the heavier any forest-related decision falls back on them, the lighter it is supposed to lean on its biophysical and environmental footings. Thus, how should a market fit into turning this logic upside-down? Put differently, how could ecosystem services, such as carbon sequestration, be priced biophysically rather than monetarily? What does that mean? How would it work?

The answers to whether prices can be equated with value or considered only indirect means of measuring values [13] rest upon the proposition that economic value should not be reduced to an ultimate one dimensional gauge, as held by the labour, utility and energy theories of value[3]. Economic value should be thought of as not bearing a single substance out of which it should be drawn or within which it should be found. In a biophysical sense, value is limited to the degree to which an item contributes to an objective or condition in an ecosystem [16]. For instance, the biophysical value of a tree species could come from its contribution to controlling soil erosion in steeply sloped areas [16]. Elsewhere, the same tree species might be worth for, say, sequestering carbon. So, although the tree species has not changed, the substance or content from which its value is drawn has.

Furthermore, just like man cannot reach too deep into the material microcosm [13], he is likely unable to search too deeply for the one and only source of economic value. According to the Heisenberg principle of indeterminacy in quantum physics, and given the high complexity of microstructures, the probability of, say, building a living cell from scratch is extremely small [13]. By the same token, the probability of drawing economic value out of a single underlying source is very low.

Within the economy's biophysical realms, useful goods and services should be valued by their usefulness[4] rather than by their exchange properties. What is meant by "usefulness" builds on Daly's [18] notion of *ultimate efficiency*, defined as the ratio of service to throughput. The services yielded by the stocks of artefacts are the ultimate benefit of economic activity. The throughput is the inevitable cost of maintaining the stocks of people and artefacts. Thus, in the

3 Dating back to classical times, the labour theory of value paradoxically stated that only when land (natural resources) is (are) running out is the maximum value reached. Therefore, any natural resource that gets into the market and thus acquires exchange value is dying out or close to extinction. So, exchange value lays bare that there is less of that natural resource than there was when it held no value at all! Later on, the neoclassical utility theory displaced the economic value to an even odder container: consumption. In the "utility world", the economic value was placed in the individual preferences for commodities. As preferences were a subjective matter, the economic value then turned out to rest upon the abstract basis of utility. From then on, the economic analysis has been cut off from its biophysical roots [14]. But as early as 1883, S. Podolinsky pioneered the idea of associating energy with value, as claimed by energy theories of value. These theories aimed at substituting energy for money as the only source of value. However, money is not particularly correlated with energy content, because there is no direct equivalence between low entropy and economic value [13][15]. For instance, the monetary value per unit energy content of a diamond is extremely large compared to the monetary value per unit energy content of a lump of coal. Nonetheless, if all indirect energy flows were to be tracked down and accounted for, the discrepancy between diamonds and coal might not be so great [15]. By and large, economists have rebutted these energy evaluation methods because of the fear that economics might end up turning into a branch of thermodynamics [14]. Furthermore, as Georgescu-Roegen [13] once pointed out, the economic process has only two flows: an input flow of low entropy and an output flow of high entropy, namely, waste. If the balance sheet of value should be set on the basis of these inputs and outputs, one "would arrive at the absurd conclusion that the value of the low entropy flow on which the maintenance of life itself depends is equal to the value of the flow of waste, that is, zero" (p. 284). This paradox only vanishes by acknowledging that the true "product" of the economic process is not a material flow, but a psychic flux — the enjoyment of life. Thus, the economic value has both psychic (neoclassical) and physical (classical) roots. An entropy-oriented, energy theory of value would only account for the supply side of the process and neglect the corresponding demand side [9]. That is why the proposition of a shadow price for natural resource inputs, on one hand, and for the waste sinking capacity of the environment, on the other, results from an economic rather than from an energy theory of value [14].

4 Although nature might well have other values — existence and bequeath — than use ones, non-use values are harder to estimate. Obviously, the use value is the one arising from the real — direct, indirect or optional — use of a given resource, whether in the present or in the future. On the other hand, the existence value is simply related to the existence of specific riches. The bequeath value measures the benefit that any individual obtains from knowing that, in the future, other people will also be able to benefit from the resource they have been left [17]. First and foremost, both non-use values essentially depend on estimating the preferences of future generations, which is not that easy to foreshadow. Moreover, non-use values resemble very much the intrinsic value of nature, which was claimed by Deep Ecology followers. According to them, nature had to be preserved for itself, rather than for satisfying the well-being of present and future generations. The intrinsic value is fully separated from any use value, even in the future [14]. Once these non-use concepts are difficult to apply, environmental goods and services are taken up thereafter in their usefulness sense.

final analysis, the stock of physical wealth is an accumulated flow of throughput, which is a cost to be minimised [18].

Likewise, if carbon removal is the service to be used, then it must be valued by the biophysical cost of delivering carbon removal stocks. Yet if not technically estimating price or value, a method that estimates costs should at least be a fairly good approximation to price and value, when markets are in equilibrium [16]. As known, whenever a commodity has a much higher value than its cost of production, its profits will be higher. The commodity will then be produced with increasing marginal costs until cost just equals price and profits are zero. The opposite happens when the commodity cost is much higher than its value. Since the commodity profits are lower, it will not be produced. The marginal costs then decrease until cost and price are equal.

As real markets are seldom in equilibrium, cost and price are expected to diverge, so that commodities could become overpriced or underpriced. The *shadow price* $\varepsilon = M \div Z$, calculated by the BESF model, is the biophysical sign of that distortion. It informs how far or close the growth rates of the economy's emissions (ln k) and of the forest stocks to remove them (ln λ) are from each other (Figure 2.a). As explained before, the greater λ is, the larger Z will be, which, *ceteris paribus*, makes ε go down (appreciate); on the other hand, the greater k is, the larger M will be, which, *ceteris paribus*, makes ε go up (depreciate). The optimal balance between k and λ defines not only ε^*, in Figure 2.a, but also the optimal level of carbon conservation (Z^*) and compensation (M^*), in Figure 2.b, as well as the optimal growth ($G(X_t)$) of and demand (h_t) for carbon removal stocks (X_t), in Figure 2.c. Hence, any bio-economic distortion in carbon prices is communicated by the bio-economic exchange rate ε.

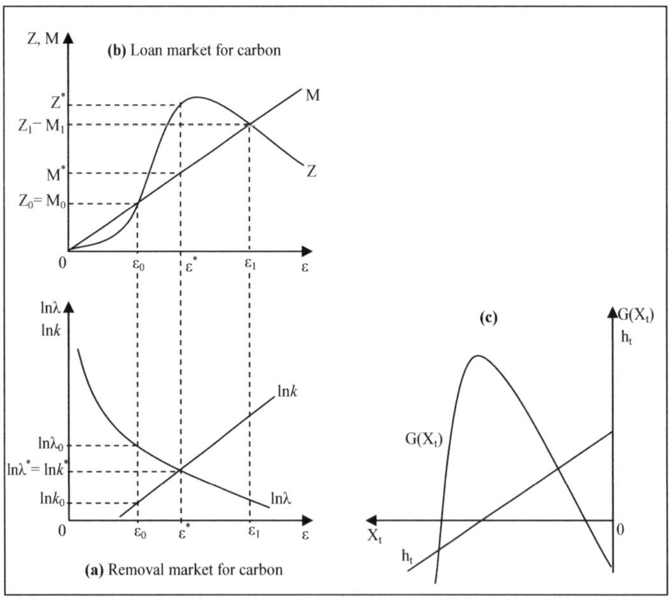

Figure 2. Loan and removal markets for carbon

For instance, given the hypothetical functional forms shown in Figure 2, ε^*, found in Figure 2.a, requires carbon conservation (Z^*) to be greater than carbon compensation (M^*), in Figure 2.b. Similarly, as ε^* was the outcome of a specific value taken on by λ and k, this bio-diversity ratio (λ^*) and economic growth rate (k^*) will then be applied to yield the curves depicted by Figure 2.c. It is these curves that set optimal deforestation targets and rates if forestland is split into deforested (u) and conserved (v) patches.

3. Model assumptions

Figure 2.c draws on bio-economic models, such as forestry and fishery models, which are concerned with the age and size of their biomass stocks (trees and fish) [3][19]. As the growth of a forest depends on the age of its trees, forestry models consider the time at which the maximum biological growth — or Maximum Sustainable Yield (MSY) — will be reached. In contrast, since fish growth depends on the size instead of on the age of fish stocks, fishery models involve the stock size at which the Maximum Economic Yield (MEY) will be attained [3].

Despite their different biological and economic emphasis, both such models are concerned with the output (trees and fish) provided by the corresponding natural source (forests and oceans). Neither of them, though, cares about the *input* — environmental service — supplied by the natural *sink* upon which the corresponding resource stock grows. Therefore, in the BESF model, the biomass stock is an *input* rendering an environmental service (carbon removal), instead of an *output* yielding biological (MSY) and economic (MEY) gains [3].

In this case, the stock harvested ($\hat{h}_t$) stands for the *throughput* of emissions removal — or the environmental cost of storing in the forest biomass the emissions given off by the production and consumption of the economy's output. Provided that emissions from economic growth have to be removed by natural standing forests, the supply of removal forest stocks ($G(X_t)$) need to meet an upper boundary (K_h) somewhere, since tree growth typically has an upper bound [3].

4. Model parameters and variables

Such an upper bound (K_h) depends both on *space* (λ) and *time*-related (k) variables (Table 1 and Table 2). On one hand, K_h is a function of λ_t (Eq. (3)) — the spatial distribution of forest biomass X_t(Eq. (1)), in GtC, between $j = 1,..., n$ conserved (**v**) and deforested (**u**) sinks at each time period $t = 1,..., m$, where $\mathbf{u} = (u_1, u_2,..., u_m) = (x_{11}, x_{21},..., x_{m1})$; $\mathbf{v} = (v_1, v_2,..., v_m) = (x_{1n}, x_{2n},..., x_{mn})$; $\mathbf{u} < \mathbf{v}$; x_{tj} = carbon-equivalent emissions by sources at time t to be stored (removed) by sink j. On the other hand, the distribution of forest biomass is highly affected by economic activity and the rate at which land-use changes take place. In other words, λ_t depends on k_j (Eq. (12)) — which measures emissions from deforestation or carbon emissions of sink j given off by its economic growth over time.

As time goes by, the ratio of energy changes (ln k) to land changes (ln λ), measured by ε (Eq. (14)), indicates in each period how much emissions from economic growth can be removed by each hectare of forestland [20]. Whereas economic growth (k) demands increasing removal of stocks, the supply of rising stocks is constrained by biophysical limits on growing and maintaining standing forests (λ).

Building on standard bio-economic (forestry and fishery) models, it is assumed that such a biological growth constraint follows a logistic pattern [3]. So as to hook the economy to its natural strings, emission flows ($\hat{h}_t$) must be capped (K_h) rather than the growth of removal forest stocks ($G(X_t)$), which already are by nature. Capping emissions from deforestation, though, implies that, at some former time T, when deforestation was negligible, there was a maximum level of removal stocks, X_K, associated with that least emission release (K_h).

t periods (emission sources)	j = 1,..., 2 removal sinks* (**u** < **v**)		X	Z	λ
	u	**v**			
1	x_{11}	x_{12}	X_1	Z_1	λ_1
⋮	⋮	⋮	⋮	⋮	⋮
m	x_{m1}	x_{m2}	X_m	Z_m	λ_m
V	V_1	V_2	$V = X$	Z	λ
M	M_1	M_2	M	ε	
k	k_1	k_2	k		

(*) j = **u** = the smallest carbon biomass stock sink or the *deforested* sink; j = **v** = the largest carbon biomass stock sink or the *conserved* sink. It is required that $\varepsilon > 0$, because, in biophysical terms, there cannot be a negative accountancy [10]. Therefore, so that $Z > 0$, j sinks must be displayed on an increasing biomass stock order. Likewise, so that $M > 0$, $M_j = |x_{mj} - x_{1j}|$.

Source: [3]

Table 1. Algebraic emission-removal matrix

$$X_t = \sum_{j=1}^{n} x_{tj} \tag{1}$$

$$Z_t = v_t - u_t = x_{tn} - x_{t1} \tag{2}$$

$$\lambda_t = v_t / u_t = x_{tn} / x_{t1} \tag{3}$$

$$\begin{aligned} \lambda &= 1 + (Z / V_1) \quad \text{(a)} \\ \ln \lambda &= Z / V_1 \quad \text{(b)} \end{aligned} \tag{4}$$

$$V = \sum_{j=1}^{n} V_j \tag{5}$$

Dimension	Variable	Description	Equation
Space	X_t	Total removal stock at time t	(1)
	Z_t	Exports, at time t, of removing services across the space and j sinks	(2)
	λ_t	Bio-diversity ratio of deforested to conserved forestland at time t	(3)
	λ	Source-sink system's bio-diversity ratio	(4)a, (4)b
	V	Total spatial bio-economic value of the j removal sinks	(5)
	M	Total spatial imports (across all sinks)	(6)
	ψ_t	†Biophysical overshoot rate at time t	(7)
Time	X	Total removal stock over time	(8)
	Z	Total exports over time	(9)
	V_j	Bio-economic value of sink j arising from its long-run supply of biomass stocks for emission removal	(10)
	M_j	Imports (loans) of removing services per j sink through time	(11)
	k_j	Deforestation rate at sink j	(12)
	k	Source-sink system's deforestation rate	(13)a, (13)b, (13)c
Space-time	ε	Bio-economic exchange rate	(14)
	ψ	Long-run overshoot rate	(15)

(†) k_t = variable k (Eqs. (13)a-c) at time t; k^* = optimal value for k.

Source: [3]

Table 2. Variables and equations of the BESF model

$$M = \sum_{j=1}^{n} M_j = \Delta X = | X_m - X_1 | \tag{6}$$

$$\psi_t = k_t / k^* \tag{7}$$

$$X = \sum_{t=1}^{m} X_t \tag{8}$$

$$Z = \sum_{t=1}^{m} Z_t = \Delta V = | V_n - V_1 | \tag{9}$$

$$V_j = \sum_{t=1}^{m} x_{tj} \tag{10}$$

$$M_j = |x_{mj} - x_{1j}| \quad (11)$$

$$k_j = x_{mj} / x_{1j} \quad (12)$$

$$\begin{aligned} k &= 1 + (M / X_1) && \text{(a)} \\ \ln k &= M / X_1 && \text{(b)} \\ k &= X_{t-1} / X_t && \text{(c)} \end{aligned} \quad (13)$$

$$\varepsilon = \frac{M}{Z} = \frac{\ln k \times X_1}{\ln \lambda \times V_1} = \frac{\Delta X}{\Delta V} \quad (14)$$

$$\psi = \sqrt[m]{\prod_{t=1}^{m} \psi_t} \quad (15)$$

The parameter K_h is an algebraically found macro-scale bound to emissions. Actually, it is the value taken on by the emission demand function $\hat{h}_t$ when $X_t = X_K$. Hence, the first step to set K_h is to find X_K, which is arrived at through vector algebra (Eq. (16)). Eq. (16) fulfils a twofold ideal requirement for sustainability:

a. *Maximum economic efficiency* ($k_j = k$), so that the rates of economic growth or return even out across the sinks [3][24];

b. *Perfect ecological efficiency* ($\varepsilon = 1$), so that neither further reallocation nor redistribution of emissions removal takes place across the sinks [3][15][25].

Theoretically, these conditions not only allow the source-sink system to simultaneously reach its economic and ecological sustainability, but also require it to remain sustainably stable. Therefore, the stock level X_K represents the "bio-economic cost" of achieving a stable state of sustainability. Rather than a target to be complied with, it translates into the space-time needed to make k stable ($k_j = k$) and $\varepsilon = 1$ [3].

$$2\begin{pmatrix} 1 & \frac{\bar{k}^{-1} + \bar{\lambda}^{-1}}{2} \\ \frac{\bar{k} + \bar{\lambda}}{2} & 1 \end{pmatrix} \begin{pmatrix} -\bar{u}_t \\ \bar{v}_t \end{pmatrix} = \begin{pmatrix} 0 \\ 0 \end{pmatrix} \quad (16)$$

In Eq. (16), the bar over the symbols stands for the corresponding mean values in the last row of Table 4. However, since the distribution of forestland (λ) is known for every period and the release of emissions from deforestation (k) instead is to be limited, $\bar{k}$ must be found by taking $\bar{\lambda} = 7.975$, $\bar{u}_t = 24$ and $\bar{v}_t = 182.40$ out of Table 4. By so doing, it turns out that $\bar{k} \approx 7.23$ and $\bar{k}^{-1} \approx 0.1383$. Next, by substituting these values into Eq. (17), in which, again, $\bar{u}_t = 24$ and $\overline{X}_t = 206.40$ come from Table 4, $v_K = 1319.20$ and $X_K = 1492.78$ are finally found.

Time periods (t)	Years	Deforest. rates (10^3 km^2)	Deforested area (10^3 km^2)[a]	Deforest. of the original forest[b]	Forest remnant (10^3 km^2)[c]	υ_t (GtC)[d]	u_t (GtC)[e]
1	1988	21.05	376.7	8.4%	4107.3	190.58	16.35
2	1989	17.77	401.4	9.0%	4082.6	189.43	17.42
3	1990	13.73	415.2	9.3%	4068.8	188.79	18.02
4	1991	11.03	426.4	9.5%	4057.6	188.27	18.51
5	1992	13.79	440.2	9.8%	4043.8	187.63	19.10
6	1993	14.90	440.2	9.8%	4043.8	187.63	19.10
7	1994	14.90	470.0	10.5%	4014.0	186.25	20.40
8	1995	29.06	497.1	11.1%	3986.9	184.99	21.57
9	1996	18.16	517.1	11.5%	3966.9	184.06	22.44
10	1997	13.23	530.1	11.8%	3953.9	183.46	23.01
11	1998	17.38	547.5	12.2%	3936.5	182.65	23.76
12	1999	17.26	564.7	12.6%	3919.2	181.85	24.51
13	2000	18.23	583.0	13.0%	3901.0	181.01	25.30
14	2001	18.17	601.1	13.4%	3882.8	180.16	26.09
15	2002	21.39	622.5	13.9%	3861.4	179.17	27.02
16	2003	25.25	647.8	14.4%	3836.2	178.00	28.11
17	2004	27.42	675.2	15.1%	3808.8	176.73	29.30
18	2005	18.85	694.0	15.5%	3789.9	175.85	30.12
19	2006	14.11	708.2	15.8%	3775.8	175.20	30.73
20	2007	11.53	719.7	16.1%	3764.3	174.66	31.23
21	2008	12.91	732.6	16.3%	3751.4	174.06	31.79
22	2009	7.46	740.1	16.5%	3743.9	173.72	32.12
23	2010	7.00	747.1	16.7%	3736.9	173.39	32.42
24	2011	6.42	753.5	16.8%	3730.5	173.09	32.70
25	2012	4.57	758.0	16.9%	3725.9	172.88	32.90

(a) According to [21], until 1997, and thereafter calculated from the previous deforested area plus the yearly deforestation rate. (b) Figures in this column are obtained by dividing the deforested area by estimates that consider 4483.972×10^3 km^2 = 4.483972×10^6 km^2 as, approximately, the original forested area of Brazil's 5×10^6 km^2 Legal Amazon [22]. (c) Figures obtained by calculating 4483.972×10^3 km^2 – deforested area $\times 10^3$ km^2. (d) "For the region's forests as a whole, the mean biomass loading (t/ha) for pre-logging biomass (weighted by the area of each eco-region present) is estimated at 464 t/ha" [22]. So the figures in this column result from 464 t/ha × forest remnant × 10^3 km^2 = 464 t/10^{-2} km^2 × forest remnant × 10^3 km^2 = (464 × forest remnant × 10^5) t = (464 × forest remnant × 10^5) 10^{-9} Gt. (e) The loading (biomass per hectare) for pre-logging biomass of forests cleared in 1990 (weighted by the deforestation rate in each state) is calculated at 434 t/ha. As before, the figures in this column result from 434 t/ha × forest remnant × 10^3 km^2 = 434 t/10^{-2} km^2 × forest remnant × 10^3 km^2 = (434 × forest remnant × 10^5) t = (434 × forest remnant × 10^5) 10^{-9} Gt [22].

Source: [23]

Table 3. Forest conservation and deforestation in the Brazilian Legal Amazon (1988-2012)

$$\begin{pmatrix} 1 & \bar{k}^{-1} \\ \bar{k} & 1 \end{pmatrix} \begin{pmatrix} \bar{u}_t \\ v_K \end{pmatrix} = \begin{pmatrix} \overline{X}_t \\ X_K \end{pmatrix} \tag{17}$$

5. Removal demand function ($\hat{h}_t$)

In an emissions-saving (low carbon) economy, a seesaw balance between deforestation and conservation of carbon stocks is expected to hold. At lower deforestation rates, the savings (conservation) of biomass stocks outstrip their consumption. But as economic growth increases, conservation savings of biomass stocks decline, while their consumption by deforestation goes up. Thus, in order to account for these offsetting effects at any time, the observed removal demands for deforestation and conservation are respectively determined by:

$$h_t(u_t) = \bar{u}_t - u_t \tag{18}$$

$$h_t(v_t) = \bar{v}_t - v_t, \tag{19}$$

where $\bar{u}_t$ and $\bar{v}_t$ are mean values for either biomass stock, displayed in the last row of Table 4.

By summing Eqs. (18) and (19), the total observed demand for removal per period is obtained:

$$h_t = \bar{x}_t - x_t, \tag{20}$$

where $x_t = u_t + v_t$ and $\bar{x}_t = \bar{u}_t + \bar{v}_t$.

The underlying assumption of Eqs. (18)-(20) is the *Permanent Income Hypothesis* (PIH), put forward by the American economist Milton Friedman in the late 1950. According to this hypothesis, consumption is smoothed in the long run, so as to handle short-run fluctuations in income. Income fluctuates because of its transitory component, which consists of unforeseen additions or subtractions likely to cancel out in the long run [26][27]. In an emissions avoiding world, where carbon storage sounds like the economy's earnings, conservation of carbon stocks adds to income, whereas deforestation lessens it. So, Eq. (20) amounts to the permanent component of income in the long run, which accounts for a smoother path based on consumption expectations ($\bar{x}_t$) rather than on current consumption (x_t). The regression[5] of the results for Eq. (20) on the observed values for X_t, in Table 5, yields the estimated long-run mean of removal demand (Eq. (21)).

$$\begin{array}{lll} \hat{h}_t = & 206.121 - & 0.999X_t \\ \text{t-stat.} & 497.034 & -496.808 \\ \text{sig.t.} & 0.0000 & 0.0000 \end{array} \tag{21}$$

5 All statistical estimations were performed in SPSS Statistics 17.0.

http://dx.doi.org/10.5772/57288

Time periods	Years	Deforestation (GtC)	Conservation (GtC)	Total biomass stock (GtC)	Deforest. demand (GtC)	Conserv. demand (GtC)	Total biomass demand (GtC)	Obs. Exports	Obs. Imports	ε	Econ. growth rate	Bio-div. ratio
(t)		(u_t)	(v_t)	(X_t)	Eq. (18)	Eq. (19)	Eq. (20)	(Z_t) Eq. (2)	(M) Eq. (6)	Eq. (14)	(k) Eq.(13c)	(λ_t) Eq. (3)
1	1988	16.35	190.58	206.93	7.65	-8.18	-0.53	174.23	–	–	1.00036	11.657
2	1989	17.42	189.43	206.85	6.57	-7.03	-0.45	172.01	2.2181	12.9	1.00020	10.874
3	1990	18.02	188.79	206.81	5.98	-6.39	-0.41	170.77	1.2392	7.3	1.00016	10.477
4	1991	18.51	188.27	206.78	5.49	-5.87	-0.38	169.77	1.0058	5.9	1.00020	10.174
5	1992	19.10	187.63	206.74	4.89	-5.23	-0.34	168.53	1.2392	7.4	1.00000	9.821
6	1993	19.10	187.63	206.74	4.89	-5.23	-0.34	168.53	0.0000	0.0	1.00043	9.821
7	1994	20.40	186.25	206.65	3.60	-3.85	-0.25	165.85	2.6760	16.1	1.00039	9.131
8	1995	21.57	184.99	206.57	2.42	-2.59	-0.17	163.42	2.4336	14.9	1.00029	8.575
9	1996	22.44	184.06	206.51	1.55	-1.66	-0.11	161.62	1.7960	11.1	1.00019	8.202
10	1997	23.01	183.46	206.47	0.99	-1.06	-0.07	160.45	1.1674	7.3	1.00025	7.974
11	1998	23.76	182.65	206.41	0.23	-0.25	-0.02	158.89	1.5610	9.8	1.00025	7.687
12	1999	24.51	181.85	206.36	-0.51	0.55	0.04	157.34	1.5499	9.9	1.00027	7.420
13	2000	25.30	181.01	206.31	-1.31	1.40	0.09	155.71	1.6367	10.5	1.00026	7.154
14	2001	26.09	180.16	206.25	-2.09	2.24	0.14	154.07	1.6312	10.6	1.00031	6.906
15	2002	27.02	179.17	206.19	-3.02	3.23	0.21	152.15	1.9212	12.6	1.00037	6.632
16	2003	28.11	178.00	206.11	-4.12	4.40	0.28	149.89	2.2672	15.1	1.00040	6.331
17	2004	29.30	176.73	206.03	-5.31	5.68	0.37	147.42	2.4626	16.7	1.00027	6.031
18	2005	30.12	175.85	205.97	-6.13	6.55	0.42	145.73	1.6924	11.6	1.00021	5.838
19	2006	30.73	175.20	205.93	-6.74	7.20	0.47	144.46	1.2670	8.8	1.00017	5.700
20	2007	31.23	174.66	205.90	-7.24	7.74	0.50	143.43	1.0356	7.2	1.00019	5.592
21	2008	31.79	174.06	205.86	-7.80	8.34	0.54	142.27	1.1594	8.1	1.00011	5.475
22	2009	32.12	173.72	205.84	-8.12	8.68	0.56	141.60	0.6703	4.7	1.00010	5.409
23	2010	32.42	173.39	205.82	-8.43	9.01	0.58	140.97	0.6286	4.5	1.00009	5.348
24	2011	32.70	173.09	205.80	-8.71	9.31	0.60	140.39	0.5763	4.1	1.00007	5.293
25	2012	32.90	172.88	205.78	-8.90	9.52	0.62	139.98	0.4105	2.9	–	5.255
Mean		**24.00**	**182.40**	**206.40**							**1.00023**	**7.975**

Table 4. Observed deforested, conserved and demanded biomass in the Brazilian Legal Amazon (1988-2012)

Time periods	Years	X_t (GtC)	h_t (GtC)	$\hat{h}_t$ (GtC)	$g(v(X_t))$ (GtC)	$\hat{F}(X_t)$ (GtC)	$G(X_t)$ (GtC)	$\hat{G}(X_t)$ (GtC)
(t)			Eq.(20)	Eq.(21)	Eq.(23)	Eq.(25)	Eq.(26)	Eq.(27)
T	?	1492.78	—	-1285.17	0.0000	0.00	-1492.78	1.25×10^7
1	1988	206.93	-0.53	-0.5983	0.5370	1431.25	1224.33	1215.00
2	1989	206.85	-0.45	-0.5242	0.5376	1252.76	1045.91	1039.41
3	1990	206.81	-0.41	-0.4829	0.5379	1153.21	946.40	941.33
4	1991	206.78	-0.38	-0.4493	0.5382	1072.51	865.73	861.74
5	1992	206.74	-0.34	-0.4080	0.5386	973.18	766.45	763.70
6	1993	206.74	-0.34	-0.4080	0.5386	973.18	766.45	763.70
7	1994	206.65	-0.25	-0.3187	0.5393	759.12	552.47	552.05
8	1995	206.57	-0.17	-0.2374	0.5400	564.94	358.38	359.66
9	1996	206.51	-0.11	-0.1775	0.5405	421.94	215.44	217.72
10	1997	206.47	-0.07	-0.1385	0.5409	329.13	122.67	125.48
11	1998	206.4139	-0.02	-0.0864	0.5413	205.20	-1.22	2.17
12	1999	206.36	0.04	-0.0347	0.5418	82.34	-124.02	-120.23
13	2000	206.31	0.09	0.0199	0.5422	-47.19	-253.50	-249.45
14	2001	206.25	0.14	0.0744	0.5427	-176.08	-382.33	-378.21
15	2002	206.19	0.21	0.1385	0.5433	-327.61	-533.80	-529.81
16	2003	206.11	0.28	0.2141	0.5439	-506.05	-712.16	-708.66
17	2004	206.03	0.37	0.2963	0.5446	-699.41	-905.44	-902.84
18	2005	205.97	0.42	0.3528	0.5451	-832.02	-1037.99	-1036.24
19	2006	205.93	0.47	0.3951	0.5455	-931.14	-1137.07	-1136.09
20	2007	205.90	0.50	0.4296	0.5458	-1012.07	-1217.97	-1217.69
21	2008	205.86	0.54	0.4683	0.5461	-1102.58	-1308.43	-1309.03
22	2009	205.84	0.56	0.4907	0.5463	-1154.85	-1360.68	-1361.82
23	2010	205.82	0.58	0.5117	0.5465	-1203.84	-1409.66	-1411.33
24	2011	205.80	0.60	0.5309	0.5466	-1248.73	-1454.53	-1456.72
25	2012	205.78	0.62	0.5446	0.5468	-1280.69	-1486.47	-1489.04

Table 5. Observed forest data and estimated demand ($\hat{h}_t$) and supply ($\hat{G}(X_t)$) functions of the BESF model

6. Removal supply function ($\hat{G}(X_t)$)

By substituting, in Eq. (21), $X_K = 1492.78$, found by Eqs. (16) and (17), for X_t, the least level of emissions demand $K_h = -1286.16$ comes out. This value is used as the lower limit of a logistic function (Eq. (23)) in the following constrained optimisation problem:

Objective function:

$$\min_{X_t} \sum_t S_t = \min_{X_t} \sum_t [g(v(X_t)) - \hat{h}_t] \tag{22}$$

Constraints:

Constraint I:

$$\hat{h}_t = 206.121 - 0.999 X_t \tag{21}$$

Constraint II:

$$g(v(X_t)) = \frac{1}{(1 / K_h) + b_0 b_1^{X_t}}$$

$$g(v(X_t)) = \frac{1}{(1 / -1286.16) + 0.07075414 \times 1.01593216^{X_t}} \tag{23}$$

Constraint III:

$$g(v(X_t)) \geq \hat{h}_t, \tag{24}$$

where S_t stands for the instantaneous surplus arising from the gap between removal growth rates ($g(v(X_t))$) and removal consumption rates ($\hat{h}_t$). Because of Eq. (24), $S_t \geq 0$, that is, at any time, the rate at which removal stocks are supplied must be greater than or equal to the rate at which they are demanded. In Eq. (23), the parameter K_h is found by Eqs. (16), (17) and (21), whereas the parameters b_0 and b_1 are provided by GAMS-IDE (General Algebraic Modelling System – Integrated Development Environment), version 24.1.2 (http://www.gams.com/download/), in which the optimisation programme described by Eqs. (21)-(24) was run.

The optimal values for $g(v(X_t))$ are displayed in Table 5. They are now employed to estimate the future supply of removal stocks ($\hat{F}(X_t)$) given the existing ones (X_t), used up in the present. The estimated variable $\hat{F}(X_t)$ stands for an *outflow-inflow ratio*, defined by logistically constrained rates of demand (numerator) and supply (denominator) of removal stocks (Eq. (25)). The difference between $\hat{F}(X_t)$ and X_t is the removal supply per time period ($G(X_t)$). Starting from the values for $G(X_t)$, found by Eq. (26) and displayed in Table 5, it is possible to estimate the removal supply function (Eq. (27)), whose values are also displayed in Table 5.

$$\hat{F}(X_t) = \frac{\partial \hat{h}_t / \partial X_t}{g(v(X_t))} = \frac{\hat{h}_t (K_h - \hat{h}_t)}{g(v(X_t))} \tag{25}$$

$$G(X_t) = \hat{F}(X_t) - X_t \tag{26}$$

$$\hat{G}(X_t) = 5.727X_t^2 - 244006.276$$
$$\text{t-stat.} \quad 588.143 \quad -588.763 \qquad (27)$$
$$\text{sig.t.} \quad 0.0000 \quad 0.0000$$

7. Avoiding deforestation *versus* stock maintenance approach

All removal market functions are drawn in Figure 3. They look like those in Figure 2.c, and their interplay shows how long emissions from deforestation can be removed by natural forest stocks. In deforestation-conservation settings, the cost of maintaining a high economic growth rate (ln k) would be an abrupt fall in the rate at which natural removal stocks grow (ln λ). As demonstrated by Eq. (3), in Table 2, and by the figures in the last column of Table 4, the value of λ indirectly defines the quantities of removal consumption (h_t) from deforestation (u_t) or removal savings from conservation (v_t). Thus, $\hat{h}_t$ (Eq. (21)) hinges on that sink distribution, and so does $\hat{G}(X_t)$ (Eq. (27)), whose estimation ultimately relies on K_h — the lower bound of removal demand. As a matter of fact, λ is the critical variable drawing the bottom line in carbon removal markets.

The impacts of λ on the equilibrium of removal markets are shown in Table 6. The equilibrium scenarios checked, but the BESF one, follow the standard fishery analysis [28]. In the context of deforestation and forest conservation, though, there are important remarks regarding equilibrium conditions. To begin with, although the MSY equilibrium might apply to unnatural, even-aged stands, it is not likely to suit the dynamics of conservation of natural forests and deforestation. It is known that in natural forests, the wide frequency and age range of tree species points to a biomass yielding function that does not reach a maximum sustainable yield. Mathematically, this is translated by taking the first derivative of Eq. (27) and making it equal to zero. Thus, it is found that $X_{MSY} = 21303.15$ GtC, which is a prohibitively high level of removal stocks, provided their logistic upper bound is $X_K = 1492.78$. The same holds for the restricted access (RA) stock level, which is found to be slightly lower ($X_{RA} = 21303.06$ GtC).

Although in standard bio-economic (fishery) analysis, the latter is expected to be higher than the former, it must be borne in mind that, in a *compensation approach* — such as *avoiding deforestation* —, the demand for removal stocks grows with deforestation (Figure 4.a). Therefore, conservation requires $X_{MSY} > X_{RA}$, since a smaller *compensating* stock means that less deforestation occurred.

On the other hand, Figure 3 shows that the greater the stock, the smaller its demand is, because the stock can only grow when its depletion is low. Thus, as is clear so far, Figure 3 illustrates a rather different standpoint, namely, a *stock maintenance approach* (Figure 4.b). It has been argued that such a *conservation approach* favours the largest forest countries, like Brazil. Unlike in small forest countries, with only tiny remnants of forest left, in countries where large expanses of forest remain standing, stock maintenance represents a much greater carbon service than does avoiding deforestation [21].

In this regard, the equilibrium conditions in Table 6 had to be adjusted to fall within a stock maintenance rationale. Since, in Figure 3, the slopes of the functions $\hat{G}(X_t)$ (Eq. (27)) and $\hat{h}_t$ (Eq. (21)) are never expected to be equal — as the columns labelled "rate of return" and "rate of depletion" in Table 6 also show —within the feasible region ($X_{0A} \leq X_t^* \leq X_K$), RA equilibrium only calls for maximising some positive level of Y_t, which could otherwise be warranted if the functions $\hat{G}(X_t)$ and $\hat{h}_t$ were to have the same slopes somewhere. However, as the outcomes in Table 6 show, a positive Y_t can be accomplished with a negative value for $\hat{G}(X_t)$, provided that, in absolute terms, this is smaller than that for $\hat{h}_t$. Since $\hat{G}(X_t) < 0$ is environmentally threatening, BESF equilibrium becomes a more stringent condition, because it calls both for positive Y_t and $\hat{G}(X_t)$. By minimising this positive level of Y_t, then, the BESF equilibrium ensures that, for a removal demand function like h_t, in Figure 4.b, and $\hat{h}_t$, in Figure 3, the stock level satisfying this will lie slightly beyond (greater than) K — the stock level at which $\hat{G}(X_t) = 0$ — and further beyond (greater than) X_{OA} — where $\hat{G}(X_t)$ becomes equal to $\hat{h}_t$, but turns out to be negative.

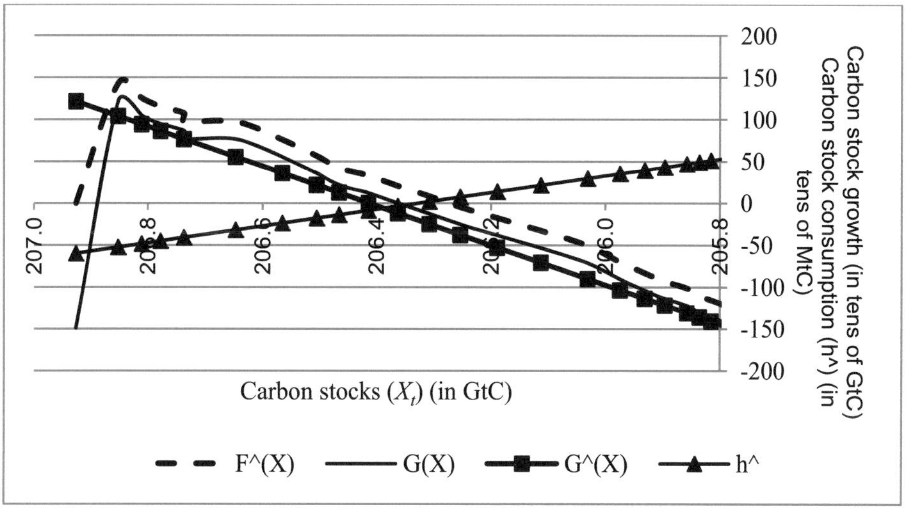

Figure 3. Removal market functions

The open access stock level (X_{OA}) is the smallest, yet just slightly smaller than the others. As expected from standard theory on renewable resources, the economic rent at this level is zero. At any other equilibrium point, it is non-zero and positive. But it is the highest at BESF, which minimises the throughput of maintaining stocks by requiring the supply of their emissions removal services ($\hat{G}(X_t)$) always to be positive.

Finally, by comparing the optimal stock levels (X_t^*), in Table 6, with the observed ones (X_t), in Table 5, it can be inferred when each equilibrium scenario must have occurred. It is worrying to ascertain that all of them are already gone somewhere between 1998 and 1999.

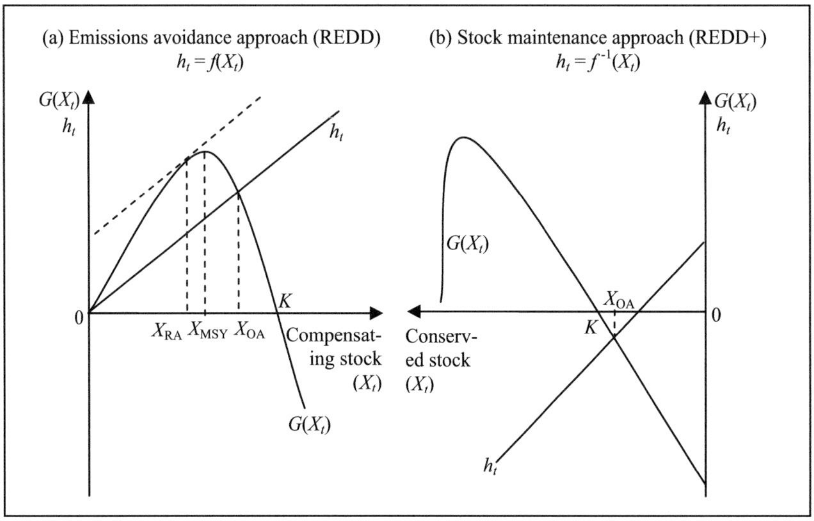

Figure 4. The BESF model functions and the REDD approaches

Equilibria[a]	Equilibrium conditions[b]	Optimal[c] stock (X_t^*) (in GtC)	Removal supply $(\hat{G}(X_t^*))$ Eq. (27) (in GtC)	Removal demand (h_t^*) Eq. (21) (in GtC)	Rate of return[d] $\left(\frac{d\hat{G}(X_t^*)}{dX_t^*}\right)$	Rate of depletion[e] $\left(\frac{d\hat{h}_t^*}{dX_t^*}\right)$	Economic rent $(Y_t = \hat{G}(X_t^*) - \hat{h}_t^*)$
BESF	min $Y_t > 0$ $\hat{G}(X_t) > 0$ $X_{OA} \leq X_t^* \leq X_K$	206.412933	→ 0.000	-0.086	2.364	0.999	> 0.086
K	$\hat{G}(X_t) = 0$	206.412933	0.000	-0.086	2.364	0.999	0.086
MSY	$\frac{d\hat{G}(X_t)}{dX_t} = 0$ $X_{OA} \leq X_t^* \leq X_K$	—	—	—	—	—	—
RA	max $Y_t > 0$ $X_{OA} \leq X_t^* \leq X_K$	206.412933	-3.49×10^{-10}	-0.086	2.364	0.999	< 0.086
OA	$\hat{G}(X_t) = \hat{h}_t$	206.412898	-0.08	-0.08	2.364	0.999	0.000

(a) BESF = Bio-Economic carbon Sequestration by Forests; K = steady-state equilibrium; MSY = Maximum Sustainable Yield; RA = Restricted Access equilibrium; OA = Open Access equilibrium. (b) In standard fishery models, however, X_{RA} is found where $d\hat{G}(X_t) / dX_t = d\hat{h}_t / dX_t$. (c) Provided by GAMS-IDE 24.1.2. (d) First derivative of Eq. (27) = $11.454X_t$. However, so that $\hat{G}(X_t)$ and $\hat{h}_t$ can be plotted together, the barter ratio between them is 10 GtC of $\hat{G}(X_t)$ per 10 MtC of $\hat{h}_t$, as the vertical axis of the graph in Figure 3 indicates. More simply, this barter ratio can be expressed as 1 GtC : 1 MtC, which means 10^3 MtC : 1 MtC. Therefore, so that rates of return and depletion can be compared with one another, the former must be multiplied by 10^{-3}. (e) First derivative of Eq. (21).

Table 6. Removal market equilibrium analysis

8. Aggregate emissions demand or removal supply function ($\hat{\lambda}(\hat{\varepsilon})$)

Perhaps these worries could have been dismissed before, if the variable regulating land-use changes (λ) had not been overlooked. As seen, ε (Eq. (14)) — the bio-economic exchange rate — is defined as the ratio of ecological debt — excess demand for removal services (supply of emissions) — to ecological credit — excess supply of removal services (demand for emissions). Although it is an underlying variable, it stands for the *shadow price* measuring, along an optimal path through time, the marginal bio-economic value of the forestland asset [19]. When the speed k of economic activity drives deforestation, forestland shrinks and thus ε is expected to rise (depreciate).

Since ε critically and ultimately depends on λ and k, it must, to begin with, be expressed in terms of them. Methodologically, this can be first accomplished by fixing $k=\bar{k}=1.00023$, given in the last row of Table 4. Then this rate is assumed to hold for every year according to the following rule:

$$X_{t-1}/X_t' = \bar{k} = 1.00023, \tag{28}$$

where X_t' is the stock level that would be observed in column X_t of Table 5, if $k=\bar{k}=1.0023$.

As λ changes, so will ε — whose calculation follows Eqs. (11) and (6), for M (imports); (2) and (9), for Z (exports); and (14), for ε proper. These outcomes must be ordered pair-wise, according to increasing values of λ. The objective of this disposition is to check how ε is affected by changes in λ. Next, an equation for ε, as a long-run function (thus, bearing no t index) of λ, is estimated:

$$\begin{array}{lll} \hat{\varepsilon}(\lambda) = \exp\left(-0.131 + \dfrac{1.096}{\ln\ln\lambda}\right) & & \\ \text{t-stat.} \quad -243.522 & 1048.581 & \\ \text{sig.t.} \quad 0.0000 & 0.0000 & \end{array} \tag{29}$$

Finally, the results for $\hat{\varepsilon}(\lambda)$, in Eq. (29), are used as inputs (independent variables) to arrive at an equation for λ as a long-run function of ε:

$$\begin{array}{lll} \ln\hat{\lambda}(\hat{\varepsilon}) = \exp\left(-1.127 + \dfrac{2.766}{\hat{\varepsilon}}\right) & & \\ \text{t-stat.} \quad -133.522 & 214.563 & \\ \text{sig.t.} \quad 0.0000 & 0.0000 & \end{array} \tag{30}$$

Eq. (30) stands for the *aggregate emissions demand* or *removal supply* function in the long run.

9. Aggregate emissions supply or removal demand function ($k(\hat{\varepsilon})$)

By a similar procedure, the functional relationship between ε and k, as well as between k and ε, can then be calculated. This time, though, the variable made fixed is $\lambda = \bar{\lambda} = 7.975$, given in the last row of Table 4. This value is kept unchanged for each two consecutive years, to either of which Eqs. (1) and (3) apply:

$$x_{t,j} + x_{t,j+1} = X_t \qquad (1)\text{a}$$

$$x_{t,j+1} / x_{t,j} = \bar{\lambda} = 7.975 \qquad (3)\text{a}$$

By substituting Eq. (3)a into Eq. (1)a, it turns out that:

$$X_t = (1 + \bar{\lambda}) x_{t,j}, \qquad (1)\text{b}$$

which can be correspondingly replaced in Eq. (13)c to find the new value of k, at each two consecutive periods, when $\lambda = \bar{\lambda} = 7.975$ and thence remains constant. Again, the resulting ε requires Eqs. (11), (6), (2), (9) and (14). Also, like before, so as to inquire into the effects of changing k on ε, these variables are taken pair-wise on an increasing order of k values. Lastly, an equation for ε, as a long-run function (thus, bearing no t index) of k, can be estimated:

$$\begin{array}{ll} \hat{\varepsilon}(k) = & 6434.098 \ln \ln k \\ \text{t-stat.} & 3921.720 \\ \text{sig.t.} & 0.0000 \end{array} \qquad (31)$$

Now, the outcomes for $\hat{\varepsilon}(k)$, in Eq. (31), are used as inputs (independent variables) to obtain an equation for k as a long-run function of ε:

$$\begin{array}{lll} \hat{k}(\hat{\varepsilon}) = 0.99999998 & + 1.554575 \times 10^{-4} \hat{\varepsilon} & \\ \text{t-stat.} \quad - & - \\ \text{sig. t.} \quad - & - \end{array} \qquad (32)$$

Eq. (32) stands for the *aggregate emissions supply* or *removal demand* function in the long run.

10. Macro-bio-economic scenarios

Based on Eqs. (30) and (32), REDD, CDM and BAU scenarios are tested to understand how well conservation (REDD) and compensation (CDM) strategies can do as compared with business-as-usual (BAU) ones (Table 7 and Table 8). The most useful results shown by Table 8 are those displayed in its last two columns. They make clear how much the natural forest and the economy are expected to grow annually, through 25 years, in each scenario.

It is noteworthy that a stringent conservation scenario, such as REDD1, requires an optimal value for λ ($\hat{\lambda}^*_{REDD1}$= 5.787) that is not too far from its observed mean value, displayed in the last row and column of Table 4 ($\bar{\lambda}$=7.975). Thus, the allowed annual deforestation rate through 25 years (Eq. (33)) is 3.18% p.a. (last column of Table 8). This figure might sound startling when compared, for instance, with the deforestation rate in the Brazilian Legal Amazon for a single year: just between August 2012 and August 2013, this rate reached 20% [31]! However, neither would more "economic growth-driven" strategies (CDM and BAU's) stand such a high annual deforestation rate. Nearly all of them (last four rows and last column of Table 8) would allow for a yearly deforestation rate of about 7.8%. On the other hand, a 100% rate of deforestation reduction, even spread over 25 years (scenario REDD2), would render no more than an economic growth rate as low as 0.00149% (last column of Table 8). These numbers help shed some light on the feasibility of the targets set by deforestation reduction programmes [29] (Figure 7). Figure 5 shows that the supply of emissions ($\hat{k}$) is nearly *perfectly inelastic* to the shadow price ε, whereas the removal of them ($\hat{\lambda}$) dramatically falls with the rise of ε. Although at some high value of ε, the demand for removal also becomes almost inelastic to price changes, this only happens at very low levels of existing removal stocks, when thus $\hat{\lambda} \to 1$ (Eq. (3)) and the share of forest conservation approaches that of deforestation (v → u) or becomes even smaller (v < u).

Scenario	Constraints	Objective-function (W or W')[a]
REDD1	$\widehat{\ln\lambda} > \ln\hat{k}$; $\widehat{\ln\lambda} = 1.755682$[b]	$\max W = \widehat{\ln\lambda} - \hat{k}$
REDD2	$\widehat{\ln\lambda} > \ln\hat{k}$; $\widehat{\ln\lambda} = 1.028114$[c]	$\max W = \widehat{\ln\lambda} - \hat{k}$
CDM	$\widehat{\ln\lambda} = \ln\hat{k}$	$\max/\min W = \widehat{\ln\lambda} - \hat{k}$ or $W' = \hat{k} - \widehat{\ln\lambda}$
BAU1	$\widehat{\ln\lambda} < \ln\hat{k}$; $\hat{k} = 1.8092514$[d]	$\max W' = \hat{k} - \widehat{\ln\lambda}$
BAU2	$\widehat{\ln\lambda} < \ln\hat{k}$; $\hat{k} = 2.6658363$[e]	$\max W' = \hat{k} - \widehat{\ln\lambda}$
BAU3	$\widehat{\ln\lambda} < \ln\hat{k}$; $\hat{k} = 3.3863549$[f]	$\max W' = \hat{k} - \widehat{\ln\lambda}$

(a) Because, as shown by Table 4, k is much smaller than λ, the greater of them must be scaled down through logarithms to make them comparable. (b) Growth rate for λ that would smooth, over 25 years (1988-2012), the accumulated deforestation reduction rate defined by a 7-year programme, from 2007 to 2015, for reducing deforestation in the Brazilian Amazon [29]. The annual reduction rates for every period t = 1,..., 7 are, respectively, 25%, 25%, 30%, 40%, 50%, 75% and 100%. The "capitalisation" (multiplication) of all these rates yields 7.4648438, which amounts to the full figure to be reached in 7 years. This 7-year time is factored into a 25-year period, thereby yielding 3.5714286 sub-periods, over which the deforestation reduction rate accumulated during 7 years is spread according to its geometric mean $\left(\sqrt[\frac{25}{7}]{7.4648438}\right)$. (c) Growth rate for λ that would amount to a 100% deforestation reduction in 25 years. It is calculated by $\sqrt[25]{2}$, where 2 is a rate of growth that is worth 100%. (d) Value needed to yield an annual rate of economic growth of 2.40% during 25 years. This rate is the one projected for Brazil's GDP growth in 2013 by the Brazilian Central Bank, in its latest Focus Report [30]. (e) Value needed to yield an annual rate of economic growth of 4.00% during 25 years. (f) Value needed to yield an annual economic growth rate of 5.00% during 25 years.

Table 7. Scenario analysis

Scenario	$\hat{\varepsilon}^*$ (× 10^{-4})	$\hat{\lambda}^*$	$ln\hat{\lambda}^*$	$\hat{k}^*$	$ln\,\hat{k}^*$	$\hat{\lambda}^*$% p.a.†	$\hat{k}^*$% p.a.†	Allowed def. rate δ^*% p.a. (Eq.(33))
REDD1	1.637	5.7874	**1.755682**	1.000254	0.000254	7.27520	0.00102	3.1804
REDD2	2.395	2.7958	**1.028114**	1.000372	0.000372	4.19819	0.00149	6.7997
$\ln\hat{\lambda}=\hat{k}$	2.453	2.7193	**1.000381**	**1.000381**	0.000381	4.08267	0.00153	6.8623
CDM	2464.70	1.3832	**0.324368**	1.383156	**0.324368**	**1.30592**	**1.30592**	7.8350
BAU1	5205.61	1.3829	0.324176	**1.809251**	0.592913	1.30515	**2.40000**	7.8352
BAU2	10715.70	1.3828	0.324087	**2.665836**	0.980518	1.30479	**4.00000**	7.8353
BAU3	15350.53	1.3827	0.324062	**3.386355**	1.219754	1.30469	**5.00000**	7.8353

(†) Growth rates per annum (p.a.). Figures under these column captions respectively come from $[(\sqrt[25]{\hat{\lambda}^*}) - 1] \times 100$ and $[(\sqrt[25]{\hat{k}^*}) - 1] \times 100$. (‡) Provided by GAMS-IDE 24.1.2.

Table 8. Optimal (*) results‡ from scenario analysis

$$\delta^* = [\sqrt[25]{(\bar{\lambda} - \hat{\lambda}^*)} - 1] \times 100, \quad \text{with } \bar{\lambda} > \hat{\lambda}^* \tag{33}$$

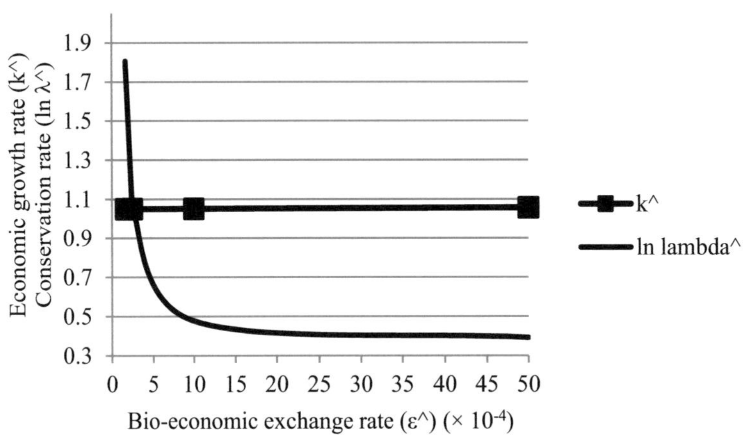

Figure 5. Aggregate equilibrium for conservation and deforestation in the Brazilian Legal Amazon

11. Removal trade

Now, the optimal values for ε brought to light in Table 8 can be used in Table 9 to define, as in Figure 2.b, the amount of removal loans across the space (Z) and over time (M). Of course, this previously requires that both Z and M are estimated as long-run functions of ε. However, as shown in Table 4, it is precisely ε that arises from observable Z and M; not the other way around. So, what must be estimated first and foremost is an equation in which ε works as a long-run function of observable Z and M.

$$\hat{\varepsilon}(Z) = \exp(1.428 \times Z) \tag{34}$$

t-stat. 26.544

sig. t. 0.0000

$$\hat{\varepsilon}(M) = 6.643 M^{0.926} \tag{35}$$

t-stat. 67.364 35.514

sig. t. 0.0000 0.0000

Then, the outcomes for $\hat{\varepsilon}(Z)$ and $\hat{\varepsilon}(M)$, in Eqs. (34) and (35), are used as inputs (independent variables) to obtain the equations for Z and M as long-run functions of ε:

$$\hat{Z}(\hat{\varepsilon}) = \exp\left(0.895 - \frac{4.154}{\hat{\varepsilon}}\right) \tag{36}$$

t-stat. 259.616 −133.481

sig. t. 0.0000 0.0000

$$\hat{M}(\hat{\varepsilon}) = 0.129 \hat{\varepsilon}^{1.08} \tag{37}$$

t-stat. — —

sig. t. — —

Scenario	$\hat{\varepsilon}$ (× 10^{-4})	ln $\hat{Z}$ (× 10^{2})	ln $\hat{M}$	β= ln $\hat{Z}$ - ln $\hat{M}$	$\hat{Z}$(GtC) Eq.(36)[a]	$\hat{M}$(GtC) Eq.(37)	B= $\hat{Z}$ - $\hat{M}$
Min B	0.940	-3.523403	-2.114586	-1.408817	0.0295	0.1207	**-0.0912**
REDD1	1.637	-1.642841	-1.515764	-0.127077	0.1934	0.2196	-0.0262
B = 0	**1.799**	-1.413446	-1.413446	0.000000	**0.2433**	**0.2433**	**0.0000**
REDD2	2.395	-0.839176	-1.104520	0.265344	0.4321	0.3314	0.1007
$\ln \hat{\lambda} = \hat{k}$	**2.453**	-0.798110	-1.078638	0.280528	0.4502	0.3401	0.1101
Max B	**5.454**	0.133360	-0.215882	0.349242	1.1427	0.8058	**0.3368**
B = 0	**10.621**	0.503872	0.503872	0.000000	**1.6551**	**1.6551**	**0.0000**
CDM	2464.70	0.893315	6.386668	-5.493353	2.4432	593.87	-591.43
BAU1	5205.61	0.894202	7.194149	-6.299947	2.4454	1331.62	-1329.17
BAU2	10715.70	0.894612	7.973880	-7.079267	2.4464	2904.10	-2901.66
BAU3	15350.53	0.894729	8.362075	-7.467345	2.4467	4281.57	-4279.12

(a) Actually, these figures are in hundreds of GtC, so that, as indicated by Table 4, their order of magnitude can be compared to that of $\hat{M}$. (b) Results provided by GAMS-IDE 24.1.2.

Table 9. Scenario analysis[b] for conservation and compensation loans

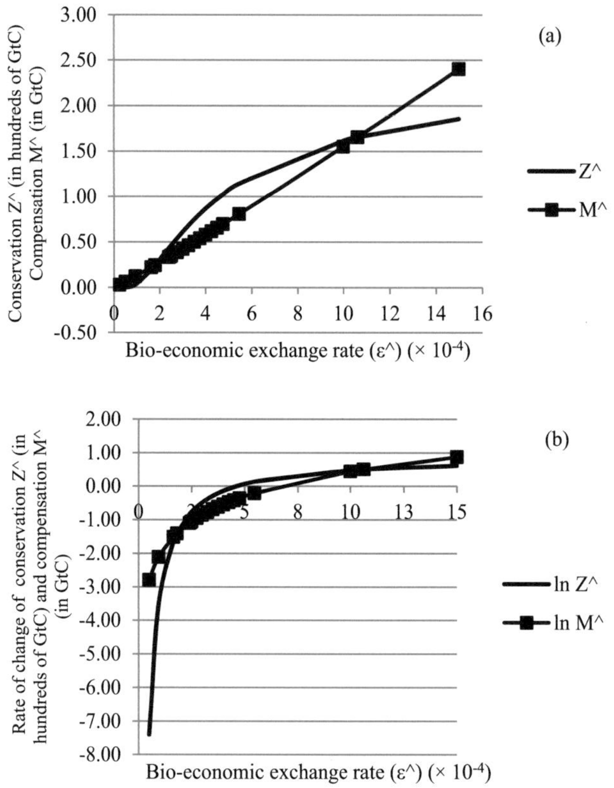

(†) Determined by GAMS-IDE 24.1.2.

Figure 6. Optimal path† for conservation ($\hat{Z}$) — REDD — and compensation ($\hat{M}$) — CDM — loans (a) and rates (b) in the long run

Eqs. (36) and (37) define, respectively, the optimal path of *conservation* (REDD) and *compensation* (CDM) to be loaned out in the long run (Figure 6.a). Since conservation is the surplus of carbon removal services, they can be *exported* (lent) to somewhere else. On the other hand, when these missing services have nevertheless been already used, some compensation is due. However, compensating for environmental services that were already lacking before implies *importing* (borrowing) them from somewhere. In this sense, *conservation* (exports) defines an *excess supply* of removal services, while *compensation* (imports) is caused by an *excess demand* for removal services. If these services are to be loaned out, the sink yielding conservation *lends* them, whereas the sink owing compensation *borrows* them.

In any event, the demand for these loans, as shown in Figure 2, lies behind disaggregate (Figure 3) — periodical — and aggregate (Figure 5) — long run — removal needs. As long as *B* (last column of Table 9) stands for the balance of carbon trading loans, it is interesting to note that a stringent scenario, such as REDD1 (Table 7), yields more ecological debt ($\hat{M}$) than ecological credit ($\hat{Z}$). This balance only turns out to be positive in the interval where $1.799 \leq \hat{\varepsilon} \leq 10.621$. *B* reaches its maximum in this interval, when $\hat{\varepsilon} = 5.454$. Later, it starts dropping, reaching zero again. From then on, it becomes increasingly negative, and neither the CDM nor BAU scenario can turn it over anymore.

Actually, CDM and BAU1 are but very loose scenarios. When they come into play, the ecological credit region ($1.799 \leq \hat{\varepsilon} \leq 10.621$) has already been left behind. Therefore, in spite of the high values for $\hat{\varepsilon}$ set by CDM and BAU scenarios, these prices appear to be too low to prevent $B < 0$. Although CDM requires that $\ln \lambda = \ln k$ (Table 7 and Table 8), existing ecological conditions demand, rather, that $\ln \lambda > \ln k$.

If these conditions are overlooked, no further rise in $\hat{\varepsilon}$, as in the BAUs scenarios, can prevent B from keep falling. That likely occurs because the environmental capacity of providing removal services has been already overshot. As it turns out, $\hat{\varepsilon}$ cannot be set like usual prices can; instead, it is set by ecological conditions, as though ecology were guiding the economy [32].

12. Target setting

Table 10 focuses on scenario REDD1, from Table 8, in order to demonstrate how an economy can be ecologically guided. Of course, the figures in Table 10 do not account for the real picture. Rather, they relate to a 25-year deforestation period (1988-2012), to show how things would look if it had followed out the optimal path suggested by the conservation scenario REDD1. The percentages in the last column of Table 10 were merged in such a way that the 25-year period is divided into 7 time lags. By so doing, REDD1 reduction path, although applying to a different time period (1988-2012), can be compared to the same time length of the deforestation reduction programme tailored by [29]. The comparison is shown in Figure 7.

Nevertheless, it can still be asked why the percentage deforestation rate along the optimal path in Table 10 (15.59%) does not match that in the last column of Table 8 (3.1804%). Even though the former is referred to in deforested area terms, in km^2, while the latter, in carbon biomass, in GtC, the proper calculation (Eq. (39)) shows that the values would be equal. Accordingly, the percentage deforestation rates displayed in Table 10 can be reckoned either in carbon biomass (GtC) or in area units (km^2). Actually, the underlying reason for the mismatch between deforestation rates in Table 10 and in Table 8 is that the rates in the former are *bounded* (Eq. (33)), whereas those in the latter are *unbounded*. This difference can be grasped from combining Eqs. (1) and (3) to yield:

$$X_t = u_t + \lambda_t u_t = (1 + \lambda_t) u_t \tag{38}$$

From Eq. (38) and from the definition of the deforestation rate (δ_t) given in note *i* of Table 10, it turns out that:

$$\delta_t = \frac{u_t}{X_t} = \frac{u_t}{(1 + \lambda_t) u_t} = \frac{1}{1 + \lambda_t} \tag{39}$$

If Λ_t, in Eq. (39), takes on whatever optimal value displayed in Table 8, then this equation can be compared with Eq. (33). Yet now, unlike in the former equation, in the latter $\hat{\lambda}^*$ has an upper bound ($\bar{\lambda}$= 7.975). Since this ceiling stands for a mean value for the whole period (1988-2012), it is not surprising that δ^* in Eq. (33) is lower than that arising from Eq. (39) — displayed in the antepenult column of Table 10.

The deforested area in Table 10 is given by Eq. (39). Its size follows closely — yet throughout a 25-year period — the size of forestland that, according to Table 3, has been cut down since 2005 (t = 18). However, while in the observed data of Table 3, the size of deforested land keeps on getting bigger, in the optimal path of Table 10, it increasingly shrinks.

t	Year	u^a (GtC)	v^b (GtC)	X_t (GtC)	Z_t^c (GtC)	M_t^d (GtC)	ε ($\times 10^{-4}$)	u^e (10^3 km^2)	v^f (10^3 km^2)	X_t^g (10^3km^2)	Remaining deforest.[h] (10^3 km^2)	Def. rate[i](%)	Def. rate[j] (km^2)	Deforest. targets[k] (%)
1	1988	30.49	176.44	206.93	145.95	0.023890	1.637	702.46	3802.57	4505.03	16806.21	15.59	—	4.01
2	1989	30.48	176.38	206.85	145.90	0.023881	1.637	702.21	3801.21	4503.42	16104.00	15.59	251.55	4.18
3	1990	30.47	176.34	206.81	145.87	0.023877	1.637	702.07	3800.45	4502.52	15401.93	15.59	140.54	4.36
4	1991	30.46	176.31	206.78	145.85	0.023873	1.637	701.96	3799.83	4501.79	14699.98	15.59	114.06	4.56
5	1992	30.46	176.28	206.74	145.82	0.023868	1.637	701.82	3799.07	4500.89	13998.16	15.59	140.54	4.77
6	1993	30.46	176.28	206.74	145.82	0.023868	1.637	701.82	3799.07	4500.89	13296.35	15.59	0.00	5.01
7	1994	30.45	176.20	206.65	145.76	0.023858	1.637	701.51	3797.43	4498.94	12594.83	15.59	303.49	5.28
8	1995	30.43	176.13	206.57	145.70	0.023848	1.637	701.24	3795.93	4497.17	11893.60	15.59	275.99	5.57
9	1996	30.42	176.08	206.51	145.66	0.023841	1.637	701.03	3794.83	4495.86	11192.57	15.59	203.68	5.89
10	1997	30.42	176.05	206.47	145.63	0.023837	1.637	700.90	3794.12	4495.02	10491.67	15.59	132.40	6.26
11	1998	30.41	176.00	206.41	145.59	0.023831	1.637	700.72	3793.16	4493.88	9790.94	15.59	177.03	6.68
12	1999	30.40	175.96	206.36	145.55	0.023825	1.637	700.55	3792.21	4492.75	9090.40	15.59	175.77	7.16
13	2000	30.40	175.91	206.31	145.52	0.023818	1.637	700.36	3791.20	4491.56	8390.04	15.59	185.62	7.70
14	2001	30.39	175.87	206.25	145.48	0.023812	1.637	700.18	3790.20	4490.38	7689.86	15.59	185.00	8.35
15	2002	30.38	175.81	206.19	145.43	0.023805	1.637	699.96	3789.02	4488.98	6989.90	15.59	217.88	9.10
16	2003	30.37	175.75	206.11	145.38	0.023796	1.637	699.70	3787.63	4487.33	6290.20	15.59	257.12	10.01
17	2004	30.35	175.68	206.03	145.32	0.023786	1.637	699.42	3786.12	4485.54	5590.78	15.59	279.28	11.12
18	2005	30.35	175.63	205.97	145.28	0.023780	1.637	699.23	3785.08	4484.31	4891.55	15.59	191.93	12.51
19	2006	30.34	175.59	205.93	145.25	0.023775	1.637	699.09	3784.30	4483.39	4192.46	15.59	143.69	14.29
20	2007	30.34	175.56	205.90	145.23	0.023771	1.637	698.97	3783.66	4482.63	3493.49	15.59	117.44	16.67
21	2008	30.33	175.53	205.86	145.20	0.023767	1.637	698.84	3782.95	4481.79	2794.65	15.59	131.49	20.00
22	2009	30.33	175.51	205.84	145.18	0.023764	1.637	698.76	3782.54	4481.30	2095.89	15.59	76.02	25.00
23	2010	30.32	175.49	205.82	145.17	0.023762	1.637	698.69	3782.15	4480.84	1397.20	15.59	71.29	33.34

t	Year	u [a] (GtC)	v [b] (GtC)	X_t (GtC)	Z_t [c] (GtC)	M_t [d] (GtC)	ε (x10⁻⁴)	u [e] (10³ km²)	v [f] (10³ km²)	X_t [g] (10³km²)	Remaining deforest.[h] (10³ km²)	Def. rate[i](%)	Def. rate[j] (km²)	Deforest. targets[k] (%)
24	2011	30.32	175.48	205.80	145.16	0.023759	1.637	698.62	3781.80	4480.43	698.58	15.59	65.36	50.00
25	2012	30.32	175.46	205.78	145.15	0.023758	1.637	698.58	3781.55	4480.13	0.00	15.59	46.55	100.00
Total								**17508.67**	**94778.08**					

(a) $u_t = X_t \div (1 + \hat{\lambda}^*)$. (b) $v_t = X_t - u_t$. (c) Eq. (2). (d) Eq. (14): $M_t = \varepsilon \times Z_t$. (e) $(u_t \div 434) \times 10^{-4}$, according to note *e* in Table 3. (f) $(v_t \div 464) \times 10^{-4}$, according to note *d* in Table 3. (g) X_t (in km²) = u_t (in km²) + v_t (in km²). However unrealistic it may sound, the total area "shrinks" to account for the loss of forestland implied by substituting the carbon stored in deforested sites for that stored in natural forests. (h) For $t = 1$ (1988), $\Sigma_t u_t$ (in km²) = 17508.67 – u_1 (in km²); for $t > 1$, (remaining deforestation)$_{t-1}$ – u_t (in km²). (i) u_t (in km²) ÷ X_t (in km²). (j) $(u_{t-1} - u_t)$ (in km²) x 10³. (k) For $t = 1$ (1988), u_t (in km²) ÷ $\Sigma_t u_t$ (in km²); for $t > 1$, u_t (in km²) ÷ (remaining deforestation)$_{t-1}$.

Table 10. Deforestation targets and rates from optimal results in the REDD1 scenario ($\hat{\lambda}^*$ = 5.7874) (Table 8)

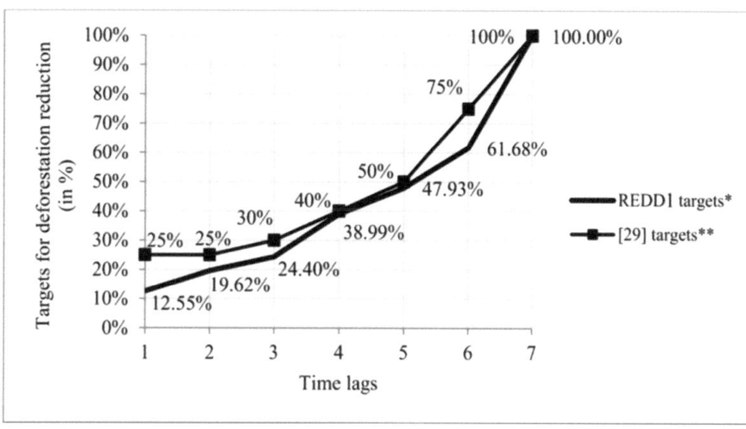

(*) During 1988-2012. (**) From 2007 to 2015 [29].

Figure 7. Targets for reducing deforestation in the Brazilian Legal Amazon

13. Conclusion

The analysis carried out so far has demonstrated that, where policy climate and deforestation are concerned, carbon conservation (REDD) and compensation (CDM) entail a trade-off that cannot be overcome by monetary mechanisms. Instead of money, the underlying variable which forest value rests upon is the bio-diversity ratio (λ). Although it is typically a space-based measure, this ratio also accounts for forestland distribution over time. When λ is affected by the demand of removal stocks (h) set off by the emissions growth from the economy, the amount of compensation (M) for these emissions over time is assumed to feed on the conservation (Z) of carbon savings carried out at each period.

Conservation surplus can be loaned out to afford compensation demands from an economy producing growing emissions. However, such demands must be halted somewhere, otherwise the supply of removal forest stocks ($G(X_t)$) will hit a ceiling and fall short of delivering enough conservation. As shown by Figure 6 and Table 9, it is likely there also is a biophysical limit to conservation, which should prevent this environmental service from being further encouraged by incentives such as monetary payments (PES). When conservation is driven too far, as in the REDD1 scenario (Table 8 and Table 9), it might end up turning the balance of carbon trading loans negative. On the other hand, when it is traded off against compensation, as in the CDM scenario, it deteriorates the balance of carbon trading loans even further. Quite often, economy-wise price setting and policy-making grow apart from ecological conditions.

Figure 2, Figure 3, Figure 5 and Figure 6 show how ecology could economically guide the economy [32]. If this linkage should be fully accomplished, setting targets for reducing deforestation would be grounded in biophysical, rather than in economic or political bearings.

Shifting from a deforestation avoidance approach to a forest stock maintenance one would certainly be a step forward. While the former carries a misleadingly uneconomic meaning, the latter brings forest conservation to the economic foreground. As avoiding deforestation usually implies forgoing profit-making activities, it mistakenly underlies monetary rewards. Thus, a great deal of the REDD mechanism draws heavily upon them. However, from a bio-economic standpoint, they are most likely to become romantic red roses, whose purchase will, rather than mend a broken heart, make money melt into thin air.

Author details

Valny Giacomelli Sobrinho*

Address all correspondence to: giacomelliv@yahoo.com.br

Department of Economics, Federal University of Santa Maria (UFSM), Santa Maria, Rio Grande do Sul, Brazil

References

[1] Linhares-Juvenal T. REDD and the Challenge of Protecting the Global Forest Cover. In: Serôa da Motta R., Hargrave J., Luedemann G., Gutierrez MBS. (eds.) Climate Change in Brazil — Economic, Social and Regulatory Aspects. Brasília: IPEA; 2011. p319-328.

[2] Kanninen M., Murdiyarso D., Seymour F, Angelsen A., Wunder S., German L. Do Trees Grow on Money? The Implications of Deforestation Research for Policies to Promote REDD. Bogor, Indonesia: CIFOR; 2007.

[3] Giacomelli Sobrinho V. Carbon Bio-Economics and Forests — Getting the BESF out of Climate Policy. In: Kheradmand H. (ed.) Climate Change — Socioeconomic Effects. Rijeka: InTech; 2011. p401-424.

[4] Martin RM. Deforestation, Land-Use Change and REDD. Unasylva 2008; 59(230): 3-11.

[5] Angelsen A., editor. Analyzing REDD+ — Challenges and Choices. Bogor, Indonesia: CIFOR; 2012.

[6] Polanyi K. A Grande Transformação — As Origens de Nossa Época. Rio de Janeiro: Campus; 2000.

[7] Whiteman A. Money Doesn't Grow on Trees: A Perspective on Prospects for Making Forestry Pay. Unasylva 2003; 54(212): 3-10.

[8] Prince's Charities' International Sustainability Unit. Interim REDD+ Finance — Current Status and Ways Forward for 2013-2000. London: The Prince of Wales' Charitable Foundation; 2012. http://www.pcfisu.org/the-princes-rainforests-project/interim-redd-finance (accessed 8 September 2013).

[9] Daly HE., Farley, J. Economia Ecológica — Princípios e Aplicações. Lisboa: Instituto Piaget; 2004.

[10] Soddy F. The Rôle of Money — What It Should Be, Contrasted with What It Has Become. London: Routledge; 1934.

[11] Tobin J. Money. In: Durlauf SN., Blume LE. (eds.) Monetary Economics. 2nd. ed. Hampshire, UK: Palgrave Macmillan; 2010. p224-241.

[12] Instituto do Homem e Meio Ambiente da Amazônia. IMAZON: Mapas: Desmatamento Acumulado na Amazônia Legal 2007-2012: Desmatamento na Amazônia Legal até 2012. http:// www.imazon.org.br/mapas/desmatamento-acumulado-2007-2010/desmatamento-amazonia-legal-acumulado-ate-2012/view (accessed 25 September 2013).

[13] Georgescu-Roegen N. Entropy, Value, and Development. In: Georgescu-Roegen N. The Entropy Law and the Economic Process. Cambridge, US, London, UK: Harvard University Press; 1999. p276-315.

[14] Faucheux S., Noël JF. Economia dos Recursos Naturais e do Meio Ambiente. Lisboa: Instituto Piaget; 1995.

[15] Ayres RU. On the Life Cycle Metaphor: Where Ecology and Economics Diverge. Ecological Economics 2004; 48: 425-438.

[16] Costanza R. Value Theory and Energy. Encyclopedia of Energy 2004; 6: 337-346.

[17] Pearce D., Moran D. O Valor Económico da Biodiversidade. Lisboa: Instituto Piaget; 1994.

[18] Daly HE. The Economics of the Steady State. The American Economic Review 1974; 64: 15-21.

[19] Clark CW. Mathematical Bio-Economics — the Mathematics of Conservation. 3rd. ed. Hoboken, US: John Wiley & Sons; 2010.

[20] Wackernagel M., Rees W. Our Ecological Footprint — Reducing Human Impact on the Earth. Gabriola Island, Canada: New Society Publishers; 1996.

[21] Fearnside PM. Forests and Global Warming Mitigation in Brazil: Opportunities in the Brazilian Forest Sector for Responses to Global Warming under the "Clean Development Mechanism". Biomass and Bioenergy 1999; 16: 171-189.

[22] Fearnside PM. Greenhouse Gases from Deforestation in Brazilian Amazonia: Net Committed Emissions. Climatic Change 1997; 35: 321-360.

[23] Instituto Nacional de Pesquisas Espaciais. INPE: Amazonia: PRODES: Taxas Anuais 1988 a 2012. http://www.obt.inpe.br/prodes/prodes_1988_2012.htm (accessed 20 September 2013).

[24] Common M. Environmental and Resource Economics — an Introduction. 2nd. ed. New York: Addison Wesley Longman; 1996.

[25] Ayres RU. The Second Law, the Fourth Law, Recycling and Limits to Growth. Ecological Economics 1999; 29: 473-483.

[26] Houthakker HS. The Permanent Income Hypothesis. The American Economic Review 1958; 48: 396-404.

[27] Meghir C. A Retrospective on Friedman's Theory of Permanent Income. The Economic Journal 2004; 114: F293-F306.

[28] Perman R., Ma Y., McGilvray J., Common M. Natural Resource and Environmental Economics. 3rd. ed. Harlow, England: Pearson Education; 2003. http://rapidlibrary.com/files/ perman-natural-resource-and-environ mental-economics-3rd-edition-pdf_ulczebmccni89on.html (accessed 19 November 2012).

[29] Instituto Socioambiental. ISA, Greenpeace, Instituto Centro de Vida. ICV, The Nature Conservancy. TNT, Conservação Internacional, Amigos da Terra – Amazônia Brasileira, Instituto do Homem e Meio Ambiente na Amazônia. IMAZON, WWF-Brasil. Pacto pela Valorização da Floresta e pelo Fim do Desmatamento na Amazônia Brasileira. http://www.socioambiental. org/banco_imagens/pdfs/doc-pacto%20desmatamento%20zero%20SUM%20ONGs%20FINAL. pdf (accessed 25 September 2013)

[30] Banco Central do Brasil. BCB. Sistema de Metas para a Inflação: Sala do Investidor: Focus: Focus — Relatório de Mercado. http://www.bcb.gov.br/pec/GCI/PORT/readout/R20130920.pdf (accessed 25 September 2013).

[31] Martins H., Fonseca A., Souza Jr. C., Sales M., Veríssimo A. Boletim do Desmatamento (SAD) Agosto de 2013. Boletim Transparência Florestal da Amazônia Legal 2013: 12. http://www. imazon.org.br/publicacoes/transparencia-florestal/transparencia-florestal-amazonia-legal/boletim-do-desmatamento-sad-agosto-de-2013-1 (accessed 25 September 2013).

[32] Giacomelli Sobrinho V. Bio-economics and ecosystem services: how ecology can economically guide the economy. In: Bürgi M. (ed.) Frontiers in Historical Ecology: Abstracts: proceedings of the International Conference of the International Union of Forest Research Organisations (IUFRO), Swiss Federal Institute for Forest, Snow and Landscape Research (WSL), 27 August – 2 September 2011, Birmensdorf, Switzerland. Birmensdorf: WSL; 2011. http://www.wsl.ch/ publikationen/pdf/11114.pdf

Chapter 2

CO_2 Utilization: A Process Systems Engineering Vision

Ofélia de Queiroz F. Araújo,
José Luiz de Medeiros and Rita Maria B. Alves

Additional information is available at the end of the chapter

http://dx.doi.org/10.5772/57560

1. Introduction

The development of economies results in increased energy consumptions, as observed in Figure 1. In the coming decades, the supply of such expanding demand will remain based on fossil fuels technologies. Expanding the share of renewable energy (*e.g.*, biofuels in the case of transportation fuels) would require massive investments in creating a new infrastructure, which would eventually raise the standards to a new economic order entirely based on renewables in the near future. On the other hand, the current scenario involves the announcement of large proven reserves of non-conventional gas and oil and expansion of installed infrastructure of production and refining.

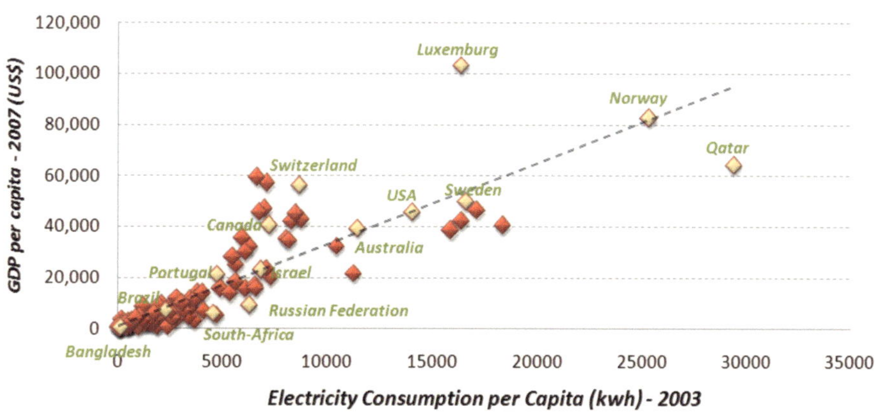

Figure 1. Economic development and electricity consumption. [graph constructed with data available at hdr.undp.org]

Fossil fuel based energy is recognized as carbon emitter. The challenge of the century is thus to expand energy supply in a carbon-constrained economy. According to the World Economic Forum (King, 2010), no truly low-carbon technology will be able to penetrate the mass market in the short term. The use of installed fossil processing infrastructure, with co-processing of biomass and fossil feedstock and capturing and utilization of emitted CO_2 is the "escape route" for a moderate transition from the present to a long-term sustainable future. In this context, putting a price on carbon will gradually build the road to a greener tomorrow. Meanwhile, bio-based products are a realistic supplement to fossil-based products, but they cannot mitigate the rising demand for fossil fuels.

According to an IEA Technology Roadmap, the manufacture of only 18 chemicals account for 80% of the energy demand in the chemical industry and 75% of its greenhouse gas (GHG) emissions. The study concludes that, "in the short to medium term (to 2025), steady progress in implementing incremental improvements and deploying best practice technologies (BPT) could provide substantial energy savings and emissions reductions compared to business as usual". "A step change in the sector's energy consumption and GHG emissions would require the development of 'game changer' technologies, such as sustainable biomass feedstocks and hydrogen from renewable energy sources which have not yet reached commercial maturity." (IEA, 2013)

With this prospect, CO_2 utilization in the short term should allow parallel production routes based on BPT, driven by emission-capture-utilization synergies. In this sense, production and conversion of synthesis gas (SYNGAS) exhibits the highest potential for medium term of commercial success. Nevertheless, it is worth noting that, while the utilization of CO_2 has potential to reduce greenhouse gas emissions into the atmosphere, CO_2 has disadvantages as a chemical reactant due to its relative significant chemical inertness. This inertness is the underlying reason why CO_2 has broad industrial application as solvent (supercritical CO_2), as fire and pest extinguisher, and as a non-toxic amusement additive in the food industry. From the standpoint of building a low-carbon economy, each potential use of CO_2 as reactant has a customarily high energy requirement, entailing associated energy-related GHG emissions. Obviously, such GHG emissions should not exceed the yield of chemical conversion of CO_2. Reverse water gas shift (RWGS) and dry reforming to yield SYNGAS and CO_2 hydrogenation to methanol are the most prominent CO_2 conversion alternatives to high volume chemical commodities. On the other hand, it is reasonable to expect that CO_2 conversion and biomass-based processing alternatives will coexist for a while with fossil-based technologies. Figure 2 illustrates the concept of co-existing fossil and biomass feedstock, having SYNGAS generation as the integration phase. In this scenario, SYNGAS can be obtained from coexisting fossil and green feedstock via gasification (of biomass, coal and heavy residues), steam reform and dry (CO_2) reform of natural gas, which, in turn, will also coexist with downstream nonconventional conversion routes – e.g., Fischer-Tropsch (FT) and methanol to olefins (MTO) – to fuels and chemical intermediates to supply the installed petrochemical industries.

In countries with large bioresources as well as significant oil and natural gas production infrastructure (e.g., USA, Canada, Mexico, China, Russia, Australia, Argentina and Brazil), fossil feedstock (mainly nonconventional oil and gas) will present the greatest challenge to bio-

http://dx.doi.org/10.5772/57560

mass. In Brazil, for instance, gas supply capacity in Santos Basin increased from 600 thousand m^3/d, in 2009, to 22.2 million m^3/d in 2013 (PETROBRAS, 2013). The identified risks associated to such expansion are, beside depletion of natural resources, increased CO_2 emissions.

Therefore, CO_2 capture, transportation and utilization must be included in the scene since CO_2 stands conceptually as a renewable feedstock. In this sense, Figure 3 illustrates CO_2 emissions capture and utilization associated to the production and refining of fossil fuel. It is worth noting that, in addition to the variety of alternative technologies, other factors influence the conception of capture and utilization of CO_2 for the production of chemicals.

Rostrup-Nielsen and Christiansen (2011) present four recognizable trends in next generation successful plants for producing chemical commodities: (a) location of cheap raw materials; (b) economy of scale; (c) highly integrated process plants; and (d) low CO_2 footprint (t CO_2/t product) Consequently, the production chain and the associated GHG emissions are complex and require a system view for optimal decision making on technologies for production of energy and chemicals taking into account the CO_2 perspective.

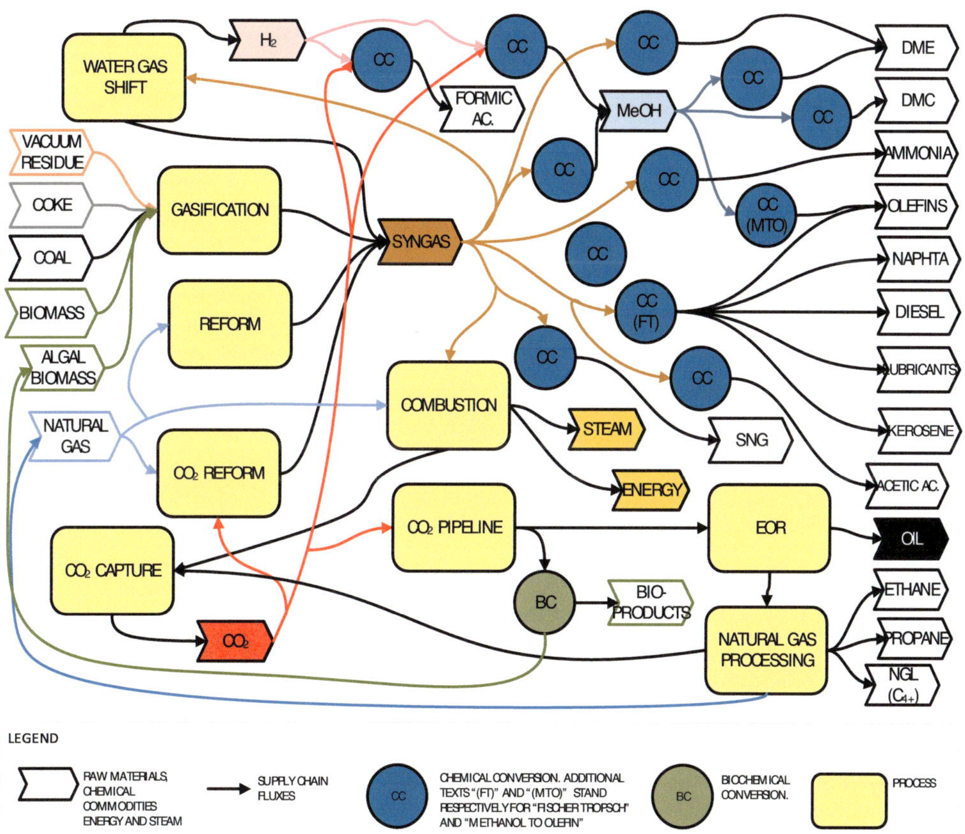

Figure 2. Coexistence of fossil and biomass feedstock uses: CO_2 capture and utilization dropped into existing production infrastructure

Chemical Process Systems Engineering (cPSE) deals with the set of basic unit operations involved in turning raw materials into products via chemical and bio-chemical processes. In a wider definition, PSE is concerned with the optimization of decision-making process for creation and operation of chemical supply chains. Such integrated framework embraces product and process design, manufacture and distribution of chemical products with multiple and conflicting technical, economic, energetic, environmental and social objectives.

With this vision, c-PSE of industrial use and reuse of CO_2 supply chain is assessed in a life cycle approach of technological alternatives involving integration of CO_2 capture (separation), CO_2 transportation and CO_2 chemical/biochemical utilization. Apart the conventional utilizations of CO_2 listed above, there are three new potential chains of CO_2 utilizations, namely: (i) production of chemicals via chemical conversion (CC); (ii) enhanced oil recovery (EOR) and carbon geological storage (CGS); and (iii) conversion via algae and microalgae to biomass or biochemical conversion (BC).

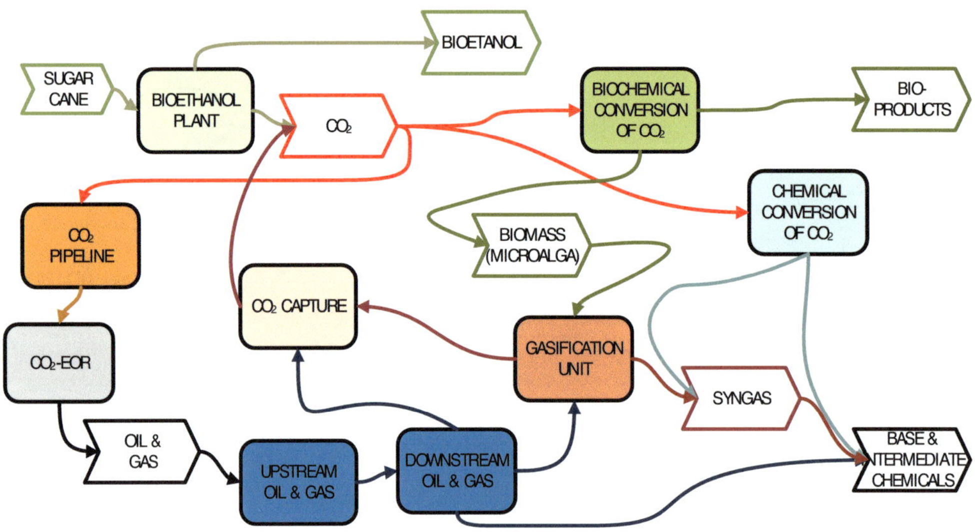

Figure 3. CO_2 capture and utilization in the oil & gas supply chain

EOR and CGS constitute, under present conditions, the only CO_2 processing chain that has some steps ready to be put into operation at high scales. These encompass some separation technologies for CO_2 post-combustion capture, CO_2 compression and CO_2 transportation via long pipelines. EOR and CGS can be reinforced if oxy-fuel technologies finally attain industrial maturity, which may occur within the next 5 decades. Pre-combustion technologies are also potential powerful contributors to EOR-CGS systems, which should be fully developed within the next 5 decades. Pre-combustion relies on some separation technologies also present in post-combustion alternatives.

BC via solar photosynthesis is a promising but still incipient package of technologies that have yet to be proven feasible in large scales. The main drawback is the impressive footprint and liquid hold-up of algae cultivation and processing plants, which have to comply with the high dispersion and periodicity of solar light and small concentration of biomass in growing medium (always below 10g/L). Besides, the use of non-solar light energy is totally out of question due to the photosynthesis efficiency limitations of green life forms (always below 10%, i.e., only a maximum of 10% of the incident light, already with appropriate wavelength, is biochemically usable to bio-convert CO_2).

CC context encompasses novel approaches extending the technology scenario for CO_2 supply chain via chemical conversion to benign, stable compounds for long-term storage or to value-added products like plastics, chemical intermediates and oxygenated octane enhancers like DMC – dimethyl-carbonate (Souza et al., 2013) for reuse.

The cPSE vision of CO_2 treats routes of CC and BC (and BC integrated with CC, i.e., BCC) to useable products and fuels, not as substitute, but rather as a complement to EOR and CGS. The present text is oriented to analyze CO_2 as a profitable feedstock, i.e., its potential use as feedstock to chemicals and fuels within economic applications, besides its relevant use in EOR. CGS is not considered an economic application; instead, it constitutes only a plausible, secure, destination of the excess of carbon, i.e., all CO_2 that has to be discarded because it is not dispatched to any economic use.

Among the routes for CO_2 utilization, BC is a natural choice as photosynthesis yields biomass, allowing the production of bioproducts, biofuels and chemicals through downstream CC processing routes. In fact, biomass gasification is the most flexible technology for dropping into conventional downstream CC routes. This integrated option configures BCC – biochemical and chemical conversion of CO_2.

The beneficial use of alkaline wastes or metallic ions to convert CO_2 via neutralization of alkaline wastes, or reaction of CO_2 with metallic ions to form less soluble carbonates that can be removed from produced water (oil & gas industry) is also a relevant CC application of CO_2. Lastly, this study presents the EOR use of CO_2, i.e., by injection into depleting oil or gas fields to maximize hydrocarbon production.

Finally, analysis of routes that undertake CO_2 reduction must take into account the life cycle of the processes in order to assess whether additional CO_2 production occurs beyond the amount abated from atmospheric emissions. This is precisely the case of CC of CO_2 into fuels and chemicals that always requires high energy input, normally derived from burning fossil fuels, entailing further associated GHG emissions. In this context, new or mature cPSE solutions should comply with the triple objective of sustainability, namely: economically feasible, environmentally benign and socially beneficial, in a supply chain approach. In connection with this, the chapter presents the CO_2 Capture Cycle as well as promising alternatives of its reutilization.

2. The thermodynamics of pure CO_2

The phase behavior of pure CO_2 (Figure 4) exhibits particularities when compared with common light species of natural gas (NG) like CH_4, C_2H_6 and C_3H_8. In general grounds, CO_2

has a fluid phase behavior similar to ethane (C_2H_6) with a very similar Critical Point (CP) temperature. The pronounced distinctive characteristic is its high Triple Point (TP) (Figure 4) temperature comparatively with light hydrocarbons. This means that solid CO_2 (dry ice) can be easily encountered below -56.6°C and above 5.2bar, if the original processing stream is rich enough in CO_2. This freeze-out of solid is not observed with the light hydrocarbon species unless below -182°C, which is about 20°C below the lowest temperature that can be achieved in LNG processing (i.e., the Normal Boiling Point of CH_4). In other words, without CO_2, the coldest NG processing (LNG plant) does not have solid formation. The phase behavior of pure CO_2 (Figure 4) is characterized by two larger (and infinite) continents corresponding to gas and solid states, and a finite intermediate liquid continent extending between the TP and CP temperatures. The three continents are two-dimensional (2D) objects due to the Phase Rule, which stipulates two degrees of freedom for a pure species at one-phase condition. One-dimensional (1D) equilibrium boundaries – SLE, SVE and VLE lines – are positioned between two neighboring continents. Their 1D nature is also a consequence of the phase Rule, which stipulates one degree of freedom for a pure species at a two-phase condition. SLE – solid-liquid (or solid-fluid) equilibrium line – is an endless line extending from the TP to indefinitely high pressures, characterized by the coexistence of solid and liquid CO_2. SVE – solid-vapor equilibrium line – is characterized by coexistence of solid and gas CO_2 ending at the TP and lower bounded by the absolute 0K. VLE – vapor-liquid equilibrium line – is a finite locus between TP and CP where liquid and gas CO_2 coexist. Just above the liquid continent there is a somewhat indefinite supercritical fluid zone (SCF). The SCF is a fusion of gas and liquid behaviors at high pressures above the CP pressure and high temperatures above the CP temperature. The SCF is characterized by high densities and high compressibility.

The contact point for the three continents is the Triple Point TP, an invariant three-phase point (zero degrees of freedom) by the Phase Rule. The VLE locus ends at the Critical Point CP where the differences between liquid and vapor vanish. The CP is also an invariant point with zero degrees of freedom, but, contrary to the TP, the CP is a single phase point satisfying two extra criticality conditions. The phase behavior of CO_2 depicted in Figure 4 can be enriched if density (kg/m^3) is put on the third axis as shown in Figure 5. In Figure 5, densities of gas phase are calculated via the Peng-Robinson (PR) equation of state (EOS), whereas liquid and solid phase densities (saturated or not) are calculated via the correlations presented by Span and Wagner (1996) and Trusler (2011), respectively. In Figure 5, the saturation lines SLE, SVE and VLE are also depicted, where the VLE line is presented with its gas and liquid branches merging at the CP. Figure 5 reveals that CO_2 can be found with densities well above the density of water (1000 kg/m^3), either as a solid or as high pressure dense liquid or SCF fluid.

The PR-EOS is a very simple thermodynamic relationship that can be used to predict gas and liquid properties of pure CO_2 and of CO_2 rich mixtures. The PR-EOS is certainly not the most accurate state relationship to address the fluid properties of CO_2 (Genesis, 2011; see also the EOS presented by Span and Wagner, 1996), but it does represent the best compromise between simplicity of use and accuracy for CO_2 rich systems without water, either in single fluid phase or in two coexisting fluid phases (VLE), at low or high pressures and densities (Li, 2008; Li and Yan, 2009; Genesis, 2011). According to Genesis (2011), the PR-EOS produces errors for

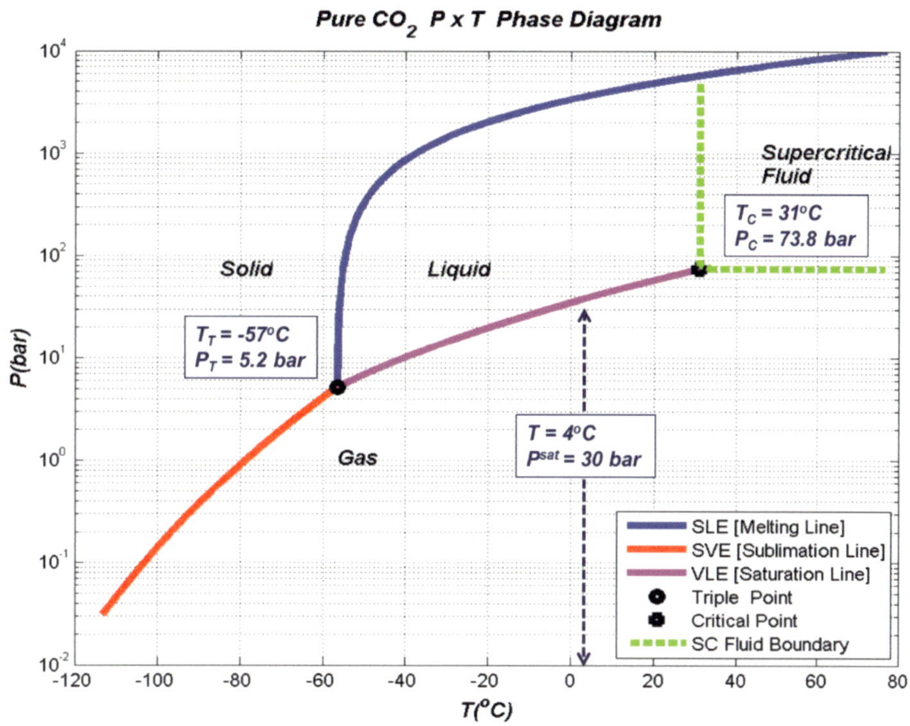

Figure 4. Phase behavior of pure CO_2 [T_C, P_C critical point; T_T, P_T triple point]

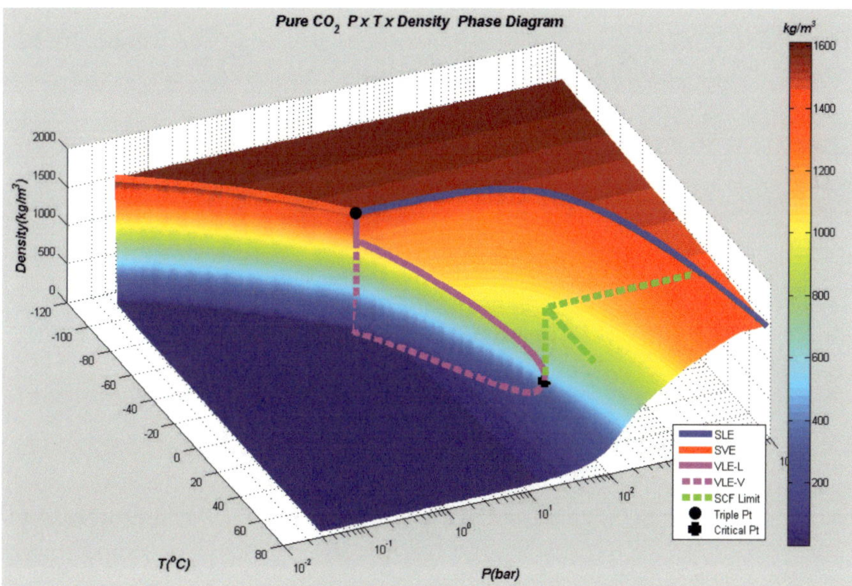

Figure 5. Density x Pressure x Temperature of Pure CO_2 [with TP and CP points, VLE, SVE & SLE lines and SCF domain]

compressibility factor (Z), molar enthalpy (H) and sound speed (C) below 1% for CO_2 subcritical gas phases and between 5% and 20% for subcritical liquid phases. Near the CP, the errors in Z, H, C can be higher but the uncertainties of experimental values are also higher. In the SCF domain, the PR errors for Z, H, C fall between 1% and 15%. In all the aforementioned cases, errors are relative to Span-Wagner EOS for fluid CO_2. The PR-EOS also seems adequate to describe the supercritical fluid (SCF) domain of CO_2 and its rich mixtures near critical transitions. This means that PR-EOS can be used to address any property (density, enthalpy, entropy, exergy, etc) of liquid and vapor phases of CO_2 and its mixtures with NG species under moderate errors, which are compatible with engineering applications (Li, 2008 and Li and Yan, 2009; Genesis, 2011). The PR-EOS is presented in Eq. (1), while its classical mixing rules follow in Eqs. (2) and (3). PR-EOS component parameters are given in Eq. (4) to (6) where Tc_i, Pc_i, ω_I and nc are, respectively, critical temperature, critical pressure, acentric factor for component i, and the number of components. K_{ij} represents the binary interaction parameter (BIP) for species i and j which is symmetric, can be used as zero in some cases, and is not necessary for pure CO_2. In Eqs. (1) to (3), V, T,P,R and N_i represent volume (m^3), temperature (K), pressure (bar), ideal gas constant ($8.314.10^{-5}$ $bar.m^3/mol.K$) and the mol number of species i.

$$P = \frac{NRT}{V - Nb} - \frac{N^2 a}{V^2 + UVNb + W(Nb)^2} \quad , \quad U = 2, \quad W = -1 \tag{1}$$

$$Nb = \sum_{i=1}^{nc} N_i b_i \tag{2}$$

$$N^2 a = \sum_{i}^{nc} \sum_{j}^{nc} N_i N_j \sqrt{a_i} \sqrt{a_j} \sqrt{\Phi_i(T)} \sqrt{\Phi_j(T)} \left(1 - K_{ij}\right) \quad , \quad (K_{ij} = K_{ji}) \tag{3}$$

$$a_i = 0.45724 \frac{R^2 Tc_i^2}{Pc_i} \quad , \quad b_i = 0.07780 \frac{RTc_i}{Pc_i} \tag{4}$$

$$\Phi_i(T) = \left(1 + g(\omega_i)\left(1 - \sqrt{T / Tc_i}\right)\right)^2 \tag{5}$$

$$g(\omega_i) = 0.37464 + 1.54226\omega_i - 0.26992\omega_i^2 \tag{6}$$

If dimensionless terms in Eq. (7) are used in Eq. (1), the classic Z cubic form results in Eq. (8):

$$Z = \frac{PV}{NRT} \quad , \quad B = \frac{PNb}{NRT} \quad , \quad A = \frac{PN^2 a}{(NRT)^2} \tag{7}$$

http://dx.doi.org/10.5772/57560

$$Z^3 - (1 + B - UB)Z^2 + (A + WB^2 - UB - UB^2)Z - AB - WB^2 - WB^3 = 0 \tag{8}$$

3. The CO_2 capture cycle

According to Oi (2010), CO_2 removal from process streams at an industrial scale has occurred since about 1930, mainly from natural gas and from industrial gases at high pressures for ammonia and methanol production. Like any separation unit in any kind of process, the CO_2 capture technology has to be judiciously chosen as it may undermine the profitability, controllability, safety, and simplicity of the plant. Nevertheless, even when properly selected, separations usually raise concerns like heat and mechanical energy consumption, increased utility use, carbon emission, chemicals demands, size, weight, footprint, construction restraints, operational hazards, etc.

The CO_2 capture cycle encompasses two or three main unit operations, which have to separate CO_2 from the gas mixture and send it to an appropriate destination. First, there is the CO_2 transfer across the gas phase into the medium that contains the binding material: a solvent or an adsorbent or a selective barrier. Second, there is (or not) the regeneration of the binding medium with concomitant CO_2 release. Third, there is the compression and cooling of the captured CO_2 because it has to be handled at high density or as a liquid.

Technologies for CO_2 capture from gas streams include chemical absorption (e.g., aqueous ethanolamines and aqueous K_2CO_3), physical absorption (e.g., propylene carbonate, selexol and rectisol), physical adsorption, membrane permeators, membrane contactors, cryogenic distillation and hybrid technologies (e.g., membrane permeator followed by ethanolamine absorption). Among those, the most relevant technologies include membrane permeation and chemical absorption with aqueous ethanolamines, the later standing as the most mature CO_2 capture technology from gas streams.

3.1. Chemical absorption with aqueous alkanolamines

With regards to the solvent, the advantages of the chemical absorption of acid gases – CO_2 and H_2S – by aqueous alkanolamines are well-known: the former are weak acids and the later are weak alkali, such that they reversibly bind at low temperature and high acid gas fugacity and subsequently untie at higher temperatures and low fugacity, leading to efficient acid gas stripping, which regenerates the solvent. The relevant variable in the liquid phase is the solvent loading (in mol of acid gas per mol of amine), which expresses the degree of conversion of amine in the solvent. Typically, the loading assumes values in the range 0 to 1.2 mol/mol amine.

AGWA (Acid Gas, Water and Amines) systems is a convenient denomination (de Medeiros et al., 2013a; de Medeiros et al., 2013b) of such reactive vapor-liquid equilibrium (RVLE) systems containing Acid Gas, Water and Amines. Amines are understood to be the common alkanolamines like monoethanolamine (MEA, MW=61), diethanolamine (DEA, MW=105), methyldiethanolamine (MDEA, MW=119) and 1-amino-2-propanol (AMP, MW=75).

MEA is a benchmark co-solvent for CO_2 capture, with: (i) satisfactory absorption capacity; (ii) fast kinetics; (iii) miscible with water in all proportions, and (iv) low cost. On the other hand, MEA is problematic in terms of solvent regeneration, because it exhibits: (i) the highest energy load per unit of stripped gas; (ii) corrosion and chemical/thermal degradation concerns; (iii) non-negligible evaporation losses due to its low boiling point; and (iv) high reactivity entailing low H_2S/CO_2 selectivity. As a secondary amine, DEA is less reactive than MEA but it has the following comparative advantages: (i) lower energy per unit of stripped gas; (ii) more resistant to degradation; (iii) less corrosive; and (iv) less volatile. MDEA exhibits the lowest reactivity with CO_2 and the greatest resistance to degradation. When compared to DEA and MEA, MDEA presents the following advantages: (i) the highest equilibrium capacity of acid gas absorption (in the case of CO_2, nearly two times the capacity of primary amine); (ii) lowest regeneration cost (does not form stable products with CO_2); (iii) high H_2S/CO_2 selectivity thanks to its low reactivity with CO_2; (iv) lowest enthalpies of reaction and lowest regeneration heat per unit of stripped gas; and (v) negligible losses due to very low vapor pressure. Alkanolamines with steric hindrance like AMP show reduced carbamate stability and the methyl adjacent to the amine group may affect absorption capacity and/or its rate. AMP exhibits absorption capacity and stripping heat similar to MDEA, but a faster reaction with CO_2 during the capture step (Medeiros et al., 2013b).

AGWA chemical reactions are really three-reactant transformations in the liquid phase with 1:1 amine to acid gas mol ratio. Water is necessary at high mol ratio to amine (8:1 or 7:1), otherwise the reaction simply does not evolve sufficiently, leaving non-solvated amine unconverted (low loading). This is the reason why MEA concentration in the solvent is upper bounded at 30%w/w in water (or 11.2%mol MEA + 88.8%mol H_2O), and 50%w/w in the case of MDEA (or 13.1%mol MDEA + 86.9%mol H_2O). Figure 6 presents a schematic of the chemical absorption system for CO_2/H_2S capture. AGWA absorption reactions take place in the colder higher pressure absorption column, whereas in the hotter lower pressure stripper column, AGWA reactions are reverted, thus releasing free acid gas and regenerating amine at the bottom. The main consumption of heat occurs in the reboiler in the bottom of the stripper column, where water is vaporized breaking the liquid phase association of acid gas, water and amine. Consequently, typical heat consumptions of MEA strippers lay between 167kJ and 200kJ per mol of stripped CO_2 (or between 3.8 GJ and 4.5 GJ/tonne of stripped CO_2). These figures are impressive high values, equaling 4 to 5 times the molar heat of vaporization of water per mol of stripped CO_2.

The literature presents several modeling approaches for absorption and stripping with AGWA systems. The most common approach involves cumulatively ionic species within ideal solution, ideal gas vapor phase, reversible chemical kinetics and rate-based interfacial mass transfer (de Medeiros et al., 2013b). This kind of approach is classical and is more adequate to low pressure and to dilute AGWA systems as in CO_2 capture from combustion gases. For high capacity and high pressure AGWA with rich CO_2 natural gas systems, high loadings and high heat effects may appear. For such AGWA systems de Medeiros et al. (2013a, 2013b) proposed a molecular Chemical Theory approach where molecular complex species are formed in the liquid phase via chemical equilibrium reactions having as reactants real AGWA species CO_2,

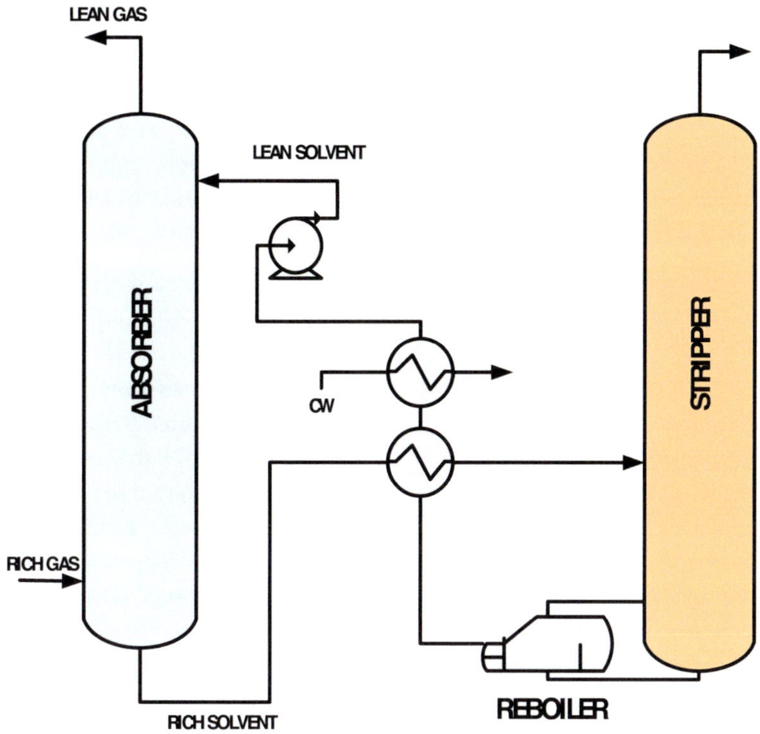

Figure 6. CO_2 capture by chemical absorption

H_2S, H_2O, MEA, DEA, MDEA and AMP. Each complex species is created by reacting three real species: an acid gas, water and an amine, as shown in Eq. (9). These complexes are reversibly created during the absorption step and destroyed during the stripping step. The main advantages of the molecular chemical theory of de Medeiros et al. (2013b) are:

i. Theory is a Reactive Vapor-Liquid Equilibrium (RVLE) framework that employs only VLE AGWA data for model tuning via implicit statistical procedures. Model can be tuned with low or high pressure data, appropriate either for the absorption column (low temperature, high pressure) or for the stripping column (high temperature, low pressure).

ii. AGWA RVLE is, in fact, an ionic scenario dominated by weak ions and several unknown solvated complexes. Thus, the limited knowledge to handle all the possible ions is circumvented via such nonionic, nonvolatile complexes.

iii. Model tuning requires only AGWA VLE data, much more available than non-equilibrium counterparts.

iv. The nonionic assumption is proposed to mimic the weakness of AGWA electrolytes, all created by incomplete dissociations.

v. All thermodynamics properties (of both vapor and liquid phases) in the formalism of de Medeiros et al. (2013b) are calculated via cubic EOS (Soave-Redlich-Kwong - SRK and Peng-Robinson - PR), because all species considered are purely molecular.

vi. The formalism can be used with both acid gases CO_2 and H_2S and with blends of alkanolamines, which are used for conjugating desirable qualities like good reactivity of MEA with lower costs of regeneration and corrosion of MDEA. Such blends are promising in terms of gains relatively to individual amines.

The AGWA formalism of de Medeiros et al. (2013b) was calibrated with VLE AGWA data from Literature. These data configure a large database of AGWA-VLE with 1331 runs shown in Figure 7. This database includes several runs with alkanolamines (MEA, DEA, MDEA and AMP), water and two acid gases or solutes (CO_2 and H_2S) at pressures ranging from 0.1 bar to 30 bar and temperatures from 25°C to 140°C. The calibration parameters correspond to chemical equilibrium constants belonging to chemical reactions that convert acid gas, water and amine into molecular complex species in liquid phase as shown in Eqs. (9). These chemical reactions are chemical equilibrium (ChE) reactions, which evolve to the right when CO_2 and H_2S are absorbed by the solvent and to the left when CO_2 and H_2S are stripped from the solvent by the action of heat and low pressure. In other words, the set of chemical equations in Eq. (9) can reproduce both phenomena absorption and solvent regeneration. This is a physically sound approach that allows, among other things, to estimate heat effects that occur in these processes like the release of heat during gas absorption and the absorption of heat during stripping of CO_2 and H_2S.

$$
\begin{array}{l}
CO_2(g) + H_2O(g) + MEA(g) \xleftrightarrow{K_1} CO_2 - H_2O - MEA(Liq) \\
CO_2(g) + H_2O(g) + MDEA(g) \xleftrightarrow{K_2} CO_2 - H_2O - MDEA(Liq) \\
CO_2(g) + H_2O(g) + DEA(g) \xleftrightarrow{K_3} CO_2 - H_2O - DEA(Liq) \\
CO_2(g) + H_2O(g) + AMP(g) \xleftrightarrow{K_4} CO_2 - H_2O - AMP(Liq) \\
H_2S(g) + H_2O(g) + MEA(g) \xleftrightarrow{K_5} H_2S - H_2O - MEA(Liq) \\
H_2S(g) + H_2O(g) + MDEA(g) \xleftrightarrow{K_6} H_2S - H_2O - MDEA(Liq) \\
H_2S(g) + H_2O(g) + DEA(g) \xleftrightarrow{K_7} H_2S - H_2O - DEA(Liq) \\
H_2S(g) + H_2O(g) + AMP(g) \xleftrightarrow{K_8} H_2S - H_2O - AMP(Liq)
\end{array} \quad (9)
$$

The parameter estimation of the AGWA model of de Medeiros et al. (2013b) involves estimating ChE constants for the chemical reactions in Eq. (9) at selected temperatures. Thus, several experimental AGWA VLE points at a chosen temperature are extracted from the AGWA database in Figure 7 and are processed via a maximum likelihood algorithm to estimate the ChE constants. The estimation algorithm also assures that the set of nonlinear constraints of each selected experiment is satisfied. The set of constraints of a given experiment is shown in Table 1 with Eqs. (10) to (16) including: (i) Real species – CO_2, H_2S, H_2O, MEA, DEA, MDEA, AMP – mass balance (RMB); (ii) mol fraction normalization for each phase (SXY); (iii) definition partial pressures of solutes (acid gases) (PPS); (iv) definition of loadings of solutes (LDG); (v) VLE of Real species (VLE); (vi) ChE of Complex formation in Eq. (9) (CHE). The nomenclature used in Table 1 is shown in Table 2, including the matrix of stoichiometric coefficients of Eq. (9) in Eq. (10).

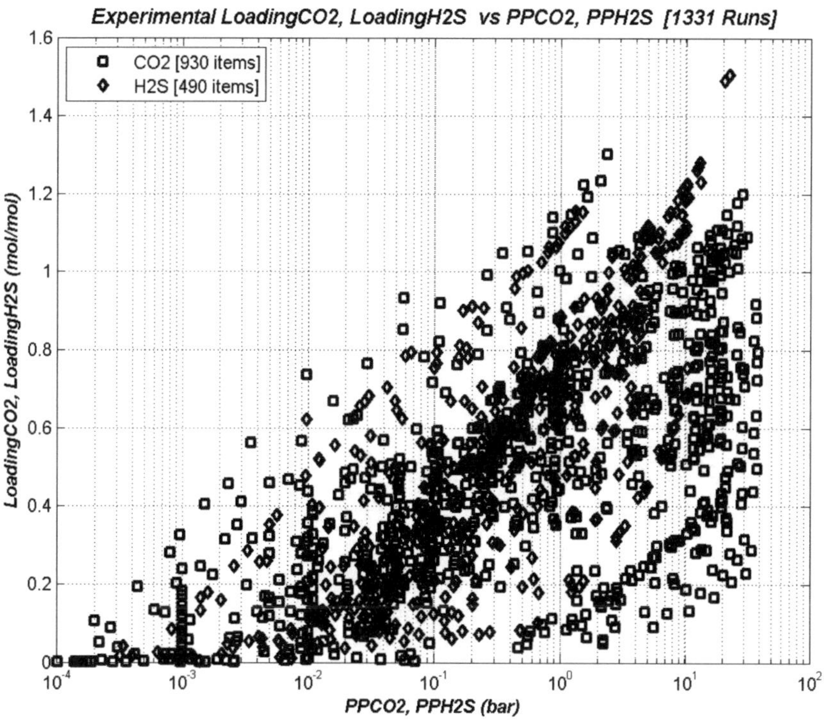

Figure 7. AGWA Database used to estimate parameters in the Chemical Theory Model of de Medeiros et al. (2013b) [PPCO2, PPH2S – partial pressures (bar) of CO_2 and H_2S]

Figures 8 and 9 depict some results after calibration of the AGWA model of de Medeiros et al. (2013b) with the database of AGWA experiments in Figure 7. Both figures refer to parameter estimation using 320 experimental AGWA data points at 40°C. Figure 8 presents experimental versus predicted loadings (mol/mol amine) of solutes CO_2 and H_2S at 40°C, whereas Figure 9 presents experimental versus predicted partial pressures (bar) of solutes CO_2 and H_2S at 40°C. As can be observed, there is a satisfactory agreement between experimental values and predicted counterparts.

Predicted values are calculated solving the set of constraints of the AGWA model in Eqs. (I) to (VIII) with all thermodynamic properties (e.g. vapor and liquid fugacities) estimated with SRK-EOS. The estimated parameters are the eight ChE constants in Eq. (9) and (VIII) at the corresponding temperature of 40°C.

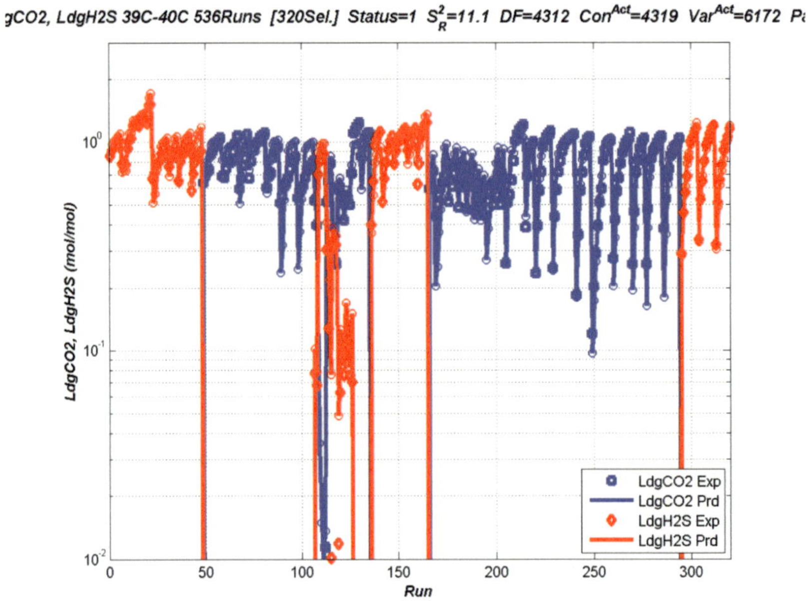

Figure 8. Experimental vs predicted Loadings of CO_2 (blue) and H_2S (red) after calibrating the AGWA model at 40°C with 320 data points (de Medeiros et al., 2013b) [LdgCO2, LdgH2S – Loadings of CO_2 and H_2S (mol/mol amine)]

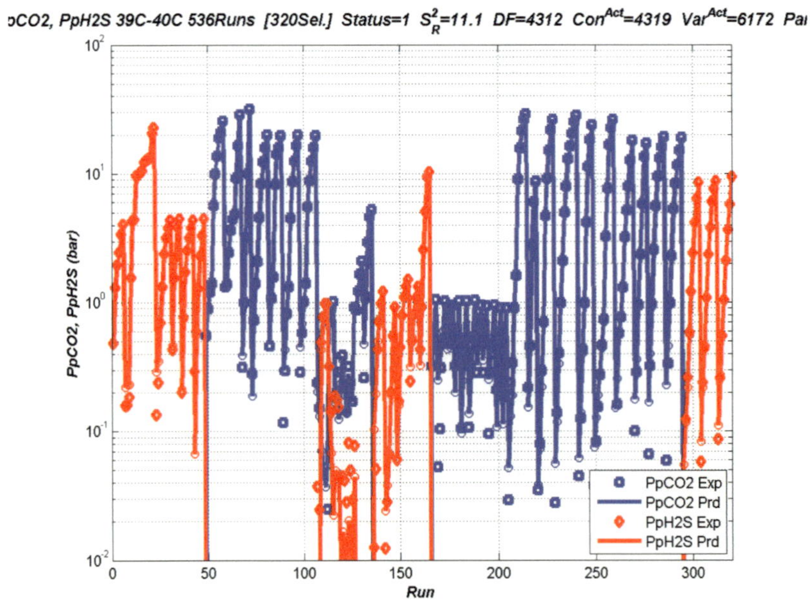

Figure 9. Experimental vs predicted partial pressures of CO_2 (blue) and H_2S (red) after calibrating the AGWA model at 40°C with 320 data points (de Medeiros et al., 2013b) [PpCO2, PpH2S – Partial pressures of CO_2 and H_2S (bar)]

http://dx.doi.org/10.5772/57560

Constraints [Label]	Expression	Eq.
n Real Species Balances [RMB]	$\underline{N} + L\underline{\underline{\Pi}}\underline{X}_C - L\underline{X} - V\underline{Y} = \underline{0}$	I
Normalizations [SXY]	$\sum_{j=1}^{n} X_j + \sum_{k=1}^{n_C} X_{Ck} - 1 = 0$	II
	$\sum_{j=1}^{n} Y_j - 1 = 0$	III
n_S *Solute Partial Pressures* [PPS]	$\underline{P}_S - P\underline{\underline{S}}_S\underline{Y} = \underline{0}$	V
n_S *Solute Loadings* [LDG]	$(\underline{1}^T \underline{\underline{S}}_A \underline{N})\underline{\alpha}_S - \underline{\underline{S}}_S(\underline{N} - V\underline{Y}) = \underline{0}$	VI
n VLE of Real Species [VLE]	$\ln \underline{\hat{f}}^V - \ln \underline{\hat{f}}^L = \underline{0}$	VII
n_r *Reactions ChE* [CHE]	$\ln \underline{X}_C + \underline{\underline{\Pi}}^T \ln \underline{\hat{f}}^L - \ln \underline{K}(T) = \underline{0}$	VIII

Table 1. Set of Constraints of AGWA VLE Experiments [real species (*n=7*): CO_2, H_2S, H_2O, MEA, DEA, MDEA, AMP], [solutes (*ns=2*): CO_2, H_2S], [Complexes (*nr=8*): see Eq. (9)]

Symbol	Definition
$\underline{\alpha}_S$	Vector of solute loadings (mol/mol of amine)
$\underline{\hat{f}}^V$, $\underline{\hat{f}}^L$	Vectors of Real species fugacities in vapor and liquid phases (bar)
$\underline{K}(T)$	Vector of ChE constants of Eq. (9)
L	Liquid phase mole number
n_S, n, n_r	Numbers of solutes (=2), Real species (=7) and Complex species (=8)
$\underline{N}$	Vector of total mole number of Real species in the experiment
$\underline{P}_S$	Vector of solute partial pressures (bar)
P	Pressure (bar)
$\underline{\underline{\Pi}}$	Matrix of stoichiometric coefficients of Eq. (9), shown in Eq. (17).
$\underline{\underline{S}}_A$, $\underline{\underline{S}}_S$	Selection matrices for amines and solutes with sizes 4 X 7 and 2 X 7
V	Vapor phase mole number
$\underline{X}, \underline{Y}$	Vectors of mol fractions of Real species in liquid and vapor phases
$\underline{X}_C$	Vector of mol fractions of Complex species in liquid phase

Table 2. Nomenclature used in Table 1

$$\underline{\underline{\Pi}} = \begin{bmatrix} -1 & -1 & -1 & -1 & 0 & 0 & 0 & 0 \\ 0 & 0 & 0 & 0 & -1 & -1 & -1 & -1 \\ -1 & -1 & -1 & -1 & -1 & -1 & -1 & -1 \\ -1 & 0 & 0 & 0 & -1 & 0 & 0 & 0 \\ 0 & -1 & 0 & 0 & 0 & -1 & 0 & 0 \\ 0 & 0 & -1 & 0 & 0 & 0 & -1 & 0 \\ 0 & 0 & 0 & -1 & 0 & 0 & 0 & -1 \end{bmatrix} \qquad (10)$$

The huge heat consumption of MEA strippers (167 kJ to 200kJ/mol of stripped CO_2) forces investigation on more efficient alternatives of solvent regeneration. Wagener and Rochelle (2010) recognize as a "monumental task" reducing energy penalty of CO_2 capture from coal-fired power plants (approximately 30%). They presented an analysis of various stripper configurations, concluding that increasing complexity improves performance at the cost of higher capital and operational expenditures, i.e., an optimal scheme should exist. The alternative configurations include simple stripping with vapor recompression, multi-pressure, double matrix, split product, internal exchange, and flashing feed. Wagener and Rochelle (2010) concluded that operating with multiple pressure levels reduces the energy requirement as "equivalent work" (including reboiler duty, pumping and heat exchangers) in 33.6 kJ/molCO$_2$ captured, with an optimal lean loading of 0.375 mol/mol amine. Moreover, they claim that the arrangement benefits from stripping at high pressure, whilst improves reversibility when returning to atmospheric conditions for the absorber. Wagner and Rochelle (2011) revisited several configurations with varying levels of complexity and reported that an inter-heated column and a simple stripper required 33.4 kJ/mol CO_2 and 35.0 kJ/mol CO_2 of equivalent work, respectively, at their optimum lean loadings.

3.2. Membrane permeators

The first membrane modules were developed as planar films. However, such arrangement has a low ratio of membrane transfer area per equipment volume. Presently, the majority of modules for CO_2 gas separation are manufactured in hollow-fiber or spiral wound configurations. Figure 10 shows a schematic of a hollow-fiber membrane permeator module and the selective transport of molecules across the hollow-fiber membrane.

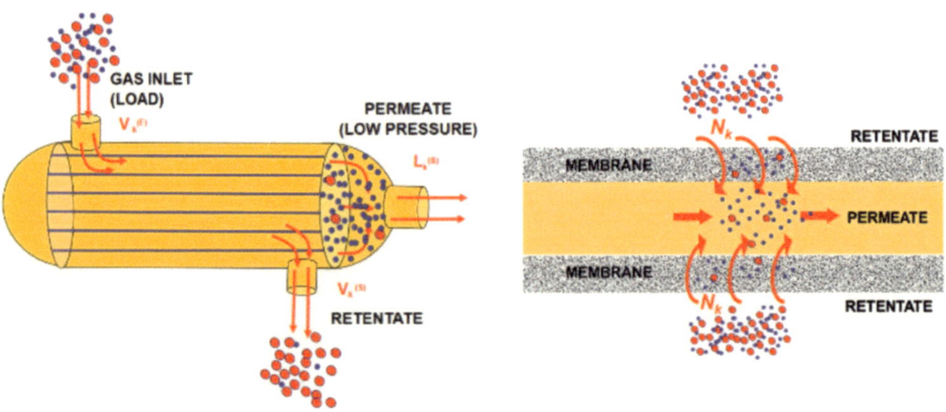

Figure 10. Hollow-fiber membrane permeator and the transfer through a hollow-fiber

The membrane acts selectively against gas diffusion from the high pressure side (the retentate) to the low pressure side (the permeate) creating separation. The membrane – either in hollow-fiber or spiral wound configurations – is composed by two polymeric structures: a permselective dense (nonporous) skin over a thicker microporous substrate. Inside the permeator gas coming from the high pressure retentate stream solubilizes into the dense skin and diffuses to the low pressure side creating an almost adiabatic expansion with consequent fall of temperature via Joule-Thomson effect. Such fall of temperature may create problems to the proper functioning of the module due to eventual condensation of less volatile species that are retained and accumulate in the retentate. This condensation is undesirable and can damage the membrane plastic material.

Ho and Wiley (2005) evaluated the economic performance of membrane separation for low pressure (flue gas applications) and high pressure (e.g., natural gas processing). The authors concluded that the highest share in capital cost was due to the compression phase (around 80%) while membrane and the respective shells exhibited 10% of the investment. For natural gas processing, however, membrane has the largest share of CAPEX (about 62%), as the compression is an existing stage of natural gas processing. Moreover, Ho and Wiley also report an alternative multi-stage configuration in order to obtain a CO_2 rich permeate in a second membrane stage, which demands a recompression step corresponding to 30% of the total CAPEX.

In oil and natural gas deep-water offshore rigs, an issue in the processing of large volumes of NG with high CO_2/CH_4 ratio *vis-à-vis* climate concerns, involves large capacity CO_2 separators, whose targets are exportation of saleable NG from the retentate product via long pipelines and feeding enhanced oil recovery (EOR) systems with hyperpressurized CO_2 from the permeate product. In this scenario, robustness, angular indifference, modularity, and compactness also influence the selection of a separation technology. In this context, membrane permeation batteries are usually favored against the more traditional absorption columns with amine solvents (Medeiros et al., 2013a). On the other hand, large permeation batteries also have their own shortcomings, mostly related to the permselective dense (nonporous) skin over the microporous substrate. The permselective skin forces the existence of a high ΔP and exhibits low fluxes which means low capacity per unit area and high consumption of power for permeate recompression and/or permeate recycles. It also has CO_2/CH_4 limited selectivity and demands feed necessarily dew-point conditioned (to prevent gas condensation in the retentate provoked by the decrease of temperature associated with permeation) and continuous inspection looking for membrane bursts.

A model for hollow-fiber permeators can be built by writing mass, momentum and energy conservation principia for both permeate and retentate. Additionally, an appropriate model for trans-membrane flux transport is also necessary. Typically, trans-membrane flux models are written as products of a permeancy coefficient and a driving force term. Each species has a permeancy coefficient for a given membrane and given conditions of temperature, pressure and gas composition, but, in general, permeances are assumed independent of gas composition and pressure. The driving force term is usually expressed as a difference of component fugacities between the retentate and permeate sides. In this context and assuming a co-current compressible flow of both permeate and retentate, Nakao et al. (2009) proposed a stationary hollow-fiber membrane permeator rigorous model which can be used to predict CO_2 separation from CO_2 rich NG. The model of Nakao et al. (2009) builds a spatial 1-dimensional axial description where species mass balances, energy and momentum balances are rigorously written for both permeate and retentate. All permeate and retentate thermodynamic properties (enthalpies, densities, fugacities, etc) are rigorously calculated via PR-EOS, Eq. (1). The model is too involved to be discussed here in detail, but the clarity and usefulness of it results are worth presenting in the context of CO_2 capture. Figure 11 depicts a natural gas feed named GBS2 (from Basin of Santos, Brazil) with 2.79554 MMNm^3/d at 53bar and 38°C with, initially, 25.32%mol CO_2 and 55.64%mol CH_4. Figure 11 depicts the VLE locus of GBS2 and the heat capacity (*Cp* in kJ/mol.K) map versus *T X P* with vivid identification of the critical and supercritical neighborhoods due to the second-order transition across the critical state that is revealed by second order properties like *Cp*. In Figure 11 [A], the gas feed is located at 28°C above its dew boundary (blue). GBS2 passes through a battery of 40 horizontal modules (0.2m X 10m each) of cellulose-acetate hollow-fibers (0.5mm ID), with CO_2 and CH_4 permeances respectively of 1.27.10^{-8} and 4.4. 10^{-10} mol/s.m^2.Pa and 1854m^2 of permeation area per module. Permeate pressure is 4bar and the external temperature is 27°C with a heat transfer coefficient of 5W/m^2.K. Resulting profiles in the axial flow direction (10m long) for one module are shown in Figure 12. Permeate and retentate initiate contact at axial position *X=0m* and cease contact at *X=10m*. The background along the axis of flow is painted in four colors for discrimination among the four quarters of a typical module. Figure 12 [A] depicts retentate profiles of %mol CO_2 and CH_4 showing a CO_2 decrease from 25.62% to 9.3%mol and a CH_4 increase from 55.64% to 65.5%mol. Figure 12 [B] shows that the final permeate recoveries of CO_2 and CH_4 are 72% and 11.5%. Figure 12 [C] depicts permeate and retentate temperature profiles revealing a drop of retentate temperature from 38°C to 29°C, insufficient for condensation (Figure 11 [A]).

3.3. Gas-liquid contactors

Gas–liquid contactors (GLC) constitute a new and versatile kind of membrane operation for CO_2 removal from NG. A GLC unit admits a gas stream which is contacted with a liquid stream (the solvent stream) through a polymeric micro-porous membrane. The solvent phase is an aqueous solution of ethanolamines (e.g. MEA and/or MDEA) that can absorb CO_2 as occurs in an absorption column. But the difference here is that the liquid and gas phases really do not mix as they do in direct contact devices like packing towers. Assuming that the GLC is manufactured with hollow-fiber membranes, the solvent phase flows in the hollow-fiber inner space while the gas phase flows in the outside shell space, but there are also configurations where the roles of liquid and gas phases are inverted. A module of hollow-fiber GLC with gas flowing in the shell side and with solvent flowing in the inner space of the hollow-fibers is sketched in Figure 13. As can be seen, the module resembles closely a shell and tube indirect contact heat exchanger.

GLC membrane operation can outperform common nonporous permeators in terms of capacity per unit area while sustaining high CO_2/CH_4 selectivity. High fluxes are possible in GLC because the membrane does not have a dense skin in order to be selective. The underlying reason is that selectivity is imposed by the solvent in the inner (permeate) side cutting the necessity of high ΔP across the membrane. As the reader can see, terms permeate and retentate are also used here despite some improperness, and refer, respectively, to the inner solvent flow and to the gas flow that was not transferred to the inner solvent.

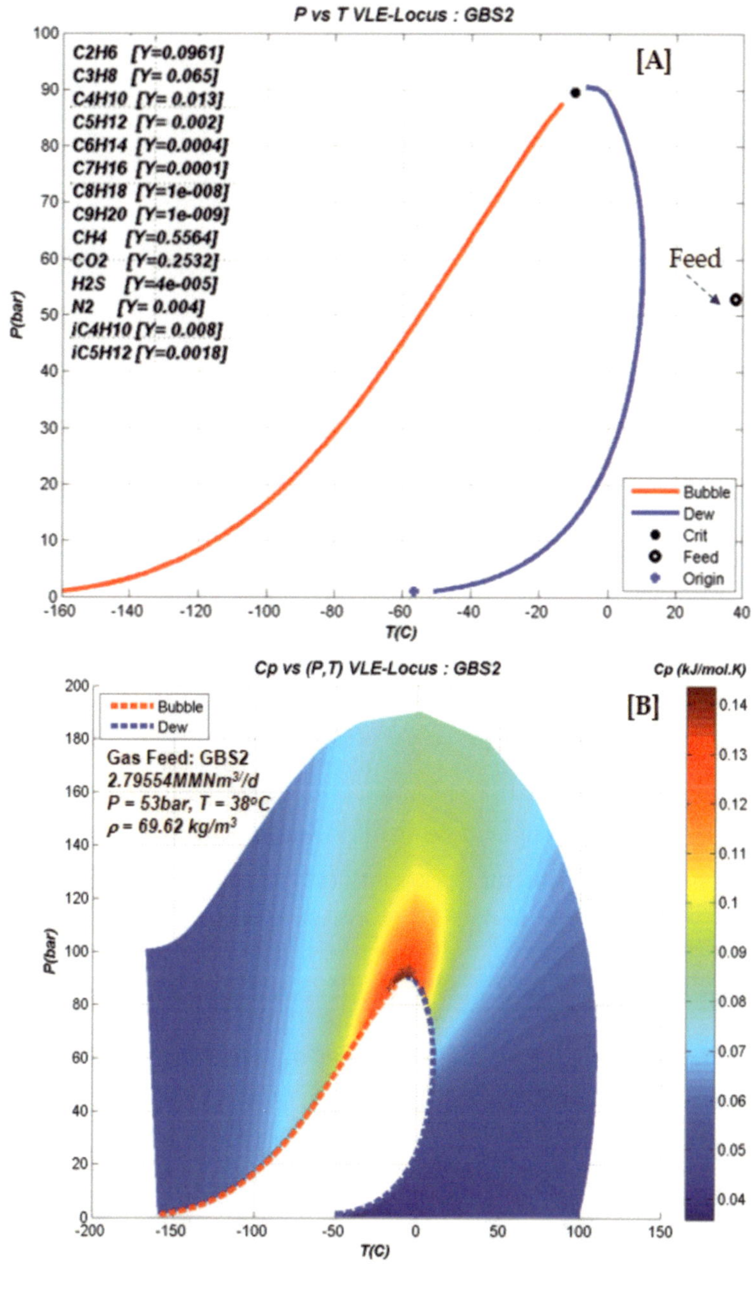

Figure 11. GBS2 a CO_2 rich natural gas (25.32%CO_2 55.64%CH_4 9.61%C_2H_6, 9.43%C_3^+) for Hollow-Fiber Permeator: [A] VLE Locus *P*(bar) X *T*(°C); [B] C_P(kJ/mol.K) vs *P*(bar) X *T*(°C)

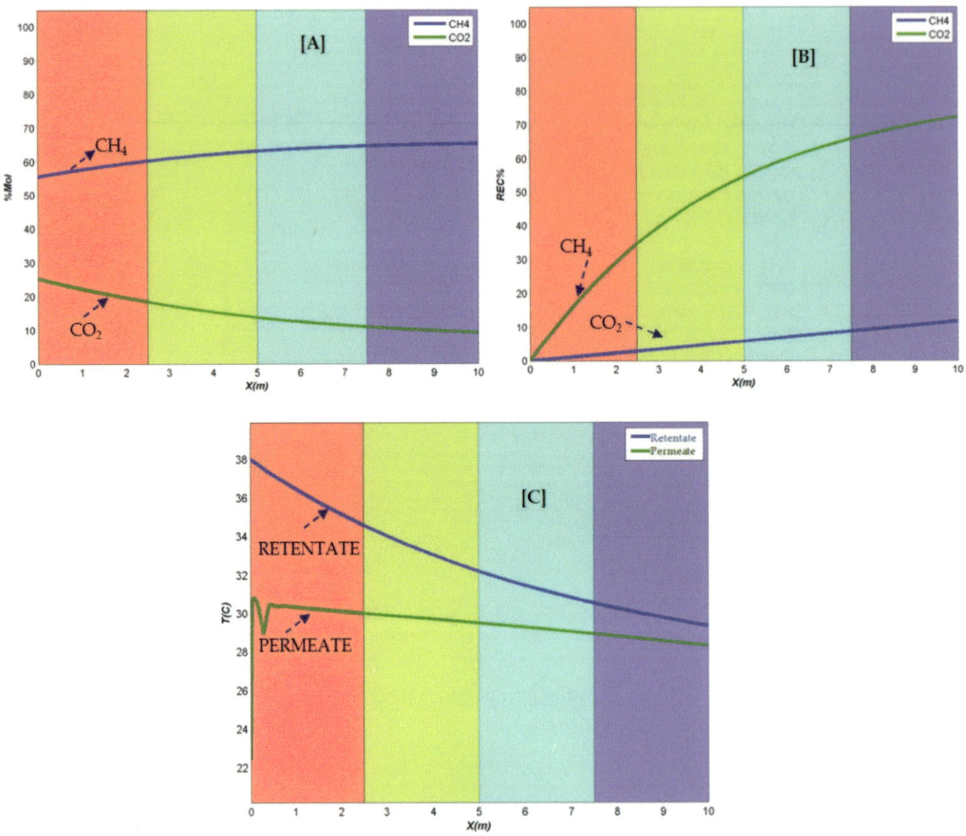

Figure 12. Profiles in a hollow-fiber module in co-current flow with feed GBS2 in Fig. 3.6 (25.32% CO_2 & 55.64% CH_4): [A] CO_2 & CH_4 %mol in retentate; [B] CO_2 & CH_4 permeate % recoveries; [C] retentate and permeate temperatures.

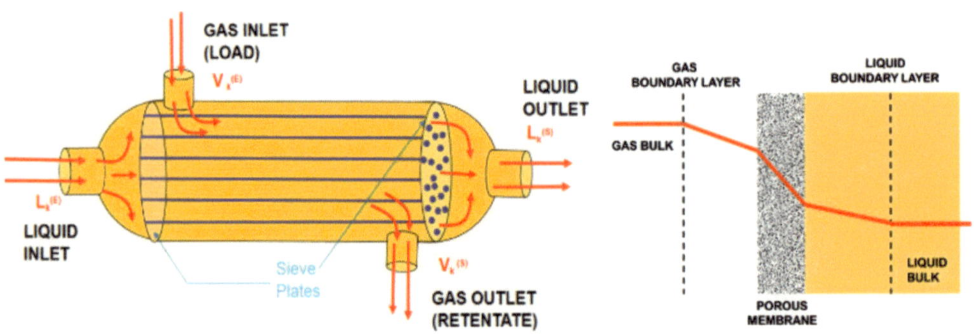

Figure 13. Hollow-fiber gas-liquid contactor with solvent in the inner space of membrane

GLC can separate CO_2 from NG offering the advantages of both membrane and absorption technologies, but leaving behind the respective shortcomings like the low flux coexisting with high difference of pressure between the membrane sides and the flooding concerns coexisting with dependence on gravity in the case of absorption columns. GLC combines polymeric membrane separation and chemical absorption, using a physically and chemically active solvent for selective CO_2 removal. However, the new aspect is that there is a membrane standing as a physical barrier against the unnecessary mixing of gas and liquid phases. Other advantages of GLC are: (i) it allows independent manipulation of liquid and gas flows; (ii) offers larger area of gas–liquid interface; (iii) modularity ; (iv) it exhibits flexibility to increase/ decrease operational scales; (v) no dew-point conditioning of the gas feed is necessary; and (v) angular indifference allowing horizontal or vertical operational arrangements. Figure 14 sketches a typical process flowsheet for operation of a GLC unit capturing CO_2 from a NG feed.

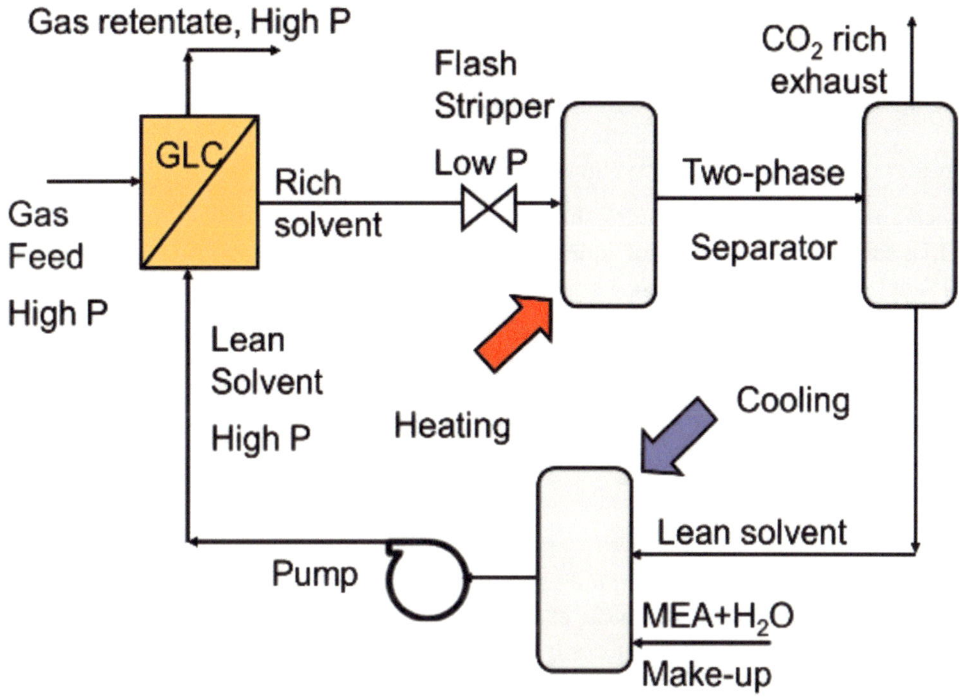

Figure 14. Typical process flowsheet for CO_2 capture with gas-liquid contactor (GLC)

Here the GLC operates t high NG pressure, but the rich solvent has to be regenerated in a low pressure stripper where CO_2 is released by the action of heating and low pressure as occurs in the second regeneration column shown in Figure 6.

A model for stationary gas-liquid hollow-fiber contactor was proposed by de Medeiros et al. (2013a) for separating CO_2 from CO_2 rich natural gas. This model is based on the AGWA theory discussed in the sub-section 3.1 (de Medeiros et al., 2013b). The model assumes a hollow-fiber contactor with co-current compressible flows of both permeate and retentate, where the permeate corresponds to the inner space inside the hollow-fibers. Permeate and retentate are separated by the membrane. The permeate is supposed in two-phase flow because the trans-membrane flux of methane will support the maintenance of a gas phase in the inner permeate flow. The permeate is, in fact, a two-phase reactive flow in continuous reactive vapor-liquid equilibrium (RVLE) because CO_2 is in reactive vapor-liquid equilibrium with water and amine via a set of chemical equations similar to Eq. (9) (without the H_2S chemical equations if the GLC is designed for CO_2 capture only).

The model of de Medeiros et al. (2013a) is based on 1-dimensional axial geometry of the GLC. Species mass balances and momentum/energy balances are written for both RVLE permeate and gas retentate. All properties of gas and liquid phases are calculated with PR-EOS. The trans-membrane flux terms are written as products of permeancy coefficients and a driving force term. Each species has a permeancy coefficient for a given membrane and given temperature conditions. The driving force term is expressed as a difference of component fugacities between the retentate and RVLE permeate sides. The major difficulty encountered in the GLC model is to represent the RVLE two-phase flow in the inner membrane space. This is accomplished by solving the AGWA VLE model (de Medeiros et al., 2012b) described in Table 1, along the entire path of the permeate in the inner space of the hollow-fibers. As an example, consider the equilibrium map in Figure 15.

This map was calculated assuming that 1 mol of natural gas (with 16.7%mol CO_2 +82.3%mol CH_4) is contacted with 1.2 mol of liquid solvent containing 14.5%mol MEA +14.5%mol MDEA +71%mol H_2O. Only the chemical reactions of CO_2 absorption by MEA (Eq. 9a) and MDEA (Eq. 9b) are considered. Each point in Figure 15 represents the resulting equilibrium vapor phase mol fraction of CO_2 (Y_{CO2}) versus T(°C) X P (bar) under reactive VLE. Clearly, the locations with low Y_{CO2} (blue) are dominated by absorption phenomena, while those with high Y_{CO2} (red) are dominated by stripping phenomena. This kind of inner RVLE solver is a key element in the construction of the GLC model of de Medeiros et al. (2013a).

3.4. The CO_2 transportation cycle

An efficient and reliable transportation system is required to displace enormous quantities of captured CO_2 to their destination site at appropriate geological formations that are capable to accommodate hundreds of billions or trillions of Nm^3 of CO_2 under stable and secure conditions. Clearly a not especially large size thermoelectric coal plant can produce something like 1 million of metric tonnes (1Mt) of CO_2 per year or about 500.10^6 Nm^3/y. Such huge capacities can only be attended by large pipeline systems operating with dense compressed fluid (liquid or dense supercritical CO_2), because the other existing alternatives – road, railroad and barge transport – simply cannot cope with dense fluid pipelines in terms of unitary cost and capacity of transportation.

In other words, despite of the existence of many options for transporting compressed (gas or liquid) CO_2 – including highway tankers, railway tankers, ships, and pipelines – it is evident that the impressive volumes that must be transported dictates that only pipelines working at high pressures, high densities and high capacities are suitable for this service. For instance, 2-3Mt/y of CO_2 have to be transported to dispose of the entire production of a single 500MW coal-fired power plant. This corresponds to transporting 230-350t/h of CO_2, just to service a single, medium-sized, client. Thus, only a network of large-scale pipelines could provide viable overland transport of such massive flow rates of CO_2. Presently, about 50Mt/y of CO_2 (equivalent to the output of 16 coal-fired power plants) are transported by 3100km of CO_2 pipelines, mainly for EOR processes in the USA and Canada (de Medeiros et al., 2008). The best example is the 808 km long, 30" diameter, Cortez Pipeline that transports 13Mt/y of CO_2 from highlands in Colorado to oilfields in Texas, USA.

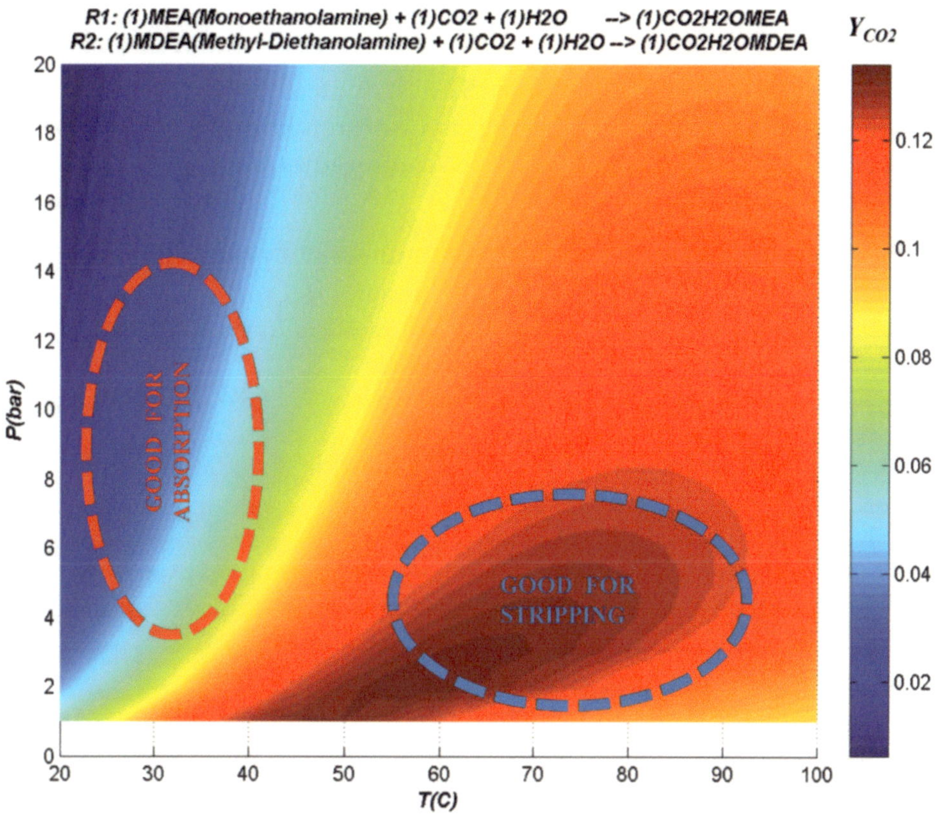

Figure 15. Equilibrium vapor phase mol fraction of CO_2 (Y_{CO2}) versus T(°C) X P (bar) under reactive VLE: 1mol of a 16.7%mol CO_2 natural gas is contacted with 1.2 mol of a liquid solvent with 14.5%mol MEA, 14.5%mol MDEA and 71%mol water.

Based on historical capital and O&M data for a 480-km long CO_2 pipeline without booster compressors, McCoy (2008) projected a fixed O&M coefficient of \$3,250/y/km for CO_2 pipelines. Considering a horizontal pipeline without appreciable elevation changes and an annualized fixed cost of 15% of capital, McCoy estimated the total unitary cost of CO_2 transportation as only \$1.16 per tonne of CO_2 per 100km. Based on a Monte Carlo sensitivity analysis, McCoy (2008) determined a range of \$0.75 to \$3.56 per tonne of CO_2 per 100km for this cost, recommending the median value of \$1.65 per tonne per 100km as a suitable estimate for investment decisions.

The design of CO_2 pipelines depends on reliable compressible flow models for dense compressible fluid near critical conditions. This model should account for thermal compressibility effects inside the fluid, i.e., temperature increases (decreases) during downhill (uphill) flow due to gravity compression (expansion).

In the same way, external heat transfer and elevation effects must be allowed. The extremely high compressibility of CO_2 near its critical state at 31°C (shown in Figures 4 and 5), leads to potential abrupt changes of velocity due to abrupt changes of density as the fluid compresses (decompresses) near the critical state. In other words, any candidate model for high capacity CO_2 pipelines must be able to calculate thermodynamic properties of dense supercritical CO_2 near its critical transition with accuracy. Such a CO_2 pipeline model has been proposed by de Medeiros et al. (2008) by solving rigorous species mass balances and energy/momentum balances along the pipeline with all thermodynamic properties given by PR – EOS or SRK – EOS.

This model is demonstrated in the simulation of a sub-sea CO_2 pipeline (Figures 16, 17 and 18) for transportation of 20 MMSm³/d of CO_2 from onshore plant to five EOR wellheads 2100m deep, 380km from the coast. The 20'' pipeline extends 380km from west to east and 320km from south to north with 700 km of length.

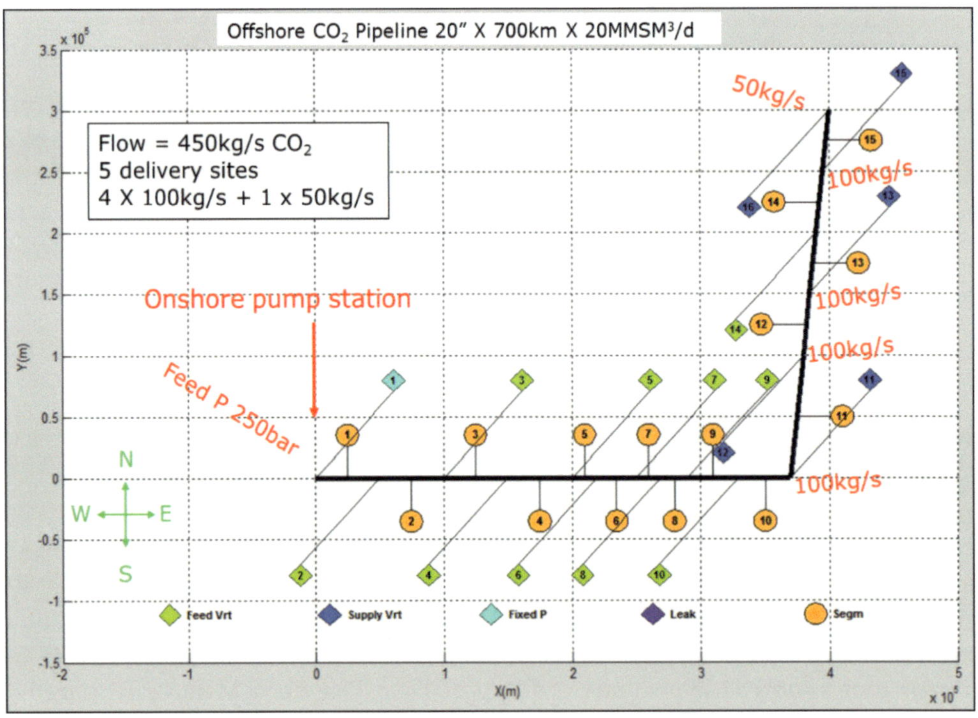

Figure 16. Hypothetical 20"X 700 km sub-sea pipeline for 20MMSm³/d of liquid CO_2 (≈450kg/s) at 250 bar from onshore plant to five offshore EOR wellheads (4X100kg/s + 1X50kg/s) 2100m deep, 380km from the coast.

As seen in Figure 17, a problematic factor was inserted between the continental shelf and the continental slope, namely, a big rift 1500m deep lies in the pipeline route. Rifts are not common in Santos Basin, Brazil, but they exist in the Norway arctic coast (Pettersen, 2011).

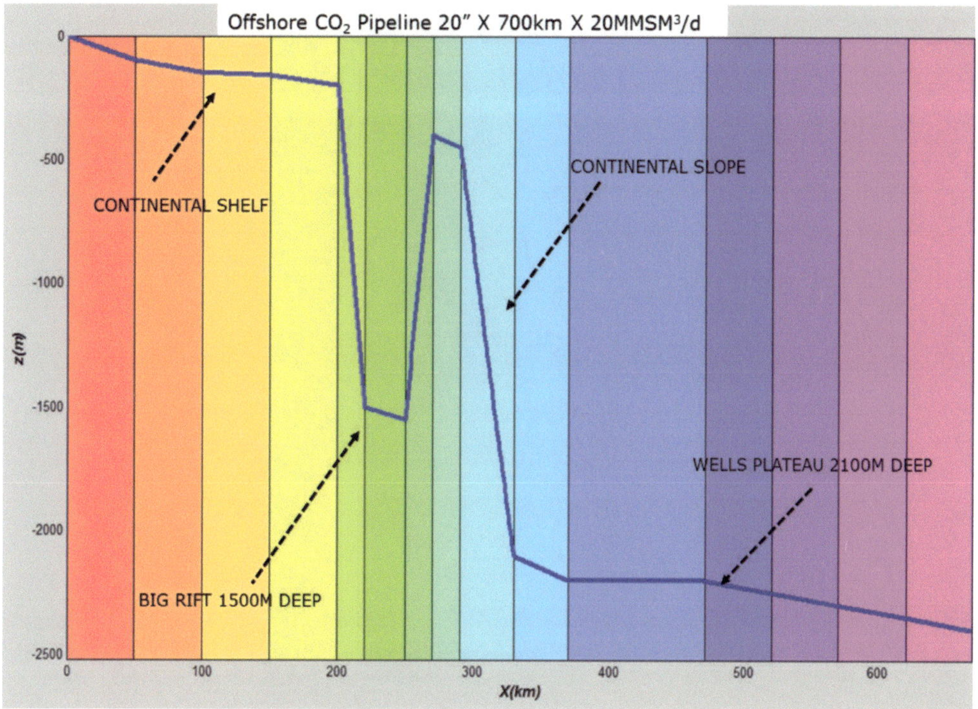

Figure 17. Elevation profile for the hypothetical sub-sea CO_2 pipeline from onshore plant to five EOR offshore well-heads 2100m deep, 380km from the coast.

4. The CO_2 utilization cycle

Although CGS has been regarded worldwide as a mitigation technology, it deals with CO_2 as a waste with an energy and an economic penalty for its disposal (Armstrong, 2012). Rather than treating CO_2 as a waste, carbon dioxide utilization (CDU) recognizes it as a raw material in chemical process to produce high added-value carbon containing products. It is also worth noting that the CDU is a complementary technology to CGS, not a competing technology, adding value to a process and thus it may help balance the costs of CGS.

While CO_2 has broad industrial application as solvent (supercritical CO_2, fire extinguishers) and in the food industry, it has disadvantages as a chemical raw material due to its low reactivity and few reactions are thermodynamically feasible. Furthermore, each potential use of CO_2 as reactant has an energy requirement that needs to be determined and must not exceed the CO_2 utilized and, although the utilization of CO_2 has been subject of research since before

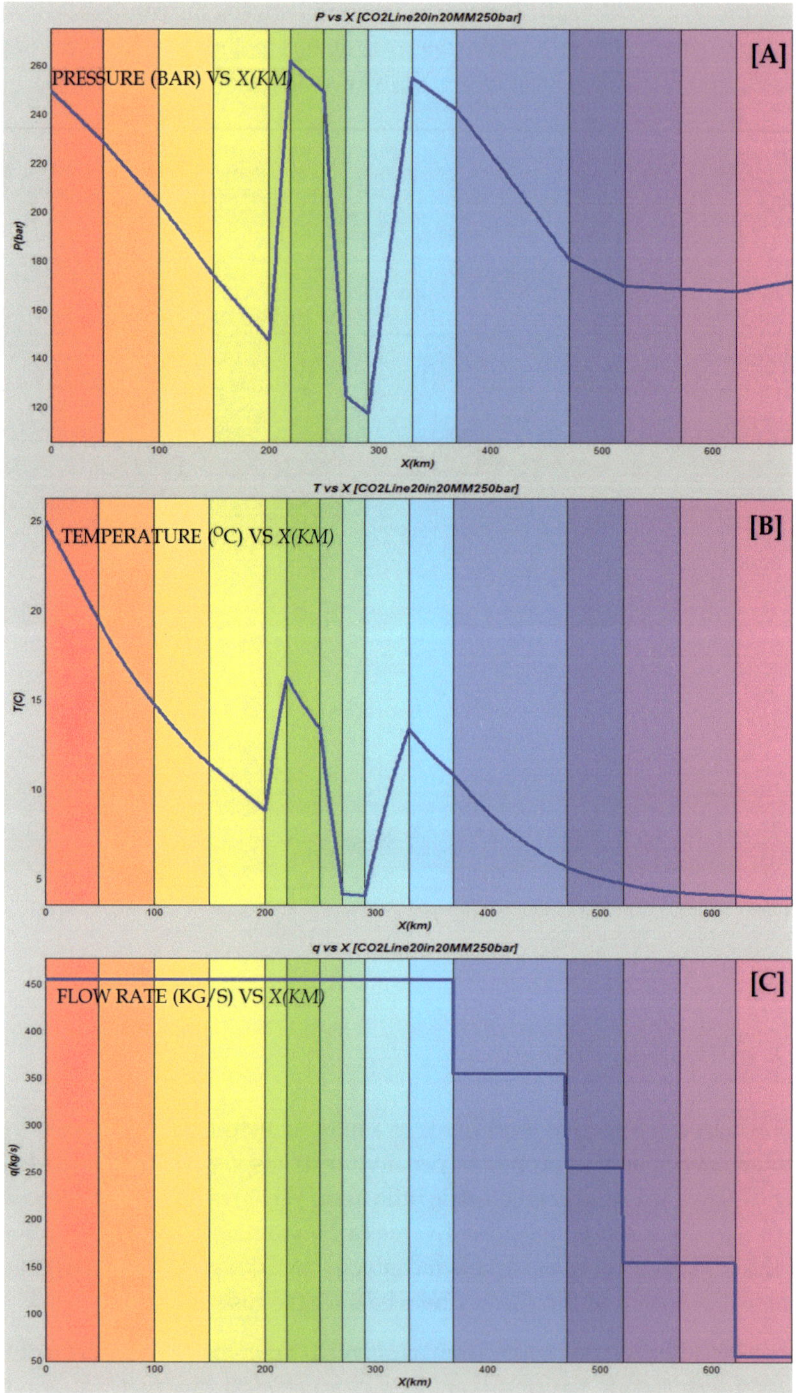

Figure 18. Calculated profiles for the hypothetical 20"X 700 km sub-sea pipeline with 20MMSm³/d (≈ 450kg/s): [A] pressure (bar); [B] temperature (°C); [C] mass flow rate (kg/s)

1970´s, there is much research still needed for CO_2 activation. Moreover, the utilization of CO_2 to cause an effective reduction in its emission into the atmosphere, must observe certain guidelines (Aresta, 2010): (i) the new process must reduce the overall CO_2 emissions; (ii) it mustbe less energy - and material - intensive with respect to the on–stream processes that it aims to replace; (iii) the new process must reduce the overall CO_2 emissions; (iv) it must employ safer and more eco-friendly working conditions; (v) it needs to be able to operate on a commercial scale and (vi) it must be economically viable.

According to Song (2006), the global market for CO_2 is estimated to be $3.2 billion/year in 2003. Carbon markets across the world were valued at 96 billion euros ($122.28 billion) in 2011 (Reuters Agency, 2012). Utilization of CO_2 by the chemical industry exists for more than one century as, for instance, the synthesis of urea (50 Mt/y) (Aresta, 1999) (Eq. 11), salicylic acid (Song, 2006) (Eq. 12) and inorganic carbonates (20Mt/y) (Aresta, 1999).

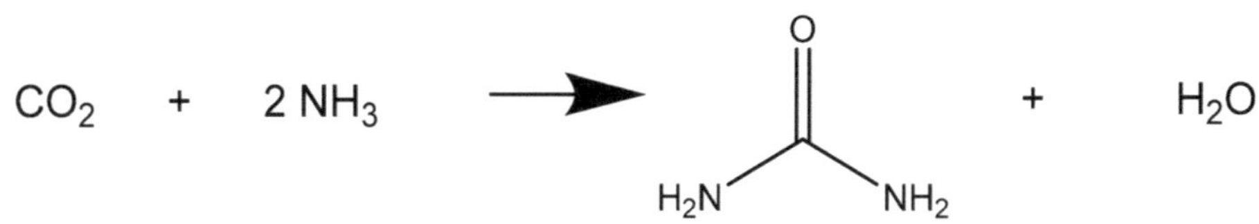

(11)

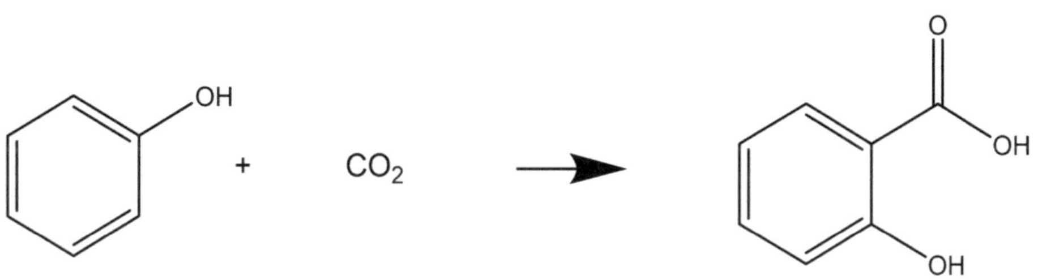

(12)

According to Song (2006), the worldwide production of urea in 2002 was about 110 million metric tonnes, which contains 51.8 million metric tonnes of nitrogen with an estimated value of US$11.5 billion. This corresponds to about 81 million metric tons of CO_2, and 22 million metric tonnes of carbon.

4.1. Thermodynamic and chemical considerations of CO_2 conversion

Chemical reactions are driven by the difference in Gibbs free energy between the products and reactants at certain conditions. The obstacle for utilizing CO_2 as feedstock to industrial processes is its low energy level - CO_2 is a highly stable molecule. Consequently, a substantial input of energy, effective reaction conditions, and often catalysts, are necessary for its chemical conversion. In other words, many reactions for CO_2 conversion involve positive change in enthalpy, requiring an energy input. There are many large-scale chemical industrial processes that are operated based on endothermic reactions in the chemical industry (e.g., ammonia Haber process).

Song (2006) states properly that it is more energy-demanding if one were to use only CO_2 as a single reactant, but it becomes easier in thermodynamically terms if CO_2 is used as a co-reactant with another substance that has higher Gibbs free energy, such as CH_4, graphite and H_2. Song (2006) illustrates this trend by the change in the reaction heat for reactions with CO_2 as the single reactant (Eq. 13) and with CO_2 as a co-reactant (Eq. 14).

$$CO_2 \longrightarrow CO + ½\,O_2 \qquad \Delta H^0 = +\,293\frac{kJ}{mol\,CO_2} \tag{13}$$

$$CO_2 + H_2 \longrightarrow CO(g) + H_2O \qquad \Delta H^0 = +51\frac{kJ}{molCO_2} \tag{14}$$

Therefore, energy input is necessary to transform CO_2 into chemicals. Four methods are possible: (i) reaction with high-energy molecules (e.g., ethylene oxide, H_2, unsaturated compounds and organometallic compounds); (ii) reaction with low energy oxidized compounds (e.g., organic carbonates), (iii) shifting chemical equilibrium towards products (via removal of a reaction product) and (iv) supplying physical energy (e.g., light or electricity) (Sakakura et al., 2007). The appropriate selection of chemical reactions makes it possible to obtain a negative Gibbs energy change.

As the carbon of the carbonyl group has an electron deficiency, CO_2 has great affinity for nucleophilic compounds and electron donors, i.e., as an anhydrous carboxylic acid, it promptly reacts with basic compounds. For instance, organometallic compounds, such as Grignard compounds, react promptly with CO_2 even at low temperatures. Reactions with CO_2 can be divided into two groups: (1) formation of a carboxylic group via a nucleophilic attack and (2) oxidative cycle addition yielding a ring of 5 members (Sakakura et al., 2007).

A relevant aspect to be considered is that utilization of CO_2 as feedstock does not necessarily contribute to the mitigation of greenhouse effects, even though CO_2 stands as a green reactant in many cases (Sakakura et al., 2007). Three points are hence relevant:

a. The chemical (or biochemical) fixation of CO_2 does not necessarily imply in reducing CO_2 emissions as its transformation requires energy, both to drive reaction (high temperatures and pressures) and separate products (separation occurs mainly at low pressures and, hence, recycling unreacted CO_2 to the reactor will require recompression at the expense of high energy input);

b. The energy demand of the world is order of magnitude higher than the amount of CO_2 fixed by chemical utilization of CO_2, and

c. In the critical phase of its life cycle, organic chemicals will emit CO_2. Nevertheless, the relevance of CO_2 as raw material stands for being a renewable feedstock, substituting the conventional fossil based routes

http://dx.doi.org/10.5772/57560

Among other utilization, CO_2 is currently used as supercritical solvent, refrigerant fluid, beverage carbonation agent, inert medium (such as fire extinguisher), pressurizing agent, neutralizing agent, gas for greenhouses, "inerting" applications to inhibit unintended chemical reactions, welding (preventing atmospheric oxygen from reacting with molten metal), food processing (suppressing aerobic bacterial activity for preservation in processes like pneumatic conveying or food storage). In any of these applications for inerting, carbon dioxide serves as a cover against atmospheric oxygen and is thus implicitly released into the atmosphere. The carbonation of beverages accounts for around $1.0x10^6$ t CO_2/y. Nevertheless, they do not constitute CO_2 sink as it is ultimately released to the atmosphere or remains in a closed loop (Ormerod et al., 1995). Hence, this study does not review such utilizations.

4.2. Supply chain considerations of CO_2 conversion

Bayer (2013) estimates that the chemical industry has over 40,000 final chemicals, produced from approximately 400 intermediate chemicals, derived from ~40 basic chemicals that, in turn, are based on 4 classic feedstocks: petroleum, natural gas, coal and biomass. The Company expands the set of feedstock with the inclusion of CO_2.

For the near and middle term time-period, i.e., next one or two decades, it is reasonable to assume that presently dominant technologies (from an economic standing point) will persist and, consequently, expanding economies' demand of energy will be met by present technologies. Consequently, GHG emission of chemical processes will expand. This same approach leads to a transition scenario to a low-carbon economy equally dominated by presently installed infrastructure.

Hence, the CO_2 utilization cycle is likely to rely on commercially mature technologies or on technologies presently in large-scale pilot or demonstration plants as *Bridge Technologies*. Therefore, only peripheral technological advances are expected in these technologies such as process intensification, enhanced selectivity and activity of catalysts, and process optimization with increased mass and energy integration.

It is worth noting that 5 chemical commodities, ammonia, methanol, ethylene, propylene and BTX dominate energy consumption and GHG emissions in the chemical industry (IEA, 2013). In the conception of co-processing of fossil feedstock, with biomass and CO_2, conversion routes to produce ammonia, methanol and olefins (e.g., ethylene and propylene) are considered (Figure 2). Although not included in Figure 2, catalytic fast pyrolysis of biomass can lead to the key aromatic compounds, Benzene, Toluene, and Xylene (BTX), with generation of paraxylene from the BTX and subsequent conversion to Purified Terephthalic acid (PTA) and PolyEthylene Terephthalate (PET).

Synthesis gas (SYNGAS), a mixture of hydrogen, carbon monoxide and CO_2, is a versatile intermediate feedstock used in the production of a number of hydrocarbons such as methanol, ammonia, synthetic hydrocarbon liquids, and as a source of pure hydrogen and carbon monoxide. Applications of these products range from petrochemical feedstock to fuels.

According to Rostrup-Nielsen and Christiansen (2011), trends in the use of SYNGAS are dominated by the conversion of inexpensive remote natural gas into liquid fuels ("gas to liquids" or "GTL") and by a possible role in a future "hydrogen economy" mainly associated with the use of fuel cells. Some relevant synthesis to the chemical industry are:

SYNGAS to Ammonia: Ammonia serves as a building block in many pharmaceuticals, fertilizers, ethanolamines, urea and cleaning products, as well as an anti-microbial agent in food processing. 50% of the world´s food production relies on ammonia fertilizers.

SYNGAS to Methanol: The main use for methanol is to produce other chemicals; about 40% is converted to formaldehyde, and further processed into plastics, plywood, paints, explosives and textiles. It is also used in anti-freeze, solvents, and fuels, and can serve as energy carrier.

SYNGAS to Hydrogen: Hydrogen generation is one of the largest energy-consuming steps in the production of the crucial chemical precursors of ammonia and methanol.

SYNGAS to Synthetic Fuels: Liquid hydrocarbons exhibit an excellent volumetric energy density and offer various opportunities for storing electric energy (Kaiser et al., 2013). Kaiser et al. (2013) point generation of SYNGAS by reverse water-gas shift (RWGS) at elevated temperatures as the first step, followed by Fischer-Tropsch (FT) synthesis. If CO is substituted by CO_2, less synthetic fuels are formed, the water-gas shift is repressed, and methane selectivity increases.

4.3. CO_2 to SYNGAS

SYNGAS is a toxic, colorless and odorless mixture. Its efficient commercial production is gaining significant attention worldwide (Raju et al., 2009) as it is a versatile feedstock to produce a variety of fuels and chemicals.

Almost any carbon source ranging from natural gas and oil products to coal and biomass can be used in the SYNGAS production. The lowest cost routes for its production, however, is natural gas (Spath and Dayton, 2003), which is also the cleanest of all fossil fuels. Furthermore, steam methane reforming is a well-established process for the production of SYNGAS and hydrogen (Gangadharan et al., 2012). The H_2/CO ratio varies over a wide range, depending on the primary feedstock and technology employed. Particular SYNGAS ratios are required depending on the chemical product desired, therefore creating flexibility for the chemical industry.

In the twofold context of avoiding emissions and standing as a renewable feedstock, carbon dioxide has been investigated as raw material in SYNGAS production. The new technologies involves CO_2: (i) reforming processes using a hydrocarbon (methane, typically) as reducing agent; (ii) using CO_2 as a co-reactant with hydrogen in the catalytic reverse water gas shift (RWGS); (iii) thermocatalytic (solar assisted) routes; (iv) electro- or photo-catalysis; (v) plasma processes, and (vi) bio-processes, e.g., by hybrid enzyme-nanoparticle systems, bioelectrochemical reduction or using a biomass char and a catalyst such as Ni/Al_2O_3.

CO_2 Reforming of CH_4 (Dry Reforming): The dry (carbon dioxide) reforming of methane has been of interest for a long time, dating back to as early as the 1920s, and was first proposed by Fischer and Tropsch (1928), but it is only in recent years that interest in it has rapidly increased for both environmental and commercial reasons (Zhang et al., 2003). Its name derives from the fact that CO_2 replaces steam in the conventional steam methane reforming process (Hartley & Tam, 2012). This reaction can be represented as

$$CH_4 + CO_2 \longrightarrow 2CO + 2H_2 \qquad \Delta H = +247\frac{kJ}{mol} \tag{15}$$

According to Hartley and Tam (2012), dry reforming utilizing CO_2 produces synthesis gas with higher purity and lower H_2/CO ratio than either partial oxidation or steam reforming. The produced SYNGAS has an H_2/CO ratio of unity without further post-reformer reactions (Zhang et al., 2003). The interest in this reforming route in recent years (Treacy and Ross, 2004, Shi et al., 2013) is due to two main reasons: (i) it produces SYNGAS with a H_2/CO molar ratio that is suitable for a variety of products including Fischer–Tropsch fuels and (ii) the reaction consumes two types of greenhouse gases, CO_2 and CH_4 (Zhang et al., 2003, Gangadharan et al., 2012). Moreover, SYNGAS production stands as the most promising alternative of CO_2 utilization as it presents flexibility of using installed infrastructure to the manufacture of important chemical commodities.

The biggest limitation to the dry reforming process appears to be the availability of a suitable catalyst. The high temperatures required to reach high conversions, due to the endothermic nature of the process, contribute to carbon deposition (both CO_2 and CH_4 give off carbon deposits), and a catalyst capable of operating at such severely deactivating conditions has not been well developed (Zhang et al., 2003, Shi et al., 2013). According to Shi et al (2013), from the viewpoint of GTL industry, developing a catalyst for CO_2 reforming of CH_4 is a challenge, because the catalyst must exhibit very high production rates to render the GTL methane reformer as small as possible. Nevertheless, progress in the development of suitable catalysts with higher activities and optimized lifetime stabilities have been reported (Bradford and Vannice, 1999, Souza and Schmal, 2003; Zhang et al., 2003, Ginsburg et al., 2005; Kahle et al., 2013; Shi et al., 2013; Zheng et al., 2013; Edwards, 1995; Wurzel et al., 2000; Nagaoka et al., 2001; Li et al., 2004). Nevertheless, there is still no process for the CO_2 reforming currently considered to be commercially feasible. However, a variation of dry reforming has been used industrially (Hartley and Tam, 2012). The CALCOR process (Teuner, 1985, Kurz and Teuner, 1990, Teuner et al., 2001) involves dry reforming of methane, optimized to reduce the hydrogen content of the product gas. Furthermore, hydrogen separation by membrane permeators produces hydrogen gas that combusts a fuel (e.g., methane) producing pure carbon monoxide. The SPARG process (promotion by poisoning) is also a dry reforming reaction process (Gunardson, 1998; O'Connor and Ross, 1998; Rostrup-Nielsen, 2006). The active catalytic sites are blocked by poisoning the feed gas with hydrogen sulfide (H_2S). The adsorption of sulfur at the catalytic sites is favored over carbon growth. The SPARG process is claimed to produce high CO content SYNGAS.

Also combined CO_2 and steam reforming systems have been operational in the industry for a number of years (Gangadharan et al., 2012). By choosing the right proportions between CH_4, water and CO_2 (3/2/1), the combination of steam and dry reforming of methane can generate SYNGAS with a H_2/CO ratio of 2, ideal, for example, for the synthesis of methanol (Rostrup-

Nielsen and Christiansen, 2011; Olah et al., 2009). This combination of steam and dry reforming was named bi-reforming.

Bi-reforming could be advantageous in the use of various natural gas sources even these containing substantial amounts of CO_2. Some natural gas as well as biogas sources contain CO_2 concentration up to 50–70%. Bi-reforming can also be used to recycle CO_2 emissions from sources such as flue gases from fossil fuel (coal, petroleum, natural gas, etc.), burning power plants, exhaust of cement factories, among other industries (Olah et al., 2013). This reaction can be represented as

$$\text{Steam reforming: } CH_4 + H_2O \longrightarrow CO + 3H_2 \quad \Delta H = +206.3 \frac{kJ}{molCO_2} \tag{16}$$

$$\text{Dry reforming: } CH_4 + CO_2 \longrightarrow 2CO + 2H_2 \quad \Delta H = +247.3 \frac{kJ}{molCO_2} \tag{17}$$

$$\text{Bi-reforming: } 3CH_4 + 2H_2O + CO_2 \longrightarrow 4CO + 8H_2 \tag{18}$$

Bi-reforming is adaptable for reforming varied natural gas (containing hydrocarbon homologues) and CO_2 sources, e.g., shale gas (Olah et al., 2013):

$$3C_nH_{(2n+2)} + (3n\text{-}1)H_2O + CO_2 \longrightarrow (3n{+}1)CO + (6n{+}2)H_2 \tag{19}$$

Numerous authors (Ashcroft et al., 1991; O´Connor and Ross, 1998; Wang et al, 200;, Jarungthammachote, 2011) have studied a similar idea, which combines dry reforming with partial oxidation. The idea again being that the combination helps overcome the endothermic requirement of dry reforming with the exothermic nature of partial oxidation, resulting in lower total energy consumption. In addition, it allows altering the H_2/CO ratio by controlling the extent to which each reaction takes place (Hartley and Tam, 2012). The combination of exothermic and endothermic reactions is called autothermal reaction (ATR). The ATR technology requires addition of CO_2 or CO_2-rich gas, in order to adjust the SYNGAS composition to the desired H_2/CO ratio.

$$\text{Dry reforming: } CH_4 + CO_2 \longrightarrow 2CO + 2H_2 \quad \Delta H = +247.3 \frac{kJ}{mol} \tag{20}$$

$$\text{Partial Oxidation of Methane: } CH_4 + 0.5O_2 \longrightarrow CO + 2H_2 \; \Delta H = -35.6 \frac{kJ}{mol} \tag{21}$$

The combined Dry Reforming and Partial Oxidation is hence:

$$CH_4 + 0.5CO_2 + 0.25O_2 \longrightarrow 1.5CO + 2H_2 \quad \Delta H = +211.7 \frac{kJ}{mol} \tag{22}$$

Integrating steam reforming and partial oxidation with CO_2 reforming could reduce or eliminate carbon formation on reforming catalyst, thus increasing catalyst life and process efficiency. Therefore, the tri-reforming, a synergetic combination of CO_2 reforming, steam reforming, and partial oxidation of methane in a single reactor for effective production of industrially useful SYNGAS (Song, 2006) could solve two important problems encountered in individual processing. Incorporating oxygen in the reaction generates heat *in situ* that could increase energy efficiency; oxygen also reduces or eliminates carbon formation on the reforming catalyst. The tri-reforming can be achieved with natural gas and flue gases using the waste heat in power plants and the heat generated *in situ* from oxidation with the oxygen that is present in flue gas (Zhou et al., 2008; Zangouei et al., 2010; Moon et al., 2004).

The tri-reforming process is presented in Eqs. (23) to (26) (Song and Pan, 2004):

$$\text{Steam reforming: } CH_4 + H_2O \longrightarrow CO + 3H_2 \quad \Delta H = +206.3 \frac{kJ}{mol} \tag{23}$$

$$\text{Dry reforming: } CH_4 + CO_2 \longrightarrow 2CO + 2H_2 \quad \Delta H = +247.3 \frac{kJ}{mol} \tag{24}$$

Partial Oxidation of Methane:

$$CH_4 + 0.5O_2 \longrightarrow CO_2 + 2H_2 \quad \Delta H = -35.6 \frac{kJ}{mol} \tag{25}$$

$$CH_4 + 0.5O_2 \longrightarrow CO + 2H_2O \quad \Delta H = -880 \frac{kJ}{mol} \tag{26}$$

Song (2006) reports experimental and computational results to support that tri-reforming produces SYNGAS with desired H_2/CO ratios (1.5–2.0) and eliminates carbon formation in the CO_2 reforming of CH_4. Song (2006) suggests that tri-reforming is especially suited to using CO_2 in concentrated sources without prior CO_2 separation, as in non-conventional (low-quality CO_2-rich) natural gas, and has been demonstrated in pilot scale in Korea.

In general, produced SYNGAS from methane reforming is converted catalytically *in situ* via one of two main routes. The first is to use Fischer-Tropsch synthesis, a process that catalytically converts SYNGAS to hydrocarbons of varying molecular weights. The second is methanol synthesis. The latter has better atomic economy, since the oxygen atom in CO is included in the product and CO_2 can be blended into SYNGAS as a reactant. However, production of methanol is very inefficient in this reaction: only 10-15% one pass conversion typically at 5.0-10.0 MPa and 523-573 K, due to the severe thermodynamic limitations of this exothermal reaction ($CO+2H_2 \rightarrow CH_3OH$) (Shi et al., 2013).

Finally, CO_2 reforming of methane can also be used as a chemical energy storage alternative and an energy transmission system (Richardson and Paripatyadar, 1990, Levitan et al., 1991; Levy et al., 1993). According to Zhang et al. (2003), in this system, solar energy is used to drive the endothermic forward reaction, and the energy thus stored can be transported via pipelines such as SYNGAS and liberated at will by the reverse reaction at any location or time. The highly endothermic reaction could be an option to store solar energy in hot regions (Zhang et al., 2013).

Reverse Water Gas-Shift (RWGS): The reverse water gas shift (RWGS) reaction has been known from over two centuries and is a well-researched and understood process for SYNGAS ratio alteration (Hartley and Tam, 2012). In fact, both the water gas shift (WGS) and the RWGS reactions are mostly used in combination with reforming of hydrocarbons to adjust the H_2/CO ratio, as shown in Eq. (27) (Song, 2006). Depending on the reaction conditions, the equilibrium for the WGS can be pushed in either the forward or the reverse direction. Efforts to explain the RWGS reaction mechanism are reported (Goguet et al., 2006, Meunier et al., 2007, Wang et al., 2013), and two main mechanisms have been proposed: the *redox* mechanism and the *associative formate* mechanism. The reversibility of the WGSR is important in the production of ammonia, methanol, and Fischer-Tropsch synthesis where the ratio of H_2/CO is critical. Many industrial companies exploit the RWGS reaction as a source of the synthetically valuable CO from cheap CO_2. In fact, catalytic RWGS reaction is the main route to produce SYNGAS from CO_2.

$$H_2 + CO_2 \longleftrightarrow CO + H_2O \quad \Delta H = +51\frac{kJ}{mol} \tag{27}$$

RWGS provides a source of hydrogen at the expense of carbon monoxide, which is important for the production of high purity hydrogen. This is a mildly endothermic reaction, as shown in Eq. (27).

Although high temperature reactions are effective for obtaining a high conversion, WGS reaction is an equilibrium-limited reaction that exhibits decreasing conversion with increasing temperature. In order to take advantage of both the thermodynamics and kinetics of the reaction, industrial scale WGS reaction is conducted in multiple adiabatic stages consisting of a high temperature shift (HTS) followed by a low temperature shift (LTS) with intersystem cooling (Byron, 2010). The initial HTS takes advantage of the high reaction rates, but is thermodynamically limited, which results in incomplete conversion of carbon monoxide and a 2-4% carbon monoxide exit composition. To shift the equilibrium towards hydrogen production, a subsequent low temperature shift reactor is employed to produce a carbon monoxide exit composition of less than 1% (Byron, 2010). A catalyst is required under these conditions because of the lower reaction rate at low temperatures. The RWGS reaction uses a variety of catalysts, including palladium, platinum on titania, copper, cobalt with manganese/ zinc oxide and rhodium with ceria (Tanaka et al., 2003, Saito and Murata, 2004, Meunier et al., 2007). Many research groups are looking at copper as a catalyst due to its effectiveness and its relatively low cost (Armstrong et al., 2013). However, there has been renewed interest in the WGSR at extreme temperatures, because of recent advances in high-temperature materials for hydrogen separation membranes (Bustamante et al., 2002).

4.4. CO_2 to methanol

Methanol is one of the most important commodity chemicals as it is used as a raw material in several intermediate chemicals and end uses. Methanol is produced industrially from SYN-GAS from natural gas or coal mainly containing CO, H_2 and a small amount of CO_2 in presence of a catalyst. Nevertheless, direct CO_2 hydrogenation has also been reported. Other non-conventional routes are electro- or photoprocesses, as well as the use of enzymes. The importance of methanol synthesis is demonstrated by widespread scientific publications of various reaction routes (Razali et al. 2012), and the development of several pilot plants to use waste carbon dioxide for methanol production.

Among the new technologies, in terms of potential for application, the CO_2 catalytic hydrogenation to methanol appears to have the highest degree of commercialization. It may be already commercially interesting when cheap sources of renewable H_2 are available, or to store excess electrical energy, as an alternative to actual systems. It is estimated that this reaction could reach the industrial stage in less than five years. This development would be pushed by experience in pilot or pre-commercial industrial plants, such as the Mitsui Chemicals Inc.'s plant (pilot in Japan capable of producing 100 t of methanol per year, and large unit expected in Singapore) and a plant by Carbon Recycling International (installed at the end of 2010) (Quadrelli et al., 2011). Mitsui's pilot plant uses CO_2 from an ethylene production plant of Osaka Works Petrochemical Complex (ADEME, 2010). It synthesizes methanol by CO_2 hydrogenation and the simultaneous water gas shift reactions. The process claims 96% selectivity (Hartley and Tam, 2009). Carbon Recycling International is capable of producing 3000 t/y of methanol (ADEME, 2010). This unit has a capacity of about 10 t of methanol from 18 t of CO_2 (Carbon Recycling International, 2009; Van-Dal and Bouallou, 2013), with CO_2 from the Svartsengi geothermal plant and an aluminum production plant. Hydrogen is generated from the electrolysis of water using a renewable source of electricity.

Methanol from SYNGAS: Synthesis gas composed of the proper ratio of hydrogen, carbon monoxide and carbon dioxide is converted to methanol. Alternatives paths to methanol are via CO from RWGS reacting with hydrogen according to Eq. (28) and via CO_2 being hydrogenated following Eq. (29).

$$CO + 2\,H_2 \longrightarrow CH_3OH \qquad \Delta H = -90.6\frac{kJ}{mol} \tag{28}$$

$$CO_2 + 3\,H_2 \longrightarrow CH_3OH + H_2O \quad \Delta H = -49.5\frac{kJ}{mol} \tag{29}$$

From Eq. (28), production of methanol involves SYNGAS production as intermediate stage. Hence, two steps are required for the manufacture of methanol: reduction to SYNGAS and reaction to form methanol. There are process variations for implementing the sequence. Before being sent to the methanol production unit, the SYNGAS must thus be subjected to the WGS reaction to enhance its hydrogen content. Alternatively, H_2 from other sources can be added. Recent efforts have been aimed at production of methanol in a one-step process without intermediate formation of SYNGAS. Homogeneous or heterogeneous catalysts are typically preferable. The conventional process occurs at relatively low pressures (5 to 10 MPa) and 210 to 350 °C employing a $Cu/ZnO/Al_2O_3$ catalyst.

Catalytic Hydrogenation Conversion of Carbon Dioxide to Methanol: The most direct and studied route to methanol from CO_2 is the catalytic conversion of CO_2 with hydrogen. Carbon dioxide hydrogenation to methanol is a relatively mature process. The main issue is the cost (and associated carbon footprint) of the H_2 necessary for the reaction. Any available energy source (alternative energies such as solar, wind, geothermal, and atomic energy) can be used for the production of needed hydrogen and chemical conversion of CO_2. The process can use lower operational pressures of 3 MPa at 240 °C. This direct CO_2 hydrogenation exhibits low conversions resulting in high volumes of recycled gas. Literature indicates that methanol is synthesized following a 3:1 hydrogen to carbon dioxide stoichiometry using catalysts of copper oxide, zinc oxide, incorporating either titania, aluminum oxide, chromium oxide and alternatively lanthanum or gallinium (Lachowska and Skrzypek, 2004, Lee et al., 2004, Stoczynski et al., 2004).

An alternative approach to the use of solid catalysts and a gas phase process is to employ the so called low-temperature methanol synthesis (LTMS) (Dixneuf, 2011). LTMS is based on the catalytic hydrogenation of methanol to formic acid (HCOOH) with subsequent etherification to methanol formate (alternative to methyl formate from SYNGAS), followed by hydrogenation of formate to two methanol molecules using Pincer-type ruthenium(II) catalyst (Balaraman et al, 2011; Dixneuf, 2011; Huff and Sanford, 2011). A liquid-phase allows CO_2 and H_2 conversion to methanol of about 95% with very high selectivity in a single pass (Olah, 2009). Waugh (2012) has published a review on catalytic methanol synthesis which includes the use of carbon dioxide as a feedstock.

Photoreduction of CO_2 to methanol: Photoelectrochemical reduction of carbon dioxide or photocatalysis generally uses semiconductors to promote reaction in the presence of sun light. The semiconductor is used as a catalyst to absorb solar energy and generate electrons and protons needed for the reduction of carbon dioxide. While hydrogenation of carbon dioxide requires high temperature and high pressure conditions, photocatalysis carries out under relatively mild conditions with advantageous energy input – sun light – a continuous and readily available source (Le, 2009).

Considerable research effort has been made on CO_2 activation by visible light photocatalysts due to the natural abundance of sunlight. Nevertheless, the efficient photoreduction of CO_2 with H_2O remains one of the most challenging tasks of environmental catalysis.

CO_2 can be reduced in water vapor or solvent by photocatalysts such as TiO_2 and ZnS. Eq. (30) describes the overall reaction.

$$CO_2 + 2H_2O \xrightarrow{h\nu} CH_3OH + 3/2\, O_2 \tag{30}$$

Due to the high energy requirements, this method is often combined with electrochemical methods via photoelectrocatalysis to drive the reaction (Hu et al., 2013). The catalysts traditionally used are transition metal complexes, TiO_2, ZnO, CdS, and functionalized metal surfaces (Yamashita et al., 1998; Kuwabata et al., 1994). A wide variety of CO_2 photoreduction has been achieved on the surface of TiO_2 under UV irradiation. The yield of photoproducts can be changed substantially under different experimental conditions such as UV wavelength, UV intensity, additives of reaction media and reactor configuration. Other variables, such as CO_2 pressure, moisture and residence time are also important in photoreducing CO_2 (Wu & Lin, 2005).

Electrochemical Production of Methanol from CO_2 and H_2O: The direct reduction of CO_2 to CH_3OH is known to occur at several types of electrocatalysts including oxidized Cu electrodes. The current stage of the technology is still very experimental. The majority of tests have been performed on a laboratory scale with a purpose of either kinetic analysis or proof-of-concept to examine product distribution for different material and condition combinations (Beck et al., 2010). An advantage of electrochemical CO_2 reduction is that unlike many other hydrocarbon processes it can occur at ambient conditions.

The electrochemical reduction of carbon dioxide to methanol is thermodynamically possible, but there seems to be no well-established technique to achieve this reaction with high current efficiencies close to 100%. Nevertheless, methanol production has been reported with the use of ruthenium: gallium arsenide and RuO_2-$Ti0_2$ mixed cathodes (Le, 2009). Cole and Bocarsly (2010) have reviewed electrochemical reduction processes, including electrochemical CO_2 conversion to methanol. Few studies have investigated the feasibility of this technology, and none has been found to provide an in-depth analysis of its potential industrial implementation.

Applications of Methanol: Methanol has traditionally been used as feed for production of a range of chemicals including acetic acid, formaldehyde and MTBE (Olah, 2009). In recent years, methanol has also been used for other markets such as production of Di-methyl-ether (DME) and olefins by the so-called methanol to olefins process (MTO) or as blendstock for motor fuels. As a liquid fuel, methanol is of interest especially for use in fuel cells (Olah, 2009).

Methanol to Olefins (MTO) explores alternative pathways to produce small olefins, in particular ethylene and propylene. Conventional steam cracker feeds are either natural gas liquids (NGL) or heavy liquids (i.e., naphtha). Ethane cracking, however, is increasing its share as feedstock. A promising alternative route is dehydration of methanol (MTO). Methanol-to-olefins (MTO) was first developed by ExxonMobil (1980s) as part of its methanol-to-gasoline (MTG) process. In the 1990s, UOP and Norsk Hydro built an MTO pilot plant in Norway. Since then, Lurgi has developed its own version of this process, methanol-to-propylene (MTP).

4.5. CO_2 to DMC

Dimethyl carbonate (DMC) is a biodegradable and nontoxic chemical acceptable environmentally as a chemical destination of CO_2. It is exempted from VOC classification and can be used as raw material for producing valuable chemicals, including aromatic polycarbonate, and qualifies as an octane booster component in gasoline and diesel. It is a safer and nontoxic substitute of well-established methylating-carbonylating hazardous chemicals like dimethyl sulfate and phosgene (Souza et al., 2013). Although DMC is presently produced on a relatively small scale, approximately 400 kt/y, its demand has grown strongly in recent times because of its green properties.

Currently, DMC is produced mainly by oxidative carbonylation of methanol (Aoussi et al., 2010). The direct methylation reaction is possible where, according to Ferreira et al. (2013), the most used catalyst is tin, employed as an oxide compound or as an organometallic complex, according to Eq. (31):

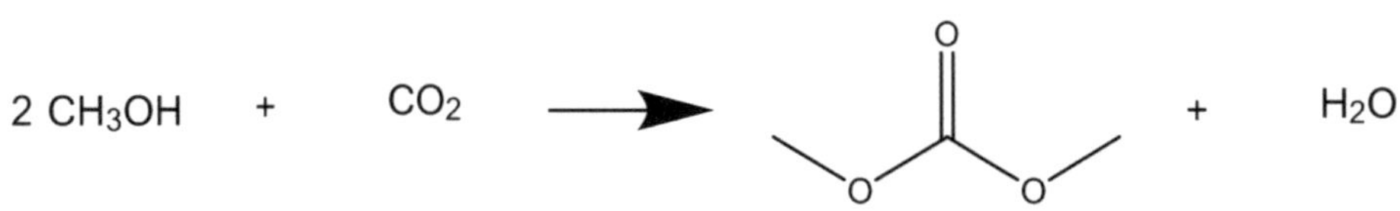

DMC

(31)

However, direct methylation presents low yields, inferior to 10%, due to the chemical inertness of CO_2 and to the deactivation of catalysts induced by water formation in the reaction (Aouissi, 2010). For large-scale production of DMC from CO_2, one route seems to be promising: the indirect route (IR) for two-step conversion of CO_2 with ethylene oxide (EO) to ethylene carbonate (EC), which then reacts with excess methanol (MeOH) giving DMC and ethylene glycol (EG) as shown in Eqs. (32) and (33).

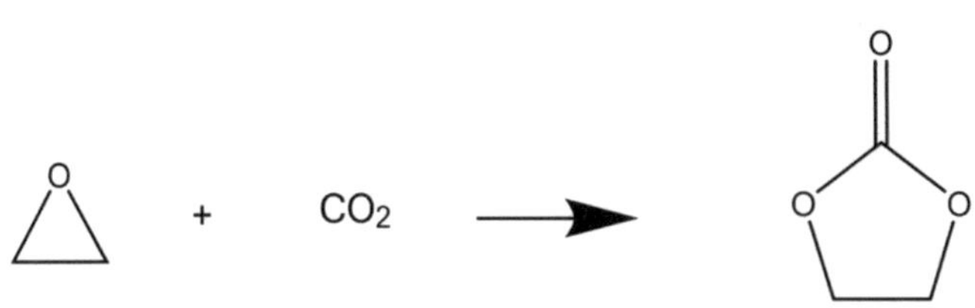

EO

EC

(32)

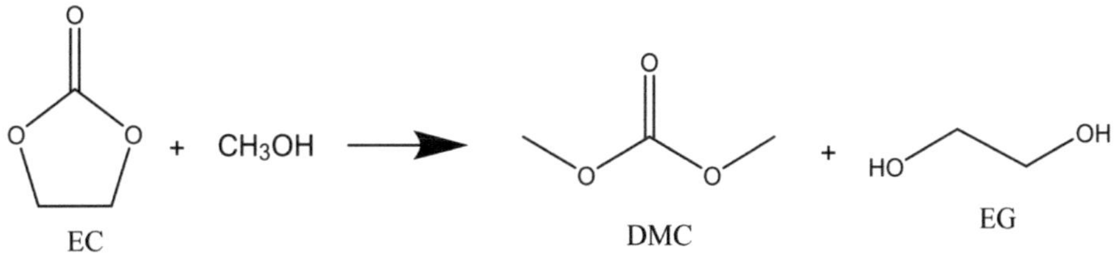

(33)

Souza et al. (2013) evaluated IR's performance from technical, economical and environmental standpoints. Accordingly, the authors proposed a process flowsheet with two serial reactors' system followed by an integrated separation section, with extractive distillation using methyl-iso-butyl ketone (IR-MIBK) and ethylene glycol (IR-EG). For environmental performance assessment, Souza et al. (2013) defined Chemical Sequestration of CO_2 (CSC) as the amount of CO_2 consumed by chemical reaction minus the amount of CO_2 emitted by heat, power and purges. Considering a production of $1.3x10^5$t/y of DMC produced, the authors reported CSC values for IR–MIBK and IR–EG of -15.9kt/y and -8.1kt/y, respectively. The negative values of CSC indicate that both alternatives are net emitters and illustrates that fixation of CO_2 in chemical products does not necessarily imply into net CO_2 reduction, and that the main aspect in CO_2 utilization as feedstock is of substituting fossil carbon source for a renewable alternative. Furthermore, Souza et al. (2013) estimated net present values for IR–MIBK and IR–EG of $\$71.5x10^6$ and $\$106.5x10^6$, respectively, and payback times of 5.5 and 4.5, respectively, concluding for economic feasibility of DMC production. It is worth noting that DMC production from CO_2 is already in use at Asahi Chemical Industry.

4.6. Biochemical conversion of CO_2

Biofixation of CO_2 with microalgae is a promising route of utilization of CO_2, as it exhibits fast growth (e.g., Picardo et al., 2013a) and produces numerous high-added-value bioproducts (Grima et al., 2003). Additionally, they do not contain lignin, a fact that renders microalgae better adapted to biochemical valorization. As advantages of CO_2 bioconversion with relation to energy crops, microalgae grow in variable climates on non-arable land with non-potable water, releasing competition with food crops, and are able to use direct flue gases as their carbon source (Fernández et al., 2012). Alternatively, to inject directly flue gases into microalgae cultures, adequate design and operation of the carbonation culture system unit are also necessary, otherwise almost all of the CO_2 fed to the culture would be released into the atmosphere. In this aspect, photobioreactors are more appropriated arrangements. Furthermore, photobioreactors provide cell requirements such as light, temperature, pH, and mixing (Fernández et al, 2012).

Chisti (2007) concluded that microalgae are the only alternative for the sustainable production of biodiesel. Accordingly, the fact that microalgae biomass is rich in lipids and are, hence, high energy density feedstock for fuels and chemicals is of relevance. Picardo et al (2013a) proposed a screening procedure for microalgae selection to meet production objectives such as SYNGAS for the production of synthetic fuels, since the biomass, or the residual biomass obtained after extraction of bioproducts, can be gasified to yield SYNGAS. An attractive alternative in this route is to employ CO_2 as oxidation agent (Butterman and Castaldi, 2007). Butterman and Castaldi (2007) report that the injection of CO_2 and H_2O in gasification increases char reactivity that results in more efficient use of the feedstock with less residual to be post-processed.

According to Grima et al. (2003), production of microalgal biomass can be carried out in fully contained photobioreactors or in open ponds and channels. Biomass productivity depends on species, operational conditions and the choice of ponds (~20g/m^2.d) or photobioreactor (~50g/m^2.d) geometry. Open-culture systems are almost always located outdoors and rely on natural light for illumination while closed photobioreactors may be located indoors or outdoors, although outdoor location is more common. Grima et al. (2003) list as biomass harvesting operations centrifugation, filtration or gravity sedimentation, which may be preceded by a flocculation step.

Microalgae contain lipids and fatty acids as membrane components, storage products, metabolites and sources of energy. Microalgae have been found to contain proportionally high levels of lipids (for some species this value can reach 50% oil by weight), with a convenient fatty acids profile and an unsaponifiable fraction allowing a biodiesel production with high oxidation stability (Grima et al., 2013). Lipid accumulation is promoted by stress, notably by nitrogen starvation (Picardo et al., 2013b).

Elemental analysis of carbon content of biomass points to ~50% (Picardo et al., 2013a), what leads to conclude that approximately 2t of CO_2 can be converted into 1t of biomass, potentially amenable to 0.2t of lipids. Its massive extension to the energy sectors constitutes a vast potential for large-volume CO_2 utilization. Fernandez et al. (2012) recognize that microalgae are not a storage strategy because the biomass produced cannot be stored for a long time. Its contribution to reducing CO_2 emissions is only possible if biofuels are produced to replace the fossil fuels use, and allowing the production of other commodities, or by-products from flue gases, which allows one to obtain revenues to mitigate the penalty of carbon capture. To illustrate the potential industrial application of microalgae, Figure 19 shows a schematic of microalgae bioconversion of CO_2 and its downstream processing in a biorefinery arrangement producing long-chain fatty acids (PUFA´s, MUFA´s and PUFA´s), biodiesel, green diesel, gasoline, biogas, urea, N_2 and propylene carbonate.

Monteiro et al. (2010) employed Pareto optimization of what was named an "industrial ecosystem" comprised of a biorefinery of microalgal biomass aiming at maximizing sustainability of the productive arrangement. The authors concluded that increasing the weight of environmental objectives against economic performance might make sectors of the proposed original superstructure of amenable processes unattractive. Therefore, the final structure of a biorefinery of microalgae depends on the priorities set for the productive complex.

4.7. Some pilot and commercial scale CO_2 utilization processes

Polycarbonate: Polycarbonate (PC) is a plastic with impact resistance and heat resistance, mainly produced (4t/y) by reacting CO and Cl_2 to form phosgene as an intermediate material. The phosgene process has a number of disadvantages, including the risk of environmental

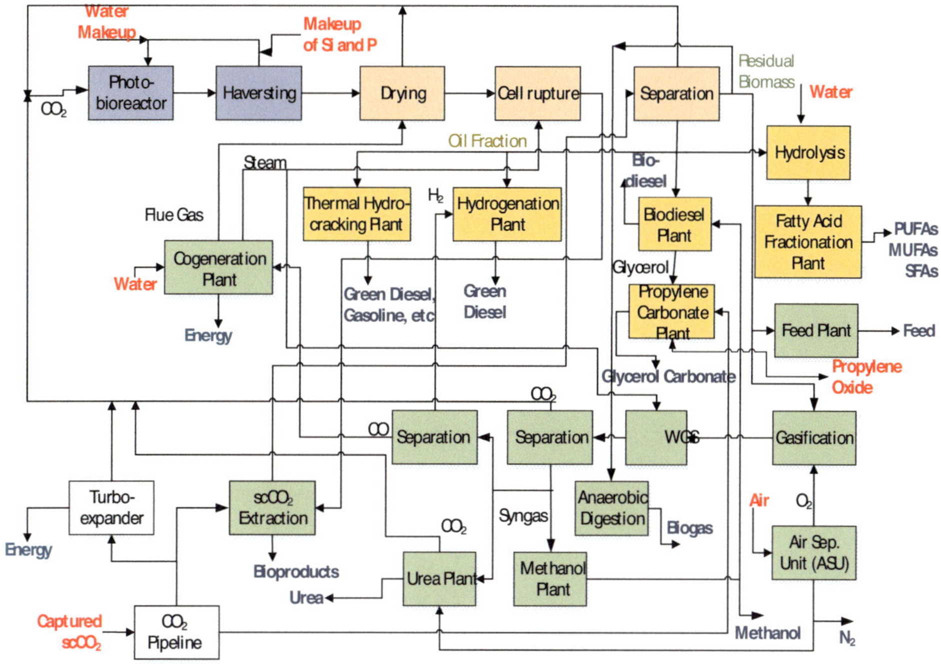

Figure 19. Bioconversion of CO_2 integrated into a biorefinery arrangement. Main external inputs are marked in red.

harm (Ushikubo, 2013). Asahi Kasei Corporation succeeded in commercializing the first non-phosgene polycarbonate production using ethylene oxide and CO_2, a by-product of ethylene oxide synthesis. The Asahi Kasei Process has as co-products high-purity monoethylene glycol (MEG). The process employs reactive distillation in the monomer production and gravity-utilized, non-agitation polymerization reactor in the melt polymerization. The monomer process consists of 3 production steps, ethylene carbonate (EC) from CO_2 and EO, dimethyl carbonate (DMC) and MEG from EC and MeOH, and diphenyl carbonate (DPC) and MeOH from DMC and phenol (PhOH). All intermediates are recycled. The by-produced PhOH is recycled to the monomer process (Sinshuke et al., 2010). According to Ushikubo (2013), the new process reduces CO_2 emissions by 0.173kg per one kg of polycarbonate. Five commercial plants using the Asahi Kasei process are operating in Taiwan (150,000 t/y), Korea (2 plants of 65,000 t/y), Russia (65,000 t/y) and Saudi Arabia (260,000 t/y) (Shinsuke et al., 2010).

Monoethylene glycol (MEG): MEG is used as an antifreeze and as a raw material for the production of polyester fibers and resins, mainly PET (Ushikubo, 2013). An expanding market share is foreseen in natural gas industry, where MEG is added into the pipeline or the gas conditioning process, either as hydrate inhibitors or for dehydration purposes to protect downstream pipelines. The pipeline can extend for thousands of kilometers and MEG is injected to inhibit hydrate formation, avoiding plugging (*e.g.*, Statoil's Snøhvit field, Pettersen, 2011). Concerning the production process, a new technology was developed by Mitsubishi Chemical Corporation with 99% selectivity while the conventional (non-catalytic) process has selectivity around 89% (Kawabe, 2010). The conventional technology produces as co-product DEG and TEG (di- and triethylene glycols, whose demand is expanding at only 2-3% as opposed to MEG expansion (world demand amounts to 17t/y). Mitsubishi technology uses a two-step catalytic synthesis: production of ethylene carbonate (EC) as intermediate followed by EC hydrolysis under almost stoichiometric condition, while the conventional hydrolysis occurs at a higher H_2O/ethylene oxide molar ratio, according to Eqs. (34) and (35).

$$EO + CO_2 \longrightarrow EC\ (\text{carbonation}) \tag{34}$$

$$EO + H_2O \longrightarrow EG + CO_2\ (\text{hydrolysis}) \tag{35}$$

The product purification is simpler (water removal and MEG purification distillation columns, while the conventional process has 4 distillation columns: water, MEG, DEG and TEG columns). Furthermore, CO_2 remains in closed loop. Ushikubo (2013) reports that several commercial plants operating with the new (and greener) technology save resources and energy and reduces the amount of wastewater and CO_2 production. It is worth noting that MEG is a co-product of the production of DMC in a process where Eq. (4.24) is replaced by Eq. (40).

Polyurethane: Bayer (2013) targets the production of polyurethane, via the utilization of CO_2 as feedstock. In this route, CO_2 is converted to polyols (HO-R-OH) which reacts with isocyanate to yield polyurethane. The conversion of CO_2 starts with its reaction with an epoxide, (propylene oxide) of higher energy content, in a catalytic route. Polyol with 30%CO_2 had 2.64 kg of equivalent CO_2 emissions. A maximum theoretical value of 43% of CO_2 can be incorporated in the polyol (Bayer, 2010). The development started in 1669, ended its laboratory scale in 2009, and is moving to industrial implementation of the named "Dream Production" with a pilot plant in Leverkusen to produce polyol, for testing purposes. In early 2013, the new method was successfully converted from the production of discrete quantities to continuous production, a key intermediate step for the industrial-scale production of CO_2-based polyurethane, which Bayer is targeting for 2015.

4.8. Emerging CO_2 utilization processes

The fixation of CO_2 into chemicals and polymers will not substantially contribute to a reduction in antropogenic GHG emissions given the current energy demand. Nevertheless, using CO_2 as a feedstock meets the requirements of sustainable development. An insight into advanced process concepts focus on chemical sequestration of CO_2 creating manufactured products from captured CO_2 with large potential markets. Integration of capture technologies into energy production schemes or oil and gas refining installations is the idea behind cPSE approach, namely, to abate chemical emissions while producing industrial products.

Formic acid: According to Armstrong et al. (2013), the amount of energy required to utilize carbon dioxide as a feedstock largely depends on the oxidation state of the intended products. The next-highest oxidation state molecules from CO_2 are formic acid (HCOOH) and carbon

http://dx.doi.org/10.5772/57560

monoxide (CO). So, carbon dioxide utilization to manufacture formic and other carboxylic acids is a relatively low-energy transformation. Formic acid has numerous applications, including food technology, agriculture, and the leather and rubber industries. Moreover, it has recently been considered as a promising candidate material for hydrogen storage and it is an important chemical with numerous applications. Moreover, formic acid has limited uses for further conversion, except reduction to methanol. The industrial methods used for its production employ CO as a raw material. Maihom et al. (2013) concluded that a first step occurs where CO_2 is hydrogenated to a formate intermediate. In the second step, the formate is further hydrogenated into formic acid. The hydrogenation of CO_2 would complete the chemical loop for hydrogen storage using CO_2. The complementary step is the catalyzed decomposition of formic acid to pure H_2 and reusable CO_2.

Carbon dioxide-based copolymers: The synthesis of organic carbonates has been one of the most widely studied areas of CDU. Typically, CO_2 is inserted into a molecule without the loss of any atoms in either the co-reactant or the gas itself (Armstrong, 2013). Carbonates are formed by the insertion of a CO_2 molecule into a guest co-reactant, typically an epoxide. Poly(propylene carbonate) (PPC), an alternating copolymer of CO_2 and propylene oxide, is one of the emerging low-cost biodegradable plastics. The fast development in catalyst design and performance improvement for PPC has created new chances for the chemical industry. In particular, high molecular weight PPC from rare earth ternary catalyst is becoming an economically viable biodegradable plastic with tens of thousands of tons produced per year, providing a new solution to overcoming the problem of high cost in biodegradable plastics (Qin and Wang, 2010). According to Qin and Wang (2010), with the continuous improvement in catalyst systems, commercialization of CO_2 copolymer is possible. The authors report industrial activities by Empower Materials producing polypropylene carbonate (QPAC®40), polyethylene carbonate (QPAC®25), polybutylene carbonate (QPAC®60), and polycyclohexene oxide(QPAC®130) on a pilot scale.

Electrochemical Reduction of CO_2. Delacourt (2010) studied the electrochemical conversion of CO_2 into SYNGAS. The driver of the proposed route is that renewable energies (e.g., solar and wind) are only alternatives to fossil fuel as they are not available on demand, thus requiring storage. Delacourt (2010) lists as a storage opportunity the conversion to liquid fuels (e.g., methanol), in which SYNGAS is the required intermediate, by converting solar energy into electricity through photovoltaic arrays, and then by using this electricity to produce fuels by electrolysis. Evolved H_2 reacts with CO_2 in a water-gas-shift reactor to make CO (and H_2O). The resulting SYNGAS is converted to methanol. Delacourt (2010) decided for a low-temperature technology (room temperature) although reported that high-temperature electrolysis (800 to 900°C) could be an attractive alternative. Because of the relatively low solubility of CO_2 in water under ambient conditions, gas-diffusion electrodes were applied to operate at higher current densities, and ion-exchange membrane was used as the electrolytic medium to limit gas crossover resulting in a decrease of the current efficiency of the electrochemical cell. Catalysts capable of reduction of CO_2 to CO at low overpotentials were selected.

Light-Driven Technologies. The rubisco enzyme is probably the most abundant enzyme of the biosphere. The fixation of CO_2 and its transfer to organic substrates in the Calvin cycle leads by way of starch to an annual production of 10^{11} t of biomass (Walther et al., 1999). With the development of catalysts able to reproduce the key steps of photosynthesis, water and sunlight would ultimately be the only needed sources for clean energy production. Light driven technologies under development include (a) photoelectrochemical cells where CO_2 present in a moistened gas stream is converted into organic molecules based on the photooxidation of water into oxygen gas O_2, protons H_+, and electrons. The conversion of CO_2 occurs at the photocathode and involves the generated protons, electrons and the "fuel" CO_2 (Kayaert et al., 2013); (b) direct water oxidation - photocatalytic water splitting - to produce H_2 and O_2 over a metal-oxide-based photocatalyst using solar energy (Maeda and Domen, 2013); (c) hydrogen-producing systems consisting of a hydrogen-evolving catalyst linked to a photosensitizer (Badura et al., 2012). Although promising alternatives, biomimetic CO_2 conversions are still in its early stage of technological development.

CO_2 Mineralization for Environmental Remediation: Lim et al. (2013) reviewed the application of carbonation to solidify or stabilize solid combustion residues from municipal solid wastes, paper mill wastes, etc. and contaminated soils, and to manufacture precipitated calcium carbonate. For instance, the red mud - a highly alkaline waste of Bayer's process - can be treated by absorption of CO_2. Machado (2012) analyzed the process of red mud carbonation with the exhausted gases from the alumina production calcinators, by developing a dynamic model representative of the mass and energy balances involved in the process, and chemical reactions occurring in the mud under carbonation. Machado (2012) was able to predict the species behavior, as well as the decrease in mud pH and the rebound phenomenon observed when the CO_2 concentration is reduced. The transient profile of the main process responses indicated a substantial reduction of CO_2 concentration in the output gas, in consequence of tons of CO_2 captured, and a significant reduction in mud pH. Concerning other environmental applications, Lim et al. (2013) report that carbonated products can be utilized as aggregates in the concrete industry and as alkaline fillers in the paper (or recycled paper) industry. Mineral carbonation consist in reacting CO_2 and Ca or Mg-bound compounds such as wollastonite (CaSiO3), olivine (Mg2SiO4), and serpentine ($Mg_3Si2O_5(OH)_4$). As a result, CO_2 is stably stored in final products such as $CaCO_3$ and $MgCO_3$. Last, the accelerated carbonation of solid wastes containing alkaline minerals such as Ca and Mg before their landfill treatment is effective for decreasing the mobility of heavy metals by adjusting pH to below 9.5 at which their solubility is lowest.

4.9. Non-conversion utilization of CO_2

CO_2 utilization that does not involve its chemical conversion is an alternative destination of captured emissions. Among such alternatives the injection of supercritical CO_2 into depleted oil wells to enhance the further recovery of oil is well established. Indeed, this is presently the only commercially viable technology adding value to large volumes to CO_2 in the order of magnitude of emissions from fossil fuel based energy generation. It has been estimated that CO_2 injection can enhance oil recovery from a depleting well by about 10 to 20 % of the original oil in place. Similarly, CO_2 can be used to recover methane from unmined coal seams. It has been estimated that, in the U.S. alone, 89 billion barrels of oil could technically be recovered using CO_2, leading to a storage of 16 Gt of CO_2 in the depleted oil reservoirs (DNV, 2011).

The use of supercritical CO_2 as a solvent in processing chemicals (e.g., flavor extraction) is also well established. New uses of supercritical CO_2 in chemical processing are emerging, and have the added benefit of reducing water usage. Supercritical CO_2 is also being explored as a heat transfer fluid for some geothermal applications. These non-conversion methods of utilization constitute a significant fraction of the total CO_2 emissions (DNV, 2011).

Enhanced Oil Recovery (CO_2-EOR): Through CO_2-EOR, oil producers inject CO_2 into wells to help sustain production in otherwise declining oil fields. The main goal of this technology is to draw more oil to the surface. In 2012, CO_2-EOR accounted for 6% of current U.S. domestic oil production. The limited CO_2 source is the main barrier to reaching higher levels of CO_2-EOR production due to insufficient supplies of affordable CO_2. With the discovery of offshore gas fields with high CO_2 contents in Brazil, there is a great opportunity to implement CO_2-EOR at those fields.

Furthermore, the offshore removal of acid gases poses a choice of onshore processing against offshore processing. Factors like safety and operability may favor onshore processing in comparison with offshore processing. The proper on land disposal of the CO_2 removed from natural gas requires the construction of CO_2 pipelines to transport CO_2 to offshore EOR applications. Another aspect is the high cost of ship hulls as shifting CO_2 removal to onshore facilities releases the weight shipped, which could overload the cost of building the required CO_2 pipelines.

In the option of onshore processing, CO_2 rich natural gas would be available as feedstock to SYNGAS production from CO_2 reforming, besides CO_2 separation and transport back to oil fields for CO_2-EOR. The current estimated cost gap for CGS from power, steel and cement plants is several times larger than the current CO_2 market price, and downward pressure on this market price is likely to increase. Investments in CO_2 reuse technologies need to be assessed as a screening procedure among potential alternatives.

5. Concluding remarks

Technologies for utilization of CO_2 amenable to commercial scales are presently a very small fraction of anthropogenic CO_2 emissions, and very endothermic due to the inertness of CO_2, what reduces their abatement potential. Furthermore, chemical and biochemical conversion of CO_2 presents a sequestration potential that is orders of magnitude lower than the CO_2 emissions associated to energy generation from fossil fuels.

Geographical synergies of CO_2 supply (power plant emissions or natural gas processing) should guide in the medium term feasible utilization alternatives. The main synergy is identified in offshore gas processing and EOR, which, due to the economic benefit, process scale and maturity, stands as the most relevant utilization route in the short to medium term. Furthermore, most of the emerging alternatives reviewed are at their early stage of technological development.

However, CO_2 stands as a promising renewable feedstock to the chemical industry, which has been limited to oil, natural gas, coal and, recently, biomass. Such as posed, SYNGAS based conversions to the downstream supply chain is a route for flexibility of raw materials. Gasification of a variety of feedstock can lead to SYNGAS. Furthermore, expanding non-conventional gas supply enforces natural gas reforming in the upstream of the chemical supply chain. CO_2 captured from emissions and natural gas processing may drop into the supply chain via Dry Reform. As SYNGAS derived products, hydrogen, methanol and synthetic fuels (e.g., olefins, naphta, diesel, lubricants and kerosene) from Fischer-Tropsch process are likely to dominate the scenario.

Additionally, methanol (MeOH) is expected to grow in relevance either as hydrogen carrier and as intermediate product such as feedstock to MeO (Methanol to Olefins) process, as well as trans-esterification agent in biodiesel and dimethyl carbonate (DMC) production processes. Nowadays, the interest in DMC has grown significantly because it is considered to be a safe and nontoxic substitute for well-established methylating and carbonylating agents (e.g., phosgene), and has potential as an oxygen-containing fuel additive. There are several technological routes to produce DMC, however, the one route considered promising for large-scale commercialization is the trans-esterification of ethylene carbonate (EC) with methanol. In this indirect route, EC is obtained by a previous reaction of CO_2 with ethylene oxide. The route yields DMC and ethylene glycol (EG) as co-products in equimolar ratio.

The use of CO_2 as a carbon source in the synthesis of chemicals, in contrast to disposal, reduces dependence on fossil fuels, generates profit and is in line with a sustainable chemical industry. However, the actual use of CO_2 corresponds to about 0.4% of the potential CO_2 suitable to be converted to chemicals (Navarro et al., 2013).

Finally, large-scale utilization of CO_2 require energy efficient CO_2 capture technologies and an expansion of CO_2 transportation infrastructure.

Acknowledgements

O. Araujo and J.L. Medeiros kindly acknowledge CNPq for scholarships and financial grants; and CAPES for grant no. 113/2008.

Author details

Ofélia de Queiroz F. Araújo[1], José Luiz de Medeiros[1] and Rita Maria B. Alves[2,3]

1 Federal University of Rio de Janeiro, Brasil

2 BRASKEM S.A., Brasil

3 University of São Paulo, Brasil

References

[1] ADEME (Agence de l'Environnement et de la Maîtrise de l'Energie), Panorama des voies de valorisation du CO_2 http://www2.ademe.fr/servlet/getDoc?cid=96&m=3&id=72052&p1= 30&ref=12441 (2010).

[2] Aouissi, A., Al-Othman, Z.A., Al-Amro, A. Gas-Phase Synthesis of Dimethyl Carbonate from Methanol and Carbon Dioxide over Co1.5PW12O40 Keggin-Type Heteropolyanion. Int. J. Mol. Sci., 11, 1343-1351 (2010).

[3] Aresta, M. Perspectives in the use of carbon dioxide. Quím. Nova, 22 (2), (1999). http://dx.doi.org/10.1590/S0100-40421999000200019

[4] Aresta, M., Aresta, Michele. Carbon Dioxide as Chemical Feedstock, WILEY-VCH Verlag GmbH & Co., (2010).

[5] Armstrong, Katy, Dawson, George and Styring, Peter. Recent Advances in Catalysis for Carbon Dioxide Utilization. The Catalyst Review, vol. 26, issue 2, 6-13, (2012).

[6] Ashcroft, A.T., Cheetham, A.K., Green, M.L.H., Vernon, P.D.F, Partial oxidation of methane to synthesis gas using carbon dioxide Nature 352, 225 - 226 (18 July 1991).

[7] Badura, A., Guschin, D., Esper, B., Kothe, T., Neugebauer, S., Schuhmann, W., Rögner, M. Photo-Induced Electron Transfer Between Photosystem 2 via Cross-linked Redox Hydrogels. Electroanalysis 20(10), 1043–1047 (2012).

[8] Bayer. Use of carbon dioxide for the production of plastics. http://www.materialscience. bayer.com/~/media/Bms/Documents/Publications/CO2-Projekte_kurz_EN.ashx. Accessed on 11/27/2013 (2013).

[9] Bayer. A Dream Comes True http://solutions.bayertechnology.com/fileadmin/user_upload/sat_pages/technologyImpulse/articles/A-Dream-Comes-True.pdf. Accessed on 11/27/2013 (2010).

[10] Beck, J., R. Johnson, R., Naya, T., Electrochemical Conversion of Carbon Dioxide to Hydrocarbon Fuels, EME 580, Spring (2010).

[11] Bradford, M. C. J., Vannice, M. A., CO_2 Reforming of CH_4, Catalysis Reviews, Vol. 41(1), 1-42, (1999).

[12] Bustamante, F., Enick, R. Rothenberger,K., Howard, B., Cugini, A., Ciocco, M. Morreale, B., Kinetic Study of the Reverse Water Gas Shift Reaction in High-Temperature, High Pressure Homogeneous Systems, Fuel Chemistry Division Preprints, 47(2), 663, (2002).

[13] Butterman, H.C., Castaldi, M.J. Influence of CO_2 Injection on Biomass Gasification. Ind. Eng. Chem. Res., 46, 8875-8886 (2007).

[14] Byron, S.R.J., Loganthan, M., Shantha, M.S., A Review of the Water Gas Shift Reaction. International Journal of Chemical Reactor Engineering 8, 1–32, (2010).

[15] Chisti, Y. Biodiesel from microalgae. Biotechnol Adv, 25, 294–306 (2007).

[16] Cole, E.B., Bocarsly, A.B., Photochemical, Electrochemical, and photoelectrochemical Reduction of Carbon Dioxide, in "Carbon Dioxide as Chemical Feedstock", M. Aresta, Editor, Wiley-VCH Verlag GmbH & Co.: Weinheim. (2010).

[17] de Medeiros, J.L., Versiani, B., Araújo, O.Q.F. A model for pipeline transportation of supercritical CO_2 for geological storage. The Journal of Pipeline Engineering, 4th Quarter, 253-279 (2008).

[18] de Medeiros, J.L., Nakao, A., Grava, W.M., Nascimento, J.F., Araújo, O.Q.F. Simulation of an Offshore Natural Gas Purification Process for CO_2 Removal with Gas Liquid Contactors Employing Aqueous Solutions of Ethanolamines. Industrial & Engineering Chemistry Research, 52, 7074-7089 (2013a).

[19] de Medeiros, J.L., Barbosa, L.C., Araújo, O.Q.F. An Equilibrium Approach for CO_2 and H_2S Absorption with Aqueous Solutions of Alkanolamines: Theory and Parameter Estimation. Industrial & Engineering Chemistry Research, 52, 9203-9226 (2013b).

[20] Delacourt, C. Electrochemical reduction of carbon dioxide and water to syngas (CO + H_2) at room temperature. Available at http://charles.delacourt.free.fr/Postdoc-manuscript_Charles-Delacourt.pdf (2010).

[21] DNV. Carbon Dioxide Utilization Electrochemical Conversion of CO_2 – Opportunities and Challenges. Available online at http://www.dnv.com/binaries/DNV-position_paper_CO2_Utilization_tcm4-445820.pdf (2011).

[22] Edwards, J.H., Potential sources of CO_2 and the options for its large-scale utilisation now and in the future. Catalysis Today, 23, 59-66, (1995).

[23] Ferreira, H.B.P., Vale, D.L., Mota, C.J.A., Miranda, J.L. Experimental Design for CO_2 Conversion into Dimethylcarbonate (DMC) using Bu_2SnO at Subcritical Conditions. Brazilian Journal of Petroleum and Gas, 6(3), 93-104, (2012).

[24] Fernández, F.G.A., González-López, C.V., Sevilla, J.M.F., Grima, E.M. Conversion of CO_2 into Biomass by Microalgae: How Realistic a Contribution May it be to Significant CO_2 Removal? Appl Microbiol Biotechnol, 96, 577–586 (2012).

[25] Gangadharan, P., et al., Evaluation of the economic and environmental impact of combining dry reforming with steam reforming of methane. Chem. Eng. Res. Des. (2012), http://dx.doi.org/10.1016/j.cherd.2012.04.008

[26] Genesis. Equation of State Prediction of Carbon Dioxide Properties. Project Kingsnorth Carbon Capture & Storage Project. CP-GNS-FAS-DRP-0001. http://www.decc.gov.uk/assets/decc/11/ccs/chapter6/6.23-equation-of-state-prediction-of-carbon-dioxide-properties.pdf, (2011).

[27] Ginsburg, J. M.; Pina, J.; El Solh, T.; de Lasa, H. I.Coke formation over a nickel catalyst under methane dry reforming conditions: Thermodynamic and kinetic models Ind. Eng. Chem. Res., 44 (14) 4846– 4854, (2005).

[28] Grima, E.M., Belarbia, E.-H., Acién Fernández, F.G., Medina, A.R., Chisti, Y. Recovery of microalgal Biomass and Metabolites: Process Options and Economics. Biotechnology Advances, 20, 491–515 (2003).

[29] Gunardson, H. Industrial Gases in Petrochemical Processing; Marcel Dekker: New York, (1998).

[30] Hartley, M., Tam, I., Non-Sequestration Utilization Options for Carbon Dioxide (CO_2), Nexant's ChemSystems PERP09/10S10, (2012).

[31] Ho, M.T., Wiley, D. E. Economic Evaluation of Membrane Systems for Large Scale Capture and Storage of CO_2 Mixtures. Report, UNESCO Centre for Membrane Science, The University of New South Wales (2005).

[32] Hu, B., Guild, C., Suib, S.L., Thermal, electrochemical, and photochemical conversion of CO_2 to fuels and value-added products, Journal of CO_2 Utilization, 1, 18–27, (2013).

[33] IEA. Technology Roadmap - Energy and GHG Reductions in the Chemical Industry via Catalytic Processes. Technology Perspectives. Available online at: http://iea.org/media/freepublications/technologyroadmaps/TechnologyRoadmapCatalyticProcessesAnnexes.pdf, (2013).

[34] Jarungthammachote, S., Combined Partial Oxidation and Carbon Dioxide Reforming Process: A Thermodynamic Study, American Journal of Applied Sciences 8 (1): 9-14, (2011).

[35] Jiang, Z.; Liao, X.; Zhao, Y., Comparative study of the dry reforming of methane on fluidized aerogel and xerogel Ni/Al2O3 catalysts. Applied Petrochemical Research, p. 1-9, (2013).

[36] Kahle, L. C. S., Roussière, T., Maier, L., Delgado, K.H., Wasserschaff, G., Schunk, S.A., Deutschmann, O., Methane Dry Reforming at High Temperature and Elevated Pressure: Impact of Gas-Phase Reactions, Industrial & Engineering Chemistry Research, 52 (34), (2013).

[37] Kaiser, P., Unde, R.B., Kern, C., Jess, A. Production of Liquid Hydrocarbons with CO_2 as Carbon Source based on Reverse Water-Gas Shift and Fischer-Tropsch Synthesis. Chemie Ingenieur Technik. Special Issue: Reaktionstechnik. 85(4), 489–499, (2013).

[38] Kawabe, K. Development of Highly Selective Process for Mono-Ethylene Glycol Production from Ethylene Oxide via Ethylene Carbonate Using Phosphonium Salt Catalyst. Catal Surv Asia 14, 111–115, (2010).

[39] Kayaert, S., Martens, J., Masschaele, K. Photo-Electrochemical Cell. United States Patent Application 20130026029, (2013).

[40] King, D. The Future of Industrial Biorefineries. World Economic Forum, (2010).

[41] Kuwabata, S., Nishida, K., Tsuda, R., Inoue, H., Yoneyama, H., Photochemical Reduction of Carbon Dioxide to Methanol Using ZnS Microcrystallite as a Photocatalyst in the Presence of Methanol Dehydrogenase, J. Electrochem. Soc. 141(6), 1498, (1994).

[42] Kuwabata, S., Tsuda, R., Yoneyama, H., Electrochemical Conversion of Carbon Dioxide to Methanol with the Assistance of Formate Dehydrogenase and Methanol Dehydrogenase as Biocatalysts. J. Am. Chem. SoC., 116, 5431-5443, (1994).

[43] Kurz, G. , Teuner, S., Calcor process for carbon monoxide production, Erdoel & Kohle, Erdgas, Petrochemie, 43(5), 171-172, (1990).

[44] Lachowska, M., Skrzypek, J., Ga, Mn, and Mg Promoted Copper/Zinc/Zirconia-Catalysts for Hydrogenation of Carbon Dioxide to Methanol, Carbon Dioxide Utilization for Global Sustainability, Elsevier, Amsterdam, (2004).

[45] Le, M.T.H., Electrochemical Reduction of CO_2 to Methanol, MSc Thesis. Graduate Faculty of the Louisiana State University and Agricultural and Mechanical College, (2011).

[46] Lee, D., Lee, J.-Y., Lee, J.S., Effects of Palladium Particle Size in Hydrogenation of Carbon Dioxide to Methanol over Pd/ZnO Catalysts, Carbon Dioxide Utilization for Global Sustainability, Elsevier, Amsterdam, (2004).

[47] Levitan, R., Levy, M., Rosin, H., Rubin, R., Closed-loop operation of a solar chemical heat pipe at the Weizmann Institute solar furnace. Solar Energy Materials vol. 24 issue 1-4, 464-477, (1991).

[48] Levy, M., Levitan, R., Rosin, H., Rubin, R., Solar energy storage via a closed-loop chemical heat pipe. Solar Energy, vol. 50, issue 2, 179-189, (1993).

[49] Li, H.. Thermodynamic Properties of CO_2 Mixtures and Their Applications in Advanced Power Cycles with CO_2 Capture Processes. Energy Processes Department of Chemical Engineering and Technology Royal Institute of Technology, Stockholm, Sweden. TRITA-CHE Report 2008:58, (2008).

[50] Li, H., Yan, J.. Evaluating Cubic Equations of State for Calculation of Vapor–Liquid Equilibrium of CO_2 and CO_2-Mixtures for CO_2 Capture and Storage Processes. Chem. Eng. and Technology/Energy Process, Royal Institute of Technology, (2009).

[51] Lim, M., Han, G.C., Ahn, J.W., You, K.S. Environmental Remediation and Conversion of Carbon Dioxide (CO_2) into Useful Green Products by Accelerated Carbonation Technology. Int. J. Environ. Res. Public Health, 7, 203-228, (2010).

[52] Li, M.W., Xu, G.H., Tian, Y.L., Chen, L., Fu, H.F. Carbon dioxide reforming of methane using DC corona discharge plasma reaction. J. Physical Chem. A., 108: 1687-1693, (2004).

[53] Machado, R.B.P. Avaliação Técnica de Processo de Carbonatação de Lama Vermelha com Gás Exausto de Calcinadores de Alumina. MSc Thesis. Graduate Program in

Chemical and Biochemical Process Technology, School of Chemistry, Federal University of Rio de Janeiro, (2012).

[54] Maeda, K. Domen, K. 7.22 – Photochemical Water Splitting Using Nanostructured Metal Oxides. Reference Module in Chemistry, Molecular Sciences and Chemical Engineering Comprehensive Inorganic Chemistry II (Second Edition). From Elements to Applications, 587–614. Vol. 7: Surface Inorganic Chemistry and Heterogeneous Catalysis, (2013).

[55] Maiohm, T., Wannakao, S., Boekfa, B., Limtrakul, J. Production of Formic Acid via Hydrogenation of CO_2 over a Copper-Alkoxide-Functionalized MOF: A Mechanistic StudyJ. Phys. Chem. C, 117, 17650–17658, (2013).

[56] McCoy, S.T. The economics of CO_2 transport by pipeline and storage in saline aquifers and oil reservoirs. PhD Thesis, Carnegie-Mellon University, Pittsburgh, USA, (2008).

[57] Meunier, F.C., Tibiletti, D., Goguet, A., Shekhtman, S., Hardacre, C., Burch, R., On the complexity of the water-gas shift reaction mechanism over a Pt/CeO_2 catalyst: Effect of the temperature on the reactivity of formate surface species studied by operando DRIFT during isotopic transient at chemical steady-state, Catalys Today, 126, 143, (2007).

[58] Moon, D.J., Ryu, J.W., Kang, D.M., Lee, B.J., Ahn, B.S., CO_2 Reforming by CH_4 over Ni-YSZ Modified Catalysts in Carbon Dioxide Utilization for Global Sustainability, Elsevier, Amsterdam, (2004).

[59] Monteiro, J.G.M.S., Silva, P.A., Araújo, O.Q.F., de Medeiros, J.L. Pareto Optimization of an Industrial Ecosystem: sustainability maximization. Brazilian Journal of Chemical Engineering, 27, 429-440, (2010).

[60] Nakao, A., Macedo, A.P.F., Versiani, B.M., Araújo, O.Q.F., and de Medeiros, J.L., Modeling of Flowcharts of Permeation through Membranes for Removal of CO_2 of Natural Gas, 10th International Symposium of Process System Engineering, PSE-2009, Computer Aided Chemical Engineering, 27, 1875-1880, ISBN 978-0-444-53472-9, Elsevier, (2009)

[61] Navarro, R., Pawelec, B., Alvarez-Galván, M. C., Guil-Lopez, R. Al-Sayari, S., Fierro, J. L. G. Renewable SYNGAS Production via Dry Reforming of Methane, in M. de Falco et al., CO2: A Valuable Source of Carbon Green Energy and Technology, pp 45-66, Spring-Verlag London, (2013).

[62] Nagaoka, K., Seshan, K., Aika, K., Lercher, J.A., Carbon deposition during carbon dioxide reforming of methane-comparison between Pt/Al_2O_3 and Pt/ZrO_2. J. Catalysis, 197: 34-42, (2001).

[63] O'Connor A.M., Ross J.R.H. The Effect of O_2 Addition on the Carbon Dioxide Reforming of Methane over Pt/ZrO_2. Catalysts. Catal. Today, 46 (2-3), 203–210, (1998).

[64] Oi, L.E. CO_2 removal by absorption: challenges in modelling. Mathematical and Computer Modelling of Dynamical Systems, 16(6), 511–533 (2010).

[65] Olah, G.A., Goeppert, A., Prakash, G.K.S., Beyond Oil and Gas: The Methanol Economy, 2nd ed., Wiley VCH, Weinheim, Germany, (2009).

[66] Olah, G.A., Goeppert, A., Prakash, Czaun, M., G.K.S., Bi-reforming of Methane from Any Source with Steam and Carbon Dioxide Exclusively to Metgas ($CO–2H_2$) for Methanol and Hydrocarbon Synthesis, J. Am. Chem. Soc., 135 (2), 648–650, (2013).

[67] Ormerod, W., Riemer, P., Smith, A. Carbon Dioxide Utilisation. IEA Greenhouse Gas R&D Programme, http://www.ieagreen.org.uk/sr4p.htm, (1995).

[68] Pettersen, J. Snøhvit field development. Available online at http://www.ipt.ntnu.no/~jsg/ undervisning/prosessering/gjester/LysarkPettersen2011.pdf, (2011).

[69] Petrobras. Visão Geral da Petrobras. Available online at: http://www.investidorpetrobras.com.br/pt/apresentacoes/visao-geral-da-petrobras.htm

[70] Pettersen, J., Snøhvit field development, TEP4520, Statoil, (2011).

[71] Picardo, M.C., de Medeiros, J.L., Monteiro, J.G.M., Chaloub, R.M., Giordano, M., Ofélia, Q.F. A methodology for screening of microalgae as a decision making tool for energy and green chemical process applications. Clean Technologies and Environmental Policy, 15, 275-291, (2013a).

[72] Picardo, M.C., de Medeiros, J.L., Ofélia, Q.F., Chaloub, R.M. Effects of CO_2 Enrichment and Nutrients Supply Intermittency on Batch Cultures of Isochrysis galbana. Bioresource Technology, 143, 242-250, (2013b).

[73] Qin, Y., Wang, X. Carbon dioxide-based copolymers: Environmental benefits of PPC, an industrially viable catalyst. Biotechnol. J., 5, 1164–1180, (2010).

[74] Raju, A.S.K., Park, C.S., Norbeck, J.M., Synthesis gas production using steam hydrogasification and steam reforming. Fuel Process. Technol. 90, 330–336, (2009).

[75] Richardson, J.T., Paripatyadar, S.A. , Carbon dioxide reforming of methane with supported rhodium Appl. Catal. 61, 293-309, (1990).

[76] Rostrup-Nielsen, J., 40 years in catalysis, Catalysis Today, 111, 4-11, (2006).

[77] Rostrup-Nielsen, J., Christiansen, L.J. Concepts in syngas Manufacture. Catalytic Series, V. 10. World Scientific, 392 pp, 2011. ISBN: 978-1-84816-567-0, (2011).

[78] Saito, M., Murata, K., Development of high performance Cu/ZnO-based catalysts for methanol synthesis and the water-gas shift reaction, Catalysis Surveys from Asia, 8, 285-294, (2004).

[79] Sakakura, T., Choi, J., Yasuda, H. Transformation of carbon dioxide. Chem Rev, 107, 2365-2387, (2007).

[80] Shinsuke, F., Isaburo, F., Masahiro, T., Kazuhiro, O., Hiroshi, H., Muneaki, A., Kazumi, H., Kyosuke, K. A Novel Non-Phosgene Process for Polycarbonate Production from CO_2: Green and Sustainable Chemistry in Practice. Catalysis surveys from Asia, 14(3-4), 146-163, (2010).

[81] Song, C., Pan, W., Tri-Reforming of Methane: A Novel Concept for Synthesis of Industrially Useful Synthesis Gas with Desired H_2/CO Ratios Using CO_2 in Flue Gas of Power Plants without CO_2 Separation, Prepr. Pap.-Am. Chem. Soc., Div. Fuel Chem., 49 (1), 128, (2004).

[82] Song, C. Global Challenges and Strategies for Control, Conversion and Utilization of CO_2 for Sustainable Development Involving Energy, Catalysis, Adsorption and Chemical Processing. Catalysis Today, 115, 2–32, (2006).

[83] Souza, M.M.V.M., Schmal, M., Methane conversion to synthesis gas by partial oxidation and CO_2 reforming over supported platinum catalysts, Catalysis Letters, 91(1–2), 11-17, (2003).

[84] Souza, L.F.S., Ferreira, P.R.R., de Medeiros, J.L.M., Alves, R.M.B., Araûjo, O.Q.F. Production of DMC from CO_2 via Indirect Route: Technical−Economical−Environmental Assessment and Analysis Sustainable Chem. Eng., dx.doi.org/10.1021/sc400279n, (2013).

[85] Span, R., Wagner, W. A New Equation of State for Carbon Dioxide Covering the Fluid Region from the Triple-Point Temperature to 1100 K at Pressures up to 800 MPa. J. Phys. Chem. Ref. Data, 25(6), (1996).

[86] Spath, P.L., Dayton, D.C., Preliminary Screening – Technical and Economic Assessment of Synthesis Gas to Fuels and Chemicals with Emphasis on the Potential for Biomass-Derived SYNGAS. National Renewable Energy Lab Golden Co., (2003).

[87] Stoczynski, J., Grabowski, R., Koslowska, A., Lachowska, M., Skrzypek, J., Effect of Additives and a Preparation Method on Catalytic Activity of Cu/ZnO/ZrO_2 System in the Carbon Dioxide Hydrogenation to Methanol, Carbon Dioxide Utilization for Global Sustainability, Elsevier, Amsterdam, (2004).

[88] Tanaka, Y., Utaka, T., Kikuchi, R., Sasaki, K., Eguchi, K., CO removal from reformed fuel over Cu/ZnO/Al2O3 catalysts prepared by impregnation and coprecipitation methods, Applied Catalysis A: General, 238, 11–18, (2003).

[89] Teuner, S., Make carbon monoxide from carbon dioxide, Hydrocarbon Processing, International Edition, 64(5), 106-7, (1985).

[90] Teuner, S. et al., The Calcor Standard and Calcor Economy Processes, in Oil and Gas European Magazine, 44-46, (2001).

[91] Treacy, D., Ross, J.R.H. The Potential of the CO_2 reforming of CH_4 as a method of CO_2 mitigation. A thermodynamic study. Prepr. Pap.-Am. Chem. Soc., Div. Fuel Chem., 49 (1), 126–127, (2004).

[92] Trusler, J. P. M. Equation of State for Solid Phase I of Carbon Dioxide Valid for Temperatures up to 800 K and Pressures up to 12 GPa. J. Phys. Chem. Ref. Data, 40 (4), 043105-1-043105-19, http://dx.doi.org/10.1063/1.3664915, (2011).

[93] Van-Dal, E.S., Bouallou, C. Design and simulation of a methanol production plant from CO_2 hydrogenation. Journal of Cleaner Production, 57, 15 October 2013, 38–45, http://dx.doi.org/10.1016/j.jclepro.2013.06.008

[94] Ushikubo, T. Green Chemistry in Japan. Chemistry International, 35(4), July-August 2013. Available online at www.iupac.org/publications/ci, (2013).

[95] Wagener, D.H.V., Rochelle, G.T. Alternative Stripper Configurations for CO_2 Capture by Aqueous Amines. AIChE Spring Meeting, (2010).

[96] Wagener, D.H.V., Rochelle, G.T. Stripper configurations for CO_2 capture by aqueous monoethanolamine. Chemical Engineering Research and Design, 89, 1639–1646, (2011).

[97] Walther, D., Ruben, M., Rau, S. Carbon Dioxide and Metal Centres: from Reactions Inspired by Nature to Reactions in Compressed Carbon Dioxide as Solvent. Coordination Chemistry Reviews 182, 67–100, (1999).

[98] Wang, W., Stagg-Williams, S.M., Noronha, F.B., Mattos, L.V.,Passos, F.B., Partial Oxidation and Combined Reforming of Methane on Ce-promoted Catalysts, Prepr. Pap.-Am. Chem. Soc., Div. Fuel Chem. 49 (1), 132-133, (2004).

[99] Wu, J.C.S., Lin, H.-M., Photo reduction of CO_2 to methanol via TiO_2 Photocatalyst, International Journal of Photoenergy, 7, 115-119, (2005).

[100] Wurzel, T., Malcus, S., Mleczko, L., Reaction engineering investigations of CO_2 reforming in a fluidized-bed reactor. Chem. Eng. Sci., 55, 3955-3966 (2000).

[101] Yamashita, H., Fujii, Y., Ichihashi, Y., Zhang, S. G., Ikeue, K., Park, D. R.,Koyano, K., Tatsumi, T., Anpo, M., Selective formation of CH_3OH in the photocatalytic reduction of CO_2 with H_2O on titanium oxides highly dispersed within zeolites and mesoporous molecular sieves, Catalysis Today 45, 221-227, (1998).

[102] Zangouei, M., Moghaddam, A.Z., Razeghi, A., Omidkhah, M.R., Reforming and Partial Oxidation of CH_4 over Ni/Al_2O_3 Catalysts in Fixed-bed Reactor International Journal of Chemical Reactor Engineering, 8. Available at: http://works.bepress.com/mohammadreza_omidkhah/2, (2010).

[103] Zhang, X., Lee, C. S.-M., Mingos, D.M.P., Hayward, D. O., Carbon dioxide reforming of methane with Pt catalysts using microwave dielectric heating. Catalysis Letters, 88(3–4), 129-139, (2003).

[104] Zhou, H., Cao, Y., Zhao, H., Liu, H., Wei-Ping, P., Investigation of H_2O and CO_2 Reforming and Partial Oxidation of Methane : Catalytic Effects of Coal Char and Coal Ash, Energy & Fuels, 22, 2341-2345, (2008).

Chapter 3

Ocean Carbon Sequestration by Direct Injection

Aaron Chow

Additional information is available at the end of the chapter

http://dx.doi.org/10.5772/57386

1. Introduction

The ocean is the largest sink of atmospheric CO_2 (about 7 petagrams (Pg) per year) (1 Pg = 1 gigaton = 10^{15} g). Dissolved CO_2 (passively entering the ocean via diffusion from the atmosphere) has already acidified the surface ocean, the most productive region of the ocean. Ocean carbon sequestration (OCS) is a method to distribute CO_2 more evenly throughout ocean depth and minimize surface ocean impacts. There are two major methods of OCS – direct injection and ocean fertilization (promoting photosynthetic fixation of CO_2 by ocean organisms). This chapter focuses only on the direct injection as a method of OCS. This chapter will first describe the physical mechanism by which CO_2 can be stored in the ocean water column at depth. It will summarize past ocean direct injection studies, and outline the effects of increased dissolved CO_2 and locally increased CO_2 partial pressure on marine organisms. It will also include a discussion of the engineering challenges of delivering CO_2 to the water column, including the selection of injection sites to minimize CO_2 outgassing to the atmosphere as well as minimizing marine life impacts. Finally, this chapter will address the legal, policy and public outreach issues that have ultimately precluded implementation of OCS using direct injection.

1.1. Motivation of Ocean Carbon Sequestration (OCS)

The ocean is presently the largest sink of atmospheric CO_2 (about 7 Pg per year) [1]. The Earth's oceans cover over 70% of the Earth's surface, and have an average depth of 3,800 m. However, dissolved CO_2 is already causing surface ocean acidification (most productive region of ocean) as it equilibrates with the atmospheric CO_2[2]. By 1994, the total atmospheric release of anthropogenic (i.e., man-made) carbon was about 244 Pg of carbon (PgC) from fossil fuel combustion, and about 140 PgC from land use change (e.g., deforestation) [3]. The oceans have absorbed about one-third of anthropogenic CO_2 (the atmosphere retained about 43%, while the oceans absorbed about 30%), leading to a decrease of surface-ocean total pH by about 0.1 units from about 8.2 to 8.1. If CO_2 emissions continue unabated the subsurface ocean total could decline by 0.7 units by 2300. To place in a geological perspective, the surface ocean pH (on a total scale) has not been below 8.1 during the past 2.1 million years. The total addition of carbon into the atmosphere is expected to be about 5000 PgC – the estimated fossil fuel reserves excluding hydrates -- in the next 500 years. This is a higher rate of carbon addition than ever experienced by the earth over a short geological time scale [3].

At the same time the ocean pH in the deep ocean has been decreasing at a relatively slower rate compared with the surface ocean [1, 4]. Rising concentrations of greenhouse gases in the atmosphere are implicated in adverse climate changes and two-thirds of the change is attributed to CO_2 [1]. Ocean carbon sequestration was conceived as a method to distribute CO_2 more evenly throughout the ocean column, especially into deep ocean waters, and minimize surface ocean impacts while the ocean CO_2 levels equilibrate with the atmosphere.

There are two major methods of OCS – direct injection and ocean fertilization (promoting photosynthetic fixation of CO_2 by ocean organisms). This chapter focuses only on the direct injection of CO_2.

2. Physical description of direct injection

2.1. Physical properties of ocean/CO_2 system

The conditions under which CO_2 can exist in a gas, liquid, solid or hydrate, and aqueous phases are depicted in the phase diagram (see Figure 1) [5] At typical ocean temperatures and pressures, CO_2 exists as a gas above 500 m depth, and a liquid below this depth. Between 500 and 2700 m depth, liquid CO_2 is less dense than seawater and would float, while below 2700 m CO_2 is denser than seawater and would sink. A solid CO_2 hydrate phase is thermodynamically stable in the ocean at low temperatures; CO_2 hydrates are discussed in Section 2.3.

2.1.1. Solubility of CO_2 in the ocean

CO_2 dissolves in ambient seawater that is not saturated with CO_2. Once dissolved, aqueous CO_2 exists in various charged forms in water according to these main reactions, known as the carbonate system [6]:

$$CO_2(aq) + H_2O = H_2CO_3(aq) \tag{1}$$

$$H_2CO_3(aq) = H^+ + HCO_3^- \tag{2}$$

$$HCO_3^- = H^+ + CO_3^{2-} \tag{3}$$

The total dissolved inorganic carbon (C_T) is defined as:

$$C_T = \left[H_2CO_3(aq)\right] + \left[HCO_3^-\right] + \left[CO_3^{2-}\right] \tag{4}$$

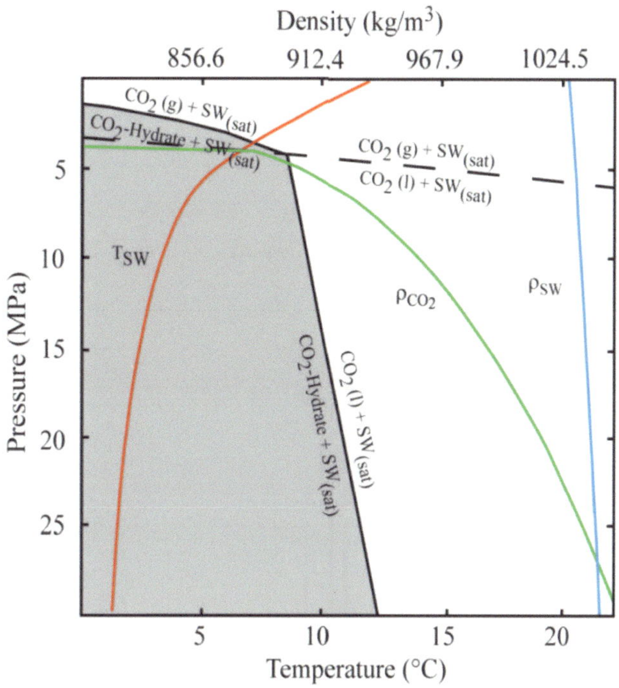

Figure 1. Phase Diagram for CO_2 in the ocean [5]. Gray area: region of stability for CO_2 hydrates; dashed line – gas/liquid phase transition for pure CO_2; red line – depth - temperature profile of the Pacific Ocean at 152°W, 40°N; green line – CO_2 density; blue line – seawater density (35.0 PSU).

The result of this is that increasing dissolved CO_2 will shift the equilibria (1) and (2) above to the right, and lower the local pH of the ambient seawater. [7]. At a typical surface seawater water pH of 8.2, the equilibrium (3) is shifted to the left with the addition of CO_2. The net result of an increase in dissolved CO_2 is the decrease of pH, an increase in HCO_3^- and a decrease of CO_3^{2-} (Figure 2).

In turn, the dissolved CO_2 causes an increase of the density of the seawater via the solute density effect [9] that has implications for the physical design of a direct injection of CO_2 into the ocean.

2.1.2. CO_2 partial pressure (pCO_2)

The partial pressure of CO_2 of a sample of water, denoted by pCO_2, is the pressure of gaseous CO_2 which, if allowed to equilibrate with water, will result in the same amount of dissolved CO_2 as observed in the sample. It is related to the solubility of CO_2, C_s, and the concentration of H_2CO_3(aq) by the following:

$$pCO_2 = \frac{[H_2CO_3(aq)]}{C_s(T,S)} \tag{5}$$

where $C_s(T,S)$ (usually expressed in μatm) is dependent on the local temperature T and salinity [10] provide empirical relations to obtain C_s.

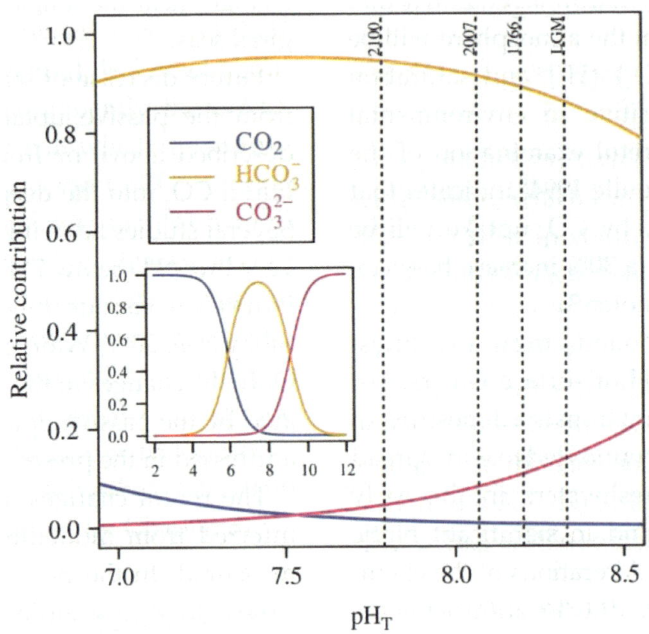

Figure 2. Bjerrum (pH_T (total scale pH) – relative speciation) plot showing the relative contributions of CO_2, HCO^{3-} and CO_3^{2-} to the dissolved inorganic carbon as a function of pH, at 15 deg C and a salinity of 35 PSU. The dashed vertical lines indicate the average open ocean surface pH_T during the Last Glacial Maximum (LGM), 1766, 2007 and 2100 (projected) [8].

As the atmospheric CO_2 concentration increases, pCO_2 levels increase in the surface ocean as it equilibrates with the atmosphere. The ratio of the relative change in pCO_2 to the relative change in C_T is known as the Revelle factor, and is inversely proportional to $[CO_3^{2-}]$. The inverse of the Revelle factor is also often termed as the buffering capacity of the ocean. According to this relation, a doubling in atmospheric CO_2 would only increase the total dissolved CO_2 by about 10%. [11]

While sea-air equilibria for most gases like oxygen occur over a time scale of days, it can take ~8 months for CO_2 to reach equilibrium at the surface, because the dissolved CO_2 in the carbonate system does not remain a dissolved gas but instead causes an increase in HCO_3^-. [4].

2.1.3. Calcium carbonate saturation state of seawater

The $CaCO_3$ saturation state of seawater is defined as Ω:

$$\Omega = \frac{\left[Ca^{2+}\right]\left[CO_3^{2-}\right]}{K_{sp}} \tag{6}$$

where $[Ca^{2+}]$ and $[CO_3^{2-}]$ are the seawater concentrations of Ca^{2+} and CO_3^{2-}, and K_{sp} is the solubility product of either calcite or aragonite (the two major forms of $CaCO_3$). If Ω for aragonite (Ω_a) for instance is greater than 1, then aragonite is supersaturated and solid aragonite would begin to precipitate; if Ω_a drops to below 1 then aragonite is undersaturated with respect to the ambient ocean and solid aragonite would begin to dissolve. Because K_{sp} increases with pressure, for both aragonite and calcite there is a transition of the saturation state from $\Omega > 1$ to $\Omega < 1$ sediments with depth [3]. The depth at which $\Omega = 1$ for a mineral is known as its saturation horizon.

2.2. Methods of direct injection

CO_2 sequestration first involves capture from their sources, of which one major type of the coal-fired power plant. The CO_2 emissions are relatively pure from coal-fired power plants and could be isolated and injected into the ocean. A typical 500 MW power plant produces about 130 kg/s of CO_2. [12]. After CO_2 capture, the CO_2 would be transported to the ocean via a pipe or ship to the ocean for direct injection. Technologies for CO_2 direct injection include: Liquid CO_2 droplets [13]; CO_2 laden seawater [9,14,15]; Solid CO_2 (dry ice) [16,17]; and CO_2 lake formation (See Figure 3)

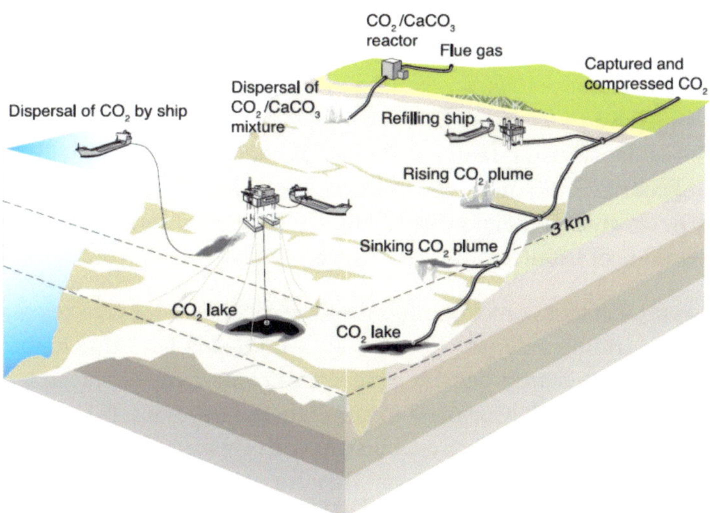

Figure 3. Ocean storage strategies (From Goddard, in [1]).

2.3. CO_2 hydrates

As shown in Figure 1, at lower temperatures (below about 5 - 10 degrees C) and at high pressure (corresponding to an ocean depth of about 400 m or greater) solid CO_2 hydrates are thermodynamically stable. CO_2 hydrates consist of molecules of CO_2 inside a cage-like structure of hydrogen-bonded water molecules [18]. They are of interest as a possible vehicle for deeper ocean carbon sequestration, because they are denser than seawater, and will sink unaided while dissolving to promote dispersion in the ocean. Pure hydrate particles are difficult to produce, but the Oak Ridge National Laboratory (ORNL) has designed a continuous CO_2-seawater co-flowing injector to create cylindrical composite particles comprised of CO_2 hydrate (negatively buoyant), liquid CO_2 (slightly positively buoyant at 1000-1500 m depths) and seawater [19].

Although CO_2 hydrates are thermodynamically stable, they will dissolve in ambient seawater upon release, because CO_2 is under-saturated in the ambient water. Field and laboratory observations confirmed that both pure hydrates and partially reacted cylindrical composite particles dissolved in the ambient seawater [20,21,23].

2.4. CO_2 droplet and hydrate studies

Numerical efforts to simulate the behaviour of CO_2 droplet plumes have included solving the full three dimensional Navier-Stokes equations in quiescent ambient sea conditions [7,22,24]. Bubble plume models calibrated using laboratory observations have also been applied to CO_2 droplet releases [25-27].

Field tests were conducted using CO_2 hydrate composite injectors [23,28]. The latest survey, with a hydrate reactor located at an ocean depth of ~1500 m, produced curved negatively buoyant cylindrical particles with diameters ~2.2 cm and lengths up to ~1 m. Applying a drag coefficient model to observed initial settling velocities and dissolution rates during the most recent survey [29,30], the hydrate conversion efficiency (percentage of liquid CO_2 converted to hydrate) in the field was ~ 15-20% resulting in particles with specific gravity 1-2% greater than seawater, which lead them to sink to a depth below discharge of roughly 100 m. Greater sinking could be achieved using larger particles. Discharging particles with a range of sizes and densities (reflecting different conversion rates) would cause differential settling resulting in spreading in the down-current and vertical directions. Furthermore, towing the source from a moving ship would contribute additional dispersion [29].

An alternative approach to enhancing mixing and vertical descent is to release a continuous stream of particles, forming a dense plume which would sink both due to the density of the particles as well as the increased density of seawater containing dissolved CO_2. An integral double plume model [25,29,31] was used to simulate the behavior of continuous streams of composite particles released to a quiescent ocean, with typical ambient stratification, at CO_2 loadings of 0.01 to 1000 kg/s. Results showed that, for a CO_2 release of 100 kg/s (roughly the emission from a 500 MW coal-fired power plant), a plume composed of 2.2 cm diameter composite particles with 16% reaction efficiency would sink about 1000 m, approximately 10 times the individual particle sinking depth. A plume composed of similar particles, but with a diameter of 5 cm, would sink about 2000 m (~5 times the individual particle depth), while plumes composed of larger particles, or particles exhibiting higher reaction efficiency, would reach the seafloor (as would the individual particles).

Two ambient effects reduce the performance of a plume: stratification and ocean currents. Plume sinking is hampered by strong ambient stratification which causes trapping of entrained seawater at intermediate depths below release. Density stratification weakens at depths below 1500 m [32], so from the perspective of reduction of plume trapping, regions of the ocean deeper than 1500 m are potentially favourable for depositing CO_2 [33].

3. Environmental impacts/challenges

Some of the concepts relevant to the impacts of OCS by direct injection (e.g. ocean acidification) are presented in this section. The reader is directed to [1,11] for a more detailed and comprehensive summary of the causes and effects of ocean acidification.

3.1. Long term stability of dissolved CO_2 in the ocean

Investigations and estimation of the long term stability is described in greater detail in [1]. Numerical ocean models indicate that placing CO_2 in the deep ocean would isolate most of the CO_2 from the atmosphere for several centuries, but over longer times the ocean and atmosphere would equilibrate.

Relative to direct atmospheric release, direct injection of CO_2 into the ocean could reduce the rise and peak of atmospheric CO_2 levels over the next several centuries. After several centuries, the CO_2 released in the ocean would be transported back to the ocean surface and interact with the atmosphere again. However, in the new equilibrium, most (66% to 85%) of the injected CO_2 would still remain in the ocean despite contacting the atmosphere [1].

Generally, carbon injected in the deep ocean would equilibrate with the atmosphere over a time scale of 300 to 1000 years, based on radiocarbon and other tracer dating to estimate the age of the deep seawater. The estimated age of the North Pacific deep water is 700 – 1000 years, while the North Atlantic deep water is estimated to be only about 300 years old. A large number of numerical three dimensional ocean general circulation models were used to study CO_2 retention. The models generally predict a higher retention time with a deeper injection depth (isolation of CO_2 from the atmosphere is nearly complete for 100 years with an injection depth of 3000 m). Consistent with the radioactive tracer dating, many of the models suggest that the Pacific Ocean would retain a larger fraction than the Atlantic Ocean. However, the models vary greatly in their predictions on the actual time taken for CO_2 injected at a particular site to once again make contact with the atmosphere [1, 34].

Additionally, other geochemical factors may affect these predictions. For example, a higher ocean temperature, as well as a higher dissolved inorganic carbon concentration may lead to a lower efficiency for the ocean to absorb additional CO_2. (See [11]).

3.2. Potential pH and carbonate system changes from added CO_2

As described in Section 2.1, ocean acidification has been occurring since the Industrial Revolution. This section describes the effect of continued ocean acidification on the ocean's carbon cycle and marine ecosystems. Between 1991 and 2006, North Pacific ocean pHT showed a decrease of 0.06 units over the upper 500 m of ocean. In the Iceland Sea, the trend of pH_T decrease between 1985 and 2008 in the surface ocean was 0.0024 units per year, with a corresponding decrease in Ω_a of 0.0117 units per year. The decline in pH_T below 1,500 m in the Iceland Sea was one-quarter of that on the surface, with a corresponding decrease in Ω_a at 0.0009 units per year [4].

Another consequence of the increased dissolved CO_2 in the ocean, as described in Section 2.1.1, is the increase of HCO_3^- and a decrease of CO_3^{2-} in the ocean. The decreased CO_3^{2-}in turn leads to the decrease of the local value of Ω in the ocean. As there is a transition from saturation to undersaturation from $\Omega = 1$, this means that the saturation horizons for both aragonite and calcite would both become less deep with time [3]. The decrease in Ω_a caused the aragonite saturation horizon (ASH), the interface between supersaturated waters above and undersaturated waters below, to rise (shoal) at a rate of 4 m per year. The decrease in Ω, and therefore the shoaling rate for the ASH, is predicted to be more pronounced near the poles, and more severe in the Arctic Ocean than the Southern Ocean, partly because the polar oceans have lower initial concentrations of CO_3^{2-} [4].

It was proposed [3] that the addition of CO_2 followed by global increase in surface temperature can be compared to that which occurred during the Paleocene-Eocene Thermal Maximum (PETM, ~55 million years ago). During PETM, about 3000 PgC was added to the over an estimated 6000 years. However, the current estimate for expected total anthropogenic carbon addition is a larger rate of carbon input over a shorter period of time, about 5000 PgC over about ~500 years. The next highest global carbon addition was experienced by the earth during the Paleocene-Eocene Thermal Maximum, (~55 million years ago) where about 3000 PgC was added over ~6000 years. During the PETM, the effects of ocean acidification on surface calcifying organisms was limited, but the conditions of the PETM were not identical to the predicted future scenario, notably in that the carbon input rate was still much slower than the modern anthropogenic carbon addition. Nevertheless, studies of the PETM may inform future predictions of the behavior of ocean marine life with a large increase of atmospheric CO_2. [3,4].

3.3. Effect of pCO_2 increase on organisms

Effects of elevated CO_2 levels and acidified seawater on marine organisms are explained in in more detail in [1,11,35,36].

At acute levels CO_2 has a narcotic effect on animals and causes respiratory distress and death. The work of [37 – 41] that model the lowered pH on passive marine organisms such as zooplankton that spend varying times in and out of a CO_2 plume, and found that minimizing the local dissolved CO_2 and pH drops will reduce the mortality rate.

Non-lethal effects have also been observed due to hypercapnia (elevated CO_2 exposure) [42-44]. Tamburri et al. [42] have observed the narcotic effects of increased CO_2 levels on mobile deep sea animals in the field; they also observe that while many tend to avoid CO_2 plumes, some may risk the narcotic effects to obtain food. They note [42] that increased partial pressure of carbon dioxide will also have a detrimental effect on marine organisms, such as causing slow respiratory distress and inducing a narcotic effect on fish. Passive marine animals may experience depressed ion exchange capability and metabolism when exposed to lower, chronic levels CO_2. Some studies also show slowed growth in mussels and corals, as well as developmental effects on some marine larvae and eggs (brittle stars and bivalves) [35].

The primary effect of acidified seawater exposure by organisms is acidosis, the decrease of pH in body fluids. Intracellular and extracellular processes have been shown to be disrupted when seawater pH drops to a range of about 6.0 – 7.8. Many marine animals counter acidosis by increasing bicarbonate ion production (e.g. in the gills) [35]. Barry et al. report that organisms that have weaker control of their internal fluid chemistry, and that rely on passive molecular diffusion for gas exchange such as sponges, echinoderms, may have greater sensitivity to ocean acidification [45].

Some organisms may adapt to hypercapnia (elevated CO_2) better than others [46]. For example, tropical fishes, as they live closer to the edge of oxygen limitation than temperate fishes, may make them more sensitive to the combined effects of ocean temperature and ocean acidification than their temperate counterparts. For example, studies on acutely exposed tropical cardinal fishes to 1 week of pCO_2 of 1000 μatm resulted in decreases of aerobic scope and critical swimming speeds by about 40 – 50%, but a similar study conducted for Atlantic cod after 12 months of exposure to both 3000 and 6000 μatm did not result in any significant change in swimming capacity.

High CO_2 levels (up to a pCO_2 of 16,000 ppm [47] have also been observed in ocean bottom waters and marine sediments where there are high rates organic matter oxidation and low rates of mixing with the overlying seawater. Under these conditions, high CO_2 concentrations are often accompanied by low O_2 concentrations. Near the surface at night, respiratory fluxes in some relatively confined rock pools of the intertidal zone can produce high CO_2 levels. [1]. Portner et al. [46] report that high pCO_2 is found in oxygen minimum layers. They report that elevated pCO_2 is linked to acid-base regulation and respiration in fish. However, they also report that coastal and mid-water animals (both pelagic and benthic) regularly experience a large range of pCO_2 values (500 to 9400 μatm) in estuaries [46]. "These patterns suggest that in some environments, organisms have evolved to tolerate relatively wide pH oscillations and/or low pH values." [1]

Organisms such as the Humboldt squid, although thought not to be able to adapt physiologically to future changes to the oceans oxygen balance, have been observed to thrive in oxygen minimum layers which tend to have low pH and are undersaturated with respect to calcium carbonates [46].

Deep sea ecosystems depend on sinking particles of organic carbon, made by photosynthesis near the ocean, settling down through the water. Most species living in the deep sea display very low metabolic rates [48, 49], especially in oxygen minimum layers [51]. Organisms living in the deep seawaters have adapted to the energy limited environment by conserving energy stores and minimizing energy turnover. Turley et al. also suggest the depletion of oxygen as a contributing factor to the increased prevalence of harmful algal blooms, though the link between anthropogenic CO_2 and algal blooms remains controversial [36].

Finally, as many marine organisms synthesize and depend on calcium carbonate structures (e.g. shells), the implication of a lowered CO_3^{2-} and Ω in the ocean is the potential for reduction of their habitats.

As a guide, [1] uses a pH drop of 0.1 units as the threshold pH drop for insignificant marine life impact; it is also within the observed natural variability in the ocean. The US Environmental Protection Agency proposed that the threshold for open waters at depths greater than the euphotic zone, the pH value should not drop more than 0.2 pH units outside the range of natural variation [11]. [39] shows that some theoretically modeled scenarios of carbon dioxide releases (for example, releasing sinking CO2 hydrates from a fixed or moving source at 1,500 m, injecting 10 to 1000 kg/s) would result in local pH drops within this guideline threshold in the vicinity of the release point. Others (e.g. Rockstrom et al.) have introduced the concept of planetary boundaries, and for CO_2 they have proposed a threshold carbonate ion concentration. As a first estimate, they proposed that the oceanic aragonite saturation state Ω_a be maintained at 80% or higher of the average global pre-industrial surface seawater level of 3.44 [50]. As with [1], these planetary boundaries are guides for a sustainable global environment, and (with the exception of the US Environmental Protection Agency for pH) have not been implemented as a regulatory threshold.

3.3.1. Comparison with naturally occurring ocean CO_2 vents

In the ocean, hydrothermal vents are submarine volcanic structures that act as natural sources of CO_2 in the ocean. These have been observed as potential natural analogues of OCS direct injection points. Field observations of hydrothermal vents have shown large fluctuations of pCO_2 (up to 80,000 ppm), over 100 times that observed in typical deep seawater). Over time, the vents have sustained organisms that are specially adapted to living in elevated pCO_2 conditions [52].

Observations near hydrothermal vents have shown that ocean acidification reduced biodiversity below a mean pH_T of 7.8 [53]. While Echinoderms are notably absent from habitats with naturally high CO_2 levels such as hydrothermal vents and shallow CO_2 vents off the coast of Italy [53], sponges appeared to tolerate these same sites.

As observed in [45], "[h]owever, while commonly the literature contains results of short term studies of organism physiology and survival, they may not be indicative of eventual long term consequences of ocean acidification."

4. Engineering feasibility/challenges

4.1. Site selection for injection

As described in [1,54], to date there are no publications dedicated to site selection for direct ocean injection. Although numerical models have predicted CO_2 retention time as a function of the injection location, they have not consistently agreed on any individual location for direct injection. The only agreement appeared to be that a larger depth of injection would result in a longer isolation of CO_2 from the atmosphere [34]. In contrast, [55] presented a study of site selection for deep sea geological storage, highlighting the potential of storage in basalt aquifers along particular seismic and aseismic oceanic ridges. This section therefore discusses factors that should be considered site selection criteria based on to be considered when selecting a site for OCS. Environmental goals of site selection include reducing the likelihood of outgassing, and minimizing acute impacts to ocean organisms, as described in Section 3. Additional considerations include the costs of OCS, applicable international policies (such as regulations regarding disposal and cross border transport) – these factors are presented in Sections 4.2 and 4.3.

4.2. Cost of OCS

Costs were estimated for ship transport of liquid CO_2 to an injection platform, with CO_2 injection from a vertical pipe, or a ship trailing an injection pipe, to water at 3000m [1]. The cost estimate of ocean storage is the sum of three major components: tank storage of CO_2 onshore awaiting shipping; the shipping of CO_2; and direct injection of CO_2 into the ocean (either via an ocean platform, a moving ship, or a pipeline). The estimated sum of the three components (including an assumption of 3% CO_2 emissions from boil off and fuel consumption) is 11.9 and 13.2 US$/ton CO_2 net stored from shipping to 100 km and 500 km offshore, respectively [56]. Cost estimates presented do not include transport of CO_2 onshore.

The cost for transporting CO_2 from a power plant located at the shore through a pipeline running on the sea floor to an injection nozzle was also estimated in [56]. CO_2 captured from a pulverized coal fired power plant with a net generation capacity of 600 MWe is transported either 100 or 500 km by a CO_2 pipeline for injection at a depth of 3000 m at a cost of 6.2 US $/ton CO_2 net stored (100 km case) to 31.1 US$/ton CO_2 net stored (500 km case). Other technical challenges that may not be accounted for include: residual chemicals, metals, minerals and oils that may be released during drilling activities; and the fact that liquefied CO_2 is highly corrosive, requiring that piping for CO_2 delivery would require anti-corrosion coatings, which themselves may pose contamination issues [35].

There are no published cost estimates specific to the production of a CO_2 lake on the sea floor; however, given the dominance of pipeline costs, it is reasonable to assume it to be similar to deep water injection. [1,56].

5. Policy issues/challenges

Since offshore OCS is likely to take place in international waters, several international environmental agreements may apply, mainly those that aim to minimize potential risk s to the marine environment. The main international treaties are the Law of the Sea, the London Convention, London Protocol, and the OSPAR Convention. A succinct background of these treaties is taken directly from [57]:

"International marine environment protection was established in 1972 with the London Convention to regulate the dumping of wastes and other matter at sea. In 1982, this field was extended through the adoption of the United Nations Convention on the Law of the Seas (UNCLOS). Being an overarching construction, UNCLOS does not contain detailed operative provisions on most maritime issues; rather, it provides a framework for all areas, including marine protection, and allows other, more targeted treaties to fill in the gaps…With regard to marine pollution, global standards are set by the Convention on the Prevention of Marine Pollution by Dumping of Wastes and other Matter, signed in London in 1972 (London Convention). Beneath the London Convention exist several regional agreements that cover specific areas of the ocean [Also listed in [57]]. The most widely known of these is OSPAR, the Convention for the Protection of the Marine Environment of the North-East Atlantic. OSPAR is also notable as its regulations on marine pollution are markedly stricter than those of the London Convention, and its decisions are legally as opposed to politically binding on its Contracting Parties."

5.1. 1996 London protocol

UN Convention on Climate Change encouraged the use of the oceans as a reservoir for CO_2, but the UNCLOS (in force since 1994) did not give clear guidance on OCS [1]. With respect to CO_2 storage, the original London Convention (with 80 contracting parties, and in force since 1975) only applied to storage by aircraft and vessels and platforms in the water column. As a result, the London Convention did not apply to storage of CO_2 in the seabed or the water column itself [57].

In November 1996, the London Protocol was established that prohibited the disposal of "industrial waste" into international waters. The list of prohibited substances that were categorized as "industrial waste" were contained in Annex I of the London Protocol. However, in 1996 the London Protocol did not give an opinion whether CO_2 was categorized as a "waste material generated by manufacturing or processing operations" [1]. The London Protocol entered into force March 2006 [57].

5.2. OSPAR convention

In 1992, the OSPAR Commission for the Protection of the Marine Environmental of the North-East Atlantic, was formed which unified the 1972 Oslo and 1974 Paris Conventions. It brought together the governments of Belgium, Denmark, Finland, France, Germany, Iceland, Ireland, Luxembourg, the Netherlands, Norway, Portugal, Spain, Sweden, Switzerland and the United Kingdom, together with the European Community (EC). It is considered the most compre-

http://dx.doi.org/10.5772/57386

hensive and strict legal framework governing the marine environment. As mentioned above, the contracting parties are legally bound to OSPAR's decisions [57,58].

5.3. 2007 OSPAR amendments for CO_2 storage and implications to OCS

In June 2007, the OSPAR issued two amendments to the OSPAR Convention: the first, a decision to prohibit the storage of CO_2 streams in the water column or on the sea bed in the Northeast Atlantic; and the second, a decision to allow the storage of CO_2 in subsea sediments. [59,60]

In the first amendment, OSPAR stated that CO_2 storage in the water column or on the sea bed "is not a sustainable storage option, is likely to result in harm to living resources and marine ecosystems and is thus neither a viable solution with regard to mitigating climate change nor compatible with the aims of the [OSPAR] Convention." However, in the first amendment, OSPAR indicated that ocean storage of CO_2 in the water column or on the seabed is nevertheless still under consideration in international forums. [59]

The OSPAR amendments provided a framework for its contracting national governments to develop permitting programs for CO_2 storage. For example, a list of the minimum items required in an offshore CO_2 storage permit included: a description of the project, including injection rates; types, amounts and sources of CO_2; the location of the facility; characteristics of the geological formation; methods of transport; and a risk management plan, with monitoring and verification measures, mitigation steps and a site closure plan [60].

In July 2011 the contracting parties of the OSPAR Convention ratified the 2007 Amendments to allow for CO_2 storage in subsea geological formations [58].

5.4. 2007 Amendment to the London Protocol

In 2007, an amendment to the London Protocol (Annex 1) allowed for storage of CO_2, if the disposal is into a sub-seabed geological formation, if CO_2 streams are "overwhelmingly" carbon dioxide, and as long as no wastes are added. This amendment provided that CO_2 streams may only be considered if [60,61]:

1. disposal is into a sub-seabed geological formation;
2. they consist overwhelmingly of CO_2. They may contain incidental associated substances-derived from the source material and the capture and sequestration processes used; and
3. no wastes or other matter are added for the purpose of disposing of those wastes or other matter.

The amendments to Annex 1 entered into force on 10 February 2007. In contrast to the OSPAR Convention Amendment that only covered the Northeast Atlantic, the 2007 London Protocol Amendment specifically prohibited direct injection of CO_2 for OCS for all London Protocol contracting parties.

5.5. 2009 London Protocol amendment for transboundary transport of CO_2

Article 6 of the London Protocol (on the export of wastes or other material) was largely interpreted by the contracting parties as prohibiting the export of CO_2 from a contracting party for injection into sub-seabed geological formations. In 2009 Article 6 was amended to allow for cross-border transportation of CO_2. [61]

As of 2011, there were 40 contracting parties to the London Protocol. Of these parties, 27 must also accept the 2009 amendment to Article 6 for it to enter into force. However, not all of the parties have been interested in offshore CO_2 storage or cross-border movement of CO_2, and have placed the ratification of Article 6 as a low priority. Cross-government cooperation will probably be required for ratification to occur. In some countries, the ratification may also be pending other laws and regulations that need to be changed for carbon storage and sequestration in general [61].

Therefore, although geological carbon sequestration in the ocean has been approved in principle the OSPAR Convention and even the London Protocol, the Article 6 amendment may continue to pose a policy barrier to OCS deployment in the foreseeable future.

5.6. Public outreach: Lessons from Hawaii

It is noteworthy that no field studies demonstrating OCS at a significant scale have been conducted so far prior to its prohibition through the 2007 London Protocol and OSPAR Convention amendments. The largest scale attempt at demonstrating OCS was the Hawaii CO_2 direct injection experiment This section outlines the failure of the Hawaii experiment mainly fuelled by a lack of early public outreach, and outlines some lessons learnt from the Hawaii project. [62,63]

In 1997 the US Department of Energy, the New Energy and Industrial Technology Development Organization of Japan (NEDO) and the Norwegian Research Council (NRC) signed an agreement to conduct experiments to evaluate the behaviour of liquid CO_2 releases in to the ocean. While the project was announced in Kyoto in 1997, with a high profile to demonstrate the sponsors' commitment to CO_2 mitigation, few resources were subsequently invested in public outreach.

The project scientists and sponsors selected an area off the coast of the Big Island of Hawaii to conduct the pilot CO_2 study, based on technical feasibility and existing research infrastructure. However, they did not gauge the public perception prior to site selection. The local population only learnt of the injection project planned in their waters when it was first published in a newspaper article. In an area where the ocean is viewed as a major natural resource, the public perceived of the "dumping" of CO_2 as a violation, and strongly opposed its continuation.

Eventually the pilot injection project was abandoned in Hawaii. In order to salvage the project, scientists attempted to instead conduct an injection study in Norway. However, here the actions of Greenpeace stopped any further testing, thus precluding completely any chance of field scale testing of direct-injection OCS.

Although the introduction of OCS was initially high profile the sponsors' commitment to CO_2 mitigation, few resources were subsequently invested in public outreach. Moreover, the sponsors largely did not include the public in their decision to site the pilot injection experiment in Hawaii, nor did they factor public perception of potentially conducting a CO_2 injection experiment in an area where the ocean is viewed as a major environmental resource.

Reiner (2008) cited the US National Institute of Standards and Technology (NIST) 2002 workshop for "Best Practices for Communication of Science and Technology to the Public" as a resource that offered key recommendations for public outreach, including:

- Illustrating both the scientific process and unresolved scientific questions, rather than showing a result based on a black box model
- Using scientists in a public education role for a range of audiences (children and adult)
- Incorporating a wide source of knowledge from others so as to avoid parochialism
- Presenting the issues from the audiences' point of view
- Using face-to-face communications whenever possible to foster trust
- Reaching out beyond the science-attentive public, e.g. presenting at shopping malls, to disadvantaged youth
- Using multimedia and illustrations
- Providing press releases in forms easily usable to time-strapped journalists
- Avoiding perceptions of environmental injustice, bias or conspiracy
- Including the public from the start to avoid charges of "sneaking up on us" or less than full disclosure
- Avoiding letting the project become a "political football" by creating vocal supporters within the community

In addition, Reiner suggested that early outreach to the public via the internet during developmental stages of a project is important before the project becomes newsworthy and receives attention from mainstream media outlets. [63].

The lack of outreach is reflected in the low level of understanding that has remained among the public, as well as relatively low public acceptance of carbon capture and sequestration (both geologic and ocean). Reiner summarized the European Commission's survey of the public from 25 countries of the European Union (the Eurobarometer) that showed that, in 2007 (at the same time as the OSPAR convention and London Protocol amendments), only 21% of those surveyed have heard of carbon capture and storage (geologic or ocean), compared with 53% for hydrogen energy and cars, 41% for fuel cells, and 44% for geothermal energy [63]. In the US in 2004, only 2.5% of 1200 respondents in a web-based survey had previously heard of carbon sequestration. In 2007, Palmgren et al. surveyed 126 community respondents, who ranked OCS less favourable than geological carbon sequestration. Both carbon sequestration options were less favourable to the respondents than nuclear power [64].

6. Conclusions

Whether CO_2 is introduced intentionally, or passively diffusing from the atmosphere to the ocean, the ocean is and will remain the largest sink of anthropogenic CO_2. In addition to climate change implications of elevated atmospheric CO_2, a further impact is the acidification of the ocean. Effects of increased acidity and pCO_2 in organisms include respiratory distress (but some deep sea organisms take advantage of the CO_2/O_2 balance). There is also a risk of a reduced habitat as calcium carbonate stability zones decrease. However, further study is required to determine the variability of responses among marine species.

This chapter presented several methods by which direct injection of CO_2 into the ocean could be introduced. Some injection technologies were developed that would theoretically, combined with proper siting of injection points, cause a relatively minor impact to marine ecosystems. Some pilot scale field studies began that would have provided more information about environmental impacts, but they were nixed due to public opposition stemming from a lack of extensive and continuous public outreach from the onset. Since 2007, international policies began to prohibit direct discharge of CO_2 into the ocean, while favouring deep sea geological sequestration. CO_2 leaks (e.g. in the form of droplets [65]) from geological structures to the ocean water column are however still possible [35], so continued research and studies about the mechanisms of CO_2 leakage and the effects of increased dissolved carbon in the ocean continues to be an important topic of study for carbon sequestration.

Acknowledgements

This chapter contains work funded by Ocean Carbon Sequestration Program, Biological and Environmental Research (BER), U.S. Dept. of Energy (grant number DE-FG02-01ER63078), the National Energy Technology Laboratory, U.S. Dept. of Energy (grant number DE-FG26-98FT40334) and the Martin Family Fellows for Sustainability.

Author details

Aaron Chow

Berkeley Research Group, Waltham, Massachusetts, USA

References

[1] Intergovernmental Panel on Climate Change (IPCC) (B. Metz B., Davidson O. eds. Carbon Dioxide Capture and Storage: A Special Report of IPCC Working Group III, Cambridge University Press, Cambridge UK; 2005

[2] Caldeira, K., Wickett, ME. Anthropogenic carbon and ocean pH. Nature 2003;425, 365-365.

[3] R. Zeebe R., Rigwell A. Past changes in ocean carbonate history. In: Gattuso J-P., Hansson L., eds; Ocean acidification. Oxford; 2010 p.21-40.

[4] Orr JC. Past and recent changes in the ocean carbonate system. In: Gattuso J-P., Hansson L., eds; Ocean acidification. Oxford; 2010 p.41-66.

[5] Bigalke N., RehderG., Gust G. Experimental investigation of the rising behavioe of CO_2 droplets in seawater under hydrate-forming conditions. Environmental Science and Technology 2008;(42)5241 – 5246.

[6] Morel FMM., Hering JG. Principles and Applications of Aquatic Chemistry. Wiley-Interscience, 1993.

[7] Alendal G., Drange H. Two-phase, near field modeling of purposefully released CO_2 in the ocean. Journal of Geophysical Research, 2001;106(C1):1085–1096.

[8] Lavigne and Gattuso. In: Gattuso J-P., Hansson L., eds; Ocean acidification. Oxford; 2010.

[9] Haugen P., Drange H. Sequestration of CO_2 in the deep ocean by shallow injection. Nature 1992;357(28):1065–1072.

[10] Dickson G., Goyet C., editors. Handbook of Methods for the Analysis of the Various Parameters of the Carbon Dioxide System in Sea Water. ORNL/CDIAC-74. 1994.

[11] Gattuso J-P., Hansson L., eds; Ocean acidification. Oxford; 2010.

[12] Herzog HJ., Adams EE., Auerbach D., Caulfield J. Environmental impacts of ocean disposal of CO_2. Energy Conversion and Management, 1996;37(6-8):999–1005.

[13] Brewer PG., Peltzer ET., Walz P., Aya I., Yamane K., Kojima R., Nakajima Y., Nakayama N., Haugan P., Johannessen T. Deep ocean experiments with fossil fuel carbon dioxide: creation and sensing of a controlled plume at 4 km depth. Journal of Marine Research, 2005;63(1):9–33.

[14] Adams EE., Golomb D., Zhang XY., Herzog HJ. Confined release of CO_2 into shallow sea water. In Direct Ocean Disposal of Carbon Dioxide, TERRAPUB 1995; Tokyo p. 153–164.

[15] Sato T., Kajishima T., Naguosa R. CO_2 sequestration at sea by gas-lift system of shallow injection and deep releasing. Environmental Science and Technology, 2000:34:4140–4145.

[16] Nakanishi N., Ohsumi T., Shitashima K. Sequestering of CO_2 in a deep ocean. Technical Report, Central Research Institute of Electric Power Industry, 1994; 1646 Abiko, Abiko-City, Chiba 270-11, Japan.

[17] Caulfield JA., Adams EE. Auerbach DI., Herzog HJ. Impacts of ocean CO_2 disposal on marine life: II. probablistic plume exposure model used with a time-varying dose-response analysis. Environmental Modeling and Assessment 1997; 2:345–353.

[18] Sloan ED. Gas hydrates: Review of physical/chemical properties. Energy & Fuels, 1998;12(2):191–196.

[19] Lee S., Liang L., Riestenberg D., West OR., Tsouris C., Adams E. CO_2 hydrate composite for ocean carbon sequestration. Environmental Science and Technology, 2003; 37:3701–3708.

[20] Rehder G, Kirby SH., Durham WB, Stern LA., Peltzer ET., Brewer PG. Dissolution rates of pure methane hydrate and carbon dioxide hydrate in undersaturated seawater at 1000m depth. Geochimica et Cosmochimica Acta 2004;68:285–292.

[21] Holder G., Mokka L., Warzinski R. Formation of hydrates from singlephase aqueous solutions and implications for ocean sequestration of CO_2. In: Preprints of Spring 2000 National Meeting in San Diego, CA. ACS Division of Fuel Chemistry, 2001.

[22] Chen B. A numerical prediction on plume structure of liquid CO_2 in the ocean - a near field model. In: Proceeding of the 5th International Conference on Greenhouse Gas Control Technologies. 2001; p. 417–422, 2001.

[23] Tsouris C., McCallum S., Aaron D., Reistenberg D., Gabitto J., Chow A., Adams E. Scale-up of a continuous-jet hydrate reactor for CO_2 ocean sequestration. AiChE Journal 2007; 53(4):1017–1027.

[24] Sato T., Hama T. Numerical simulation of dilution process in CO_2 ocean sequestration. In: Proceedings of the 5th International Conference on Greenhouse Gas Control Technologies. 2001; p. 475–480.

[25] Crounse B., Wannamaker EJ., Adams EE.. Integral model of a multiphase plume in quiescent stratification. Journal of Hydraulic Engineering, 2007;133(1):70–76.

[26] Socolofsky SA., Adams EE. Multi-phase plumes in uniform and stratified crossflow. Journal of Hydraulic Research, 2002;40(6):661–672.

[27] Socolofsky SA., Adams EE. Liquid volume fluxes in stratified multiphase plumes. Journal of Hydraulic Engineering, 2003;129(11):905 – 914.

[28] Riestenberg, D., Tsouris C., Brewer PG., Peltzer ET., Walz P., Chow A., Adams E. Field Studies on the Formation of Sinking CO_2 Particles for Ocean Carbon Sequestration: Effects of Injector Geometry on Particle Density and Dissolution Rate and Model Simulation of Plume Behavior. Environmental Science and Technology, 2005:39, 7287-7293.

[29] Chow A., Adams EE., Israelsson PH., Tsouris, C. Carbon dioxide hydrate particles for ocean carbon sequestration. Energy Proceedia, 2009;1(1), 4937-4944.

[30] Chow AC., Adams, EE. Prediction of Drag Coefficient and Secondary Motion of Free-Falling Rigid Cylindrical Particles with and without Curvature at Moderate Reynolds Number. Journal of Hydraulic Engineering 2011; 137(11)1406 – 1414.

[31] Wannamaker EJ., Adams EE. Modeling descending carbon dioxide injections in the ocean. Journal of Hydraulic Research, 2006; 44(3):324–337.

[32] R. Schlitzer. Electronic atlas of WOCE hydrographic and tracer data now available http://www.ewoce.org. Eos Trans. AGU, 2000; 81(5):45.

[33] Chow A., Adams EE. Particle laden flows through an inverted chimney with applications to ocean carbon sequestration. Environmental Fluid Mechanics 2012; 12(1) 3 – 21.

[34] Orr, JC. Modelling of ocean storage of CO_2---The GOSAC study, Report PH4/37, International Energy Agency, Greenhouse Gas R&D Programme, Cheltenham, UK, 2004; 96 pp.

[35] Blackford J., Widdicombe S., Lowe D., Chen B. Environmental risks and performance assessment of carbon dioxide (CO_2) leakage in marine ecosystems. In: Maroto-Valer MM., editor. Developments and innovation in carbon dioxide (CO_2) capture and storage technology. 2010, CRC Press.

[36] Turley C., Blackford J., Harman-Mounford N., Litt E., Llewellyn C., Lowe D., Miller P., Nightingale P., Rees A., Smyth T., Tilstone G., Widdcombe S.. Carbon uptake, transport and storage by oceans and the consequences of change. In: Hester RE., Harrison RM., Carbon Capture, Sequestration and Storage. Royal Society of Chemistry, 2010.

[37] Auerbach I., Caulfield JA., Adams EE., Herzog HJ. Impacts of ocean CO_2 disposal on marine life: I. a toxicological assessment integrating constant-concentration laboratory assay data with variable concentration field exposure. Environmental Modeling and Assessment, 1997;2:333–343.

[38] Caulfield JA., Adams EE., Auerbach DI., Herzog HJ.. Impacts of ocean CO_2 disposal on marine life: II. Probablistic plume exposure model used with a time-varying dose-response analysis. Environmental Modeling and Assessment, 1997;2:345–353.

[39] Israelsson PH., Chow AC., Adams EE. An updated assessment of the acute impacts of ocean carbon sequestration by direct injection. International Journal of Greenhouse Gas Control 2010;4(2): p. 262-271.

[40] Israelsson PH. Studies of Lagrangian Modeling Techniques with Applications to Deep Ocean Carbon Sequestration. PhD thesis. Massachusetts Institute of Technology; 2008.

[41] Sato, T. Numerical Simulation of Biological Impact Caused by Direct Injection of Carbon Dioxide in the ocean. Journal of Oceanography 2004;60, 807-816.

[42] Tamburri MN., Peltzer ET., Friederich GE, Aya I., Yamane K., Brewer PG. A field study of the effects of CO_2 ocean disposal on mobile deep-sea animals. Marine Chemistry 2000; 72:95–101.

[43] Kita J., Watanabe Y. Impact assessment of high-CO_2 environment on marine organisms. In: Proceedings of the Eighth International Conference on Greenhouse Gas Control Technologies, 2006.

[44] Kikkawa T., Sato T., Kita J., Ishimatsu A. Acute toxicity of temporally varying seawater CO_2 conditions on juveniles of japanese sillago (Sillago japonica). Marine Pollution Bulletin, 2006; 52:621–625.

[45] Barry JP, Widdicombe S, Hall-Spencer JM. Effects of ocean acidification on marine biodiversity and ecosystem function. In: Gattuso J-P., Hansson L., eds; Ocean acidification. Oxford; 2010, p.192-209.

[46] Pörtner, H.-O., Gutowska, M., Ishimatsu, A., Lucassen, M., Meizner, F. and Seibel, B.A. (2010). Effects of ocean acidification on nektonic organisms. In: Gattuso J-P., Hansson L., eds; Ocean acidification. Oxford; 2010, p.-154-175.

[47] Knoll AH., Bambach RK., Lykkeboe G. Blood gas transport in the cephalopod, Sepia officinalis. Journal of Experimental Biology 1982;99:331-8.

[48] Childress, JJ. Are there physiological and biochemical adaptations of metabolism in deep-sea animals? Trends in and Evolution Ecology 1995;10(1), 30-36.

[49] Riebesell U., Tortell, PD. Effects of ocean acidification on pelagic organisms and ecosystems. In: Gattuso J-P., Hansson L., eds; Ocean acidification. Oxford; 2010, p. 99-121.

[50] Rockström, J., Steffen W., Noone K., Persson Å., Chapin FS, Lambin E., Lenton TM, Scheffer M., Folke C., Schellnhuber H., Nykvist B., De Wit CA., Hughes T., van der Leeuw S.,. Rodhe H., Sörlin S., Snyder PK., Costanza R., Svedin U., Falkenmark M., Karlberg L., Corell RW., Fabry VJ, Hansen J., Walker B., Liverman D., Richardson K., Crutzen P., Foley. M. Planetary boundaries:exploring the safe operating space for humanity. Ecology and Society 2009.;4(2): 32. [online] URL: http://www.ecologyandsociety.org/vol14/iss2/art32/

[51] Seibel BA., Thuesen EV., Childress JJ., Gorodezky LA. Decline in pelagic cephalopod metabolism with habitat depth reflects differences in locomotory efficiency. Biological Bulletin 1997;192(2) 262-278.

[52] Vetter EW., Smith CR. Insights into the ecological effects of deep ocean CO_2 enrichment: the impacts of natural CO_2 venting at Loihi seamount on deep sea scavengers. Journal of Geophysical Research 2005;110(C09S13).

[53] Hall-Spencer JM., Rodolfo-Metalpa R., Martin S. Volcanic carbon dioxide vents show ecosystem effects of ocean acidification. Nature 2008; 454:96-9.

[54] Herzog H., Caldeira K., Reilly J. An issue of permanence: assessing the effectiveness of ocean carbon sequestration. Climatic Change 2003;59(3), 293-310.

[55] Goldberg D., Slagle AL. A global assessment of deep-sea basalt sites for carbon sequestration. Energy Proceedia 2009;1:3675-3682.

[56] Akai M., Nishio N., Iijima M., Ozaki M., Minamiura J., Tanaka T.. Performance and Economic Evaluation of CO_2 Capture and Sequestration Technologies. Proceedings of the Seventh International Conference on Greenhouse Gas Control Technologies, 2004.

[57] International Energy Agency (IEA) and Organisation for Economic Co-operation and Development (OECD). Legal Aspects of Storing CO_2, 2007.

[58] OSPAR Commission for the Protection of the Marine Environment of the North-East Atlantic, Press release. Ratification of OSPAR carbon capture and storage measures. October 28, 2011. http://www.ospar.org/html_documents/ospar/news/ospar_pr_11_ratification_of_ccs_measure_en.pdf. Accessed August 6, 2013.

[59] 4th Kyoto, Japan. Oslo and Paris Commission (OSPAR). OSPAR Decision 2007/1 to Prohibit the Storage of Carbon Dioxide Streams in the Water Column or on the Seabed. 25-29 June, 2007.

[60] International Energy Agency (IEA), 2008. CO_2 capture and storage: a key abatement option. 266 pp. http://www.iea.org/publications/freepublications/publication/name,35329,en.html. Accessed October 16, 2013.

[61] International Energy Agency (IEA). Carbon Capture and Storage and the London Protocol. Options for Enabling Transboundary CO_2 Transfer. 2011.

[62] De Figueiredo MA., Reiner DM., Herzog HJ. Ocean carbon sequestration: A case study in public and institutional perceptions. Proceedings of the Sixth International Conference on Greenhouse Gas Control Technologies, 2002.

[63] Reiner DM. A Looming Rhetorical Gap: A Survey of Public Communications Activities for Carbon Dioxide Capture and Storage Technologies. 2008; http://www.eprg.group.cam.ac.uk/wp-content/uploads/2008/11/eprg08012.pdf. Accessed Accessed August 6, 2013.

[64] Palmgren CR., Morgan MG., de Bruin WB., Keith DW. Initial Public Perceptions of Deep Geological and Oceanic Disposal of Carbon Dioxide. In: Wilson EJ, Gerard. D., eds. Carbon Capture and Sequestration, Blackwell Publishing, Ames, Iowa; 2007.

[65] Chen B., Nishio M, Song Y., Akai M. The fate of CO_2 bubble leaked from seabed. In Gale J, Herzog H., Braitch J., eds. Greenhouse Gas Control Technologies 9, Proceedings of the Ninth International Conference on Greenhouse Gas Control Technologies (GHGT-9), Energy Proceedia 2009:1:4969-4976.

Chapter 4

Estimation of Regional Carbon Storage Potential in Mangrove Soils on Carmen Island, Campeche, Mexico

Julia Griselda Cerón-Bretón,
Rosa María Cerón-Bretón,
Jesús Jaime Guerra-Santos and
Atl Victor Córdova-Quiroz

Additional information is available at the end of the chapter

http://dx.doi.org/10.5772/57055

1. Introduction

Warming because of the increasing concentration of greenhouse gases is a global concern. Of all greenhouse gases, CO_2 has the greatest contribution to global warming. CO_2 can be stored in forests through biogeochemical processes governing the exchange between the atmosphere and forestry systems. Although forests are carbon sources, they can also mitigate global warming through carbon sequestration in different plant ecosystems, known as sinks, where carbon is accumulated by absorbing atmospheric CO_2. Carbon is assimilated and stored, both in live biomass (stems, branches, leaves and roots) and dead biomass (litter, wood waste, organic matter from soils and forestry products), and oxygen is released to the atmosphere during this process. Therefore, forests play an important role in the global carbon cycle [1].

IPCC has pointed out the importance of carbon sequestration by vegetation as a low-cost choice to reduce the atmospheric CO_2 content. Tropical forests store almost 50% more carbon than forests located outside the tropics [2]. Among the world's most productive wetlands are the mangroves, which fix and store large amounts of carbon in soils with unique characteristics of salinity, under anoxic and acidic conditions, with frequent flooding. The fact that these systems are productive under extreme conditions has been a point of interest to the scientific community. Therefore, correct management of tropical forests and wetlands constitutes an opportunity to store carbon. In the specific case of mangrove soils, where decomposition rates are low and their ability to store carbon is high, mangrove forests are an attractive alternative for carbon sequestration. Organic matter decomposition constitutes the main flux of carbon and nutrients in most terrestrial ecosystems. In the case of tropical forests, this is regulated by several factors: vegetation species, morphology, C:N ratio, climate, flooding frequency, moisture, salinity, temperature, and so on [3-4].

Mangrove forests have a special adaptation capacity since they can tolerate oxygen deficit, high levels of salinity and different flooding patterns. Mangroves like many other plants have the ability to change the physical and chemical properties of soils in which they occur [5]. The fast-growing young trees absorb about 30% more carbon than mature wood, but an old-growth forest generally stores more carbon in the soil, groundwater and surface vegetation than a plantation of trees with the same size. Latitude, climate, species diversity and other biological factors also affect carbon fluxes in forests [3]. It has been reported that degradation of fresh organic matter is slower in anaerobic environments (such as mangroves), which additionally involves oxidizing agents such as nitrate and microorganisms like bacteria and fungi [6], which only operate when the tannin concentrations are low, because they inhibit growth [7-8]. The presence and abundance of macro invertebrates such as amphipods (crabs) and isopods (cochineal) also accelerates the breakdown of tissues by direct consumption of leaves [6].

Carbon storage in estuarine wetlands is an efficient process with minimal release of greenhouse gases [9-11]. Although these wetlands cover only about 5% of the earth's surface, they contain around 40% of global soil organic carbon. Mangrove forests cover vast extensions along the coastal zone in tropical regions. Mexico is the fifth country in the world in terms of the size of mangrove cover (655,667 ha). However, this surface is being lost at a yearly rate between 10,000 and 40,000 ha, mainly due to aquaculture, agriculture, deforestation and change in land use. Mangroves in Mexico are distributed within coastal lagoons and deltaic systems of the Gulf of Mexico (Veracruz, Tabasco and Campeche States) and the Pacific Ocean (Baja California, Sonora, Nayarit, Oaxaca and Chiapas) [12-14], with some coastal lagoons having ephemeral mouths which open during the rainy season or by action of the fishermen [5]. In the state of Campeche, mangrove cover accounts for 29.98% of the total mangrove cover in the country, with an extension of 196,552 ha [15]. Along the shoreline of Carmen Island, there are four main mangrove species: *Rhizophora mangle, Laguncularia racemosa, Avicennia germinans* and *Conocarpus erectus*. It is common to find them in associations, as a result of a succession process depending on the tidal levels that flood them but establishing clear control of one prevailing species or associations of two or three species.

Estimation of the dynamic of net fluxes of carbon between forests and the atmosphere is currently an open question in the main discussion forums about climate change. Sequestration and emission processes comprise a system with four main carbon sinks: vegetation, decomposition of organic matter, soils and forestry products. Since these sinks are interrelated, it is necessary to undertake a comprehensive systemic analysis. However, this is particularly difficult for tropical forests, where data about carbon content in vegetation and soils are scarce or unavailable. In spite of the importance of mangrove forests as carbon sinks, most of the research on carbon storage is focused on terrestrial ecosystems and little attention has been given to this type of ecosystem. Although in Mexico there are 113 Ramsar sites, the location of specific carbon sinks and their potential to sequester atmospheric carbon remains poorly defined, and there are uncertainties concerning carbon stock in wetlands at the local and

http://dx.doi.org/10.5772/57055

regional scale. In order to complete carbon storage inventories and to reduce these uncertainties, it is necessary to obtain more local and regional information to support environmental policies to promote conservation and restoration projects as mitigation strategies. Therefore, this study focused on estimating regional carbon storage potential in mangrove soils in six sites uniformly distributed along Carmen Island in the region of Terminos Lagoon, in Campeche, Mexico, considering three climatic periods (cold fronts, called "north season", dry season and rainy season) and at two different soil depths (0-30 and 30-60 cm) during 2009 and 2010.

2. Sampling procedure

2.1. Study area description

The study area is located within the Atasta peninsula in southeastern Mexico in Campeche State, within the natural protected area called "Laguna de Terminos". This area has a warm-humid climate with rains occurring mainly in summer [Am(f)], according to the Köppen classification modified by García [16]. The geomorphology is characterized by marshes and wetlands, with altitudes between 0 to 20 masl, and the annual mean temperature is between 21 to 24.7°C.

Soils in this region are clayey with high fertility and high organic matter content, associated with dominant vegetation of red mangrove (*Rhizophora mangle*), black mangrove (*Avicennia germinans*), white mangrove (*Laguncularia racemosa*) and buttonwood mangrove (*Conocarpus erectus*) [17]. Climate conditions in this zone show three well-defined periods: dry season (from March to May), rainy season (from June to October) and "north season" (from November to February).

Sampling campaigns were carried out between February 2009 and October 2010 in the three climatic periods. Based on visual inspections, transects were established in a representative area of mangrove forest in each sampling site. They were selected to assure representative regional samples, taking into account the type of vegetation, easy access and the hydrology. The locations of the selected areas for this study are shown in Figure 1:

1.-Puerto Rico (P), located at 18° 36′ 55″ N and 91° 56′ 35″ W. Altitude: 11 masl. A great portion of this area has been deforested to establish farms. The remaining mangrove areas present in this zone correspond to the basin [18]. Three sampling zones were selected (P1, P2, and P3), which were located inland, with minimum ground slope and slow water renewal. Accumulated floodwaters in depressions in these sites correspond to cycling of organic matter and nutrients in a closed ecosystem [19]. Individuals of *Avicennia germinas, Rhizophora mangle* and *Laguncularia racemosa* were registered in this site, with dominance of *Avicennia germinans.*

2.-Xicalango (XC), located at 18° 37′ 02″ N and 91° 58′ 20″ W, with dominance of *Avicennia germinans* and *Conocarpus erectus*. Three sampling zones were selected (XC1, XC2 and XC3). The first one was completely flooded, the second one was dry and the last one was partially flooded.

3.-Nuevo Campechito (NC), located at 18° 38′ 28″ N and 91° 57′ 29″ W, with an association of *Conocarpus erectus, Laguncularia racemosa* and *Rhizophora mangle*. Three sampling zones were selected (NC1, NC2 and NC3). During the three climatic periods, the first one was flooded all the time, the second one was partiallyflooded and the last one was dry.

4.-Estero Pargo (EP), a 5.3 km long tidal channel on the lagoon side of Carmen Island, located at 18° 39′ 02.8″ N and 91° 45′ 46.9″ W, with an association of *Rhizophora mangle-Laguncularia racemosa* and *Conocarpus erectus*. This site only was flooded by rainfall during rainy and "north" seasons. Three sampling zones were selected (EP1, EP2 and EP3), which were moderately flooded during the dry and rainy seasons.

5.-Bahamitas (BH), located at the border of Términos Lagoon. This site is within the natural protected area named "Términos Lagoon", a Ramsar site since 2004. This site is located at 18° 41′ 57.1′′ N and 91° 41′ 50.7′′ W, with an association of *Rhizophora mangle*, *Laguncularia racemosa* and *Avicennia germinans*. Again, three sampling zones were selected (BH1, BH2 and BH3). This site was always flooded (received regular tidal inundation and was covered with fresh water during the "north" and rainy seasons).

6.-Puerto Rico 2 (R), located at 18° 36' 56.1 "N and 91 ° 54' 33.6" W. The sampling zones selected were R1, R2 and R3. This site shows a great diversity of tropical vegetation, including *Rhizophora mangle, Avicennia germinans, Laguncularia racemosa,* and *Conocarpus erectus*.

2.2. Sampling methodology

The field study was carried out at six sites with three sampling zones for each one, between February 2009 and October 2010. In all sampling sites, the mangrove forest is subject to anthropogenic exploitation and can be considered as perturbed. We performed sampling campaigns to assess seasonal changes in carbon storage, considering three climatic periods: "north" season, dry season and the beginning of the rainy season, with six sampling campaigns. Not all sampling sites had comparable tidal influences during the study period. Some of them were always flooded (received regular tidal inundation and were covered with fresh water during "north" and rainy seasons), whereas other sites only were flooded because of the occurrence of rainfall during rainy and "north" seasons. Three sampling plots of 4 m x 12 m in each site were selected considering free access to the zone, risks, mangrove distribution by species and disturbances (Figure 1, Table 1). Based on visual inspections of the study area, transects were established in a representative area of the mangrove forest, locating three sampling points of 1 m^2 approximately for each of three sampling zones for all sites.

Duplicate soil samples were collected from the ground to 30 and 60 cm depth using a 193.3 cm^3 soil sampler in an area of 48 m^2[20]. The corer was carefully inserted into the soil and pushed down to 0.3 and 0.6 m. Because sampled soils are typically moist, the corer has a one-way check valve that creates a vacuum inside the liner as it is pushed into the soil, and when the device is pulled out it creates a suction force that retains the sample in the tube. This sampling method used was as proposed by Grossman and Reinsch [21] and Bernal and Mitsch [22] for sampling wet or inundated soils.

http://dx.doi.org/10.5772/57055

A total of 326 soil samples with replicates were taken. After extraction, each core was labeled and, sealed using tight-fitting end caps, and sent for laboratory analysis.

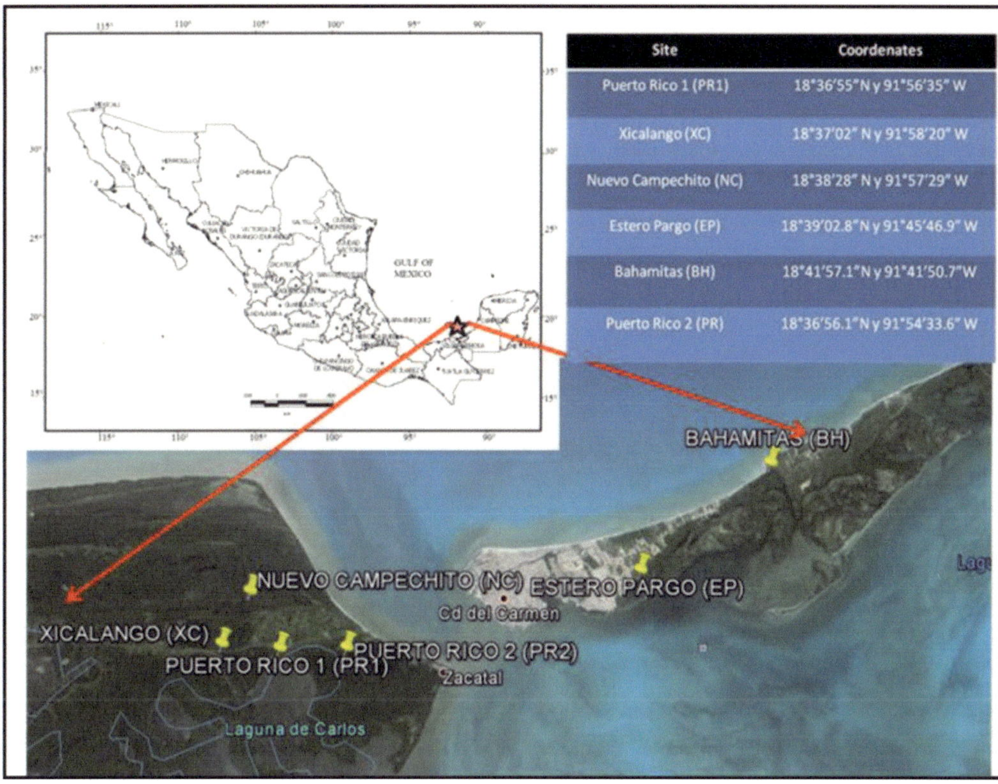

Figure 1. Location of the sampling sites.

3. Analytical procedure

Free water was drained away and all biomass and solid materials (shells, roots, leaves, and so on) were removed. Then the samples were ground, dried at room temperature and sieved to pass through a 2 mm mesh. Organic carbon (OC) was quantified by using the ignition method and organic matter (OM) was determined by warming samples to 550 °C during 4 h[22] and the content of organic carbon was estimated by multiplying by a factor of 0.4 [23]. Total nitrogen (NT) was determined according to the semimicro-Kjeldahl method [24]. By this method, 0.5 g of soil sifted through 0.250 mm is weighed and then digested with a mixture of catalyst and sulfuric acid (H_2SO_4), after which it is distillated with NaOH and titrated with H_3BO_3 using a Shiro-Tashiro mixture as indicator [24].

To determine bulk density (Da), we used the test tube technique [25], by which a dry sample is passed through a 2 mm sieve, then a 50 ml plastic test tube is weighed and then 20 to 50 g of sifted soil is added. After this, the sample is placed on a firm surface and then it is hit 30 times per second with a rubber mallet in a vertical trajectory from 0.20 to 0.30 m. Finally, the volume and sample weight are registered [25]. Electrical conductivity (CE) was measured with a CL35 conductivity meter by using a 1:5 soil/water solution [26]. Soil pH was measured with a Thermo Orion model 290A pH meter by using a 1:2 soil/water solution [27]. Texture determination was carried out by the Bouyoucos method using a 5% sodichexametaphosphate solution as dispersant [28].

To estimate the carbon storage rate (CA), the following equation was used:

CA= CO% Da Pr,

Where CA= carbon storage rate, CO% = organic carbon content, Da= density and Pr= soil depth [29].

Descriptive, comparative and relational statistical analyses were performed for carbon storage rate, sampling site, sampling depth and climatic period.

4. Results

4.1. Relative humidity

Figures 2a-2c present the relative humidity (RH) values at 30 and 60 cm depth for all the sampling zones for north, dry and rainy seasons, respectively. Puerto Rico 1, Estero Pargo and Bahamitas showed the highest relative humidity whereas Nuevo Campechito and Xicalango showed the lowest values. Most of the sampling zones showed higher RH at 30 cm depth. Seasonal variation for all the sampling zones showed higher values of RH for the rainy season and the lowest values for the north season.

4.2. Soil texture

Bahamitas (BH) and Estero Pargo (EP) showed sandy texture, whereas the texture was sandy clay loam for Nuevo Campechito (NC), and Xicalango (XC). Puerto Rico 1 (P) and Puerto Rico 2 (R) showed sandy clay texture.

4.3. pH and electrical conductivity

The mean values for pH and electrical conductivity for all sampling sites during the three climatic periods are presented in Figures 3a-3c and Figures 4a-4c. pH ranged from 6.7 to 7.5 for Bahamitas, and from 6.89 to 7.12 for Estero-Pargo. This suggests that neutral soils are dominant in this study area. Significant differences were not found among the three climatic periods and sampling depths in BH. Soils in this site can be considered moderately to strongly saline. Electrical conductivity values were higher during the dry season for BH1 and during the rainy season for BH2 and BH3, where *Rhizophora mangle* and *Avicennia germinans* are the dominant species. The highest values for electrical conductivity in EP was found at 30 cm

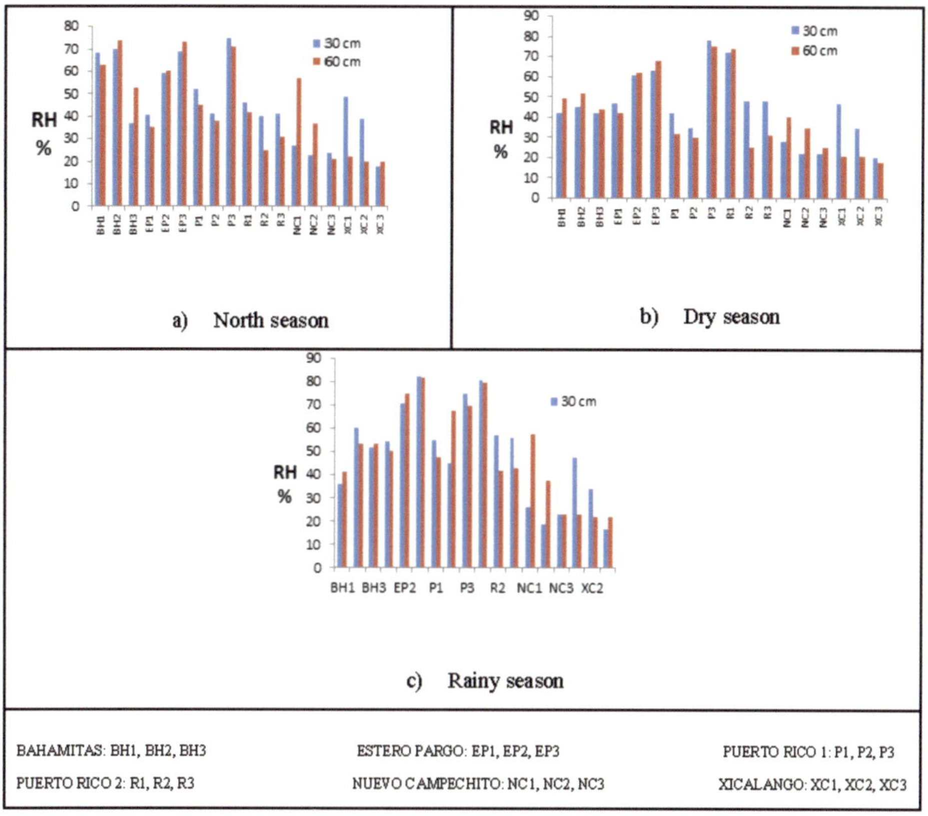

Figure 2. Relative humidity at 30 and 60 cm depth for all the sampling zones during north season (a), dry season (b) and rainy season (c).

depth, being higher during the dry season in EP1 and EP3, where *Conocarpus erectus* and *Rhizophora mangle* are the dominant species.

The pH values in Xicalango ranged from 7.96 to 8.35, whereas in Nuevo Campechito, the pH ranged from 6.55 to 8.46, suggesting that the soils are slightly alkaline in these sites and indicating the presence of soluble salts. Soils in these sites were acid, with the lowest pH values (6.5) and high values of electrical conductivity. No significant differences were found for electrical conductivity in NC and XC at 30 and 60 cm depth (ANOVA, $P< 0.05$). The highest values were found during the dry season, being highest for NC at 60 cm and highest for in XC at 30 cm depth.

Electrical conductivity in Puerto Rico did not show a clear pattern of variation regarding sampling depth. Marine aerosols due to its proximity to the coast probably influence its high salinity. Likewise, the low permeability in Puerto Rico soil promotes water accumulation, increasing sodium concentrations and contributing to low micro biota activity in these soils. Moreover, the hydrological characteristics of this mangrove forest are similar to those typical from a basin, characterized by heavy flooding, little or no tidal contact and high salinity. Soil

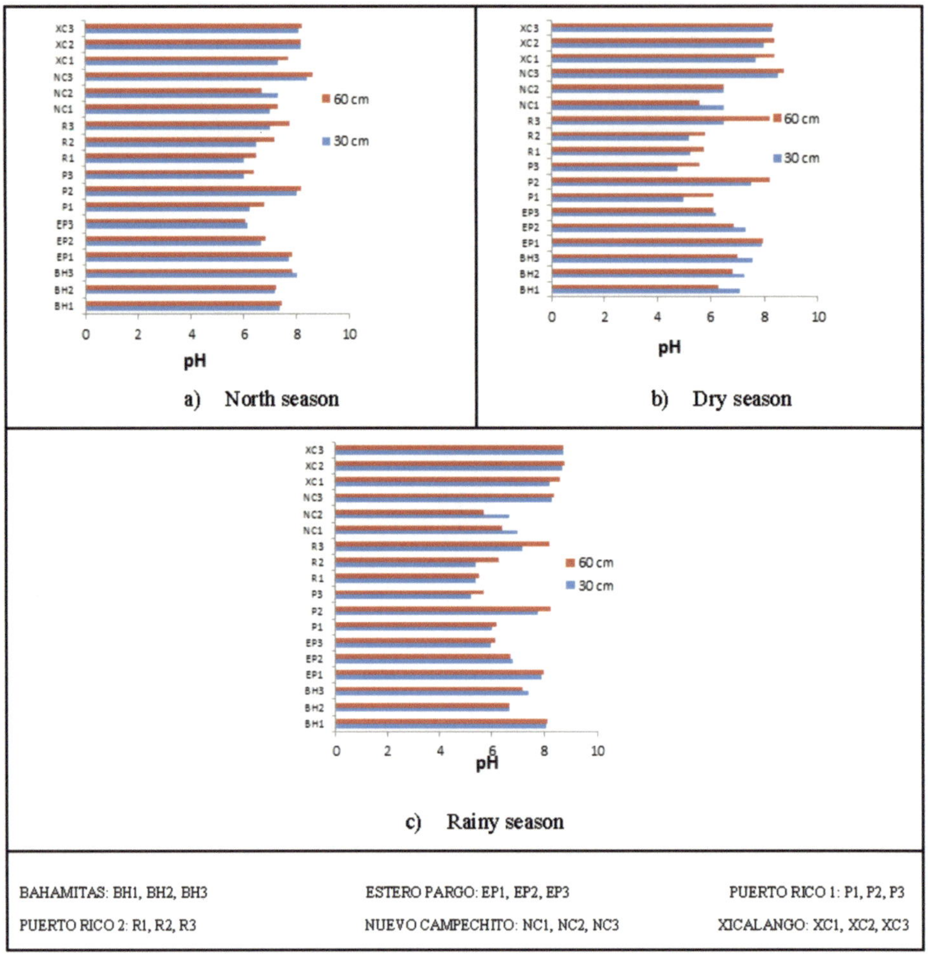

Figure 3. pH at 30 and 60 cm depth for all the sampling zones during north season (a), dry season (b) and rainy season (c).

pH variations in Puerto Rico were significant at the different depths and climatic periods (Tukey test, $P<0.05$). Soils in these sampling zones were more acidic at 30 cm depth, being more alkaline during the north season.

4.4. Organic carbon and organic matter content

The organic carbon concentrations for the three sampling zones in Bahamitas and Estero Pargo during the north, dry and rainy seasons are reported in Figures 5 a and 5b. The organic carbon (OC) content ranged from 4.76 to 15.73% for Bahamitas, and from 2.81 to 19.7% for Estero-Pargo. The mean carbon storage levels were 23.16 and 23.08 Kg C m-2 for Bahamitas and Estero-Pargo, respectively. Figures 6 a and 6b show the organic matter content (%) for the three sampling zones in Bahamitas and Estero Pargo during the north, dry and rainy seasons. In these sites, the long periods of tidal flooding (sites were flooded throughout the sampling

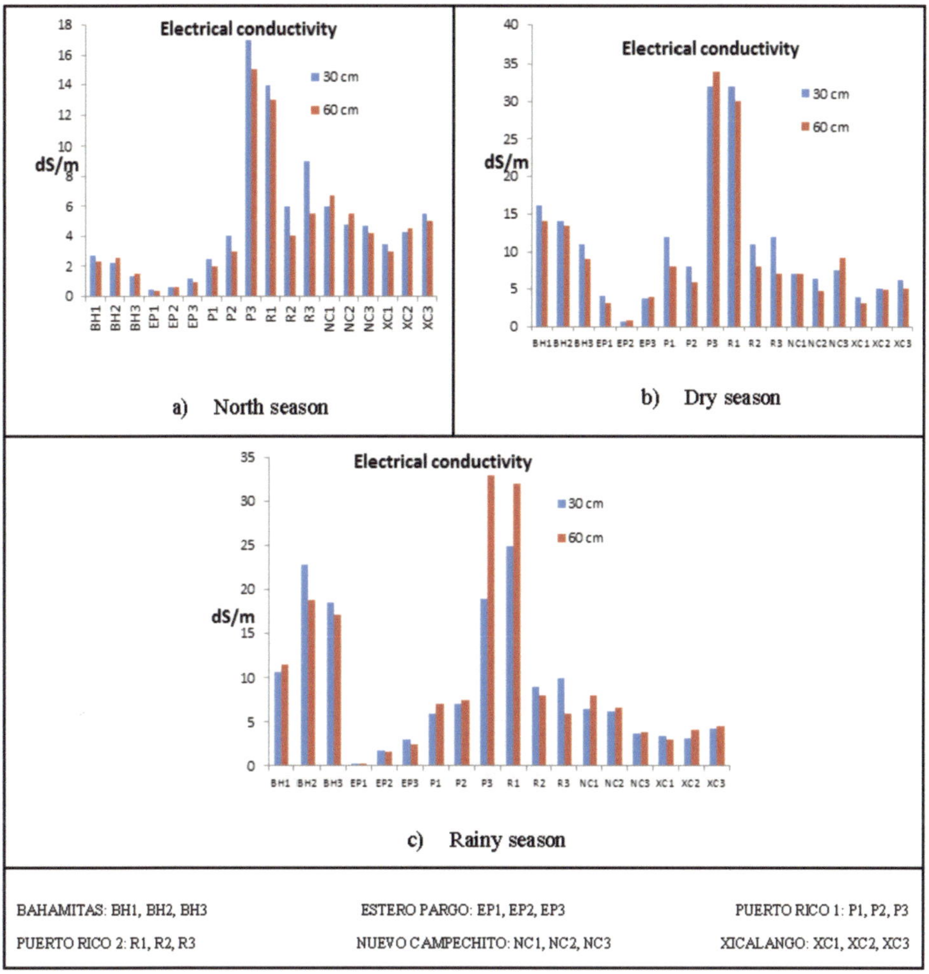

Figure 4. Electrical conductivity at 30 and 60 cm depth for all the sampling zones during north season (a), dry season (b) and rainy season (c).

period) maintained anoxic conditions (below 10 cm of depth) and high organic matter and organic carbon contents. This explains why the highest values were found in the surface layer. This pattern is a general phenomenon observed in forests. The accumulation of organic matter is enhanced in sites with abundant rainfall or with deficient drainage. The highest values for organic matter, organic carbon and carbon storage were found during the north (when heavy rains occurred) and dry seasons. During the rainy season in these sites, accumulation of organic matter increases, but decomposition is slow and the accumulation remains until the north and dry seasons, resulting in values slightly lower during the rainy season in comparison with subsequent seasons. During the dry season, salts and dissolved organic carbon concentrate with increasing evaporation, whereas during the rainy season, pore waters were diluted with rain and runoff waters, resulting in lower organic carbon concentration.

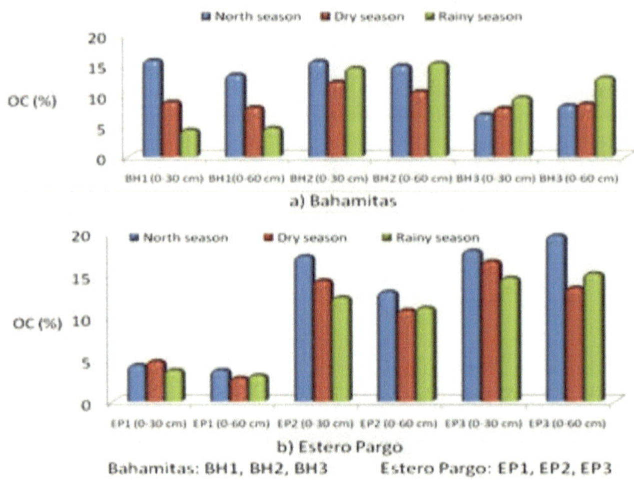

Figure 5. Organic carbon content at 30 and 60 cm depth for all the sampling zones during the three climatic periods for a) Bahamitas and b) Estero Pargo.

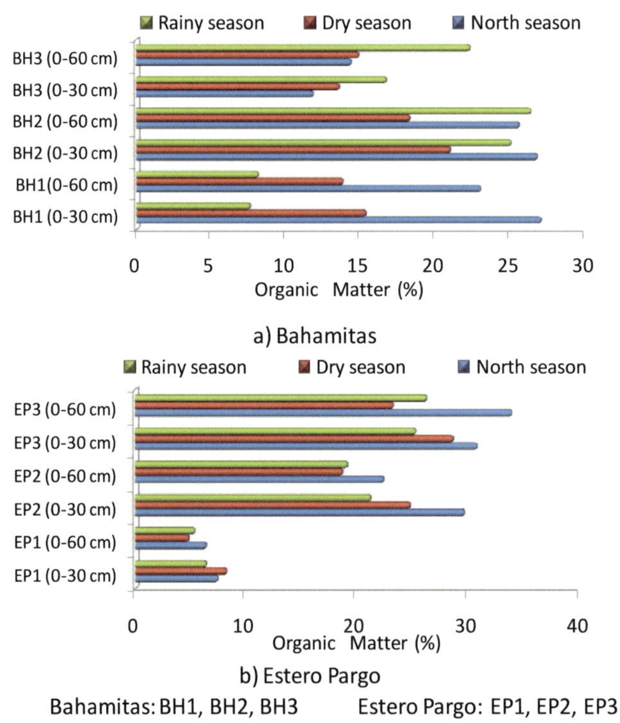

Figure 6. Organic matter content (%) at 30 and 60 cm depth for all the sampling zones during the three climatic periods for a) Bahamitas and b) Estero Pargo.

http://dx.doi.org/10.5772/57055

The organic matter content for the three sampling zones in Nuevo Campechito and Xicalango during the north, dry and rainy seasons are shown in Figures 7a and 7b. The organic matter and organic carbon concentrations were higher at 30 cm depth. The same was observed for the organic matter content (%) for the three sampling zones in Nuevo Campechito and Xicalango during the north, dry and rainy seasons (Figures 8a and 8b). Organic matter ranged from 5.2 to 5.6% for Nuevo Campechito, and from 3.4 to 3.77% for Xicalango. The mean organic carbon contents were 72.78 Kg C m^{-2} for Nuevo Campechito and 79.29 Kg C m^{-2} for Xicalango. These high organic carbon contents were associated with the high productivity of the mangrove species that prevail in these sites, where litter accumulation causes high organic matter levels in the soils. In addition, the flooded conditions that prevailed during the sampling period maintained anoxic conditions that enhanced the carbon storage. Carbon storage was higher in live biomass in comparison with the soils. The forestry inventory revealed that most of the trees in these sites were younger than 7 years. Since it is well known that young trees store 30% more carbon than mature trees, this suggests that young individuals in growing stage in these sites store more carbon in live biomass than in soils. The organic matter and organic carbon contents were slightly higher in Nuevo Campechito than in Xicalango. Individuals of *Rhizophora mangle* with higher diameter and height were found in Nuevo Campechito in comparison with Xicalango. This species contributes a great quantity of organic matter due to litter falling on flooded soils, resulting in a low decomposition rate that increased the organic matter and carbon concentrations. On the other hand, in Xicalango there were zones where *Conocarpus erectus* was the prevailing species of mangrove, where litter was deposited on sandy soils with low moisture, resulting in a slower accumulation process.

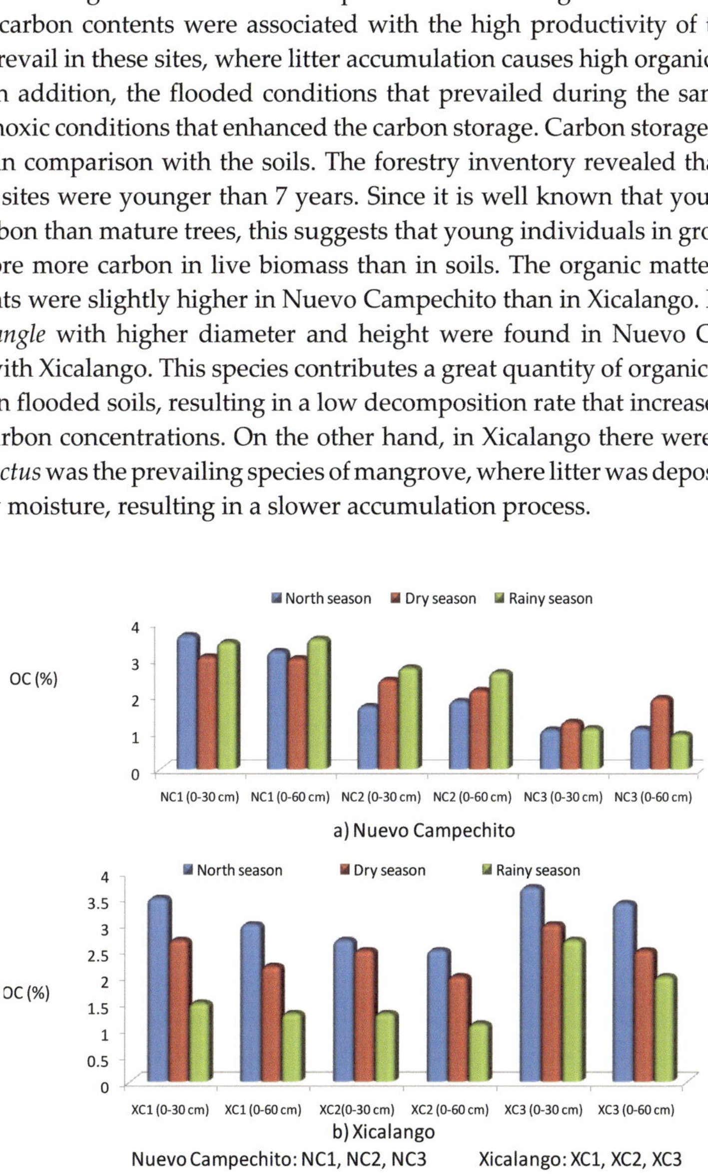

Figure 7. Organic carbon content at 30 and 60 cm depth for all the sampling zones during the three climatic periods for a) Nuevo Campechito and b) Xicalango.

Figures 9a and 9b show the result of the organic carbon content of the three sampling areas in Puerto Rico 1 and 2 in Puerto Rico for the north, dry and rainy seasons. In Puerto Rico, organic carbon ranged from 4.8 to 23.16% at 30 cm depth and from 2.99 to 18.61% at 60 cm depth for P1 and R1 sampling zones. Higher organic carbon concentrations were found in P3 and R1, reaching almost 14% of the content for the three climatic periods. P1, P2, R2 and R3 together accounted for almost 10% of the organic carbon content. The OC content diminished as the sampling depth decreased for all the sampling zones of this site. No significant differences were found for OC in the three climatic periods. These sites showed an organic carbon content between 7 and 11%, 0.2-0.35% of nitrogen, a C:N ratio from 41-47, CE and a carbon storage rate from 160 to 220 Kg C m^{-2}. The carbon storage rate decreased slightly with depth. Long periods of flooding maintained anoxic conditions. This could explain the finding of highest organic carbon and organic matter concentrations at 30 cm depth in this site.

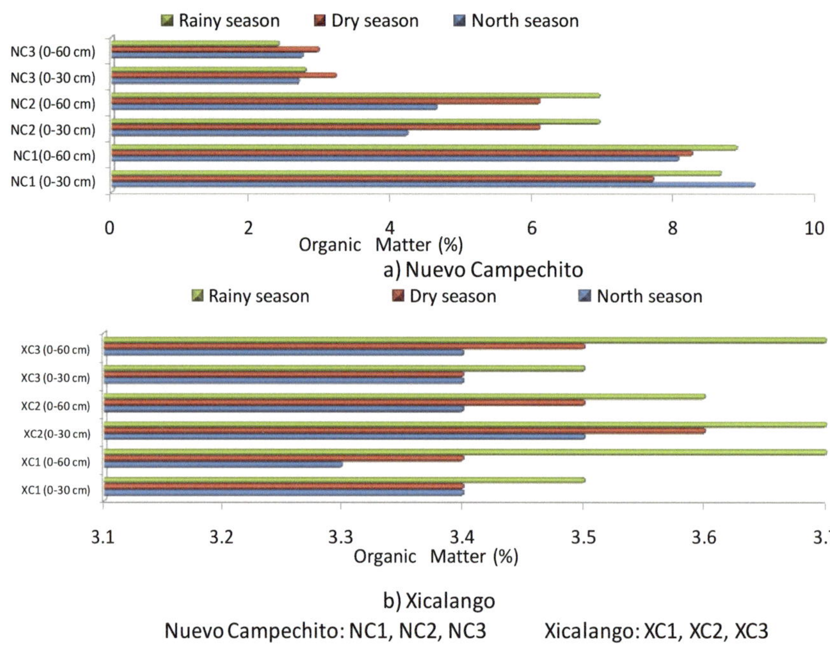

Figure 8. Organic matter content (%) at 30 and 60 cm depth for all the sampling zones during the three climatic periods for a) Nuevo Campechito and b) Xicalango.

The carbon storage in Bahamitas showed significant differences between the different climatic periods, with the highest values being found during the north season.The highest values found were for BH1 at 30 cm depth, for the three climatic periods, and for BH2 and BH3, during the dry season. However, for BH2 and BH3, the highest values found were at 60 cm depth during the north and rainy seasons. Carbon storage in Estero Pargo showed the highest values for EP2 and EP3, for the three climatic periods, where *Laguncularia racemosa* and *Rhizophora mangle* are the dominant species. the The lowest values for carbon storage for all three climatic periods were found in EP1, where the dominant species is *Conocarpus erectus*.

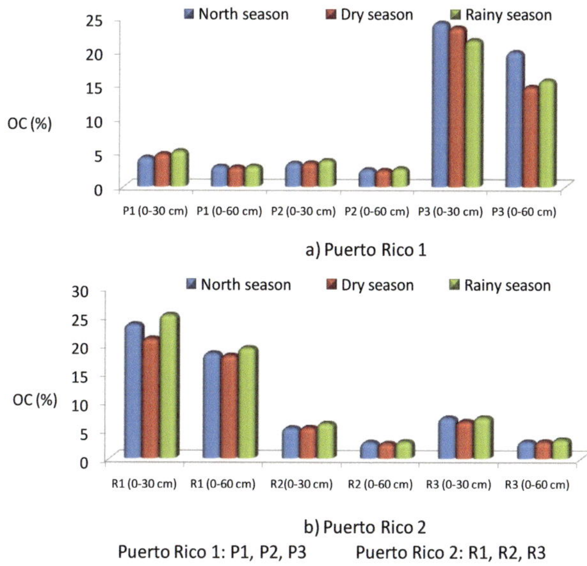

Figure 9. Organic carbon content at 30 and 60 cm depth for all the sampling zones during the three climatic periods for a) Puerto Rico 1 and b) Puerto Rico 2.

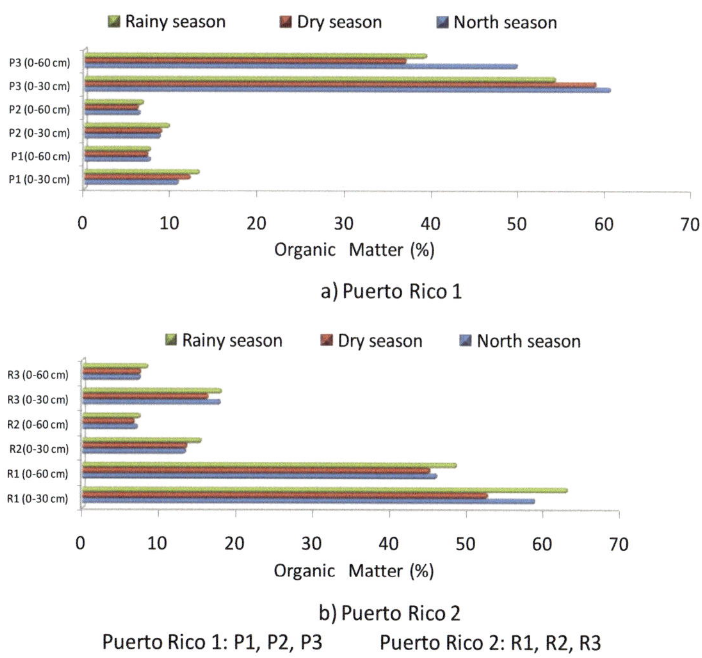

Figure 10. Organic matter content (%) at 30 and 60 cm depth for all the sampling zones during the three climatic periods for a) Puerto Rico 1 and b) Puerto Rico 2.

Site/Land Use	carbon storage (kg C m^{-2}).	Reference
Gahanna woods, Ohio, USA/isolated and forested wetland.	3.03	[22]
Old Woman Creek, Ohio, USA/riverine flow-through wetland.	2.77	
La Selva, Costa Rica/isolated and forested wetland.	0.43	
Earth University, Costa Rica/slow-flowing slough wetland.	1.67	
Palo Verde, Costa Rica/riverine flow-through wetland.	1.36	
Southern California, USA/coastal lagoon and wetland complex.	0.033	[30]
Okinawa Island, Japan/pioneer mangrove stand.	5.73	[31]
Tabasco, Mexico/ Red and white mangrove stand.	47.2-82.2	[9]
Southeast Australia/disturbed wetland soils.	6.61	[32]
Southeast Australia/undisturbed wetland soils.	11.19	
Sydney, Australia/coastal wetlands.	13.9	[33]
Tropical and subtropical China/marsh vegetation	40.0	[34]
Brazil/Oxisol.	12.0-24.0	[35]
Brazil/agricultural soils.	2.0-10.0	[36]
Bahamitas, Campeche, Mexico/neutral and sandy soils, perturbed mangrove forest.	23.16	This study
Estero Pargo, Campeche, Mexico/neutral and sandy soils, perturbed mangrove forest. Estero-Pargo.	23.08	
Nuevo Campechito, Campeche, Mexico/sandy clayloam, associations of *Rhizophora mangle-Laguncularia racemosa-Conocarpuserectus*.	70.00	
Xicalango, Campeche, Mexico/sandy clay loam, associations of *Avicenniagerminans-Laguncularia racemosa-Conocarpuserectus*	55.00	
Puerto Rico, Campeche, Mexico/sandy clayloam, associations of *Avicenniagerminans-Rhizophora mangle-Laguncularia racemosa.*	190	

Table 1. Comparison of carbon storage rate found in this work with those obtained in other sites.

Conocarpus erectus and *Laguncularia racemosa* are the dominant species in NC3 and NC1. Carbon storage in soils of NC was lower in NC1 and NC2, which have associations of *Rhizophora mangle, Conocarpus erectus* and *Laguncularia racemosa*. In contrast, XC2 and XC3 did not show significant differences between each other. Carbon storage in soils of NC3 and XC1 were similar to those reported in forests with associations of *Rhizophora mangle-Avicennia germinans* (143.3 y 122.2 t C ha^{-1}) in French Guiana [29]. XC showed the highest values of carbon storage, ranging from 74.10±4.16 to 119.50±20.40 t C ha^{-1}, with significant differences ($p<0.05$) with NC. This behavior

http://dx.doi.org/10.5772/57055

is due to the high productivity of species associated with this type of soil, resulting from defoliation and incorporation of high organic matter content in these soils.

There were significant differences for carbon storage in Puerto Rico 1 and Puerto Rico 2, showing the highest values in P3 and R2, with 325 and 320.50 t ha^{-1}, respectively, followed by R3 with 141.56 t ha^{-1} and finally P2 with 101.78 t ha^{-1}. In general, carbon storage in all sampling zones in the six studied sites showed significant differences regarding sampling depth.

In mangrove forests, the stand age is a determinant factor that influences the amount of organic carbon in the soil regardless of season. Comparing our results with carbon storage data obtained in other sites (Table 1), we can suggests that sandy and neutral soils as in our study area and associations of red and white mangrove have good potential of carbon sequestration, considering that our mangrove individuals were from young to mature in reproductive age. We can expect this potential to increase in the coming years.

5. Conclusions

Climatic season and soil type (vegetative community present and hydrogeomorfology) were the most important variables in this study. Concentrations of carbon in tropical soils tend to decrease with depth. This behavior indicates that a small fraction of carbon that is being introduced into the soil remains there, being typical of tropical forests where organic matter and nutrients do not accumulate because they are quickly used by biotic systems. It can be concluded that the accumulation of organic matter and carbon storage are determined by the rate of decay rather than the production rate of organic matter. The combination of anaerobic conditions on site and productivity of the system tends to cause the soils that remain flooded most of the time to be highly organic. High rates of carbon sequestration in this study indicate that conservation efforts to protect wetlands would have high benefits for the mitigation of global warming through the regulation of atmospheric carbon concentrations in this area.

Author details

Julia Griselda Cerón-Bretón[1*], Rosa María Cerón-Bretón[1], Jesús Jaime Guerra-Santos[2] and Atl Victor Córdova-Quiroz[1]

*Address all correspondence to: jceron@pampano.unacar.mx

1 Universidad Autónoma del Carmen (UNACAR), Chemistry School, Col. Benito Juárez, Ciudad del Carmen, Campeche, Mexico

2 Centro de Investigación de Ciencias Ambientales (Environmental Sciences Research Center), Col. Benito Juárez, Ciudad del Carmen, Campeche, Mexico

References

[1] Ordoñez A. Estimación de la captura de carbono en un estudio de caso. Instituto Nacional de Ecología. SEMARNAT. México, D.F.1999.

[2] IPCC. The Scientific basis, climate change. Houghton JT., Ding Y, Griggs DJ,Noguer M, Van der Linden PJ, Dai X.,Maskell K, Johnson CA (eds). Cambridge University Press. Cambridge, UK.2001.

[3] Brown S, Lim B,Schlamadinger, B. Evaluating approaches for estimating net emissions of carbon dioxide from forest harvesting and Wood products. Report of a meeting sponsored by the IPCC held in Dakar, Senegal, 5-7 May, 1998.

[4] Romero, LM, Smith TJ,Fourqurean JW. Changes in mass and nutrient content of Wood during decomposition in a South Florida mangrove forest. Journal of ecology2005; 93: 618-631.

[5] López PJ and Ezcurra E. Los manglares de México: una revisión. Madera y Bosques Número especial2002; 27- 51.

[6] Middleton BA and Mckee KL. Degradation of mangrove tissues and implications for peat formation in Belizean island forests. Journal of Ecology 2001; 89: 818-828.

[7] Tam NFY, Wong YS, Lan CY and Wang LN. Litter production and decomposition in a subtropical mangrove swamp receiving wastewater. Journal of Experimental Marine Biology and Ecology 1998; 226: 1-18.

[8] Lee SY. Tropical mangrove ecology: Physical and biotic factors influencing ecosystem structure and function. AustralianJournal of Ecology1999; 24: 355-366.

[9] Moreno EC, Guerrero AP, Gutiérrez MC,Ortíz CA, Palma DJL. Los manglares de Tabasco, una reserva natural de carbono. Madera y bosques2002; Número especial: 115-128.

[10] Steinke TD, Barnabas AD,Somaru R. Structural changes and associated microbial activity accompanying decomposition of mangrove leaves in Mgeni Estuary. South African journal botany1990; 56: 39-48.

[11] Olsen, M.W; Frye, R.J; Glenn, E.P. 1996. Effect of salinity and plant species on CO2 flux and leaching of disolved organic carbon during decomposition of plant residue. Plant and soil 179: 183-188.

[12] Segura Z D y Ramírez G P. Estructura de los manglares de la Laguna de Panzacola. En: Resúmenes del XI Congreso Mexicano de Botánica1990, Oaxtepec, Morelos p.128.

[13] Ramírez G P. y Segura Z D. Ordenación de la vegetación de manglar de la Laguna Panzacola, Chiapas. Los sistemas litorales Universidad Autónoma Metropolitana-Iztapalapa- UNAM 105-113.1994.

[14] Montes C, Castillo ACGS y López P J. Distribución del manglar en cuatro sistemas lagunares de la costa de Chiapas, México. Boletín de la Sociedad Botánica de México 1999; 64:25-34.

[15] CONABIO, 2008. Manglares de México. Comisión Nacional para el Conocimiento y uso de la Biodiversidad. México, D.F.

[16] García, E. 1973. Modificaciones al Sistema de Clasificación Climática de Köppen. Instituto de Geografía, UNAM, México, D.F.

[17] INE, 1997. Programa de Manejo del área de protección de flota y fauna "Laguna de Términos"; SEMARNAT.

[18] Lugo A E andSnedaker SC. The ecology of mangroves. Annual Review of Ecology and Systematics 1974:39-64.

[19] Twilley RW, Lugo A E and Patterson-Zucca C. Litter Production and Turnover in Basin Mangrove Forests in Southwest Florida. Ecology 1986;67(3): 670-683.

[20] Robadue R, Calderon A, Oczkowski LB, and CepedaMF. Characterization of the Region of the Laguna de Términos Campeche, Mexico. Draft for discussion. 2004.

[21] Grossman RB and ReinschTG. Bulk density and linear extensibility. In: Dick, W.A. (ed.). Methods of soil analysis: Physical methods. Madison: SSSA, 2002: 201-208.

[22] Bernal B and Mitsch J. W. A comparison of soil carbon pools and profiles in wetland in Costa Rica and Ohio. Ecological Engineering 2008; 34:311-323.

[23] Heiri O, LotterAF and LemckeG. Loss on Ingnition as a method for estimating organic and carbonate content in sediments: reproducibility and comparability of results. Journal of paleolimnology2001; 25: 101-110.

[24] Craft BE, SenecaD, and BroomeSW. Loss on ignition and kjeldahl digestion for estimating organic carbon and soils: Calibration with dry combustion". Estuaries 1991; 14: 175-179.

[25] RyanJ, Estefan G and RashidA. Soil and Plant Analysis Laboratory Manual. International Center for Agricultural Research in the Dry Areas (ICARDA), Islamabad, Pakistan. 172p. 2001.

[26] Naidoo G. Factors Contributing to Dwarfing in the Mangrove Avicennia marina. Annals of Botany 2006; 97: 1095–1101.

[27] Joshi H and Ghose M. Forest structure and species distribution along soil salinity and pH gradient in mangrove swamps of the Sundarbans. Tropical Ecology 2003; 44(2): 197-206.

[28] BrahimNM, BernouxD, BlavetB and GallaliT. Tunisian soil organic carbon stocks. Int. J. Soil Sci., 2010; 5:34-40.

[29] Batjes N H. A global data set of soil pH properties", Tech. Rep. 27, Int. Soil Ref. and Inf. Cent., Wageningen, Netherlands. 1995.

[30] Fromard F, Puig H, Mougin E, Marty G, Betoulle JL, Cadamuro L. Structure above-ground biomass and dynamics of mangrove ecosystems: new data from French Guiana. Oecologia1998; 115:39–53.

[31] Brevik EC, Fenton TE and Horton R. Effect on soil electrical conductivity as measured with the Geonic EM-38, Precision Agriculture 2004; 5: 145-152.

[32] Webb A. Pre-clearing soil carbon levels in Australia. National carbon Accounting System. Technical Report No. 12. AustralianGreenhouseOficce, Canberra, 204 p.2002.

[33] Zhang L and Qiguo Z. Organic carbon content and distribution in soils under different land uses in tropical and subtropical China. Plant and Soil2001; 231: 175-185.

[34] Howe AJ, Rodríguez JF, Saco PM. Surface evolution and carbon sequestration in disturbed and undisturbed wetland soils of the Hunter estuary, southeast Australia. Estuarine, Coastal and Shelf Science 2009;84: 75-83.

[35] Lal R. World cropland soils as a source or sink for atmospheric carbón. Advances in Agronomy2001; 71: 145-191.

[36] Kjerfve BL, Drude de LacerdaC, RezendeE and Coelho OvalleAR. Hydrological and hydrogeochemical variations in mangrove ecosystems. In: A. Yañez-Arancibia and A.L. Lara-Dominguez (eds.). Ecosistemas de Manglar en América Tropical. Instituto de Ecología A.C. México, UICN/ORMA, Costa Rica, NOAAA/NMFS Silver Spring, M.D. USA:71-82. 1999.

Chapter 5

Long-Term Crop Rotation, Tillage, and Fertility Effects on Soil Carbon and Nitrogen in Dry-Seeded, Delayed-Flood Rice Production Systems

Jill M. Motschenbacher, Kristofor R. Brye,
Merle M. Anders, Edward E. Gbur,
Nathan A. Slaton and Michelle A. Evans-White

Additional information is available at the end of the chapter

http://dx.doi.org/10.5772/57064

1. Introduction

Rice (*Oryza sativa* L.) production systems occupy 156 million hectares of soil around the world (USDA-FAS, 2012) and rice is considered the world's most important staple food crop due to its prevalence and societal longevity. As a result of human activities, such as cultivation, domestication, dispersal, and diversification, rice is now grown in more than 100 countries across the latitudinal span of 40°S to 53°N (Chang, 2003), accounting for 20% of annual worldwide grain production (USDA-FAS, 2012). In the United States, 82% of the land area used for domestic rice production (1.1 million hectares) occurs in Arkansas (44%), California (22%) and Louisiana (16%), but smaller quantities of rice are also produced in Texas (8%), Mississippi (7%), and Missouri (6%; USDA-NASS, 2012). Rice production in the Mississippi Delta region of the United States is primarily drill-seeded during dry conditions as opposed to wet- or transplant-seeded during flooded conditions. These are the two most common methods used in most rice-producing areas of the world (De Datta, 1981).

Drill-seeded rice production systems are unique from other row-crop production systems in that the crop is grown under flood-irrigated conditions from about one month post-emergence until a few weeks prior to harvest, where the upper-most part of the soil profile is nearly to completely saturated (Norman et al., 2003). Drill-seeded rice production systems are also unique from wet- or transplant-seeded rice production systems because they involve shorter periods of saturated soil conditions and longer periods of dry soil conditions as a result of delayed flooding and time allowed between cropping periods. Since the cycling of waterlogged and dry soil conditions has been known to greatly influence the storage and cycling of soil organic matter (SOM; IPCC, 2007), drill-seeded rice production systems can play an influential role in the biogeochemical cycling of carbon (C) and nitrogen (N) in agriculturally managed systems.

Soil organic matter is one of the main reservoirs of C in the biosphere (Bernsten et al., 2006). The actual C content in SOM is approximately 57% by weight (Sundermeiser et al., 2005), with the remaining components made up of oxygen (O_2), hydrogen, N, and smaller amounts of other nutrients (Bot and Benites, 2005). Follett (2001) estimated that there are about 1550 petagrams (Pg) of soil organic carbon (SOC) stored in the world's soils. This is more than twice the estimated C contained in living vegetation (560 Pg) or in the atmosphere (750 Pg; Sunquist, 1993). Information on the dynamics of SOM and the associated SOC and N storage in agricultural soils has gained interest over recent years because of its influences on global climate change and crop productivity (Majumder et al., 2007; IPCC, 2007).

The concept of global warming is primarily attributed to three C and N greenhouse gases that are present both naturally and from anthropogenic sources: carbon dioxide (CO_2), methane (CH_4), and nitrous oxide (N_2O; IPCC, 2007). According to Lal (2004), land management practices have the potential to enhance SOC accumulation, thereby easing the gaseous C load to the atmosphere and enriching the soil. The decomposition of SOM, and thus the loss of C from the soil through the conversion of SOC into the gaseous compounds of CO_2 under aerobic conditions and CH_4 under anaerobic conditions is promoted by an array of factors, such as location in the soil, fibrous composition of the substrate, nutrient content, soil texture, soil pH, moisture conditions, and soil temperature (Denman et al, 2004; Wolf and Snyder, 2003; Seiter and Horwath, 2004; Alvarez and Lavado, 1998; Bayer, 1996; Filcheva and Mitova, 2002).

The SOM content present in the soil is a reflection of a long-term balance between additions and losses, thus the SOM content is greater when there is a large annual plant productivity rate and a low decomposition rate (Weil and Magdoff, 2004). Therefore, crops that produce large amounts of above-ground biomass have a greater contribution to the overall SOM content. The length of time SOC can be stored in the soil is controlled by the chemical composition of the biomass, which in turn determines the microbial breakdown potential of the substrate (Debusk et al., 2001; Farquhar et al., 2001). In relation to nutrient content, a large portion of the total nitrogen (TN) within soil (90 to 95%; Smith et al., 1993) occurs along with SOC in the constituents of SOM, and the ratio of N available in the substrate directly affects the rate of microbial SOM decomposition (McGill and Cole, 1981; Weil and Magdoff, 2004). Thus, the C and N cycles are directly linked and the C cycle of an ecosystem and cannot be properly examined or modeled without giving reference to N cycling (McGill and Cole, 1981; Weil and Magdoff, 2004). Furthermore, the presence or absence of O_2 plays an important role in the process dynamics and management responses of decomposition systems. Generally, an abundant O_2 supply promotes rapid decomposition, whereas a deficiency in O_2 results in a substantially lower decomposition rate (DeBusk et al., 2001; Shaffer and Ma, 2001). Under extremely wet conditions, decomposition can become limited by O_2 availability (Wolf and Snyder, 2003). Decreased decomposition rates result in greater accumulations of SOM in saturated soils, which essentially increase the amount of stored SOC and TN within the soil (DeBusk et al., 2001).

Since soil moisture conditions substantially influence the availability of O_2, and consequently, the overall storage of SOC and the cycling of N in the soil, examining crop production systems that contain periods of both aerobic and anaerobic conditions is an important component in accurately estimating the global storage of SOC and TN in agricultural production systems. The most essential aspect of soil management practices is to maintain soil physical properties in a way that supports crop growth and ensures an adequate amount of the biomass gets recycled back to the soil to serve as a long-term organic nutrient source (Lal, 2007). There have been a number of studies examining the changes in SOC due to different production practices, such as crop variety, tillage, fertilization and climate. However, much of the previous research has been conducted in dryland cropping systems across the globe or in paddy-grown rice in Asia. Hence, there is a great deficiency in research pertaining to SOC and TN storage and cycling in other production systems, such as the dry-seeded, delayed-flood rice production system common in the main rice-production region of the United States.

1.1. Research objectives and hypotheses

The objective of this study was to evaluate the long-term effects of rice-based crop rotations [i.e., with corn (*Zea mays* L.), soybean (*Glycine max* L.), and winter wheat (*Triticum aestivum* L.)], tillage [conventional tillage (CT) and no-tillage (NT)], and soil fertility (optimal and sub-optimal) on SOC and TN contents, the partitioning of SOC and TN in SOM, and C:N soil ratios in the top 10 cm of a silt-loam soil after 11 years (1999 to 2010) of consistent management. Furthermore, SOC and TN contents were modeled for the 11-year period and into the future using the Century soil organic model (Parton et al., 1987; NREL, 2006). It was hypothesized that: (1) NT practices would have greater SOC and TN than that under CT over time as a result of decreased aeration and soil disturbance; (2) rotations with increased frequencies of high-residue-producing crops, such as rice, corn, and those double-cropped with wheat, would have greater SOC and TN from greater amounts of biomass returned to the soil than lower-residue producing crop rotations, such as those with greater frequencies of soybean; (3) the optimal fertility regime would have greater SOC and TN than that in the sub-optimal fertility regime over time as a result of increased biomass production related with greater nutrient inputs from fertilizer; (4) the partitioning of SOC and TN in SOM would not vary greatly as a result of treatment combinations; (5) soil C:N ratios would be greater in high-residue-producing rotations, such as rice and corn, and would be lower in rotations including soybean, as a result of crop residue inputs; and (6) the Century model outcomes are expected to be correlated with measured SOC and TN obtained in the study. The Century model is expected to show that systems with high-residue-producing crops, NT, and optimal fertility result in greater SOM in the soil, thus increasing the SOC and TN contents over time more in these treatment combinations than in others.

2. Materials and methods

2.1. Site description

This field study was conducted at the University of Arkansas' Rice Research and Extension Center (RREC) near Stuttgart (34°27' N, 91°24' W), which is located in the Mississippi Delta region of eastern Arkansas in an area known as the Grand Prairie (USACE, 2000). The geographic area is made up of silt-loam and clay-textured soils from the Southern Mississippi River Alluvium and terraces and the Arkansas River Alluvium, which are Major Land Resource Areas 131A and 131B, respectively (USDA-NRCS, 2008). The 30-yr mean monthly air temperature ranges from a minimum of 0.2 °C in January to a maximum of 33.1 °C in July, and the 30-yr mean annual precipitation is 132 cm (SRCC, 2012).

The study began in 1999 on a Dewitt silt loam (fine, smectitic, thermic, Typic Albaqualf; USDA-NRCS, 2008), which is a common soil present in the Grand Prairie region. The Dewitt series is made up of very deep, poorly drained, slowly permeable soils that formed in alluvium (USDA-NRCS, 2008). The top 10 cm are primarily dark grayish brown (10YR 4/2) in color when the soil is moist and have a moderate medium granular structure near the surface and weak medium and course subangular blocky structure in the sub-soil (USDA-NRCS, 2008).

Prior to the initiation of the study in 1999, the study area had been fallow for numerous years due to a lack of irrigation capability. Vegetation present consisted of a mixture of grasses and weeds that were mowed in the summer. In preparation for the study, the site was land leveled to a 0.15% grade in Fall 1998, which is common practice in the area to facilitate uniform distribution of flood-irrigation water (Brye et al., 2003). Top soil was first removed, piled off to the side of the area, and then spread back over the area that was leveled.

2.2. Experimental design and field treatments

The experimental design for this study was a randomized complete block, with each block partitioned as a split-strip-split plot. The blocks included four rectangular sections 76-m long by 120-m wide (9120 m^2) within the 1.9-ha experimental site. Each block was divided into two tillage treatments (CT and NT) and each tillage treatment was split into two fertility regimes (optimal and sub-optimal). Each of the tillage-fertility combinations were stripped across with six crop rotations and split across time (1999 and 2010). The six crop rotations included continuous rice (R), rice-soybean (RS), rice-corn (RC), rice (winter wheat) [R(W)], rice (winter wheat)-soybean (winter wheat) [R(W)S(W)], and rice-corn-soybean (RCS). All rotations evaluated started with rice during the first year of the study and followed the respective rotations in successive years (Table 1). There were 96 plots evaluated and each plot representing a tillage-fertility-rotation treatment combination measured 6- by 19-m.

In the CT treatment, crop residues were incorporated into the soil by disking one to two months following harvest and prior to spring planting. Spring field preparation was also followed by multiple passes with a light field cultivator (i.e., Triple-K) to achieve the desired seedbed for rice planting. In the NT treatment, crop residues were left on the surface after harvest and were

	Rotation					
Year	R	RS	RC	R(W)	R(W)S(W)	RCS
1999	Rice	Rice	Rice	Rice	Rice	Rice
2000	Rice	Soybean	Corn	Rice	Soybean	Corn
2001	Rice	Rice	Rice	Rice	Rice	Soybean
2002	Rice	Soybean	Corn	Rice	Soybean	Rice
2003	Rice	Rice	Rice	Rice	Rice	Corn
2004	Rice	Soybean	Corn	Rice	Soybean	Soybean
2005	Rice	Rice	Rice	Rice	Rice	Rice
2006	Rice	Soybean	Corn	Rice	Soybean	Corn
2007	Rice	Rice	Rice	Rice	Rice	Soybean
2008	Rice	Soybean	Corn	Rice	Soybean	Rice
2009	Rice	Rice	Rice	Rice	Rice	Corn
2010	Rice	Soybean	Corn	Rice	Soybean	Soybean

Table 1. Summary of summer crop rotations by year with all rotations tilled prior to planting in 1999 and the no-tillage treatment starting in 2000. Crops used in the rotations include rice (R), soybean (S), corn (C), and winter wheat [(W)]. Crops in parentheses were grown in the winter.

not manipulated by any means prior to planting in the spring. Rice, soybean, and wheat were sown into 19-cm rows in both tillage treatments using an Almaco NT drill (Almaco, Nevada, IA). Rice was drill-seeded at a rate of 100 kg seed ha^{-1}, soybean at a rate of 56 kg seed ha^{-1}, and wheat at a rate of 67 kg seed ha^{-1}. Corn was planted in 76-cm rows at a plant population of 79,000 seeds ha^{-1}. Fertility treatments applied were based on soil analyses from the study site at the beginning of the study in 1999 (Table 2). Each year, phosphorus (P) was applied as triple super phosphate and potassium (K) was applied as muriate of potash, which were broadcast pre-plant and pre-tillage with a hand-spreader. During the years rice was grown, zinc (Zn) was applied as zinc sulfate, which was also broadcast pre-plant and pre-tillage with a hand-spreader. Nitrogen was applied as urea with a hand-spreader pre-flood approximately one month after planting. Phosphorous, K, and Zn were incorporated into the soil in the CT treatment and were applied to the surface in the NT treatment.

Following N fertilization of the rice, a 5- to 10-cm permanent flood was established, which was maintained on all of the rice plots throughout the growth period. Corn and soybean were furrow-irrigated on an as-needed basis during the summer growth period, with the irrigation amount varying based on the rainfall received and the growth of the crop. Irrigation water originated from a nearby surface reservoir, with water additions mainly from an adjacent stream channel and some groundwater. Wheat grown during the winter was rainfed only. All recommendations for stand establishment, irrigation, weed control and pest management for rice (Slaton, 2001), soybean (Ashlock, 2000), corn (Espinoza and Ross, 2003) and wheat (Kelley,

		Soil Fertility Treatment (kg ha^{-1})	
Crop	Nutrient	Sub-Optimal	Optimal
Corn	N	224	337
	P_2O_5	67	90
	K_2O	112	168
Soybean	N	0	0
	P_2O_5	45	67
	K_2O	67	135
Rice	N	112	168
	P_2O_5	45	67
	K_2O	67	101
	$ZnSO_4$	11	11
Wheat	N	112	168
	P_2O_5	34	67
	K_2O	34	67

Table 2. Summary of the annual nitrogen (N), phosphorous (P), potassium (K), and zinc (Zn) added to corn, soybean, rice and winter wheat to comprise the optimal soil fertility treatments in a long-term, rice-based rotation study at the Rice Research and Extension Center near Stuttgart, AR on a silt-loam soil.

1999) followed the University of Arkansas Cooperative Extension Service recommendations for the duration of the study.

2.3. Soil sampling and analyses

Soil samples were collected from the top 10 cm in each tillage-fertility-rotation treatment combination prior to spring planting in March during both evaluation years. The 1999 samples were collected after land leveling had occurred and prior to any actual field treatment. The 2010 samples were collected prior to the spring planting of rice, corn or soybean. During the time of sampling in 2010, wheat was present in the R(W) and R(W)S(W) rotations, whereas the R, RS, RC, and RCS rotations were still fallow from the winter. Soil samples for both years were dried at 70°C in a forced-draft oven for 48 hours. In 2010, soil samples were acquired using a 4.7-cm diameter stainless steel cylinder. Following drying, samples were weighed for bulk density determination. In 1999, bulk density was not measured; therefore, soil particle-size analyses were conducted on all of the 2010 soil samples from the top 10 cm in order to estimate bulk density. The percentages of sand and clay measured from the 2010 samples were combined with measured SOM concentrations from 1999 to estimate bulk density for the 1999 samples on a plot-by-plot basis, using the bulk density prediction equation incorporated in the Soil-Plant-Atmosphere-Water (SPAW) model (Saxton and Rawls, 2006).

In both 1999 and 2010, total C and TN were measured by high-temperature combustion after samples were crushed to pass a 2-mm mesh screen. All measured C was assumed to be associated with the organic fraction of the soil because, upon treatment with concentrated hydrochloric acid (HCl), there was no effervescence. Therefore, all measured C is hereafter referred to as SOC. Measured SOC and TN concentrations (mg kg^{-1}) were converted to contents (kg m^{-2}) using the 10-cm sampling depth and soil bulk densities. The partitioning of SOC and TN within SOM, in addition to C:N ratios, were calculated from SOM, SOC, and TN concentrations measured in each tillage-fertility-rotation treatment combination.

When the 2010 samples were collected after 11 years of management, R and R(W) rotations had produced 11 rice crops, and the RS, RC, and R(W)S(W) rotations had produced six rice crops with five crops in the respective rotation with corn or soybean (Table 3). Furthermore, the RCS rotation had produced four rice crops, four corn crops, and three soybean crops, and the R(W) and R(W)S(W) rotations with had produced 11 winter wheat crops (Table 3). Tillage had occurred in the tilled plots five months before (late-October 2009) the collection of samples in March 2010.

Rotation	Number of Crops			
	Rice	Corn	Soybean	Wheat
Continuous Rice	11	-	-	-
Rice-Soybean	6	-	5	-
Rice-Corn	6	5	-	-
Rice-(Wheat)	11	-	-	11
Rice-(Wheat)-Soybean-(Wheat)	6	-	5	11
Rice-Corn-Soybean	4	4	3	-

Table 3. Summary of the crop rotations and the number of crops grown in the respective rotations during the 11-yr study period (1999 to 2010) at the Rice Research and Extension Center near Stuttgart, AR on a silt-loam soil. Crops in parentheses were grown during the winter.

2.4. Century model simulation

For comparison to measured data over time, simulations using the Century model were conducted for each tillage-fertility-rotation treatment combination. Century is a SOM model that simulates C and N cycling in different plant production systems based on a monthly time step (Parton et al., 1987, 1988; Parton and Rasmussen, 1994). The model works by allocating various plant components and animal excreta into different SOM pools with varying timescales of decomposition, which include the active (a few months to a few years), slow (20 to 50 years), or passive (400 to 2000 years) pools. This is done by partitioning plant residues as either structural or metabolic organic pools based on their lignin:N ratio (Shibu et al., 2006), in that plant parts with a larger ratio take longer to decompose. The model works by dividing the structural pool into lignin and cellulose components (NREL, 2006). The lignin is moved to the slow decomposition pool, whereas the cellulose and metabolic C (i.e., microbial biomass) are added to the active pool. The actual decomposition rates of the SOM from the active pool are calculated based on soil texture, soil temperature and soil moisture and the decomposition rates of the structural pool are determined based upon the lignin content of the plant material.

In the Century model, the nitrification process is not incorporated because there is no distinction made within the mineral-N pool between nitrate and ammonium (Shibu et al., 2006).

The limitations of the model in relation to rice-based cropping systems are that it was originally designed for grasslands, arable lands, forests and savanna ecosystems under aerobic conditions (NREL, 2006; Parton et al., 1988), not periodically flooded ecosystems. This has resulted in the Century model not being extensively used to simulate rice-based crop rotations. However, Bhattacharyya et al. (2007) used Century to model a jute (*Cochorus capsularis* L.)-rice-wheat rotation in West Bengal, India and Milne et al. (2008) modeled a 3-year rice-soybean, wheat-soybean rotation in eastern Arkansas, United States, in which the field was flooded for rice production once every three years. Bharracharyya et al. (2007) reported that the model overestimated SOC content, but was able to simulate trends in SOC cycling over the 30-year study period. Milne et al. (2008) reported that the model underestimated the increase of SOC content in the first year, but concluded that Century could be used to accurately predict the long-term SOC dynamics in rice-based crop rotations that are flooded every three years.

Parameters for the Century SOM model (Parton et al., 1987; NREL, 2006) simulation in this study were adjusted using a combination of recorded historic climatic data, measured plant properties and responses, and measured soil data for the top 10 cm that were adjusted to reflect a depth of 20 cm. In 1999, soil samples had only been collected from the top 10 cm, whereas 2010 samples were collected from both the 0- to 10- and 10- to 20-cm depths. In order to model trends over time using Century, which only simulates C and N cycling in a fixed depth of the top 20 cm, depth adjustments had to be made from the measured 0- to 10-cm sampling depth to reflect the top 20 cm. In order to calculate the percentage of SOC and TN that was contained in the top 10 cm of the 0- to 20-cm depth interval, the average percent content was calculated from the 2010 soil samples. Based on 192 samples from both the top 10 cm and the 10- to 20-cm depth, the top 10 cm contained an average of 66% of the SOC and 61% of the TN in the top 20 cm of soil. To facilitate comparison to measured data from the top 10 cm, output values from Century from the top 20 cm were adjusted to SOC and TN contents for the top 10 cm based on the average calculated percentages from observed data.

Different simulated model designs for field treatments were set up for each tillage-fertility-rotation treatment combination. However, the input file for all model designs used the same values for monthly average maximum and minimum air temperature (SRCC, 2012), monthly precipitation (SRCC, 2012), soil texture, soil bulk density, soil pH, and initial SOC and TN contents from 1999. This approach was justified because the entire study site was uniformly maintained and land leveled prior to the initiation of the study and SOC and TN concentrations were similar across the study site in 1999. Therefore, all model input files started with the same site and control parameters for the model. The initial values for soil texture, soil bulk density, and SOC and TN contents were fixed based on the average values across all preassigned tillage-fertility-rotation treatment combinations. Any observed differences among tillage-fertility-rotation treatment combinations over time were assumed to be a result of the imposed treatments instead of inherent differences in soil physical and chemical properties. Anaerobic conditions associated with flooding during the rice growing season, which results in slower decomposition rates, were accounted for by adjusting the drainage potential of the soil and the irrigation frequency.

2.5. Data analyses

Initial soil properties in 1999, in addition to the effects of tillage, fertility regime, crop rotation, over time (1999-2010) on SOC and TN contents, the partitioning of SOC and TN in SOM, and C:N ratios were evaluated by analysis of variance (ANOVA) using the PROC MIXED procedure in SAS® (version 9.2, SAS Institute, Inc., Cary, NC). When appropriate, means were separated using Fisher's protected least significant difference (LSD) at the 0.05 level. Modeled results for SOC and TN storage trends over time were compared to direct observations via a linear regression analyses using Minitab (version 15, Minitab, Inc., State College, PA).

3. Results and discussion

3.1. Initial soil properties

Samples collected at the initiation of the study in 1999, which were collected after land leveling and prior to the implementation of tillage, fertility, and rotation treatments, showed that soil properties in the top 10 cm were primarily uniform among preassigned tillage-fertility-rotation treatment combinations. There were no differences in SOC and TN contents, the partitioning of SOC and TN within SOM, and soil C:N ratios among preassigned tillage, fertility, and/or rotation treatments ($P > 0.05$; Table 4). Furthermore, the soil particle-size distribution did not differ among any treatment combinations when measured in 2010 ($P > 0.05$; Table 4).

Treatment Effect	BD	SOC	TN	SOC/SOM	TN/SOM	C:N
			P			
Tillage	0.595	0.345	0.139	0.382	0.683	0.452
Fertility	0.569	0.550	0.396	0.392	0.588	0.531
Fertility* Tillage	0.517	0.544	0.523	0.451	0.484	0.553
Rotation	**0.005**	0.113	0.671	0.320	0.921	0.349
Rotation*Tillage	0.316	0.999	1.000	0.873	0.811	0.993
Rotation*Fertility	0.418	0.400	0.397	0.488	0.762	0.760
Rotation*Tillage*Fertility	0.313	0.253	0.305	0.515	0.189	0.772

Table 4. Analysis of variance summary of the inherent differences of soil properties before tillage, fertility and crop rotation treatments. Soil properties and their interactions related to the inherent soil differences in the assigned treatment combinations on soil bulk density (BD), soil organic carbon (SOC) content, total nitrogen (TN) content, portion of SOC in soil organic matter (SOM), portion of TN in SOM, and carbon to nitrogen ratios (C:N) prior to any treatment being imposed in 1999. The study site was located at the Rice Research and Extension Center near Stuttgart, AR on a silt-loam soil. Treatment effects in bold are considered non-significant ($P > 0.05$).

Although there were no SOC and TN related soil property differences among preassigned treatments in 1999, there were a few minor inherent differences in estimated soil bulk density

($P = 0.005$) among preassigned rotations (Table 4). Estimated soil bulk density in the R(W) (1.44 g cm^{-3}) rotation was 2 to 4% greater than that in the R (1.41 g cm^{-3}), RC (1.40 g cm^{-3}), and RCS (1.38 g cm^{-3}) rotations, when averaged across preassigned tillage and fertility treatments. Furthermore, estimated soil bulk density was 3% greater in the R(W)S(W) and RS (both 1.42 g cm^{-3}) rotations than in the RCS rotation. Though statistically different, all differences in soil bulk density were ≤ 0.06 g cm^{-3}, which are not large enough to cause substantial differences in SOC and TN contents.

3.2. Tillage effects on SOC and TN

Similar to that hypothesized, when averaged across fertility regimes, crop rotations and time, SOC content (kg m^{-2}) in the top 10 cm was affected by tillage ($P = 0.012$), and when averaged across fertility regimes and crop rotations, TN content (kg N m^{-2}) differed over time between tillage treatments ($P = 0.019$; Table 5). Soil organic carbon content was 14% greater under NT (1.14 kg m^{-2}) than under CT (1.00 kg m^{-2}; Figure 1), but unlike what was expected, SOC content did not differ between the tillage treatments over time (i.e., 1999 to 2010; $P = 0.075$). However, as expected, TN content increased 34% from 1999 (0.10 kg m^{-2}) to 2010 (0.13 kg m^{-2}) under NT, but did not differ over time under CT (Figure 2). Consequently, NT management resulted in a net TN sequestration rate of about 0.003 kg m^{-2} yr^{-1} from 1999 to 2010.

Treatment Effect	SOC	TN	SOC/SOM	TN/SOM	C:N
	P				
Tillage	**0.012**	**0.002**	**0.007**	0.169	0.861
Fertility	0.469	0.080	0.589	0.167	0.349
Fertility* Tillage	0.538	0.678	0.319	0.322	0.377
Rotation	0.120	**0.010**	0.941	0.591	0.231
Rotation*Tillage	0.543	0.548	0.495	0.635	0.964
Rotation*Fertility	0.535	0.364	0.517	0.352	0.670
Rotation*Tillage*Fertility	0.839	0.441	0.464	0.457	0.447
Time	**0.039**	**0.036**	0.837	0.214	0.155
Time*Tillage	0.075	**0.019**	0.595	0.098	0.193
Time*Fertility	0.466	0.064	0.526	0.099	0.179
Time*Rotation	**< 0.001**	**0.003**	**0.047**	**0.037**	0.654
Time*Tillage*Fertility	0.623	0.738	0.348	0.344	0.580
Time*Tillage*Rotation	0.275	0.820	0.533	0.871	0.643
Time*Fertility*Rotation	0.505	0.302	0.464	0.235	0.898
Time*Tillage*Fertility*Rotation	0.786	0.439	0.568	0.605	0.312

Table 5. Analysis of variance summary of the effects of tillage, fertility, crop rotation and time on soil organic carbon (SOC) and total nitrogen (TN) contents, SOC and TN fractions of the soil organic matter (SOM), and carbon to nitrogen ratios (C:N) after 11 years of consistent management.

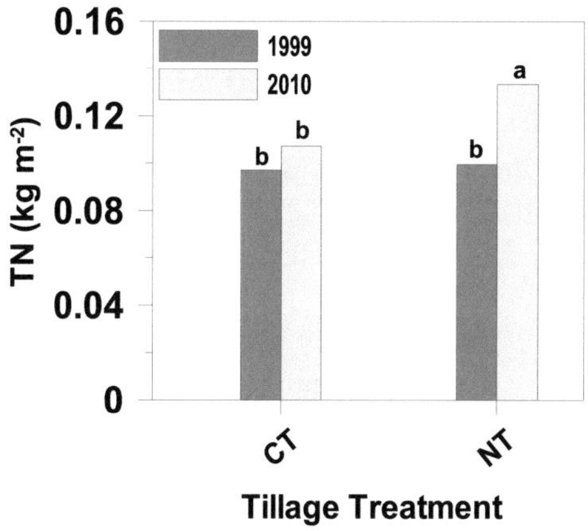

Figure 1. Tillage [conventional tillage (CT) and no-tillage (NT)] and time (1999 and 2010) effects on total nitrogen (TN) in the top 10 cm. Different letters atop bars are significantly different at the 0.05 level.

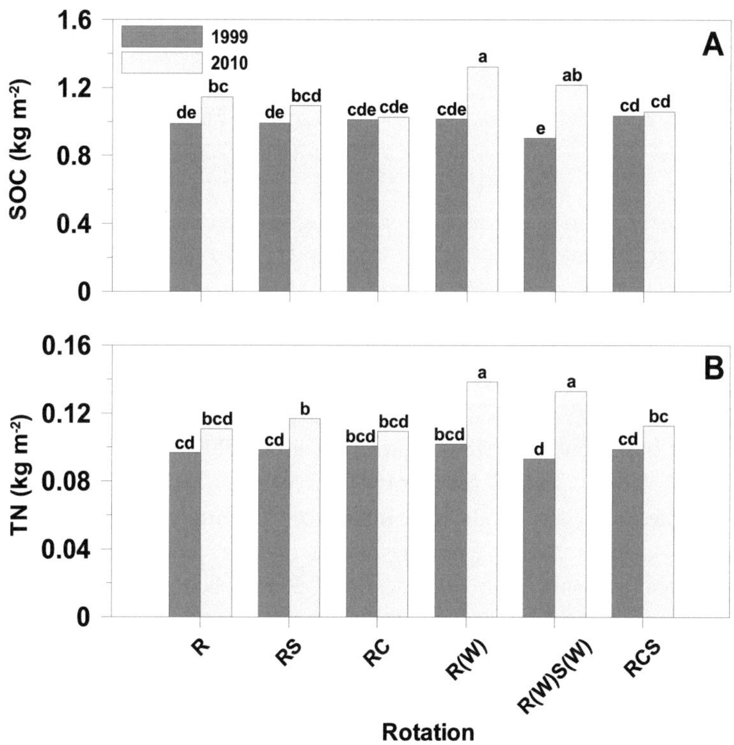

Figure 2. Rotation [rice (R), soybean (S), corn (C) and winter wheat (W)] and time (1999 and 2010) effects on soil organic carbon (SOC) [A] and total nitrogen (TN) [B] in the top 10 cm. Different letters atop bars within a panel are significantly different at the 0.05 level.

Tillage effects on SOC and TN contents appear to be directly related to SOM content differences, whereas the SOM content was 9% greater under NT (2.88 kg m^{-2}) than under CT (2.64 kg m^{-2}), when averaged across all other treatments (Motschenbacher et al., 2012). Although SOC content did not differ over time, the increase in TN content over time under NT corresponded to the 14% increase in SOM content over the same time period [1999 (2.58 kg m^{-2}) to 2010 (2.95 kg m^{-2}); Motschenbacher et al., 2012], when averaged across all other treatments. Similar to results observed in this study, greater SOC under NT as opposed to that under CT is a common occurrence, widely reported in past agronomic studies (West and Post, 2002; McCarty et al., 1998; Xu et al., 2007). Soils that have been degraded through excessive tillage tend to have less SOM due to an increased amount of exposed surface area, which facilitates aerobic decomposition (DeBusk et al., 2001). Carbon makes up more than half the mass of SOM (Montgomery et al., 2000), and it has been shown that cultivating the land influences the dynamics of SOC and, in turn, the amount of C emitted from the soil as CO_2 due to the oxidation or decomposition of SOM (Paustian et al, 1995; Reicosky et al., 1995).

Although flooded-soil conditions in rice-based crop rotations alter the environmental conditions contributing to C and N sequestration and cycling, due to slower decomposition rates and increased CH_4 emissions as opposed to CO_2 emissions, studies of long-term effects of different tillage regimes implemented in non-flooded upland soils can also provide insight on the impacts of soil disturbance from tillage. Salinas-Garcia et al. (1997) reported that a decrease in tillage increased SOM, microbial biomass C, inorganic N, and labile C and N pools compared to tilled systems in a long-term quadrennial corn-cotton (*Gossypium hirsutum*) rotation on a sandy-clay-loam soil in Texas. Furthermore, West and Post (2002) showed that among 67 non-flooded, long-term studies of various soil textures in countries throughout the world, the average soil C sequestration rate was 57 ± 14 g C m^{-2} yr^{-1} on land converted from CT to NT systems, with the exclusion of wheat-fallow systems. West and Post (2002) also estimated that C sequestration reaches a maximum between 5 and 10 years after conversion from CT to NT, and after 15 to 20 years the soil reaches a greater equilibrium C concentration.

In a study evaluating a wheat-soybean double-cropping system on a silt-loam soil in east-central eastern Arkansas, SOC content was greater under NT in the top 10 cm than that under CT after two years of continuous cultivation (Brye et al., 2006a) and C lost as CO_2 from soil respiration was 38% greater from CT than from NT (Brye et al., 2006b). Reicosky and Lindstrom (1993) attributed the large initial rate of soil surface CO_2 flux after tillage to the release of CO_2 in newly exposed soil pores and from dissolution or direct oxidation of C substrates, which further demonstrates the loss of C through increased decomposition rates as a result of aeration from tillage. Furthermore, SOM is made up of approximately 5% N, which is mineralized into ammonium (NH_4^+) during the decomposition process. Mineralized N is susceptible to removal from or translocation within the soil after nitrification through the leaching of nitrate (NO_3^-) and through gaseous losses during denitrification (Havlin et al., 2005; Schlesinger, 1997).

The reduction of tillage intensity by switching from CT to NT has been widely recognized as management practice that increases the amount of C storage in soils (Lal and Kimble, 1997; Doa, 1998; Kern and Johnson, 1993; Dick et al.,1998) and influences N cycling (Shaffer and Ma, 2001) in the near-surface soil. McCarty et al. (1998) reported increased SOC (38%), microbial biomass C (33%), TN (30%), and microbial biomass N (87%) concentrations in the top 2.5 cm of a silt-loam soil three years after the transition from CT to NT on a continuous corn rotation in the United States, whereas SOC (7%), microbial biomass C (15%), TN (6%), and microbial biomass N (35%) concentrations in the 12.5 to 20 cm depth decreased. An 18-year study of a rice-wheat rotation on a clay-loam soil in China also showed greater SOC, TN, and microbial biomass C and N concentrations in the top 5 cm under NT compared with CT, whereas SOC and TN concentrations were greater in the 5- to 10-cm and 10- to 20-cm depths under CT compared with NT (Xu et al., 2007). The results of previous studies suggest the elimination of tillage greatly influences the stratification of SOC and TN in the near-surface soil layers, whereas there are greater quantities of less-decomposed residue in the upper-most soil layers under NT and that the SOC and TN contents decrease with depth.

In contrast, SOC and TN contents are commonly unstratified and similar throughout the plow layer under CT because the mechanical mixing action of tillage distributes residue more evenly. Tillage also allows for the incorporation of SOM deeper into the soil by mixing plant residue and microbial biomass that usually remains in the upper-most layers under undisturbed conditions. This mixing action in the plow layer can result in greater SOC and TN contents in soil depths immediately below the plow depth under CT than that under NT due to the placement of the SOM near the bottom of the plow layer under full-inversion tillage (Angers and Eriksen-Hamel, 2008).

3.3. Fertility regime effects on SOC and TN

In contrast to that hypothesized, SOC and TN contents in the top 10 cm did not differ between fertility regimes imposed in the study ($P > 0.05$; Table 5). Fertilization of the soil has been well documented as being directly correlated to the quantity of biomass produced, which in turn is directly correlated with the amount of crop residues that are returned to the soil to become SOM, thus contributing to the SOC and TN pools (Follett, 2001). However, in this study the differences among optimal and sub-optimal fertility treatments (Table 2) were likely not great enough to significantly affect SOC and TN contents even after 11 years of consistent management. The lack of fertility treatment differences in SOC and TN contents between optimal and sub-optimal fertility directly correspond to the lack of differences in SOM contents that were also observed in this study, when averaged across all other treatment factors ($P > 0.05$; Motschenbacher et al., 2012). However, the lack of SOC and TN content differences under optimal and sub-optimal fertility do not mean that soil fertilizer treatments failed to increase the overall input of crop residues added to the soil when compared to the quantity that would be added upon the elimination of N-P-K treatments. For the purposes of this, only the effects of commonly recommended fertilizer rates provided to producers in the Mississippi Delta area of eastern Arkansas were evaluated (Table 2; Espinoza et al., 2007). Therefore, the design of this study did not include a control treatment in which no fertilizer was applied, as that would not be a common recommendation for field-crop production in the geographic study area.

Furthermore, there is a chance that differences in the fertility treatments over time might have existed if soil of a greater depth interval was compared. Electrical conductivity (EC) values evaluated in Motschenbacher et al. (2012) decreased substantially over the 11-year time period evaluated, which suggests that nutrient leaching beyond the top 10 cm occurred from the infiltration and movement of irrigation water. Therefore, it is reasonable to assume that some of the SOC and TN could have also leached further into the profile with the irrigation water. Therefore, differences between fertilizer treatments might have been too small to be significant in analyses from the top 10 cm, but lower soil depths could potentially show differences related to imposed fertility regimes.

Based on past studies, it can be assumed that both fertilization treatments allowed a greater quantity of above-ground biomass to be returned to the soil than would have been returned without fertilization. Halvorson et al. (2002) reported that N fertilization increased the quantity of crop residue returned to the soil in the top 30 cm of a loam soil after 12 years in two dryland cropping systems in North Dakota. The dryland cropping systems used in the study included spring wheat-winter wheat-sunflower (*Helianthus annuus* L.) and spring wheat-fallow rotations under different tillage treatments (Halvorson et al., 2002). Clapp et al. (2000) had similar findings in a 13-year study in the 15- to 30-cm depth of a silt-loam soil in Minnesota, which showed that adding N fertilizer to a corn cropping system increased SOC content when residues were returned to the soil in NT systems. On average, soil fertility management practices are estimated to add roughly 50 to 150 kg ha^{-1} of SOC to the soil every year, depending on the cropping system (Lal et al., 1998).

In the United States, most of the rice cultivars grown require 135 to 200 kg ha^{-1} of N fertilizer to produce profitable grain yields (Norman et al., 2003). Shen et al. (2007) indicated that chemical fertilizer treatments produced a net gain of 16 to 18 g SOC kg $soil^{-1}$ in rice-wheat agroecosystems in China when compared to untreated rice-wheat agroecosystems. In an 18-year study of a rice-wheat rotation in China, Xu et al. (2007) reported greater SOC and TN concentration increases in the top 10 cm when a combination of organic and inorganic N-P-K fertilization treatments were applied to a clay-loam soil than with no fertilization. Furthermore, Pampolino et al. (2008) reported that cropping systems delivering two to three rice crops per year on silty-clay or clay soils in the Philippines had greater SOC and TN concentrations in the top 20 cm with the application of N-K, N-P and N-P-K fertilization than in areas with no fertilization after 17 to 21 years of continuous production.

3.4. Crop rotation effects on SOC and TN

As hypothesized, SOC ($P < 0.001$) and TN ($P = 0.003$) contents (kg m^{-2}) in the top 10 cm were affected by crop rotation and time, when averaged across tillage treatments and fertility regimes (Table 5). The SOC content in rotations that included winter wheat [i.e., R(W) and R(W)S(W)] increased more than 30% for a sequestration rate of 0.028 kg^{-1} SOC m^{-2} yr^{-1} and the continuous rotation increased 16% for a sequestration rate of 0.014 kg^{-1} SOC m^{-2} yr^{-1} from 1999 to 2010, whereas SOC contents in the RS, RC, and RCS rotations did not differ over the 11-year time period (Figure 1A). Likewise, TN contents increased 36 to 46% in the R(W) rotation for a sequestration rate of 0.003 kg^{-1} TN m^{-2} yr^{-1} and in the R(W)S(W) rotation for a sequestration rate of 0.004 kg^{-1} TN m^{-2} yr^{-1} (Figure 1B). However, the RS rotation increased 19% for a sequestration rate of 0.002 kg^{-1} TN m^{-2} yr^{-1} from 1999 to 2010, whereas the TN contents in R, RC, and RCS rotations did not differ over time (Figure 1B). The greater increase of SOC and

TN over time in rotations with wheat could be partially due to greater quantities of annual biomass from the double-cropped rotations, as opposed to rotations that were fallow in the winter, and partially due to the presence of wheat in the ground during sampling. The presence of the wheat crop suggests there were greater concentrations of fresh root biomass and increased microbial activity in the near-surface soil compared to rotations that were fallow during the winter.

The increases in SOC and TN correspond to similar observed differences in SOM among rotations and increases in SOM content over time. Soil organic matter content increased 14% from 1999 (2.58 kg m^{-2}) to 2010 (2.95 kg m^{-2}), when averaged across all tillage, fertility and rotation treatment combinations, and SOM content was 5 to 13% greater in R(W) (3.01 kg m^{-2}) than in all other crop rotations (2.67 to 2.78 kg m^{-2}), when averaged across tillage, fertility and time (Motschenbacher et al., 2012). However, SOC contents only increased in the R(W), R(W)S(W), and R rotations and TN contents only increased in the R(W), R(W)S(W), and RS rotations over the 11-year period, whereas SOM content changes over time were unaffected by individual crop rotations (Motschenbacher et al., 2012).

Increasing the amount of SOC, and associated TN, in the soil requires the addition of enough crop residue to exceed the losses from SOM decomposition, erosion and leaching (Follett, 2001). This can be accomplished by using crop rotation systems that can influence the volume, distribution and turnover of the active and passive pools of SOC (Franzluebbers et al., 1994). Examining each year separately and averaged across tillage and fertility regimes, SOC content in the RCS (1.04 kg m^{-2}) rotation was 14% greater than that in the R(W)S(W) (0.91 kg m^{-2}) rotation at the beginning of the study in 1999, while SOC contents in all other rotations did not differ (Figure 1A). For TN content, there were no differences among crop rotations in 1999 (Figure 1B). However, after 11 years of continuous management, the R(W) (1.32 kg m^{-2}) and R(W)S(W) (1.22 kg m^{-2}) rotations had 15 to 28% greater SOC contents than rotations that included corn [i.e. RC (1.03 kg m^{-2}) and RCS (1.06 kg m^{-2}); Figure 1A]. Furthermore, the R(W) rotation had 15 to 28% greater SOC content in 2010 than that in the R, RS, RC, and RCS rotations (1.03 to 1.15 kg m^{-2}), which did not differ between one another (Figure 1A). In 2010, TN content was 14 to 27% greater in the R(W) (0.14 kg m^{-2}) and R(W)S(W) (0.13 kg m^{-2}) rotations than that in any of the other crop rotations (0.11 to 0.12 kg m^{-2}), whereas TN content among the R, RS, RC, and RCS rotations in 2010 did not differ (Figure 1B). Greater TN contents in winter-wheat rotations [i.e., R(W) and R(W)S(W)] compared to the R, RS, RC, and RCS rotations could partially be due to the timing of the last fertilizer-N application prior to the 2010 sampling date in mid-March. The R(W) and R(W)S(W) rotations received N fertilization at the beginning of March 2009 for the winter-wheat crop, whereas the last application of N fertilizer was in the previous spring (i.e., Spring 2009) for all other crop rotations (Table 2).

Dick et al. (1998) concluded that crop rotation and cover crops are an effective way to facilitate C sequestration, especially when combined with NT and optimal fertilization. After evaluating 67 long-term agricultural studies, West and Post (2002) estimated that enhancing the rotation complexity, including changing from monoculture to continuous rotation cropping, crop-fallow to continuous monoculture or rotation cropping, or increasing the number of crops in a rotation system, can sequester an average of 0.02 ± 0.01 kg C m^{-2} yr^{-1} in the top 30 cm of dryland systems. However, these results excluded the change from continuous corn to corn-soybean, which may not lead to a significant change in SOC (West and Post, 2002). The rate reported in West and Post (2002) is comparable to the SOC sequestration rate of 0.03 kg^{-1} m^{-2} yr^{-1} in the top 10 cm for rotations which included winter wheat [i.e., R(W) and R(W)S(W)] in this study compared to continuous rice (0.01 kg^{-1} m^{-2} yr^{-1}) and other rotations, which did not significantly sequester SOC over time.

In this study, crop rotation appeared to be a major contributor to the quantity of SOC present in the top 10 cm. Rice and corn are two high-residue-producing crops that are capable of producing 6.5 Mg ha^{-1} (rice) to 8.0 Mg ha^{-1} (corn) of above-ground dry matter per crop produced under adequate fertilization, whereas soybean is only capable of producing around 2.2 Mg ha^{-1} of above-ground dry matter during a cropping season (USDA-NASS, 2012). The above-ground biomass quantities produced annually can also be increased by double-cropping a system with wheat (West and Post, 2002), which is capable of producing an additional 3.3 Mg dry matter ha^{-1} per year (USDA-NASS, 2012), as opposed to keeping the rotation fallow for the winter. For this study, rotations that produced rice and wheat during a year [i.e., R(W) and R(W)S(W)] were capable of producing 9.8 Mg dry matter ha^{-1} per year, and during the years soybean and wheat were grown, the R(W)S(W) rotation produced roughly 5.5 Mg dry matter ha^{-1} per year (USDA-NASS, 2012). Therefore, a greater SOC content in the R(W) rotation after 11 years of rotation management was expected due to the consistent input of at least 23% more above-ground dry matter added annually than that from the RS, RC, and RCS rotations, and the R(W)S(W) rotation during the years soybean and wheat were produced (Figure 1A).

In addition to large above-ground biomass inputs, rice is a unique crop to include in a crop rotation because unlike all other row crops, rice is grown under nearly to completely saturated soil conditions (Norman et al., 2003), which slows the rate of SOM decomposition and affects N cycling dynamics. Thus, flooding the soil during the period rice is produced greatly affects SOM, and consequently SOC, accumulation in the soil, which is then susceptible to rapid decomposition when the field is drained for harvest and during the fallow period. Although crop rotations involving high-residue-producing crops like corn typically facilitate substantial increases in SOC, the anaerobic conditions under which rice is grown also affect the breakdown of crop residues in the soil. Witt et al. (2000) demonstrated that soils continuously cropped with flooded rice had 11 to 12% more C sequestration and 5 to 12% more N accumulation than soils which supported a dry-season, maize-flooded-rice rotation. The results were attributed to a 33 to 41% increase in the estimated amount of mineralized-C and decreased input of N from biological N fixation during the dry-season, maize-cropping period (Witt et al., 2000). Aulakh et al. (2001) reported that C sequestration in a sandy-loam soil in India was 69 to 107% greater when wheat residues were added to flooded rice. Furthermore, Aulakh et al. (2001) also showed that adding wheat residue immobilized mineral-N during the fallow period, but the amount of mineral-N increased rapidly at the start of the flooded rice season when green manure or urea-N were applied.

The increase in TN in the RS rotation over time in this study can be explained by the increased frequency of an N-fixing legume crop in the rotation, whereas the R, RC, and RCS did not result in increased TN over time (Figure 1B). This is because the *Rhizobia* bacteria present in the nodules on legume roots are able to absorb dinitrogen gas (N_2) from the air and convert it to NH_4^+, which can either be released into the soil or taken up by the plant and returned to the soil as residue-N at a later time (Havlin et al., 2005). The R(W)S(W) rotation also had a biennial rotation with soybean, but it is unclear whether the increase in TN content over time was a result of having a leguminous crop present in the rotation or related to the application of fertilizer-N in the fall. The R(W)S(W) rotation had a similar TN content as the R(W) rotation in 2010, which also had fertilizer-N applied in the fall, but did not include a leguminous crop (Figure 1B). These results are similar to those of Granatstein et al. (1987) from a 10-year study of winter wheat-legume rotations using different tillage regimes. They reported the greatest increase in TN in the top 5 cm was in rotations with greater frequencies of N-fixing legume crops in the rotation.

3.5. Partitioning of SOC and TN within SOM

Similar to SOC and TN contents, the fraction of SOM made up by SOC in the top 10 cm differed between tillage treatments ($P = 0.007$), when averaged across fertility, rotation and time. Also similar to SOC and TN contents, the fraction of SOM made up by SOC ($P = 0.047$) and TN ($P = 0.037$) differed by rotation and time, when averaged across tillage and fertility regimes, but SOC and TN fractions of SOM were unaffected by fertility regime ($P > 0.05$; Table 5).

The fraction of SOM made up of SOC was 5% greater under NT (39.7%) than that under CT (37.7%). Furthermore, the fraction of SOM made up of SOC was roughly 1.4% greater in the RS rotation in 1999 (40.7%) and R(W)S(W) rotation in 2010 (40.4%) than the R(W)S(W) rotation in 1999 (35.6%), whereas there were no differences in the SOC fraction of SOM among any other rotation-time combinations. Therefore, the fraction of SOM made up of SOC increased 13% over 11 years in the R(W)S(W) rotation, but the SOC fraction of SOM did not change over time in any other rotation evaluated. The fraction of SOM made up of TN was 2% greater in the R(W)S(W) rotation in 2010 (4.4%) than in the R, RC, R(W), R(W)S(W), and RCS rotations in 1999 and the R and RC rotations in 2010 (both years ranged from 3.7 to 3.9%). Similar to SOC, the TN fraction of SOM in the R(W)S(W) increased 21% from 1999 (3.6%) to 2010 (4.4%).

Although there were differences in the SOC and TN fractions of SOM, these differences only ranged from 1 to 5%. With the exception of the RS rotation in 1999, which had a greater fraction of SOC in SOM, only the R(W)S(W)/2010 treatment combination had greater SOC and TN fractions of SOM than a majority of the other rotation-time combinations. Furthermore, the R, RC, R(W), R(W)S(W), and RCS rotations in 1999 and the R, RS, RC, R(W), and RCS rotations in 2010 did not differ in their SOC and TN fractions of SOM. Therefore, there is a possibility that the greater SOC and TN fractions of SOM in R(W)S(W) in 2010 could be an isolated sampling discrepancy instead of a true rotation effect over time.

3.6. Soil C:N ratio

The soil C:N ratio in the top 10 cm was unaffected by tillage, fertility, rotation or time ($P > 0.05$; Table 5). Carbon:N ratios ranged from 8.1 to 14.0 in both 1999 and 2010. The C:N ratio of the soil is important because N is essential to microbial growth and reproduction, thus the microorganisms must obtain N either from the material being decomposed or an external source (i.e., inorganic N fertilizer compounds in the soil) in order to continue consuming SOM (Wolf and Snyder, 2003). If the SOM being decomposed has a large C:N ratio, then soil microbes either have to acquire N from the surrounding soil or decrease the rate of decomposition (Havlin, 2005). A common rule of thumb is that a C:N ratio of 20 to 30 in crop residues maintains an equilibrium mineral-N level in the soil, whereas any C:N ratio above 30 results in a net loss of NH_4^+- and NO_3^--N, and a C:N ratio below 20 results in a net gain of NH_4^+- and NO_3^--N (Stevenson and Cole, 1999). However, in this study there were no differences in soil C:N ratios among treatment combinations evaluated.

3.7. Century model SOC and TN contents

The linear relationship between Century-modeled and measured SOC ($P = 0.010$) and soil TN ($P = 0.024$) after 11 years of consistent management was significant, but there were relatively weak predictive relationships between modeled and measured SOC ($r^2 = 0.14$; Figure 3A) and TN ($r^2 = 0.10$; Figure 3 B) contents. For the analyses of SOC and soil TN contents, all modeled estimations for the 24 tillage-fertility-rotation treatment combinations were evaluated against the means of direct observation from 2010. The mean values of the direct observations were used to get a direct comparison between modeled and measured SOC and soil TN contents.

	SOC (kg m^{-2})								TN (kg m^{-2})							
	Measured				Modeled				Measured				Modeled			
	CT		NT		CT		NT		CT		NT		CT		NT	
Rotation	**O**	**SO**	**O**	**SO**	**O**	**SO**	**O**	**SO**	**O**	**SO**	**O**	**SO**	**O**	**SO**	**O**	**SO**
R	1.11	0.97	1.28	1.22	1.72	1.67	1.95	1.67	0.12	0.12	0.11	0.09	0.15	0.13	0.17	0.14
RS	0.96	0.93	1.24	1.26	1.42	1.38	1.38	1.34	0.14	0.13	0.10	0.10	0.12	0.11	0.12	0.11
RC	0.84	0.90	1.19	1.18	1.94	1.86	1.95	1.84	0.13	0.12	0.09	0.10	0.17	0.15	0.17	0.15
R(W)	1.20	1.21	1.45	1.43	1.38	1.38	1.43	1.42	0.15	0.15	0.13	0.12	0.12	0.12	0.13	0.12
R(W)S(W)	1.08	1.08	1.49	1.22	1.46	1.41	1.37	1.37	0.17	0.13	0.12	0.12	0.13	0.13	0.13	0.12
RCS	1.01	1.06	1.10	1.07	1.74	1.66	1.72	1.62	0.12	0.12	0.11	0.10	0.15	0.13	0.15	0.13

Table 6. Measured and Century-estimated soil organic carbon (SOC) and total nitrogen (TN) contents under different tillage [conventional tillage (CT) and no-tillage (NT)], fertility [optimal (O) and sub-optimal (SO)] and crop rotation [with soybean (S), corn (C), and/or wheat (W)] treatment combinations after an 11-yr study period (1999 to 2010). The study site was located at the Rice Research and Extension Center near Stuttgart, AR on a silt-loam soil. Values presented are the measured and model predictions from the 2010 sampling period. Crops in parentheses were grown during the winter.

A numerical evaluation of Century-estimated values and direct observations over the 11-year study period (1999 to 2010) indicated that the Century model estimated greater SOC (Figure 4A) and TN (Figure 4B) contents than that measured in 2010 (Table 6). Century-modeled SOC

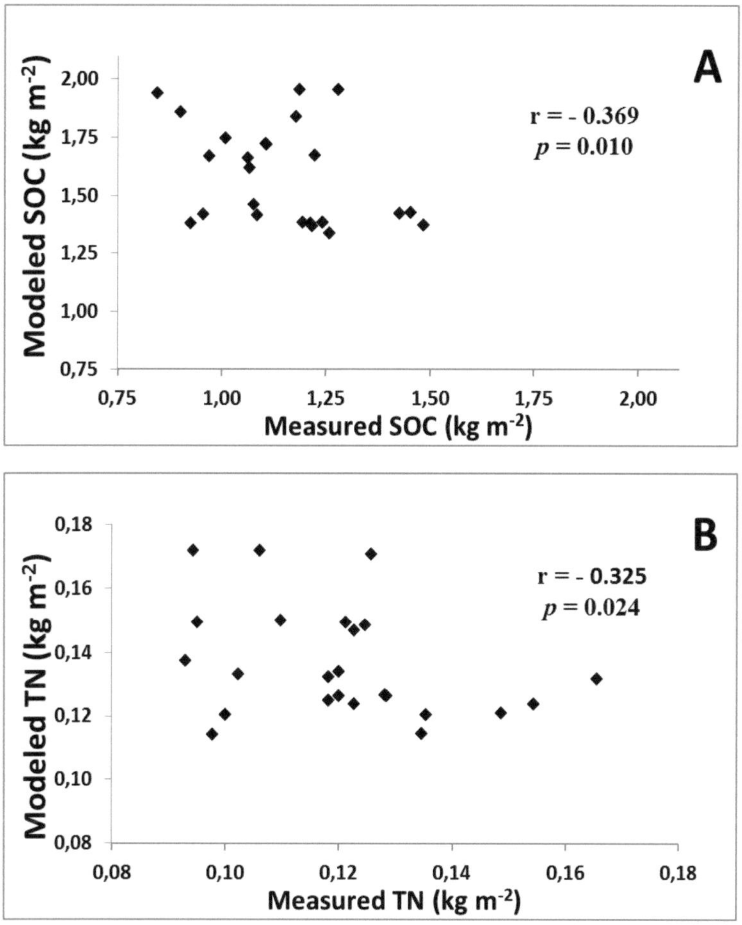

Figure 3. Century-modeled verses measured soil organic carbon (SOC) [A] and total nitrogen (TN) [B] contents in the top 10 cm after 11 years of management for 24 different tillage-fertility-rotation treatment combinations.

contents were numerically greater than measured SOC contents by 6% (0.08 kg SOC m^{-2}) to 56% (1.10 kg SOC m^{-2}) in all tillage-fertility-rotation treatment combinations except for the NT/optimal/R(W), NT/sub-optimal/R(W), and the NT/optimal/R(W)S(W) treatment combinations, which were slightly underestimated by 1% (0.01 kg SOC m^{-2}) to 8% (0.12 kg SOC m^{-2}; Table 6).

On average, Century-modeled SOC contents overestimated SOC by 26%. Estimated SOC in 8 of the 24 treatment combinations differed less than 14% from directly measured SOC in 2010, while the other 16 treatment combinations exceeded a 23% difference from measured SOC (Table 6). It appears that this overestimation by Century would increase into the future if management practices continued into the future (Figure 3A). The overestimation of SOC during a longer time period in this study is similar to modeling results reported by Bhattacharyya et al. (2007) in a jute (*Cochorus capsularis* L.)-rice-wheat rotation over a 30-year period. Bhattacharyya et al. (2007) reported that SOC was generally overestimated by 15% in the top

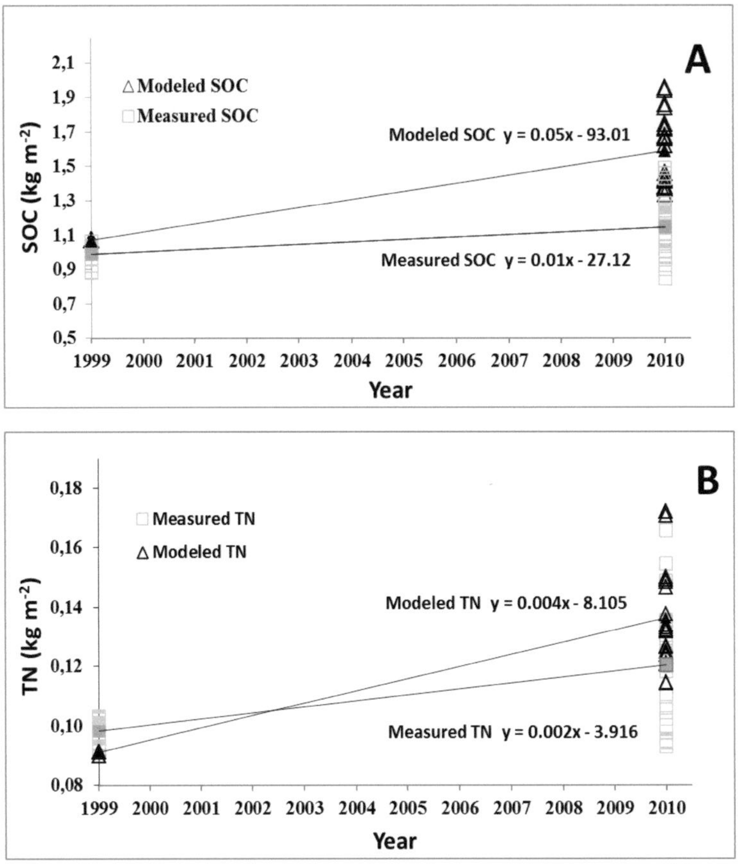

Figure 4. Century-modeled and measured soil organic carbon (SOC) [A] and total nitrogen (TN) [B] contents in the top 10 cm over time (1999 and 2010) for 24 different tillage-fertility-rotation treatment combinations. The regression equations on each graph represent the average modeled and measured SOC and TN contents across all treatment combinations over time.

20 cm when modeled in Century. Therefore, if the model were to be used for more accurate prediction in the rice-based rotations, such as those used in this study, a more in-depth evaluation and further adjustments of the Century modeling environment for flood-irrigation would be necessary.

Similar to modeled SOC, Century-modeled soil TN contents were also generally greater than the direct measurements after 11 years of management. Soil TN was greater by 1% (1.0 g SOC m^{-2}) to 45% (78 g SOC m^{-2}) in most treatment combinations except in the NT/optimal/R(W), CT/optimal/RS, CT/optimal/R(W), CT/optimal/R(W)S(W), CT/sub-optimal/RS, CT/sub-optimal/R(W), and CT/sub-optimal/R(W)S(W), which were underestimated by 1% (< 0.01 kg TN m^{-2}) to 26% (0.03 kg TN m^{-2}; Table 6). On average, the modeled soil TN content was overestimated by 10% across all tillage-fertility-rotation treatment combinations, which was a more accurate value than that of the average estimated SOC content over time. Century-modeled soil TN in 14 of the 24 treatment combinations had a range of less than 19% different from measured observations, whereas the other 10 treatment combinations exceeded a 23% difference from measured observations (Table 6). Similar to the pattern of modeled SOC, it appears that this overestimation of soil TN by Century would also continue to increase slightly over time if management practices continued into the future (Figure 3B).

Based on numeric values for modeled estimations and direct observations, it appears that Century can predict SOC and TN contents more accurately for certain tillage-fertility-rotation treatment combinations compared to other modeled treatment combinations (Table 6). The closest estimations of SOC to direct observations were for treatment combinations that included rotations with winter wheat [i.e., R(W) and R(W)S(W)] during the study period, and the closest estimations for soil TN over time occurred in treatment combinations that included the R(W)S(W) rotation (Table 6). The least accurate estimations of SOC compared to direct observations were for treatment combinations that included rotations with corn (i.e., RC and RCS), and the closest estimations for soil TN were in treatment combinations that included the R(W) rotation (Table 6). The predicted estimates of SOC and soil TN contents for rotations that included wheat could be a result of overestimating or underestimating the inputs from the winter wheat crop, which was in the ground during the time of soil sampling. The overestimation of SOC contents for rotations with corn could be a result of reduced yields at the study site compared to common yield values produced in the local geographic area, which is the yield value simulated in the Century model. This could be a result of many factors, but the more likely scenario is that the size of the research plot resulted in essentially an edge effect that would not be present in larger production fields.

The inaccuracies associated with the modeling results suggest that a few input parameters could be better adjusted in the Century model to more accurately predict SOC and TN cycling in rice-based crop rotations that are flooded for a majority of the growing season. Perhaps incorporation of the measured plant nutrient composition of the specific crop varieties grown during the study period and specific yields in each research area sampled over the 11-year period would improve the accuracy of specific input variables. Furthermore, soil sampling to a greater soil depth for more direct comparison to Century-modeled results might create not only a greater understanding of soil cycling in the near-surface soil, but also could help improve the overall accuracy of the model.

4. Summary and conclusions

This study demonstrated SOC and soil TN contents in the top 10 cm were affected by tillage, rotation and/or time after 11 years of consistent management. As hypothesized, soil TN contents increased over time under NT, but SOC contents did not differ under either tillage treatment (i.e., NT and CT) over the 11-year study period. Furthermore, soil under high-residue-producing rotations double-cropped with winter wheat [i.e. R(W) and R(W)S(W)] had greater increases in both SOC and TN contents over time compared to all other crop rotations. The only high-residue-producing, mono-cropped rotation with greater SOC over time was continuous rice, whereas SOC contents in all other crop rotations (i.e., RS, RC, and RCS) did not differ during the 11-year study period. Similar to the double-cropped rotations, the RS rotation also had an increase of soil TN over time. However, other crop rotations (i.e., R, RC, and RCS) did not differ in soil TN contents over time. In contrast to that hypothesized, fertility treatments had no effect on SOC and TN contents. The primary difference in SOC and TN contents among treatment combinations included the presence or absence of a winter wheat crop.

The Century model is a predictive tool for future environmental nutrient cycling estimations. Results from this study suggest that Century can be used to predict general trends in SOC and TN cycling in rice-based crop rotations over time, but further adjustment of the model is needed to increase accuracy. The direct measurements and the modeled estimations from this study are important because there has not been a great amount of research conducted on SOC and TN storage in dry-seeded, delayed-flooded rice production systems. A majority of C and N research in agricultural systems has either been conducted on non-flooded crops or in paddy-grown rice. Relatively few studies have actually evaluated long-term SOC and TN storage using dry-seeded, delayed-flood rice production practices commonly used in the Unites States.

This study was meant to provide a long-term evaluation of SOC and TN storage in soil used for flood-irrigated, rice-based crop rotations. Because soil moisture conditions have such a large impact on the overall storage of SOC and the cycling of N in the soil, examining crop production systems that contain periods of both aerobic and anaerobic conditions is an important component in obtaining a more accurate global estimate of SOC and TN storage in agricultural production systems. The results obtained from this study can help contribute to the ongoing effort to study the sustainability of rice cropping in the United States by enabling the direct quantification of C and N storage in the soil over time.

Acknowledgements

This study was partially funded by the Arkansas Rice Research and Promotion Board. Field assistance provided by Terry Sells, Daniel McCarty and Tara Moss Clayton is gratefully acknowledged.

Author details

Jill M. Motschenbacher[1*], Kristofor R. Brye[1], Merle M. Anders[1], Edward E. Gbur[2], Nathan A. Slaton[1] and Michelle A. Evans-White[3]

*Address all correspondence to: jmotsche@email.uark.edu

1 Department of Crop, Soil and Environmental Sciences, University of Arkansas, USA

2 Agricultural Statistics Laboratory, University of Arkansas, USA

3 Department of Biological Sciences, University of Arkansas, USA

References

[1] Alvarez, R., and R.S. Lavado. 1998. Climate, organic matter and clay content relationship in the Pampa and Chaco soils, Argentina. Geoderma 83:127-141.

[2] Angers, D.A., and N.S. Eriksen-Hamel. 2008. Full-inversion tillage and organic carbon distribution: A meta-analysis. Soil Sci. Soc. J. Am. 72:1370-1374.

[3] Ashlock, L. 2000. Arkansas soybean handbook. Handbook. MP 197. Univ. of Arkansas Coop. Ext. Serv., Little Rock.

[4] Aulakh, M.S., T.S. Khera, J.W. Doran, and K.F. Bronson. 2001. Managing crop residue with green manure, urea, and tillage in a rice-wheat rotation. Soil Sci. Soc. Am. J. 65:820-827.

[5] Bhattacharyya, T., D.K. Pal, M. Easter, S. Williams, K. Paustian, E. Milne, P. Chandran, S.K. Ray, C. Mandal, K. Coleman, P. Falloon, D.S. Powlson, and K.S. Gajbhiye. 2007. Evaluating the Century model using long-term fertiliser trials in the Indo-Gangetic Plains, India. In: E. Milne, D.S. Powlson, and C.E.P. Cerri (eds.) Soil carbon stocks at regional scales. Agric. Ecosys. Environ. 122:73-83.

[6] Bayer, C. 1996. Dinâmica da material organic em sistemas de manejo de solos. (In Portuguese.) Tese de Doutorado em Ciênia do Solo, PPG-Agronomia, UFRGS, Porto Alegre. *As cited in* C. Bayer, L. Martin-Neto, J. Mielniczuk, C.N. Pillon, and L. Sangoi. 2001. Changes in soil organic matter fractions under subtropical no-till cropping systems. Soil Sci. Soc. Am. J. 65:1473-1478.

[7] Bot, A., and J. Benites. 2005. The importance of soil organic matter. Food and Agriculture Organization of the United Nations, Rome. [online] Available at http://www.fao.org/docrep/009/a0100e/a0100e00.htm (verified 22 July, 2012).

[8] Brye, K.R., M.L. Cordell, D.E. Longer, and E.E. Gbur. 2006a. Residue management practice effects on soil surface properties in a young wheat-soybean double-cropping system. J. Sustain. Agric. 29:121-150.

[9] Brye, K.R., D.E. Longer, and E.E. Gbur. 2006b. Impact of tillage and residue burning on CO_2 flux in a wheat-soybean production system. Soil Sci. Soc. Am. J. 70:1145-1154.

[10] Brye, K.R., N.A. Slaton, M.C. Savin, R.J. Norman, and D.M. Miller. 2003. Short-term effects of land leveling on soil physical properties and microbial biomass. Soil Sci. Soc. Am. J. 67:1405-1417.

[11] Chang, T. 2003. Origin, domestication, and diversification. pp. 3-26. *In* C.W. Smith and R.H. Dilday. (eds.) Rice: Origin, history, technology, and production. John Wiley & Sons, Inc., NJ.

[12] Clapp, C.E., R.R. Allmaras, M.F. Layese, D.R. Linden, and R.H. Dowdy. 2000. Soil organic Carbon and ^{13}C abundance as related to tillage, crop residue, and nitrogen fertilization under continuous corn management in Minnesota. Soil Till. Res. 55:127-142.

[13] DeBusk, W.F., J.R. White, and K.R. Reddy. 2001. Carbon and nitrogen dynamics in wetland soils. pp. 27-53. In M. J. Shaffer, L. Ma, and S. Hansen. (eds.) Modeling carbon and nitrogen dynamics for soil management. CRC Press, Boca Raton, FL.

[14] De Datta, S.K. 1981. Principles and practices of rice production. International Rice Research Institute. John Wiley & Sons, Inc., Singapore. [online] Available at http://books.irri.org/0471097608_content.pdf (verified 21 July, 2012).

[15] Denman, K.L., G. Brasseur, A. Chidthaisong, P. Ciasis, P.M. Cox, R.E. Dickinson, D. Hauglustaine, C. Heinze, E. Holland, D. Jacob, U. Lohmann, S. Ramachandran, P.L. da ilva Dias, S.C. Wofsy, and X. Zhang. 2007. Couplings between changes in the climate system and biogeochemistry. *In*: S. Solomon, D. Qin, M. Manning, Z. Chen, M. Marquis, K. B. Averyt, M. Tignor, and H.L. Miller. (eds.) Climate Change 2007: The Physical Science Basis. Contribution of working group I to the fourth assessment report of the Intergovernmental Panel on Climate Change. Cambridge University Press, Cambridge, United Kingdom and New York, NY, USA.

[16] Dick, W.A., R.L. Blevins, W.W. Frye, S.E. Peters, D.R. Christenson, F.J. Pierce, and M.L. Vitosh. 1998. Impacts of agricultural management practices on C sequestration in forest-derived soils of the eastern Corn Belt. Soil Till. Res. 47:235-244.

[17] Espinoza, L., and J. Ross. 2003. Corn production handbook. Handbook. MP 437. Univ. of Arkansas Coop. Ext. Serv., Little Rock.

[18] Espinoza, L., N. Slaton., and M. Mozaffari. 2007. Understanding the numbers on your soil test report. University of Arkansas Extension Publication FSA 2118 [online]. Available at http://www.uaex.edu/Other_Areas/publications/pdf/FSA-2118.pdf (verified 6 July, 2012).

[19] Farquhar, G.D., M.J.R. Fasham, M.L. Goulden, M. Heimann, V.J. Jaramillo, H.S. Kheshgi, C. Le Quéré, R.J. Scholes, and D.W.R. Wallace. 2001. The carbon cycle and atmospheric carbon dioxide. *In* J.T. Houghton, Y. Ding, D.J. Griggs, M. Noguer, P.J. van der Linden, X. Dai, K. Maskell, and C.A. Johnston. (eds.) Climate change 2001: The scientific basis. IPCC, Wembley, United Kingdom.

[20] Filcheva, E., and T. Mitova. 2002. Effect of crop rotation and the composition of soil organic matter. pp. 237-263. *In* J.M. Kimble, R. Lal, and R.F. Follett. (eds.) Agricultural practices and policies for carbon sequestration in soil. CRC Press, Boca Raton, FL.

[21] Follett, R.F. 2001. Soil management concepts and carbon sequestration in cropland soils. Soil Till. Res. 61:77-92.

[22] Granatstein, D.M., D.F. Bezdicek, V.L. Cochran, L.F. Elliott, and J. Hammel. 1987. Long-term tillage and rotation effects on soil microbial biomass, carbon and nitrogen. Bio. Fert. Soils 5:265-270.

[23] Halvorson, A.D., B.J. Wienhold, and A.L. Black. 2002. Tillage, nitrogen, and cropping system effects on soil carbon sequestration. Soil Sci. Soc. Am. J. 66:906-912.

[24] Havlin, J.L., J.D. Beaton, S.L. Tisdale, and W.L. Nelson. 2005. Soil fertility and fertilizers: An introduction to nutrient management 7th ed. Pearson Education, Inc. Upper Saddle River, NJ.

[25] Intergovernmental Panel on Climate Change (IPCC). 2007. Climate change 2007: Synthesis report [online]. Core Writing Team, R.K. Pachauri, and A. Reisinger (eds.) Available at http:www.ipcc.ch/pdf/assessment-report/ar4/syr/ar4_syr.pdf (verified 23 July, 2012).

[26] Kelley, J. 1999. Arkansas wheat production and management. Handbook. MP 404. Univ. of Arkansas Coop. Ext. Serv., Little Rock.

[27] Kern, J.S., and M.G. Johnson. 1993. Conservation tillage impacts on national soil and atmospheric carbon levels. Soil Sci. Soc. Am. J. 57:200-210.

[28] Lal, R. 2004. Soil carbon sequestration impacts on global climate change and food security. Science 304:1623-1627.

[29] Lal, R. 2007. Soil Physical Properties and Erosion. pp. 165-178. *In* J.M. Kimble, C.W. Rice, D. Reed, S. Mooney, R.F. Follett, and R. Lal. (eds.) Soil carbon management: Economic, environmental and societal benefits. CRC Press, Boca Raton, FL.

[30] Lal, R., and J.M. Kimble. 1997. Conservation tillage for carbon sequestration. Nutr. Cycl. Agroecosyst. 49:243-253.

[31] Lal, R., J. Kimble, R.F. Follett, and C.V. Cole. 1998. The potential of US cropland to sequester carbon and mitigate the greenhouse effect. p. 128. Ann Arbor Press, Chelsea MI.

[32] Majumder, B., B. Mandal, P.K. Bandyopadhyay, and J. Chaudhury. 2007. Soil organic carbon pools and productivity relationships for a 34 year old rice-wheat-jute agroecosystem under different fertilizer treatments. Plant Soil 297:53-67.

[33] McCarty, G.W., N.N. Lyssenko, and J.L. Starr. 1998. Short-term changes in soil carbon and nitrogen pools during tillage management transition. Soil Sci. Soc. Am. J. 62:1564-1571.

[34] McGill, W.B., and C.V. Cole. 1981. Comparative aspects of cycling of organic C, N, S, and P through soil organic matter. Geoderma 26:267-286.

[35] Milne, E., S. Williams, K.R. Brye, M. Ester, K. Killian, and K. Paustian. 2008. Simulating soil organic carbon in a rice-soybean-wheat-soybean chronosequence in Prairie County, Arkansas using the Century model. J. Integ. Biosciences 6:41-52.

[36] Montgomery, D.R., D. Zabowski, F.C. Ugolini, R.O. Hallberg, and H. Spaltenstein. 2000. Soils, watershed processes, and marine sediments. pp. 159-194. *In* M.C. Jacobson, R.J. Charlson, H. Rodhe, and G.H. Orians. (eds.) Earth system science: From biogeochemical cycles to global change. Academic Press, San Diego, CA.

[37] Motschenbacher, J.M., K.R. Brye, M.M Anders, and E.E. Gbur. Long-term rice rotation, tillage, and fertility effects on chemical properties in a silt-loam soil. Ph.D. Diss. Chapter 4. University of Arkansas. Fayetteville, AR.

[38] National Resource Ecology Laboratory (NREL). 2006. Century soil organic matter model version 5 [online]. Available at http://www.nrel.colostate.edu/projects/century5/ (verified 23 July, 2012).

[39] Norman, R.J., C.E. Wilson, Jr., and N.A. Slaton. 2003. Soil fertilization and mineral nutrition in U.S. mechanized rice culture. pp. 331-411. *In* C.W. Smith and R.H. Dilday. (eds.) Rice: Origin, history, technology, and production. John Wiley & Sons, Inc., NJ.

[40] Pampolino, M.F., E.V. Laureles, H.C. Gines, and R.J. Buresh. 2008. Soil carbon and nitrogen changes in long-term continuous lowland rice cropping. Soil Sci. Soc. Am. J. 72:798-807.

[41] Parton, W.J., and P.E. Rasmussen. 1994. Long-term effects of crop management in wheat-fallow: II. Century model simulations. Soil Sci. Soc. Am. J. 58:530-536.

[42] Parton, W.J., D.S. Schimel, C.V. Cole, and D.S. Ojima. 1987. Analysis of factors controlling soil organic levels of grasslands in the Great Plains. Soil Sci. Soc. Am. J. 51:1173-1179.

[43] Parton, W.J., J.W.B. Stewart, and C.V. Cole. 1988. Dynamics of C, N, P, and S in grassland soils: A model. Biogeochem. 5:109-131.

[44] Paustain, K., G.P. Robertson, and E.T. Elliot. 1995. Management impacts on carbon storage and gas fluxes (CO_2, CH_4) in mid-latitudes cropland. pp. 69-83. *In* R. Lal, J. Kimble, E. Levine, and B.A. Stewart. (eds.) Soil management and the greenhouse effect. Lewis Publ., Boca Raton, FL.

[45] Reicosky, D.C., W.D. Kemper, G.W. Langdale, C.L. Douglas Jr., and P.E. Rasmussen. 1995. Soil organic matter changes resulting from tillage and biomass production. J. Soil Water Conserv. 50:253-261.

[46] Reicosky, D.C., and M.J. Lindstrom. 1993. Fall tillage method: Effect on short-term carbon dioxide flux from soil. Agron. J. 85:1237-1234.

[47] Salinas-Garcia, J.R., F.M. Hons, and J.E. Matocha. 1997. Long-term effects of tillage and fertilization on soil organic matter dynamics. Soil Sci. Soc. Am. J. 61:152-159.

[48] Saxton, K.E., and W.J. Rawls. 2006. Soil water characteristic estimates by texture and organic matter for hydrologic solutions. Soil Sci. Soc. Am. J. 70:1569-1578.

[49] Schlesinger, W.H. 1997. Biogeochemistry: An analysis of global change, 2nd ed. Academic Press, San Diego, CA.

[50] Seiter, S., and W.R. Horwath. 2004. Strategies for managing soil organic matter to supply plant nutrients. pp. 269-293. *In* F. Magdoff and R. R. Weil. (eds.) Soil organic matter in sustainable agriculture. CRC Press, Boca Raton, FL.

[51] Shaffer, M.J., and L. Ma. 2001. Carbon and nitrogen dynamics in upland soils. pp. 11-26. *In* M.J. Shaffer, L. Ma, and S. Hansen. (eds.) Modeling carbon and nitrogen dynamics for soil management. CRC Press, Boca Raton, FL.

[52] Shen, M.X., L.Z. Yang, Y.M. Yao, D.D. Wu., J. Wang, R. Guo, and S. Yin. 2007. Long-term effects of fertilizer management on crop yields and organic carbon storage of a typical rice-wheat agroecosystem of China. Biol. Fertil. Soils 44:187-200.

[53] Shibu, M.E., P.A. Leffelaar, H. Van Keulen, and P.K. Aggarwal. 2006. Quantitative description of soil organic matter dynamics: A review of approaches with reference to rice-based cropping systems. Geoderm 137:1-18.

[54] Slaton, N.A. 2001. Rice production handbook. Handbook. MP 192. Univ. of Arkansas Coop. Ext. Serv., Little Rock.

[55] Smith, J.L., R.I. Papendick, D.F. Bezdicek, and J.M. Lynch. 1993. Soil organic dynamics and crop residue management. pp. 65-94. *In* F.B. Metting, Jr. (ed.) Soil microbial ecology. Marcel Dekker, Inc., New York.

[56] Southern Region Climate Center (SRCC). 2012. Louisiana State University. Baton Rouge, LA. [online]. Available at http://www.srcc.lsu.edu/stations/index.php?action=metadata&network_station_id=030240 (verified 23 July, 2012).

[57] Stevenson, F.J., and M.A. Cole. 1999. Cycles of soil. John Wiley & Sons, Inc., New York, NY.

[58] Sundermeiser, A., R. Reeder, and R. Lal. 2005. Soil carbon sequestration – fundamentals. Ohio State University Extension Factsheet [online]. http://ohioline.osu.edu/aex-fact/0510.html (verified 23 July, 2012).

[59] United States Army Corps of Engineers (USACE). 2000. Grand prairie area demonstration project: Overview [online]. Available at http://www.mvm.usace.army.mil/grandprairie/overview/default.asp (verified 23 July, 2012).

[60] United States Department of Agriculture (USDA) Foreign Agricultural Service. (FAS). 2012. Production, supply and distribution [online]. Available at http://www.fas.usda.gov/psdonline/psdHome.aspx (verified 16 July, 2012).

[61] United States Department of Agriculture (USDA) National Agricultural Statistics Service (NASS). 2012. Crop Explorer: United States 2011 – area, yield, and production [online]. Available at http://www.nass.usda.gov/QuickStats/PullData_US.jsp (verified 15 July, 2012).

[62] United States Department of Agriculture (USDA) National Resources Conservation Service (NRCS). 2008. Web soil survey. Data from survey on November 12, 2008 [online]. Available at http://websoilsurvey.nrcs.usda.gov/app/WebSoilSUrvey.aspx (verified 23 July, 2012).

[63] Weil, R.R., and F. Magdoff. 2004. Significance of soil organic matter to soil quality and health. pp. 1-43. *In* F. Magdoff and R. R. Weil. (ed.) Soil organic matter in sustainable agriculture. CRC Press, Boca Raton, FL.

[64] West, T.O., and W.M. Post. 2002. Soil organic carbon sequestration rates by tillage and crop rotation: A global data analysis. Soil Sci. Soc. Am. J. 66:1930-1976.

[65] Witt, C., K.G. Cassman, D.C. Olk, U. Biker, S.P. Liboon, M.I. Samson, and J.C.G. Ottow. 2000. Crop rotation and residue management effects on carbon sequestration, nitrogen cycling, and productivity or irrigated rice systems. Plant Soil 225:265-278.

[66] Wolf, B., and G.H. Snyder. 2003. Sustainable soils: The place of organic matter in sustainable soils and their productivity. Food Products Press, New York.

[67] Xu, Y., W. Chen, and Q. Shen. 2007. Soil organic carbon and nitrogen pools impacted by long-term tillage and fertilization practices. Comm. Soil Sci. Plant Anal. 38:347-357.

Chapter 6

Stratification Ratio as Soil Carbon Sequestration Indicator in Oxisol and Alfisol Under No-Tillage

Ademir de Oliveira Ferreira and
Telmo Jorge Carneiro Amado

Additional information is available at the end of the chapter

http://dx.doi.org/10.5772/57063

1. Introduction

Soil organic carbon (SOC) stocks generally decrease from topsoil to subsoil layers. This is due to continuous aboveground carbon input by plant and animal residues and the absence of soil disturbance in natural ecosystems [1]. A similar phenomenon occurs in long-term no-tillage systems, promoting stratification of SOC stocks between topsoil and the adjacent soil layers [2,3,4].

The topsoil layer mediates energy, water, nutrient and gas exchanges between soil and atmosphere and therefore has vital importance for the functioning of ecosystems [2,3]. Most of these processes are regulated by soil organic matter. NT systems decrease SOC losses due to biological oxidation or soil erosion [5]. Continuous deposition of vegetal and animal residues on the soil surface and minimum soil disturbance under NT enhances SOC stabilization mechanisms, allowing increase of SOC stocks [3,4,6]. SOC accumulation in topsoil layers provides higher resistance to erosion and compaction [7], increases water infiltration and enhances aggregate and pore stability [2,8,9,10]. Thus, increase of SOC stocks, and consequently increase of the carbon stratification ratio (CSR), in the topsoil can be a good indicator of soil management quality [3,4,11].

CSR is the ratio between SOC stocks from two soil layers, usually the topsoil, with strong influence of soil and crop management practices, and the adjacent soil layer, which is less affected by these operations [2]. CSR values of 3.4, 2.0, and 2.1 were observed in NT soils from Georgia and Texas (USA) and Alberta (Canada), respectively [2]. The higher CSR values were related with agro-ecosystems which provided the maintenance or increase of soil quality. Additionally, CSR values higher than 2.0 would be unexpected in degraded soils from temperate climate regions.

CT managed soils from Alabama, Georgia, South and North Carolinas and Virginia in the USA had an average CSR of 1.4 [12]. CSR increased to 2.8 after ten years of NT adoption on the average of the evaluated soils. Three different soils from Virginia with original CSR values of 1.5 under CT, showed CSR of 3.6 after 14 years of NT adoption. Another study verified that CSR in a soil from the southeastern USA increased from 2.4 to 3.1 five years after conversion from CT to NT, reaching a CSR of 3.6 twelve years after the conversion [11].

Tropical NT soils have CSR values generally lower than temperate NT soils. CSR values of 1.48 (450 g clay kg^{-1} soil) and 1.73 (600 g clay kg^{-1} soil) were reported in Oxisols under NT from Southern Brazil [3,13]. These CSR values (SOC stock ratio of the 0-5 and 5-10 cm soil layers) had a close relationship with soil carbon sequestration rates verified in the 0-10 cm soil layer in these soils. Therefore, CSR could be used as an indicator of carbon sequestration in NT managed soils when historical data is not available for calculation of carbon sequestration rates. Furthermore, this procedure can be a tool to contribute to carbon stock inventories.

2. Material and methods

2.1. Description of the experimental areas

This study was carried out using two long-term experiments in southern Brazil. The first one was established in 1985 in Cruz Alta, state of Rio Grande do Sul, Brazil (28°33'S 53°40'W, altitude of 409 m). The local climate is subtropical humid (Cfa 2a according to the Köppen classification) with mean annual rainfall and temperature of 1,774 mm and 19.2 °C, respectively. The highest mean temperatures (30.0 °C) are registered in January and the lowest (8.6 °C) are registered in June [14]. The soil is a dystroferric Hapludox (referred to in the text as Oxisol) with 4.7% slope and predominance of kaolinite and iron oxides (63.5 g kg^{-1}) (Campos et al., 2011). The second experiment was established in 1991 in Santa Maria, state of Rio Grande do Sul, Brazil (29°43'S 53°42'W, altitude of 86 m). The local climate is subtropical (Cfa in the Köppen classification) with mean annual rainfall and temperature of 1,769 mm and 19.3 °C, respectively [14]. The soil is a dystrophic Paluedalf (referred to in the text as Alfisol) with 5.5% slope, a moderate A horizon and clay loam texture. Further soil characteristics are presented in Table 1.

The long-term experiment in the Oxisol site had a split plot design with two soil tillage systems as the main plots (i.e., conventional tillage (CT) and no-tillage (NT)) and three crop systems in the subplots without replications: a) succession R_0 - soybean (*Glycine max* L Merrill)/wheat (*Triticum aestivum* L.); b) winter rotation R_1 – wheat/soybean/oat (*Avena strigosa* Schreber)/ soybean; c) summer and winter rotation R_2- wheat/soybean/oat/soybean/oat+vetch (*Vicia sativa* L. Walp)/maize (*Zea mays* L.)/forage radish (*Raphanus sativus* var. *oleiferus* Metzg.). The

City	Soil	Layer	Carbon content	pH H_2O	Phosphorus	Potassium	Sand	Silt	Clay
		m	g kg^{-1}		mg dm^{-3}		g kg^{-1}		
Cruz Alta, RS, Brazil	Oxisol	0-0.20	19.0	4.5	19	82	310	120	570
Santa Maria, RS, Brazil	Alfisol	0-0.20	14.2	4.5	1.8	33	660	253	87

(1) Source: Adapted from [15,16,17].

Table 1. Main soil characteristics at the time of establishment of the experiments.

soil was amended with 5 Mg ha^{-1} of lime at the time of establishing the experiment in 1985. Wheat and maize received 60 and 90 kg N ha^{-1}, respectively. R_1 and R_2 crop systems received 52 and 62 kg P_2O_5 ha^{-1} yr^{-1} and 75 and 105 kg K_2O ha^{-1} yr^{-1}, respectively, in the first 15 years of the experiment [18]. Afterwards, phosphorus and potassium amendment was standardized at 50 kg P_2O_5 and K_2O ha^{-1} yr^{-1}. Further experimental details are available in [19].

The long-term experiment in the Alfisol site had a completely randomized block design with six treatments and two replications: a) maize + jack beans (*Canavalia ensiformis* DC)/soybean (M/JB); b) bare soil (BS); c) maize/fallow/soybean(M/F); d) maize/ryegrass (*Lolium multiflorum* Lam.) + vetch/soybean (M/R); e) maize + velvet beans (*Stizolobium cinereum* Piper & Tracy)/ soybean (M/VB); and f) maize/forage radish/soybean (M/FR). Details regarding the experiment are described in [17,20]. Soil acidity was corrected by the application of 3.5 Mg ha^{-1} of lime in 1991 and 2 Mg ha^{-1} of lime in 1996. The nitrogen fertilization rate for maize was 130 kg N ha^{-1} in the M/F and M/R treatments and 65 kg N ha^{-1} in the M/VB treatment. Further experimental details are described in [17]. The average carbon input values through plant residue in both long-term experiments are presented in Table 2.

2.2. Soil sampling and determination of SOC content

Soil samples were collected in 2001 (T_1) and 2007 (T_2) in the Alfisol site and in 2004 (T_1) and 2007 (T_2) in the Oxisol site. Small trenches (0.3x0.3x0.15 m, LxWxD) were opened in the plots and samples were collected with a spatula. The thickness of the sampled soil layers (0-0.05 and 0.05-0.15 m) was based on [10,11], considering that SOC stock increment due to changes in soil and crop management practices occurs mainly in the topsoil layer (0-0.05 m). Soil bulk density was determined in undisturbed soil samples collected from the same layers, by using steel rings measuring 0.05x0.04 m (diameter x height) [21]. Soil samples used for C/N analysis were air dried and root and plant residues were manually removed. Then the samples were finely ground in a porcelain mortar. SOC content was determined in soil samples collected at T_1 (2001 and 2004 in the Alfisol and Oxisol sites, respectively) by humid combustion [22,23], and by dry combustion using an elemental C/N analyzer (Flash EA 1112 Series ThermoFinnigan) for samples collected at T_2 (2007 for both sites). The SOC stocks were calculated based in equivalent

Soil	Tillage systems	Management	Carbon input
			Mg ha^{-1}
Oxisol	CT	R_0 [1]	3.54
	CT	R_1 [2]	4.48
	CT	R_2 [3]	5.04
	NT	R_0	3.90
	NT	R_1	4.82
	NT	R_2	6.00
Alfisol	NT	M/JB[4]	4.07
	-	BS[5]	0.00
	NT	M/F[6]	1.88
	NT	M/R[7]	3.76
	NT	M/VB[8]	4.51
	NT	M/FR[9]	4.10

[1] R_0 = succession soybean/wheat;

[2] R_1= wheat/soybean/oat/soybean;

[3] R_2= wheat/soybean/oat/soybean/oat+ vetch/maize/radish;

[4] M/JB= maize + Jack bean/soybean;

[5] BS= bare soil;

[6] M/F= maize/fallow/soybean;

[7] M/R= maize/ryegrass+ vetch/soybean;

[8] M/VB= maize/velvet bean; [9] M/FR= maize/forage radish/soybean.

Table 2. Mean annual carbon input through crop residues in two experimental areas.

soil masses [24], taking as references the treatments M/F and CT R_0 (T_1) for the Alfisol and Oxisol sites, respectively.

2.3. Calculation of carbon stratification ratio (CSR)

The CSR was calculated as proposed by [2], by the ratio of SOC stocks of the superficial soil layer (0- 0.05 m) and SOC stocks of the adjacent soil layer (0.05- 0.15 m) in a given treatment. The temporal variation in CSR (Δ CSR) was obtained by the difference of CSR determined in the second period (T_2) and the CSR in the first (T_1). Soil carbon sequestration rate was determined by the temporal variation of SOC stocks in each treatment in the same period (six and three years for the Alfisol and Oxisol sites, respectively).

2.4. Statistical analysis

The results were submitted to analysis of variance (ANOVA) using the software SISVAR 5.0 [25] and the means were compared by the Tukey test ($p<0.05$). The regression analysis was performed through the software JMP IN version 3.2.1 [26], using the F test ($p<0.05$).

3. Results and discussion

3.1. Carbon input and CSR in Oxisol and Alfisol

Significant linear relationships between carbon input and CSR were observed for both the Oxisol ($p = 0.003$; Figure 1a) and Alfisol ($p = 0.0004$; Figure 1b) sites. These results indicate that increase of carbon input leads to direct increase of CSR, confirming the strong influence of aboveground carbon input for SOC accumulation in the topsoil layers for both soil types.

The average CSR of the Alfisol was 1.66 ± 0.33 (Table 4), which was 21.1% higher than the CSR of the Oxisol (1.31 ± 0.25) (Table 5). The CSRs of these subtropical climate soils were lower than previously reported for temperate soils, where a NT soil from the southeastern U.S. achieved CSR of 3.6 after 12 years [11]. Higher CSR values in temperate climate soils could be related to the less C-oxidative environment in relation to subtropical or tropical climate soils. However, the CSR values observed in our study were close to the results from other Brazilian tropical soils [4,13]. A CSR value of 2.0 was suggested as a lower limit for soil quality in temperate climate regions [2], yet a CSR value of 1.5 would be more appropriate for subtropical or tropical soils [3,4,27]. Another study proposed that declining quality NT soils would result in CSR < 1.0, soils in transition from CT to NT would have CSR between 1.0 and 1.25, a consolidated NT soil would have CSR values ranging from 1.25 to 1.5, and a rising/high quality NT soil would achieve CSR > 2.0 [27]. Comparing our results with that proposed classification, treatments NT R_1 and NT R_2 in the Oxisol, and M/JB, M/VB, M/FR and M/R in the Alfisol, would be considered rising/high quality systems. However, treatments NT R_0 in the Oxisol and M/F in the Alfisol would be classified as consolidated systems, while treatments CT R_0, CT R_1 and CT R_2 in the Oxisol and BS in the Alfisol would be classified as systems under transition.

3.2. CSR and carbon sequestration rates

The relationship between CSR and C stocks (Mg ha^{-1}) was linear and significant for both the Oxisol ($p = 0.0001$; Figure 2a) and Alfisol ($p = 0.0001$; Figure 2b). Similar results were observed in others Oxisols from Brazil [4,28].

The linear and significant relationships observed between Δ CSR (0-0.05: 0.05-0.15 m) and Δ C stock (0-0.05m layer) in the Oxisol ($p = 0.0001$; Figure 3a) and Alfisol ($p = 0.0001$; Figure

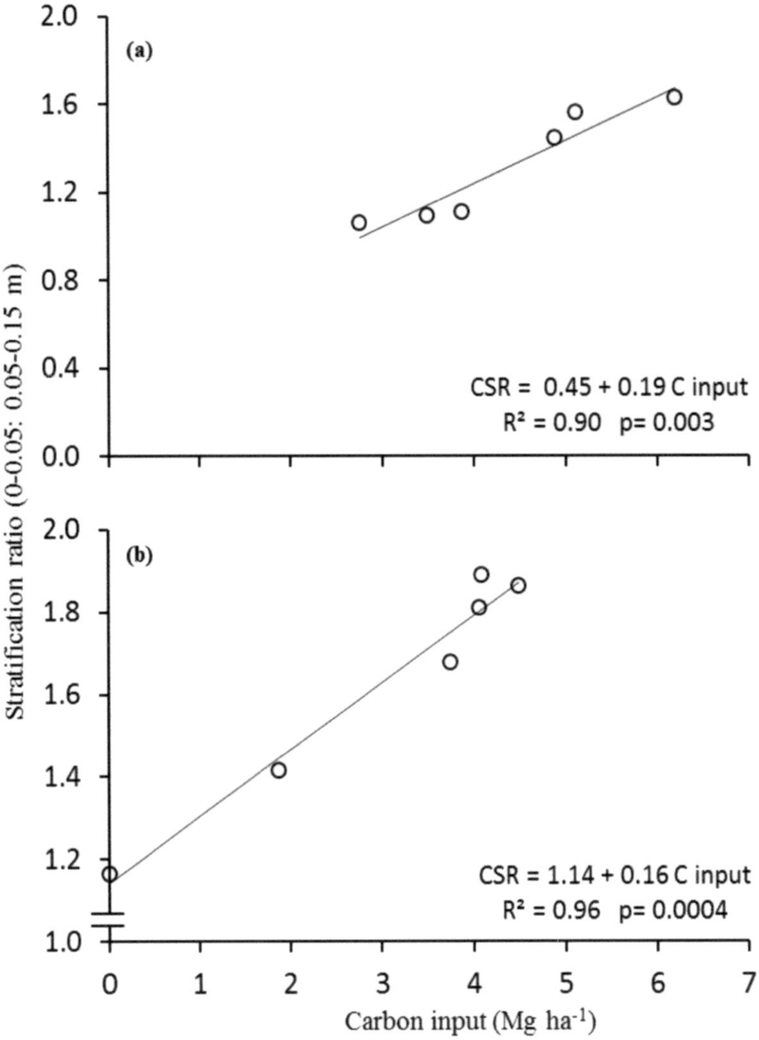

Figure 1. Relationship between annual carbon input and CSR (0-0.05: 0.05-0.15 m) under no tillage. (a) Oxisol. (b) Alfisol. CSR = carbon stratification ratio.

3b), suggest that C sequestration rates are associated with increasing SOC stratification in the soil profile [3,4,28]. Those studies related this effect with the aboveground carbon input by crop residues which are maintained on the soil surface and also with the higher SOC physical protection inside soil macroaggregates under NT. The slope of the adjusted linear equation for the Oxisol was almost twice that for the Alfisol. These results could be associated with the higher clay and Fe and Al oxides content in the Oxisol, favoring C stabilization [29]. The importance of the organo-mineral interaction for C stabilization in Oxisols has been previously reported in the literature [30,31,32]. Also, the linear relationship observed between variables indicates that both soils continue accumulating SOC.

Nevertheless, after more and 20 years, no evidence of SOC saturation was found in any of the evaluated treatments in both soils [33,34].

Both CT and NT treatments had linear relationship between Δ CSR and Δ SOC in the Oxisol site. This significant relationship (p = 0.006) indicates that high carbon input (R_2) could promote SOC accumulation even under CT (Figure 4a), while low carbon inputs leads to depletion of SOC stocks under intensive soil disturbance (R_0 and R_1). This relationship was more pronounced under NT (p = 0.0005) with higher C sequestration rates (Figure 4b). These results confirm the soil as an atmospheric CO_2 sink when conservation agriculture (NT) is associated with high carbon inputs through crop residues [16,35,36,37].

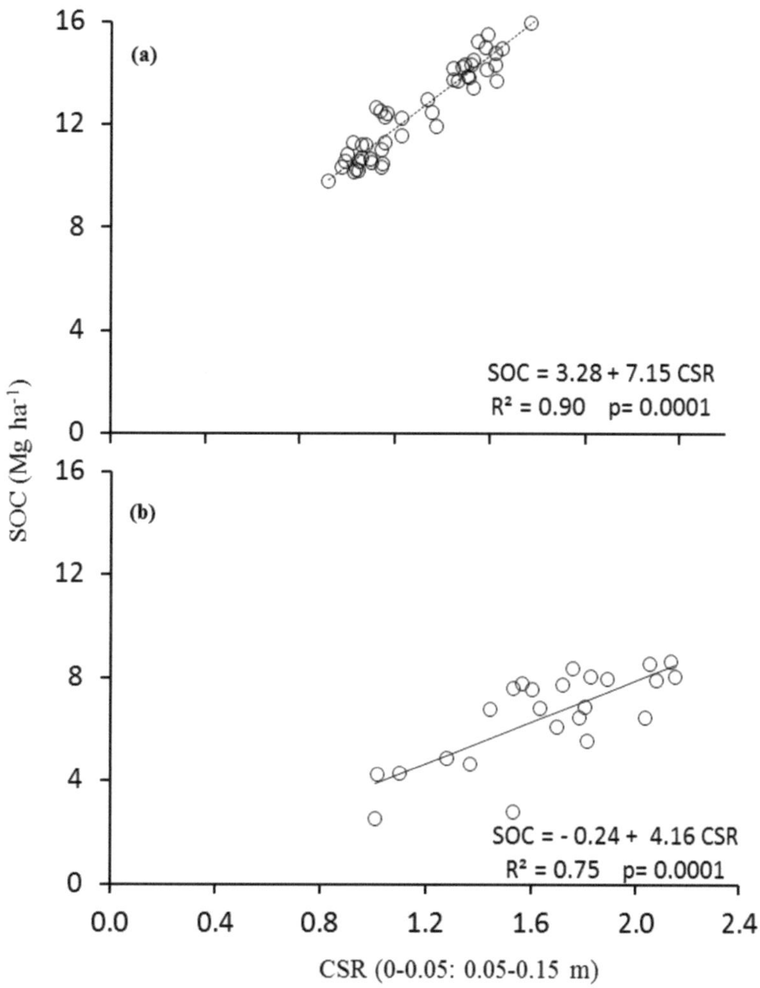

Figure 2. Relationship between CSR (0-0.05: 0.05-0.15 m) and SOC stocks in the 0-0.05 m layer. (a) Oxisol. T_1=2004; T_2= 2007 (interval of 3 years). (b) Alfisol. T_1=2001; T_2= 2007 (interval of 6 years) CSR = stratification relation.

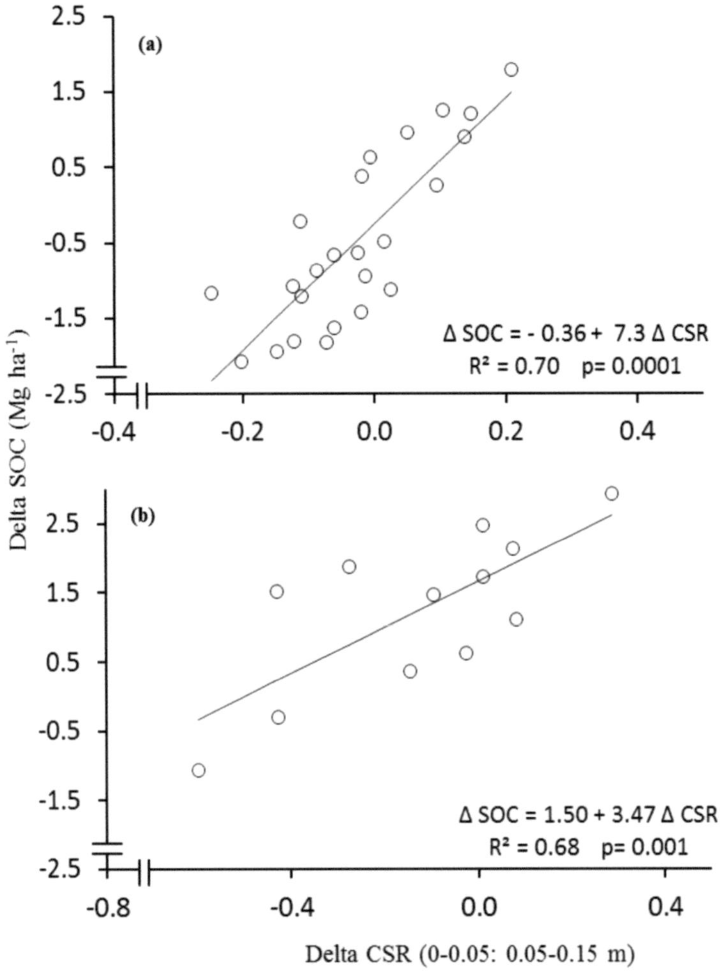

Figure 3. Relationship between the temporal CSR variation ($\Delta CSR = T_2 - T_1$) and temporal SOC stock variation (ΔSOC stock = $T_2 - T_1$) in the 0-0.05 m layer. (a) Oxisol. T_1=2004; T_2= 2007 (interval of 3 years). (b) Alfisol T_1=2001; T_2= 2007 (interval of 6 years). CSR = stratification relation.

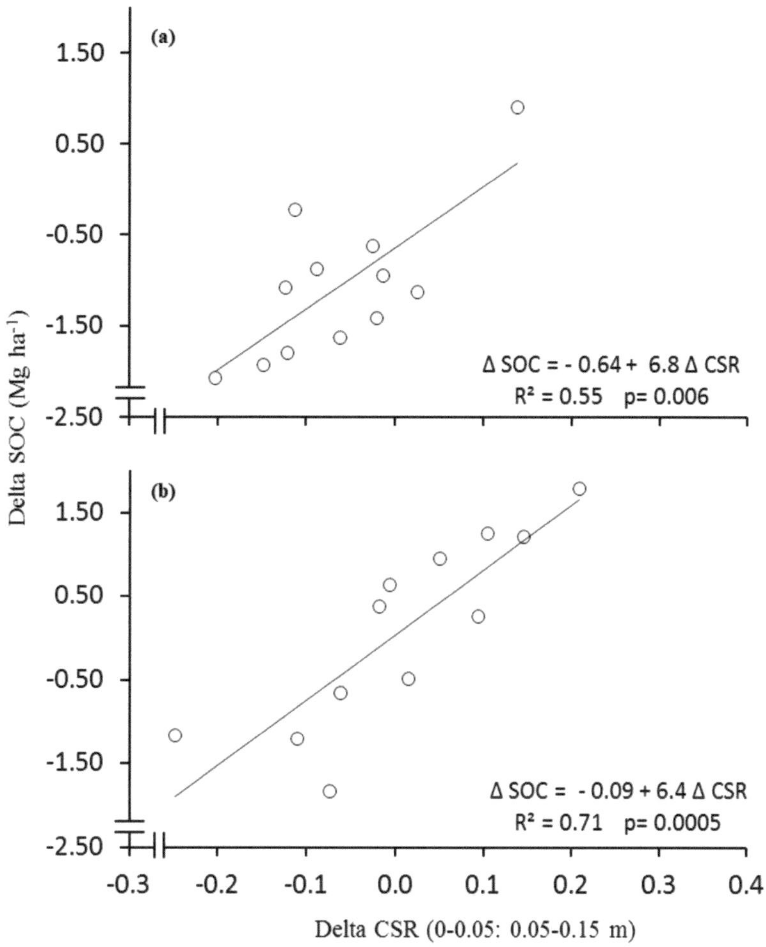

Figure 4. Relationship between the temporal CSR variation ($\Delta CSR = T_2 - T_1$) and temporal SOC stocks variation (Δ SOC stock = $T_2 - T_1$) in the 0-0.05 m layer of the Oxisol. (a) conventional tillage, (b) no tillage. CSR = stratification relation.; T_1= 2004; T_2=2007 (interval of 3 years).

4. Conclusions

The linear relationship between ΔCSR and ΔSOC in the topsoil layer indicates that SOC accumulation is related to carbon stratification in the soil profile. This relationship was more pronounced in the Oxisol than in the Alfisol. Higher CSR values were observed with the association of NT and intensive crop rotation (NT R_2) in the Oxisol. Use of cover crops promoting high carbon and nitrogen input in the soils led to higher CSR values in the Alfisol. Although the CSR values observed in subtropical soils (our study) were lower than those reported for temperate climate soils, this index was efficient for evaluation of carbon sequestration in agricultural soils.

Author details

Ademir de Oliveira Ferreira and Telmo Jorge Carneiro Amado

*Address all correspondence to: aoferreira@mail.ufsm.br

Soil Department, Federal University of Santa Maria (UFSM). Santa Maria-RS, Brazil

References

[1] Prescott C.E., Weetman G.F., Demontigny, L.E., Preston C.M., Keenan R.J. Carbon chemistry and nutrient supply in cedar-hemlock and hemlock-amabilis fir forest floors. In: McFee W.W., Kelley J.M. (eds.). Carbon Forms and Functions in Forest Soils. Madison: American Society of America; 1995. p377-396.

[2] Franzluebbers, A.J. Soil organic matter stratification ratio as an indicator of soil quality. Soil and Tillage Research 2002; 66(2) 95-106.

[3] Sá, J.C.M. & Lal, R. Stratification ratio of soil organic matter pools as an indicator of carbon sequestration in a tillage chronosequence on a Brazilian Oxisol. Soil and Tillage Research 2009; 103(1) 46-56.

[4] Ferreira, A.O., Sá, J.C.M., Harms, M.G., Miara, S., Briedis, C., Quadros Netto, C., Santos, J.B., Canalli, L.B.S., Dias, C.T.S. Stratification ratio as soil carbon sequestration indicator in macroaggregates of Oxisol under no-tillage. Ciência Rural 2012; 42(4) 645-652.

[5] Sá, J.C.M., Cerri, C.C., Dick, A.W., Lal, R., Venske Filho, S.P., Piccolo, M.C. & Feigl, B.E. Soil organic matter dynamics and sequestration rates for a tillage chronosequence in a Brazilian Oxisol. Soil Science Society of America Journal 2001; 65(5) 1476-1499.

[6] Bayer C., Amado T.J.C., Tornquist C.G., Cerri C.E.P., Dieckow J., Zanatta J.A., Nicoloso R.S., Carvalho P.C.F. Estabilização do carbono no solo e mitigação das emissões de gases de efeito estufa na agricultura conservacionista. In: Klauberg Filho, O., Mafra A.L., Gatiboni L.C. (ed.) Tópicos em ciência do solo. Viçosa: Sociedade Brasileira de Ciência do Solo, 2011. V.VII, cap.2, p55-118.

[7] Braida, J. A.; Reichert, J. M.; Reinert, D. J. & Veiga M. Teor de carbono orgânico e a susceptibilidade à compactação de um Nitossolo e um Argissolo. Revista Brasileira de Engenharia Agrícola e Ambiental 2010; 14(2) 131-139.

[8] Conceição, P.C., Amado, T.J.C., Mielniczuk, J. & Spagnollo, E. Qualidade do solo em sistemas de manejo avaliada pela dinâmica da matéria orgânica e atributos relacionados. Revista Brasileira de Ciência do Solo 2005; 29(5) 777-788.

[9] Moreno, F., Murillo, J.M., Pelegrín, F. & Girón, I.F. Long-term impact of conservation tillage on stratification ratio of soil organic carbon and loss of total and active $CaCO_3$. Soil and Tillage Research 2006; 85(1-2) 86-93.

[10] Causarano, H.J., Franzluebbers, A.J., Shaw, J.N., Reeves, D.W., Raper, R.L. & Wood, C.W. Soil organic carbon fractions and aggregation in the Southern Piedmont and Coastal Plain. Soil Science Society of America Journal 2008; 72(1) 221-230.

[11] Franzluebbers, A.J. Achieving soil organic carbon sequestration with conservation agricultural systems in the southeastern USA. Soil Science Society of America Journal 2010; 74(2) 347-357.

[12] Spargo, J.T., Alley, M.M., Follett, R.F. & Wallace, J.V. Soil carbon sequestration with continuous no-till management of grain cropping systems in the Virginia Coastal Plain. Soil and Tillage Research 2008; 100(1)133-140.

[13] Tormena, C.A., Friedrich, R., Pintro, J.C., Costa, A.C.S. & Fidalski, J. Propriedades físicas e taxa de estratificação de carbono orgânico num Latossolo Vermelho após dez anos sob dois sistemas de manejo. Revista Brasileira de Ciência do Solo 2004; 28(6)1023-1031.

[14] Moreno, J.A. Clima do Rio Grande do Sul. Porto Alegre, Secretaria da Agricultura, Diretoria de Terras e Colonização, Seção de Geografia, 1961. 46p.

[15] Campos BC. Carbon dynamics on a Rhodic Hapludox on soil tillage and crop systems. Doctoral Thesis. Federal University of Santa Maria; 2006.

[16] Amado, T.J., Bayer, C., Conceição, P.C., Spagnollo, E., Campos, B.C., Veiga, M. Potential of C accumulation in no-till soils with intensive use and cover crops in Brazil. Journal of Environmental Quality 2006; 35(4) 1599-1607.

[17] Lanzanova, M.E., Eltz, F.L.F., Nicoloso, R.S., Amado, T.J.C., Reinert, D.J. & Rocha, M.R. Atributos físicos de um argissolo em sistemas de culturas de longa duração sob semeadura direta. Revista Brasileira de Ciência do Solo 2010; 34(4) 1333-1342.

[18] Boddey, R.M., Jantalia, C.P., Conceição, P.C., Zanatta, J.A., Bayer, C., Mielniczuk, J., Dieckow, J., Santos, H.P., Denardin, J.E., Aita, C., Giacomini, S.J., Alves, B.J.R. & Urquiaga, S. Carbon accumulation at depth in Ferralsols under zero-till subtropical agriculture. Global Change Biology 2010; 16(2) 784-795.

[19] Campos, B.H., Amado, T.J.C., Bayer, C., Nicoloso, R.S., Fiorin, J. Carbon stock and its compartments in a subtropical Oxisol under long-term tillage and crop rotation systems. Revista Brasileira de Ciência do Solo 2011; 35(3) 805-817.

[20] Debarba, L. & Amado, T.J.C. Desenvolvimento de sistemas de produção de milho no sul do Brasil com características de sustentabilidade. Revista Brasileira de Ciência do Solo 1997; 21(3) 473-480.

[21] Empresa Brasileira de Pesquisa Agropecuária - Embrapa. Centro Nacional de Pesquisa de Solos. Manual de métodos de análise de solos. 2 ed. Rio de Janeiro, Embrapa Solos, p. 212, 1997.

[22] Nelson D.W. & Sommers L.E. Total carbon, organic carbon and organic matter. In: Sparks D.L., Page A.L; Helmke, P.A. & Loeppert R.H. (eds.) Methods of soil analysis: chemical methods. Part 3. Madison: American Society of America, 1996. p961-1010.

[23] Rheinheimer, D.S., Campos, B.C., Giacomini, S., Conceição, P.C. & Bortoluzzi, E.C. Comparação de métodos de determinação de carbono orgânico total no solo. Revista Brasileira de Ciência do Solo 2008; 32(1) 435-440.

[24] Ellert, B.H. & Bettany, J.R. Calculation of organic matter and nutrients stored in soils under contrasting management regimes. Canadian Journal of Soil Science 1995; 75(4) 529-538.

[25] Ferreira, D. F. Sisvar versão 5.3 (Biud 75). Sistemas de análises de variância para dados balanceados: programa de análises estatísticas e planejamento de experimentos. Lavras, MG, Universidade Federal de Lavras, 2010.

[26] Sall, J., Creighton L., Lehman A. JMP start statistics: a guide to statistics and data analysis using JMP and JMP IN software. 3rd ed. Cary: Duxbury Press; 2005. 580p.

[27] Canalli L.B.S, Sá, J.C.M., Santos J.B., Ferreira A.O., Briedis C. & Tivet F.E. Proposta de um protocolo para a validação e certificação da qualidade do SPDP relacionada a redução das emissões de CO_2. In: 12° Encontro Nacional de Plantio Direto na Palha - Tecnologia que mudou a visão do produtor, 23-25 de junho de 2010. Ponta Grosa: Federação Brasileira de Plantio Direto na Palha (FEBRAPDP); 2010. p85-94.

[28] Sá J.C.M., Seguy, L., Sa, M.F.M., Ferreira, A.O., Briedis, C., Santos, J.B. & Canalli L.B.S. Gestão da matéria orgânica e da fertilidade do solo visando sistemas sustentáveis de produção In: Prochnow L.I., Casarin V. & Stipp S.R. (eds.) Boas Práticas para Uso Eficiente de Fertilizantes. Piracicaba, SP: International Plant Nutrition Institute - Brasil (IPNI); 2010. p383-420.

[29] Denef, K. & Six, J. Clay mineralogy determines the importance of biological versus abiotic processes for macroaggregates formation and stabilization. European Journal of Soil Science 2005; 56(4) 469-479.

[30] Zinn, Y.L., Lal, R., Bigham, J.M. & Resck, D.V.S. Edaphic controls on soil organic carbon retention in the Brazilian cerrado: soil structure. Soil Science Society of America Journal 2007; 71(4) 1.215-1.224.

[31] Zotarelli, L., Alves, B.J.R., Urquiaga, S., Boddey, R.M. & Six, J. Impact of tillage and crop rotation on light fraction and intra-aggregate soil organic matter in two Oxisols. Soil and Tillage Research 2007; 95(1-2) 196-206.

[32] Fabrizzi, K.P., Rice, C.W., Amado, T.J.C., Fiorin, J., Barbagelata, P., Melchiori, R. Protection of soil organic C and N in temperate and tropical soils: effect of native and agroecosystems. Biogeochemistry 2009; 92(1-2) 129-143.

[33] Six, J., Feller, C., Denef, K., Ogle, S.M., Sá, J.C.M. & Albrecht, A. Soil organic matter, biota and aggregation in temperate and tropical soils - effects of no-tillage. Agronomie 2002; 22(7-8) 755-775.

[34] Stweart, C.E., Paustian, K., Conant, R.T., Plante, A.F., Six, J. Soil carbon saturation: concept, evidence, and evaluation. Biogeochemistry 2007; 86(1) 19-31.

[35] Bayer, C., Martin-Neto, L., Mielniczuk, J., Pavinato, A. & Dieckow, J. Carbon sequestration in two Brazilian Cerrado soils under no-till. Soil and Tillage Research 2006; 86(2) 237-245.

[36] Cerri, C.E.P., Sparovek, G., Bernoux, M., Easterling, W.E., Melillo, J.M. & Cerri, C.C. Tropical agriculture and global warming: impacts and mitigations options. Scientia Agricola 2007; 64(1) 83-99.

[37] Ferreira, A.O., Sá, J.C.M., Harms, M.G., Miara, S., Briedis, C., Quadros Netto, C., Santos, J.B., Canalli, L.B.S. Carbon balance and crop residue management in dynamic equilibrium under a no-till system in Campos Gerais. Revista Brasileira de Ciência do Solo 2012; 36(5) 1583-1590.

Chapter 7

The Classification Indices-Based Model for NPP According to the Integrated Orderly Classification System of Grassland and Its Application

Huilong Lin

Additional information is available at the end of the chapter

http://dx.doi.org/10.5772/57297

1. Introduction

Grassland accounts for 25% of the world's land area and has global significance for climate-carbon feedback [1]. As one of the most widespread ecosystem types, grassland makes a large contribution to carbon sequestration and plays a significant role in regional climates and the global carbon cycle [2]. Grassland net primary productivity (NPP) refers to carbon accumulation (photosynthesis less respiration) in grassland per unit of time, and can be used as an indicator of the capacity of the grassland ecosystem to accumulate carbon and to support grazing animals. Studies of the NPP have contributed significantly to our understanding of the pattern, process and dynamics of grassland ecosystems. Hence, the accurate estimation of the NPP of grasslands is very important not only for scientifically guiding grassland management but also for understanding the global carbon cycle. Reliable long-term data on grassland NPP are urgently needed to estimate carbon fluxes. Due to their time- and labor-intensive nature, ground measurements of grassland NPP are very limited in temporal and spatial coverage. The precision evaluation of NPP estimated in regional or global scales is always very difficult. Direct measurement of grassland NPP is tedious and not practical for large areas, so it is therefore appropriate to use computer models, calibrated with existing data, to study the spatial and temporal variations of grassland NPP [3].

A number of different methods of modelling the dynamics of grassland NPP have been reported. In climate-vegetation models, empirical relationships are developed, allowing NPP to be estimated as a function of climate variables such as temperature, precipitation, and evaporation [4-12]. Process models for estimating NPP simulate a series of plant physiological processes and their response to changes in environmental factors, including photosynthesis, autotrophic respiration and transpiration [13, 14]. In spite of this, the body of knowledge of basic biological mechanisms to simulate grassland NPP is still limited. Furthermore, process models cannot be used in developing countries because long-term data required for the parameters of the models is not available. Algorithms for estimating NPP by remote sensing may be poorly parameterized if the 'calibration' estimates of NPP are based upon an inappropriate method. In this respect, climate-vegetation models have been drawing much attention and have been widely applied internationally, owing to their simplicity [4]. These models have been shown to yield 'reasonable estimates' of regional or global NPP distribution [4, 9, 14-16].

Vegetative types and their distribution pattern can be related to certain climatic types in a series of mathematical forms. Thus, the climate can be used to predict vegetative types and their distribution, or the reverse [17]. A growing number of research efforts have demonstrated the importance of climate-vegetation interaction in understanding climate sensitivity and climate change [18]. However, current climate-vegetation classification models mostly simulate the equilibrium state of vegetation types, and do not include NPP [19]. Furthermore, at present there is no vegetation-climate model based on a classification system including the NPP function. Therefore, development of climate-vegetation classification models coupled with NPP is urgently needed in order to evaluate the possible impacts of global change on terrestrial vegetation types [19]. So the focus of future research should be to master the link between NPP and grassland class, and to develop pertinent NPP models under a consistent classification system.

2. The classification indices-based model according to the Integrated Orderly Classification System of Grassland (IOCSG)

A grassland classification system, named the Integrated Orderly Classification System of Grassland (IOCSG), uses the factors Growing Degree Days (GDD), MAP (mean annual precipitation) and the moisture index (K value) to classify grassland diversity [17]. The theory behind the IOCSG has been developed in the last 60 years since it was first put forward in the 1950s, and it has achieved widespread use in China [17, 20]. Liang et al. [20] compared the change in potential vegetation distributions from 1911 to 2000 between the IOCSG, the Holdridge Life Zone [5] and BIOME4 [21]. Their results show that the IOCSG has the advantage of being simple and operational in its simulation of grassland classes, making it by far the best grassland classification system.

The Classification Indices-based Model, dubbed the Holdridge life-zone system [5], and the IOCSG [17] were originally built using eco-physiological features and a regional evapotranspiration model with the elimination of the common variable RDI (radioactive dryness index) by the chain rule. It results in a value for NPP (Mg DM ha^{-1} yr^{-1}) as a function of GDD and the moisture index (K value). Its ecological base is the IOCSG [17, 20, 22-25]. The method of integrating the classification indices of IOCSG to estimate the NPP is of the form:

$$NPP = \frac{0.1GDD \cdot L^{2}(K) \cdot e^{-\sqrt{13.55+3.17K^{-1}-0.16K^{-2}+0.0032K^{-3}}} \cdot (K^{6}+L(K)K^{3}+L^{2}(K))}{[(K^{6}+L^{2}(K)) \cdot (K^{5}+L(K)K^{2})]}$$

Where GDD is defined as the mean of positive unit-period temperatures with the substitution of zero for all unit-period values below 0 °C. The K value is the ratio of MAP to GDD, which provides an index of biological humidity conditions.

$$K = MAP / 0.1GDD, \; L(K) = 0.58802K^{3} + 0.50698K^{2} - 0.0257081K + 0.0005163874$$

The Classification Indices-based Model not only has the advantage of being simple, operational, and especially compatible with the IOCSG used to derive NPP from its class, it also responds to dynamic vegetation-climate relationships.

3. Model validation and model inter-comparison

Model inter-comparison has been used as an alternative method for indirect validation and identification of model weaknesses and inconsistencies where ground observations are lacking [26]. However, a simultaneous comparison of these climate-vegetation models to estimate grassland NPP, using a consistent classification system, has not been attempted on a regional, China-wide or global scale.

The following three performance indicators were used as the model evaluation and performance criteria: (1) the mean bias error (MBE) [24,27,28]; (2) the coefficient of variation of the root-mean-square error (CVRMSE), which is defined as the RMSE normalized to the mean of the observed values; and (3) the forecast efficiency, E [29]. The prediction is considered excellent with the CVRMSE < 10%, good if 10-20%, fair if 20-30%, poor if >30% [30]. CVRMSE is the relative difference between the simulated and observed data, while the Nash-Sutcliffe model efficiency statistic (E) [29], regarded as a measure of the overall fit between observed and predicted values, is the primary criterion.

For the purpose of comparison with model simulations, the observed or predicted NPP was expressed in g C m^{-2} per year, where 1 g carbon is equivalent to 2.2 g oven-dried organic matter [31-34].

3.1. Evaluation of seven methods to predict grassland net primary productivity along an altitudinal gradient in the Alxa Rangeland, Western Inner Mongolia, China

China's grasslands are mainly distributed in arid and semi-arid areas of West China [35]. The main grassland classes within China are Temperate meadow-steppe, Steppe and Desert-steppe [35, 36]. Helan Mountain (elevation 3556 m), located in an ecologically vulnerable area of the northwestern part of China, further enriches the diversity of vertical grassland classes. The rangeland classes change from the high to low elevation: Alpine meadow, Steppe, Desert steppe, Steppe desert, and Desert. Thus, Helan Mountain provides an ideal target system to validate the Classification Indices-based Model and test IOCSG theory, which can be done from plots to estimate landscape-scale effects, something that so far has proven to be a challenge to ecologists worldwide [37].

This study continues our efforts to develop a robust estimation capability of NPP for grassland and, specifically, verifies the grassland NPP estimates from the climate-vegetation models along an altitudinal gradient on Helan Mountain and in the surrounding desert rangeland. The specific research objectives are to: 1) compare field-measured NPP based on field-observed data among different grassland classes along an altitudinal gradient in the Helan Mountain range and the surrounding desert; 2) compare and evaluate NPP estimated from the Miami Model [6], Thornthwaite Memorial Model [38], Schuur Model [15], Chikugo Model [8], Beijing Model [12], Synthetic Model [10,11,39], and the Classification Indices-based Model [22,24,25,27,28] with NPP derived from measurements at eight sites in the region; and 3) evaluate the applicability and reliability of the seven models among these grassland classes.

3.1.1. Study area and methods

3.1.1.1. Study area and site descriptions

The study area is located along the northern slopes of the Helan Mountain and the surrounding Alxa Desert (long. 105°32′ to 105°51′, lat. 38°39′ to 39°19′; alt. 3556 m to1360 m a.s.l.), in western Inner Mongolia (Figure 1). Eight study sites were selected following annual rainfall gradient variations with respect to grassland classes at three altitudes that represent the full range of grassland vegetation, environmental, terrain and soil conditions (Figure 1). Based on the historical meteorological data among the eight study sites, GDD and the moisture index (K value) were calculated. According to the IOCSG, the grasslands at the eight sites were classified into five classes: Alpine meadow (site 1), Cold temperate-humid montane meadow (site 2), Cool temperate-sub-humid meadow steppe (site 3), Cool temperate-semiarid temperate typical steppe (site 4) and Cool temperate-arid temperate zonal semi-desert (sites 5-8), respectively (Figure 1).

3.1.1.2. Plots-based NPP collection

The total observed NPP dataset was the sum of above- and below-ground NPP, following standard methods [3,40,41], and it was based on averages of five duplicated plots at each sampling site from April to October once every month for three consecutive years (2003-2005). One hundred and twenty plot-based NPP sets from the eight study sites were obtained [25].

3.1.2. Results

3.1.2.1. Descriptive statistics of observed NPP

The gradient distribution of grassland classes varied in different NPP values among the study sites. With all data pooled together, the observed NPP values of the Cool temperate-arid temperate zonal semi-desert has an extreme degree of variation, ranging from 72.60 to 242.02 g C m^{-2} per year, with a coefficient of variation of 28.38%. The mean NPP of the

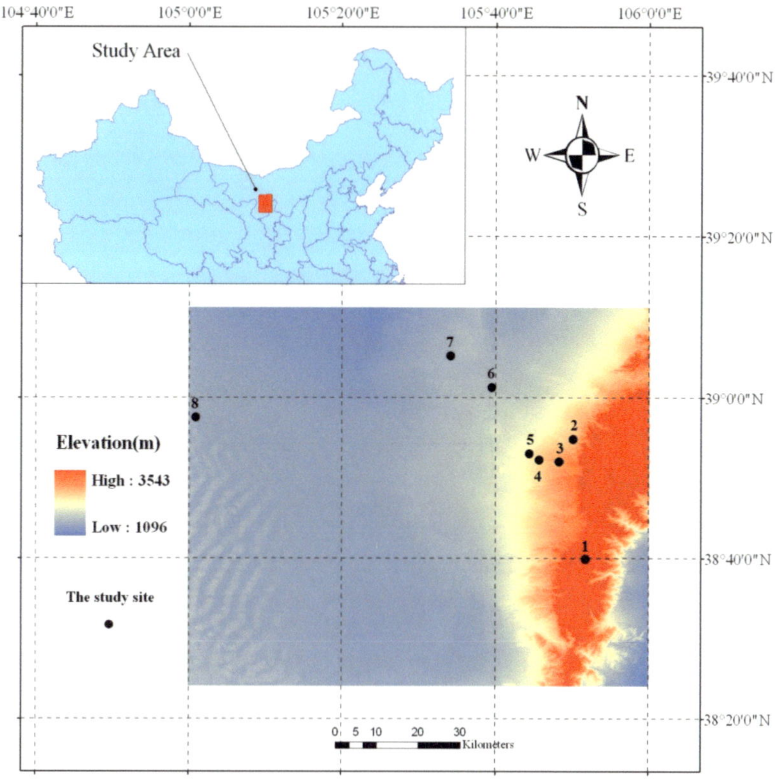

Figure 1. Location map of study area and study sites. Dark circles represent the study sites in elevation. Arabic numerals denote the study site

Cool temperate-arid temperate zonal semi-desert was the lowest (133.43 g C m^{-2} per year), followed by that of Cool temperate-semiarid temperate typical steppe (160.57 g C m^{-2} per year), Cool temperate-subhumid meadow steppe (171.94 g C m^{-2} per year) and Alpine meadow (185.88 g C m^{-2} per year). The mean NPP of Cold temperate-humid montane meadow was the largest (192.54 g C m^{-2} per year).

3.1.2.2. Comparison of the seven models in terms model performance values

The performance of each of the seven models compared and evaluated in this study is also shown graphically by plotting predicted NPP values as a function of measured values, as presented in Figure 2. The biases that each model introduces are reflected at the relative positions of the 1:1 line. Data points above the 1:1 line are over-predicted while those under the 1:1 line are under-predicted. It can be seen that the Thornthwaite Memorial Model, Beijing Model and Miami Model over-predicted grassland NPP while the Synthetic Model and Schuur Model slightly under-predicted those quantities. Considering the slopes and intercepts for the regression of predicted NPP by the Chikugo Model and Classification Indices-based Model versus observed NPP, the regression of the Chikugo Model has a slope (0.9878) approximately equal to that (0.9873) of the Classification Indices-based Model, but the intercept (11.13) of the regression line of the Classification Indices-based Model is much closer to zero than that (17.13) of the regression line of the Chikugo Model.

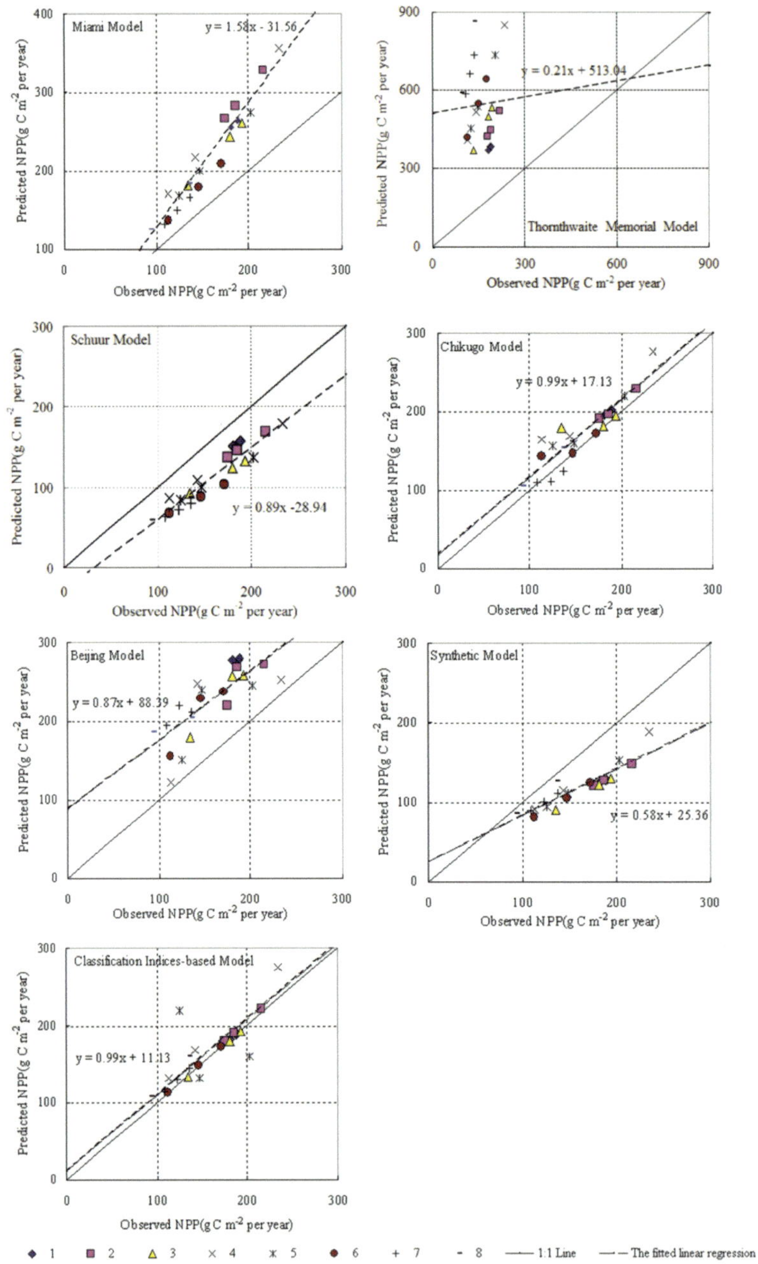

Figure 2. Predicted net primary productivity (NPP) versus Observed NPP from seven different models compared in this study. Each data point represents average of five replicates. Arabic numeral denotes the study site No..

Accordingly, the high values of E (0.67 and 0.87) in predicting NPP using the Chikugo Model and the Classification Indices-based Model indicate that these models outperform the others. The fitted linear regressions (intercepts are forced to zero) between the observed NPP and the methods are: the predicted NPP by Chikugo Model=1.09×Observed NPP, the predicted NPP by Classification Indices-based Model=1.05×Observed NPP, respectively. The results indicate that predicted NPP by the Classification Indices-based Model has a closer correlation with the observed NPP compared to the predicted NPP by the Chikugo Model. With E value of -0.27, -0.45, -1.79 and -2.61 for predicting NPP, respectively, the Synthetic Model, Schuur Model, Miami Model and Beijing Model ranked third, fourth, fifth and sixth. With E value of-113.16, the Thornthwaite Memorial Model was the least suitable. Overall, comparison of the predicted with the observed grassland NPP indicated that the CVRMSE value using the Classification Indices-based Model was less than 10%, which denotes that the prediction suitability of the Classification Indices-based Model is considered excellent, while the prediction of the Chikugo Model is considered good with CVRMSE=14.29%. As the CVRMSE value using the Synthetic Model is 28.19%, the prediction of the model is considered fair. Because of the CVRMSE values beyond 30%, the prediction using the other models is considered poor (Table 1).

Model	CVRMSE(%)	E	Ranked
Miami Model	41.84	-1.79	5
Thornthwaite Memorial Model	267.59	-113.16	7
Schuur Model	30.14	-0.45	4
Chikugo Model	14.29	0.67	2
Beijing Model	47.57	-2.61	6
Synthetic Model	28.19	-0.27	3
Classification Indices-based Model	8.98	0.87	1

Table 1. Comparison of the seven models in terms model performance values

3.1.3. Discussion

Seven different methods were compared in terms of their performance in predicting grassland NPP. The relatively high model efficiency in predicting grassland NPP using the Classification Indices-based Model and Chikugo Model indicates that these models outperform the others. The comparison of models involves the input data requirements from each model into account, as these factors affect the value of model efficiency (E), if they are optimized to maximize E. Generally, models with a large number of input variables and parameters would result in higher values of E than models with a small number of input variables and parameters. When comparing the Chikugo Model with the Classification Indices-based Model, it is evident that the Chikugo Model is more complicated in structure and highly intensive in its data requirements. The Chikugo Model provides insight into the relative importance of individual environmental variables in the evapotranspiration process [7, 11, 42, 43]. The model is a mechanistic approach to estimate grassland NPP and always slightly overestimated the actual NPP. Furthermore, the Classification Indices-based Model is based on the IOCSG and used the classification indices as independent variables. It not only takes into account dynamic grassland classes [17], but also simulates NPP of corresponding grassland classes [22, 24, 25, 27, 28]. Thus it can be very easy to implement and understand. These factors support the Classification Indices-based model as a more accurate prediction among grassland classes in this study. So the Classification Indices-based Model was found to be the best choice for the given grassland classes. The model is especially suitable for grassland NPP research and application in many developing/undeveloped countries or regions that generally lack the detailed and complex observation data required by other models (i.e., BIOME4).

The results presented in this study were not only specific to this region, but more importantly, were specific to the given grassland classes according to the IOCSG approach, which can be scaled up from plots to estimate landscape-scale effects. The ascending order of grassland NPP and its class may have significant implications for grassland succession in Chinese grassland ecosystems with predicted changes in spatio-temporal patterns of precipitation under the influence of global climate change. So there is a need to extrapolate in order to calculate the grassland's response to climate change (for instance changing temperature and rainfall). Continued climate change is expected to result in changes in temperature and precipitation, the same as the changes along the altitudinal gradient in this study area. The changing climate will in turn affect the growth of plants, and result in changes to grassland classes and subsequently NPP. The combined affect of precipitation and temperature, that is, the moisture index (K), is to some extent most important to NPP spatial distribution. Thus, the Classification Indices-based Model has the potential to evaluate the possible effects of climate change on grassland classes and their NPP in the future, in order to improve the accuracy of NPP prediction and reduce the evaluation uncertainty of the possible effects of climate change in the future.

3.2. Model validation and modelling the potential net primary productivity of grassland in China

China is located in the southeast monsoon climate of the Eurasian continent, and covers a vast territory, with complicated terrain and distinctive distribution patterns of temperature and precipitation with different climatic zones, ranging from cold temperate to tropical and from moist to extreme drought. In addition, the ascent of the Qinghai-Tibetan Plateau(the so-called third polar region of the world) leads to a special plateau climate and thus forms a special pattern of vegetation distribution, termed as 'plateau zonality of vegetation' to distinguish horizontal zonality and vertical zonality of vegetation. The vegetation distribution of China can be characterized by the latitudinal zonality, longitudinal zonality, vertical zonality, and plateau zonality [44]. Thus, China's climate is unique and forms a special pattern of vegetation distribution. Grasslands account for 41.7% of land area in China [17]. Traditionally, China's grasslands have been economically productive areas and have provided important natural resources for the nation, including meat, milk, wool and animal hides [35]. Grassland NPP can be used as an indicator of the capacity of the grassland ecosystem to accumulate carbon and to support grazing animals. The ability to accurately estimate grassland NPP is critical to our understanding of grassland dynamics [45]. But evaluation of the precision of grassland NPP estimates in China is difficult. Since it is a developing country, some mechanistic models cannot be used because long-term data required for the parameters of these models are not available in China [16]. But the required data are available for NPP-climate models, and these models have been shown to yield 'reasonable estimates' of global patterns of productivity [4, 9, 14, 16]. A growing number of research efforts [18] are focused on developing NPP-climate relationships. The Miami Model [38], the Schuur Model [15] and the Classification Indices-based Model [22, 24, 25, 27, 28] are examples of such efforts. But a simultaneous comparison using a consistent classification system has not been attempted in China.

The purpose of this section is: 1) to compared NPP estimated using the Miami Model [38], Schuur Model [15], and the Classification Indices-based Model [22,24,25,27,28] with NPP derived from measurements at 3767 sites in China, to evaluate the applicability and reliability of the Classification Indices-based Model; and 2) to simulate the spatial distribution patterns and its NPP characteristics of China's potential grassland under current climate scenarios using the IOCSG approach and NPP-climate models.

3.2.1. Data acquisition and methods

3.2.1.1. Observed NPP database

A large reference data set (n = 3767) of observations of grassland NPP with paired climatic variables was compiled for this study. This data set came from surveys conducted by the National Animal Husbandry and Veterinary Service of the Ministry of Agriculture of the People's Republic of China from 2004 to 2005. Total observed NPP dataset from the 3767 sites was the sum of above- and belowground NPP following the standard methods [3, 40] and it was based on the average of five duplicated plots at each sampling site. Locations of 3767 sampling sites were plotted using their associated geographic coordinates, as shown in Figure 3.

3.2.1.2. Climate data

Meteorological data for annual rainfall and GDD at 2348 meteorological stations, including longitude, latitude and altitude data of each meteorological station, were obtained from a database of the China Meteorological Data Sharing Service System from 1961 to 1990 (downloaded from the website: http://cdc.cma.gov.cn). The locations of the 2348 meteorological stations are shown in Figure 4. The Digital Elevation Model (DEM) was derived from the global 1km×1km DEM issued by the U.S. Geological Survey (USGS) (downloaded from the website: http://edcdaac.usgs.gov/gtopo30/ README. asp#h17). Regional data were derived from a database of 1:4 million Chinese regional data points, which was last updated in May 2004, and were used to confirm the Chinese boundary.

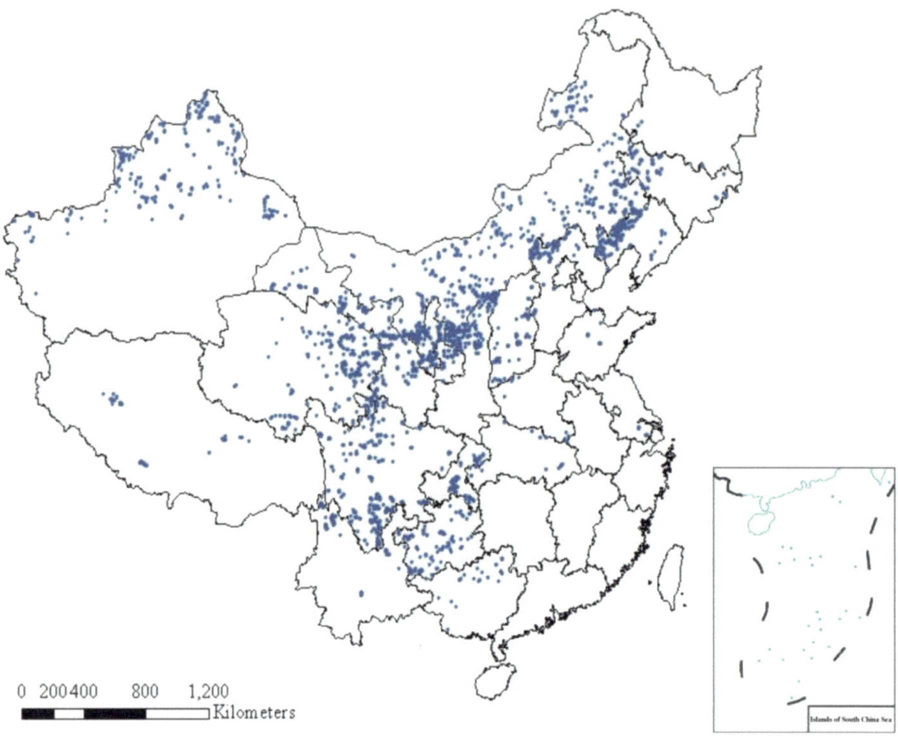

Figure 3. Locations of observation sites of grassland NPP in China.

3.2.1.3. Estimating grassland NPP and its distribution patterns in China

Spatialization of GDD and K value was created by introduction into the interpolation analysis of meteorological elements. The 1km×1km grid data of the moisture index averaged over years was obtained by the Kriging interpolation method (Figure 5). The inverse distance squared method (IDS) [46], modified by DEM in the smaller regions, was used to spatialize GDD data and to improve the interpolation precision of GDD in China [47]. By mosaicking the grid data of GDD on the actual ground of eight smaller areas and clipping the integrated interpolation area by the Chinese regionalism, we finally obtained the 1km×1km grid data of GDD (Figure 6). The 1km×1km grid data of the moisture index and GDD were used to predict the potential classes recognized by the IOCSG (Figure 7), and were then used to drive the Miami Model, Schuur Model, and the Classification Indices-based Model, respectively. To more explicitly reflect the spatial distribution patterns of potential biomes in China, and for ease of comparison and application, we merged the 42 classes of IOCSG to Grassland and Forest categories. The Grassland category consists of seven grassland super-class groups (biomes): Tundra and alpine steppe, Frigid desert, Semi-desert, Steppe, Temperate humid grassland, Warm desert and Savanna. The Forest category includes Temperate forest, Sub-tropical forest and Tropical forest (Table 2). The summed NPP has a unit of g C for pixels or Tg C (1 Tg = 10^{12}g) for each grassland super-class group (biome) or Pg C (1 Pg = 10^{15}g) for Grassland or Forest category in China.

http://dx.doi.org/10.5772/57297

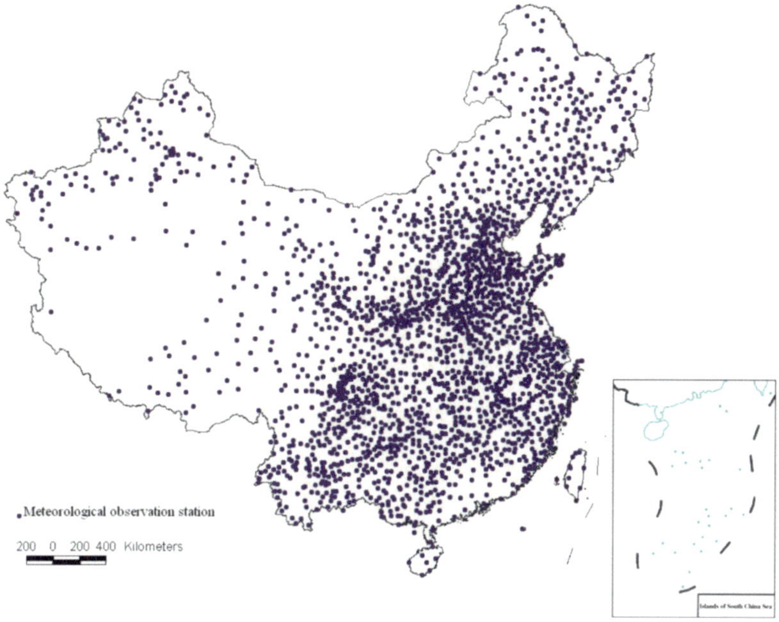

Figure 4. Spatial distribution of 2348 meteorological observation stations used for plotting spatial distribution of GDD and K value

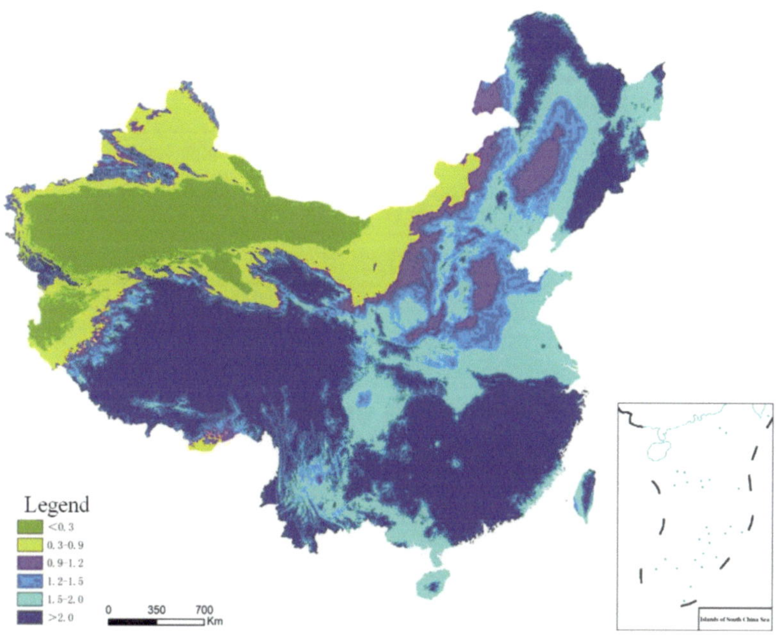

Figure 5. 1km x 1km grid map of China showing the moisture index (K value) obtained by Kriging interpolation. The legends show the zones of K value.

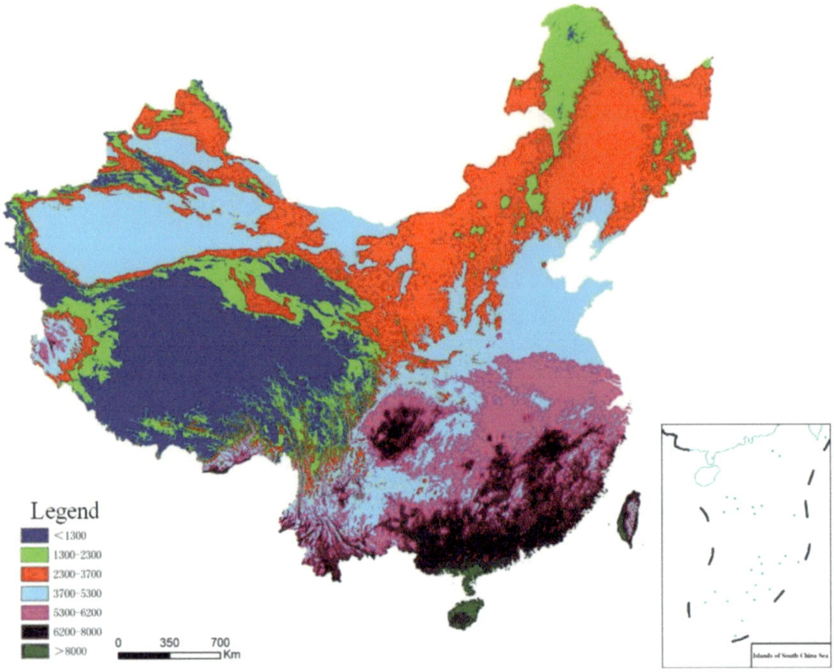

Figure 6. 1km x 1km grid map of China showing spatial distribution of GDD by DEM+ IDS interpolation method. DEM, the Digital Elevation Model; IDS, the inverse distance squared method. The legends show the zones of GDD.

Name of super-class group (biome)	Corresponding class code[a]
Tundra and alpine steppe	I A 1,I B 8, I C 15,I D 22, I E 29, I F 36
Frigid desert	II A 2, III A 3, IV A 4
Semi-desert	II B 9, III B 10, IV B 11, V B 12
Steppe	16IIC,17IIIC,18IV,19VC,25IVD
Temperate humid grassland	IID 23, IIID 24, IIE30
Warm desert	VA 5, VIA 6, VIIA 7
Savanna	VIB 13, VIIB 14, VIC 20, VIIC 21
Forest, including Temperate forest, Sub-tropical forest and Tropical forest	VD 26, VID 27, VIID 28, IIIE 31, IVE 32, VE 33, VIE 34, VIIE 35, IIF 37, IIIF 38, IVF 39,V F40, VIF 41, VIIF 42

[a] The class name refers to Figure7. Explanation

Table 2. The relationships between super-classes (biomes) and classes according to IOCSG approach

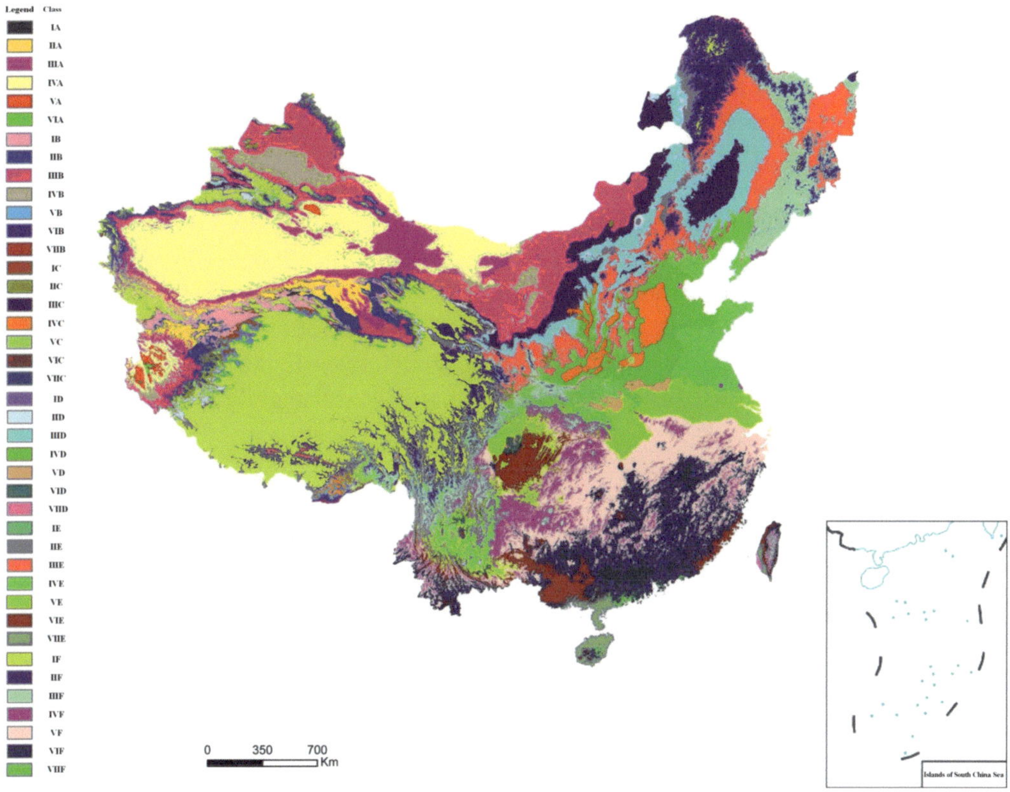

Explanation: IA 1 Frigid-extrarid frigid desert, alpine desert; IIA 2 Cold temperate-extrarid montane desert; IIIA 3 Cool temperate-extrarid temperate zonal desert; IVA 4 Warm temperate-extrarid warm temperate zonal desert; VA 5 Warm-extrarid subtropical desert; VIA 6 Subtropical-extrarid subtropical desert; VIIA 7 Tropical-extrarid tropical desert; IB 8 Frigid-arid frigid zonal semidesert,alpine semidesert; IIB 9 Cold temperate-arid montane semidesert; IIIB 10 Cool temperate-arid temperate zonal semidesert; IVB 11 Warm temperate-arid warm temperate zonal semidesert; VB 12 Warm-arid warm subtropical semidesert; VIB 13 Subtropical arid subtropical desert brush; VIIB 14 Tropical arid tropical desert brush; IC 15 Frigid-semiarid dry tundra,alpine steppe; IIC 16 Cold temperate-semiarid montane steppe; IIIC 17 Cool temperate-semiarid temperate typical steppe; IVC 18 Warm temperate-semiarid warm temperate typical steppe; VC 19 Warm-semiarid subtropical grasses-fruticous steppe; VIC 20 Subtropical-semiarid subtropical brush steppe; VIIC 21 Tropical-semiarid savanna; ID 22 Frigid-subhumid moist tundra,alpine meadow steppe; IID 23 Cold temperate subhumid montane meadow steppe; IIID 24 Cool temperate-subhumid meadow steppe; IVD 25 Warm temperate-subhumid forest steppe; VD 26 Warm-subhumid deciduous broad leaved forest; VID 27 Subtropical-subhumid sclerophyllous forest; VIID 28 Tropical-subhumid tropical xerophytic forest; IE 29 Frigid-humid tundra, alpine meadow; IIE 30 Cold temperate-humid montane meadow; IIIE 31 Cool temperate-humid forest steppe, deciduous broad leaved forest; IVE 32 Warm temperate-humid deciduous broad leaved forest; VE 33 Warm-humid evergreen-deciduous broad leaved forest; VIE 34 Subtropical-humid evergreen broad leaved forest; VIIE 35 Tropical-humid seasonal rain forest; IF 36 Frigid perhumid rain tundra,alpine meadow; IIF 37 Cold temperate perhumid taiga forest; IIIF 38 Cool temperate perhumid mixed coniferous broad leaved forest; IVF 39 Warm temperate perhumid deciduous broad leaved forest; VF 40 Warm-perhumid deciduous-evergreen broad leaved forest; VIF 41 Sub-tropical perhumid evergreen broad leaved forest; VIIF 42 Tropical-perhumid rain forest. The legend above also applies to Table 2)

Figure 7. Map of China showing the spatial distribution of potential class recognized by the IOCSG.

3.2.2. Results

3.2.2.1. Comparison between NPP observations and predictions

Figure 8. plots predicted grassland NPP versus observed NPP values. The results indicate that the Schuur Model underestimated grassland NPP. The Miami Model overestimated the large observations, while the Classification Indices-based Model overestimated as many data points as it underestimated. Considering the slopes and intercepts of the regression of predicted NPP by the Miami Model and Classification Indices-based Model versus observed NPP, the regression of the Miami Model has a slope similar to that of the Classification Indices-based Model (0.79 and 0.72, respectively), but the intercept of the regression line of the Classification Indices-based Model (64.19) is much closer to zero than is that of the Miami Model (113.40). The intercept differences suggest that the Miami Model overestimated the high values of observed NPP, whereas the Classification Indices-based Model slightly underestimated the high values.

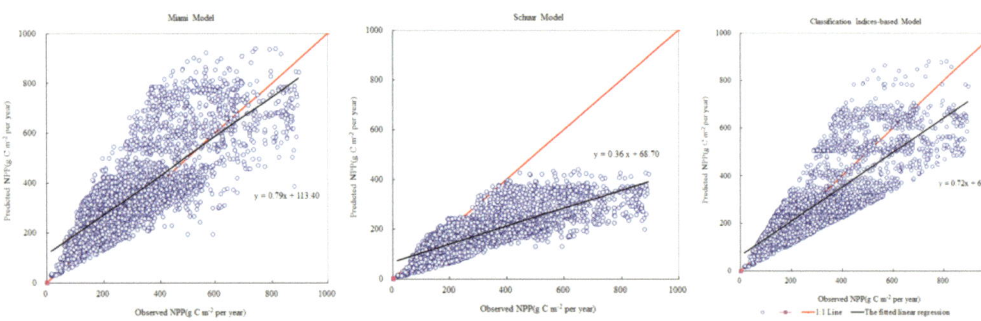

Figure 8. Performance of the Miami, Schuur and Classification Indices-based Models expressed as scatter-plots of predicted versus observed grassland NPP in China.

Comparison of the predicted with the observed grassland NPP indicated that the CVRMSE value using the Classification Indices-based Model was <10%, demonstrating that the prediction of the Classification Indices-based Model is considered excellent, while the output of the Miami Model is considered good, with CVRMSE 13.0%. As the CVRMSE value using the Schuur Model is 23.9%, its predictive ability is considered fair. The results of linear regression showed that the Classification Indices-based Model explained, on average, 65.3% (R^2) of the variation in observed NPP, while the Miami Model explained 63.0% (R^2) of the variation. With the highest E value, the Classification Indices-based Model ranked first for predicting NPP, followed by the Miami and Schuur Models, respectively, indicating that the Classification Indices-based Model can estimate grassland NPP more precisely (Table 3).

3.2.2.2. Distribution patterns of grassland NPP in China

Almost all classes (42 classes) occur in China, except for tropical-extra-arid tropical desert (VIIA7) under current climate conditions (Figure 7). The spatial distribution of grassland

Model	MBE	CVRMSE (%)	R^2	E
Miami Model	102.0	13.0	0.630	0.72
Schuur Model	191.7	23.9	0.629	0.58
Classification Indices-based Model	58.7	9.5	0.653	0.77

Table 3. The mean bias error (MBE), coefficient of variation of the root mean square error (CVRMSE), coefficients of determination (R^2), and model efficiency (E) for the comparison of the predicted with the observed grassland NPP in China.

biomes (Table 2) under current conditions is shown in Figure 9a. The Grassland category was predicted to be distributed primarily in the arid and semi-arid regions of Northern China, the Qinghai-Tibetan Plateau, and scattered throughout warm-temperate and tropical regions (Figure 9a). From north to south, grassland biomes were predicted to be mainly distributed as follows: Tundra and alpine steppe (accounting for 20.42% of Chinese land area); Cold desert (13.1%); Semi-desert (8.1%); Steppe (8.0%); Temperate humid grassland (6.0%); Warm desert (0.1%) and Savanna (0.1%) (Figure 9a and Table 4). These percentages correspond to the actual grassland distribution.

Compared with the Classification Indices-based Model and Miami Model, the Schuur Model estimated substantially lower TNPP, especially in Tundra and alpine steppe, Cold desert, Semi-desert, Steppe, Temperate humid grassland and Savanna areas (Table 4), but the simulated spatial pattern for grassland NPP is very similar (Figure 9b, c, d). Comparing the NPP maps (Figure 9b, c, d) with the data in Table 4 shows highest TNPP of grassland biomes in Tundra and alpine steppe, followed by Steppe and Temperate humid grassland, respectively. The Tundra and alpine steppe, which covers the largest area, was predicted to be mainly distributed in the Tibet Autonomous Region (94.4 million hectares), Qinghai Province (49.1 million hectares) and Xinjiang Uygur Autonomous Region (32.1 million hectares). Although Tundra and alpine steppe has lower NPP values than Temperate humid grassland on a unit area basis, the TNPP for Tundra and alpine steppe, estimated by the Classification Indices-based Model, was predicted to be nearly twice as large as that of Temperate humid grassland because Tundra and alpine steppe was predicted to cover a greater area (Table 4). The total area of Steppe and Temperate humid grassland was predicted to be 130.75 million hectares, accounting for 25.0% of the total grassland area, with nearly half of distributed on the Inner Mongolian Plateau. Savanna and Warm desert super-classes were predicted to cover less than 1.3 million hectares. Savanna was predicted to be located in southern China, particularly in Hainan Province, and Warm desert was predicted to be located in the Tarim Basin in Xinjiang Uygur Autonomous Region and the Qaidam Basin in Qinghai Province (Figure 9a). Warm desert and Savanna were predicted to have the lowest TNPP, and Semi-desert and Cold desert to have moderate TNPP values (Figure 9 and Table 4). The Semi-desert was predicted to be mainly distributed in Xinjiang Uygur Autonomous Region (33.8 million hectares) and Inner Mongolia Autonomous Region (26.4 million hectares). The Cold desert was predicted to be mainly distributed in Xinjiang Uygur Autonomous Region (84.3 million hectares), the western region of Inner Mongolia Autonomous Region (18.7million hectares) and Gansu Province (11.9 million hectares).

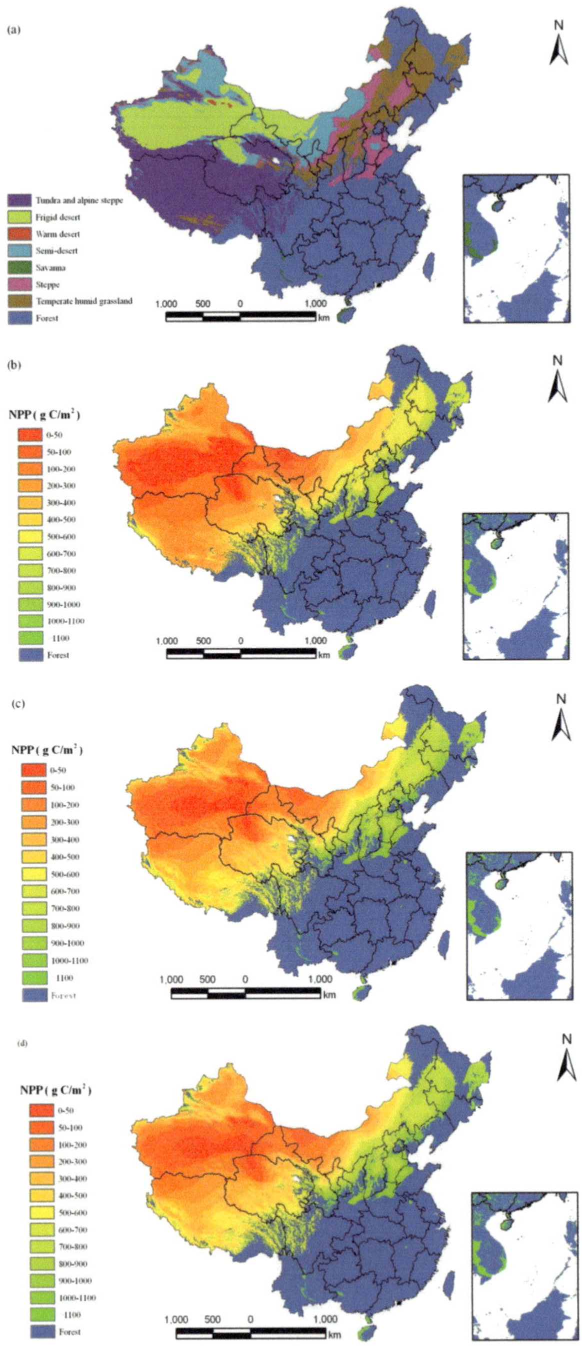

Figure 9. Spatial distribution of (a) potential grassland biomes (super-class group) and (b, c, d) geographical distribution of grassland NPP evaluated by the Miami Model, Schuur Model and Classification Indices-based Model at current climatic condition, respectively.

Biome (super-class group)	Area (million hectares)	TNPP (Tg C)		
		Miami Model	Schuur Model	Classification Indices-based Model
Tundra and alpine steppe	191.47	329.63	184.42	250.70
Cold desert	123.21	52.11	29.15	36.09
Semi-desert	76.19	110.61	61.88	93.68
Steppe	74.99	265.34	148.45	211.65
Temperate humid grassland	55.76	167.86	93.91	124.70
Warm desert	0.80	0.14	0.08	0.04
Savanna	1.25	7.34	4.11	6.71
Forest, including Temperate forest, Sub-tropical forest and Tropical forest	413.65	2529.04	1414.94	2227.83

Table 4. Total NPP (TNPP) of major China's terrestrial biomes estimated by the Miami Model, Schuur Model and Classification Indices-based Model under current climatic condition.

In conclusion, under current climate conditions, the main parts of China's grassland are predicted to be the Tundra and alpine steppe and Steppe, and they account for 50.88% China's total grassland and 63.77-63.90% China's grassland TNPP (Table 4).

3.2.3. Discussion

3.2.3.1. Model-data comparison

The performance of the different models can be compared on the basis of their error frequency distributions. Figure 10 shows the distribution of errors for the three models. Even though the pattern is to some extent controlled by the heterogeneous distribution of the validation points, the direct influence of the different models on the error structure remains evident. In the Miami Model, the error distribution (Figure 10a) shows a unimodal trend with a peak centered on a negative value (average local error of about -48 g C m^{-2} per year) and an error range from -419 to 492 g C m^{-2} per year. The largest underestimated values occur with the Schuur model, although a few overestimated predictions are also present in the same areas. In the histogram of error frequency (Figure 10b), a large single maximum is clearly centered on values of approximately 132 g C m^{-2} per year, indicating that grassland NPP is underestimated almost everywhere within the test area (Figure 8). The fact that the Schuur Model produced a model efficiency value of 0.58 is an indication that any predictions arising from this model must be treated cautiously. The Classification Indices-based and Miami models have more consistent behavior throughout the validation points, and Figure 10c indicates that the results of the Classification Indices-based Model are in good agreement with observed data (mean error 23 g C m^{-2} per year with a standard deviation of 105.6 g C m^{-2} per year) and the mean absolute error is lower than that obtained from the Miami Model.

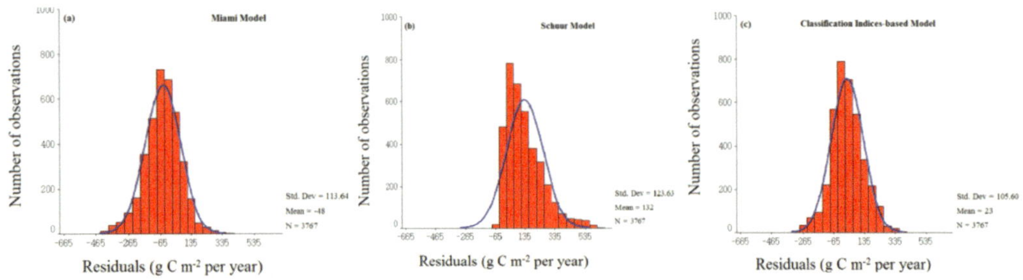

Figure 10. Frequency distribution of residuals for (a) the Miami, (b) the Schuur, and (c) the Classification Indices-based Models. Residuals are expressed as the difference between observed and predicted NPP.

The Classification Indices-based Model performs slightly better than the Miami Model, with a narrower error range and an error histogram that approximates a bell-shaped distribution (Figure 10). In general, the distribution of residuals is homogeneous, although a few larger errors occur in the Classification Indices-based model, which can compromise its reliability. Comparing the Miami Model with the Classification Indices-based Model, the Miami Model determines NPP for a particular location by comparing the minimum value of either temperature or precipitation functions. Hence, it only considers a single factor, and there are no interactions between the two variables. In contrast, the Classification Indices-based Model considers three meteorological measurements and their interactions in its formulation. The reduction in errors produced by the Classification Indices-based Model should lead to significant improvements in applications as a newer NPP-climate relationship model (Figure 10).

3.2.3.2. Grassland carbon budgets

The Classification Indices-based Model has two advantages: 1) it needs few input parameters; and 2) remote sensing data can be incorporated easily. Consequently, the Classification Indices-based Model is effective and practical for large-area application. But the simulated NPP by the Classification Indices-based Model was ideally potential value and land-use practices were not taken into account. The potential grassland, as a final state of succession which achieves the balance with its habitat, is the most stable and mature climax grassland class of the habitat without human interference, and is the trend of the regional grassland development. The study of potential grassland can substantially reveal the impact of climate on the changes of grassland patterns. It is the starting point of the vegetation-environmental classification and relationship study, as well as the key point of the global change and grassland ecosystem study [27].

The Classification Indices-based Model estimate of China's TNPP at 2.95 Pg C (Table 4) is close to the estimates of 2.65 Pg C and 2.24 Pg C by the light-use efficiency model [48] and the BEPS (Boreal Ecosystem Productivity Simulator) model [49] under current climate conditions, respectively. The Classification Indices-based Model outputs suggest that potential TNPP of grassland in China is 0.72 Pg C under current climate conditions (Table 4). China's grassland TNPP was estimated by Piao et al. [50] as 1.04 Pg C from 1980 to 1988, while Ni [51], using the carbon density method, estimated the TNPP of China's grasslands as 3.06 Pg C between 1949 and1978 and also between 1979 and 1990. This wide range in the reported potential TNPP could be due to their use of different estimation methods, inconsistent classification systems, and/or different time periods with changing grassland areas.

3.2.3.3. Human Appropriation of Net Primary Production (HANPP)

One of the central issues of sustainable development is the question of how to meet the globally growing demand for biomass in an ecologically sound way. Land use activities, primarily for agricultural expansion and economic growth, have transformed one-third to one-half of our planet's land surface from forest or grassland clearance, agricultural practices and urban expansion, which make profound impacts on ecosystem service, food production and the environment [52]. China, occupying only 7% of the world's total arable land, supports nearly 21% of the world's total human population. Agriculture is the major sector of growth in the Chinese economy and with 80% of Chinese living in villages, a large proportion of these people are still dependent upon agriculture. However, with rapid industrialization, urbanization is accelerating and leading to changes in grassland and forest land uses. With increasing scientific and political interest in regional aspects of the global carbon cycle, there is a strong impetus to better understand the carbon balance of China. This is not only because China is the largest emitter of fossil-fuel CO_2 into the atmosphere, but also because it has experienced regionally distinct land-use histories and climate trends, which together control the carbon budget of its ecosystems [53, 54]. Changes in terrestrial ecosystems through human land use and management are also important in land degradation and carbon balance studies. Environmental degradation is a major concern for China and research aimed at understanding the linkages between land use changes, socioeconomics and the biophysical variables governing environmental degradation is needed. Despite the vast amount of research conducted independently on land use changes and social indicators in China, an integrated assessment combining them to study the socioeconomic metabolic flows and their impact on land resources on a large spatial scale is absent. In this context, the concept of HANPP (human appropriation of net primary production) is an important tool for assessing the impact of anthropogenic forces through land use changes on biomass.

HANPP is an indicator of the extent to which ecosystem processes are altered by human activities. HANPP equates to the difference between the NPP of the potential natural vegetation and the amount of NPP remaining in ecosystems. The notion of HANPP represents the food supply for humans, providing the free energy for humans to sustain their metabolic activities and allowing them to perform physical work. For example, agriculture and forestry harness biomass energy for socio-economic purposes and thereby reduce the amount of NPP remaining in ecological food chains [52, 55-58]. HANPP alters the composition of the atmosphere, levels of biodiversity, energy flows within food webs and the provision of important ecosystem services [59-61]. This removal of carbon naturally comes at a cost in that increased intensities of human appropriation should directly result in less vegetative growth compared to its natural state, if all other factors are held constant, such as nutrient availability. In this study, we present a comprehensive assessment of Chinese HANPP based on vegetation modelling using our model, agricultural and forestry statistics, and geographical information systems data on land use, land cover, and soil degradation that localizes human impact on ecosystems. We find, overall, that the HANPP values suggest that more than 43.2% of the available NPP has already been modified and exploited in China. This value is comparatively higher than the range of global estimates of 20% to 40% for HANPP [62, 63]. Furthermore, compared with HANPP estimates cited for Austria [57, 59], where HANPP declined from 1980 to 1995, HANPP has increased in China. Increases in HANPP may lead to carbon fluxes from biota to the atmosphere, they may contribute to biodiversity loss, and they may result in diminished resilience of ecosystems [62]. As Chinese human population grows and with mounting political interest in increasing GDP, there is likely to be an increased requirement for more agricultural areas and further increases in HANPP over the present value. These analyses suggest policy options for slowing future growth of HANPP. It also offers a means to aid people in using the aforementioned carbon sinks to fulfill China's commitment of reducing greenhouse gases.

3.3. Model-data comparison and modelling the potential net primary productivity of global grassland

Grassland biomes account for 25% of the world's land area and have global significance for climate-carbon feedback [1], containing approximately 30% of global soil carbon stocks [64]. Ranging from the savannas of Africa to the North American prairies and the converted grasslands of Latin America and Southeast Asia, grassland ecosystems support the majority of the world's livestock and large mammals [65], besides playing a very important role in regulating the global carbon cycle and providing meat and milk for human beings.

As one of the most widespread ecosystem types, natural grassland plays a significant but poorly recognized role in the global carbon cycle [1, 2]. In order to effectively manage grassland ecosystems and maintain their sustainability, large-scale analysis and modelling of grassland NPP are needed to develop a better grasp of the spatial distribution of grasslands and their productivity [3, 4, 13, 20, 66]. Continuous monitoring of global grassland productivity has never been possible because of technological limitations. So it is necessary to use computer models, calibrated with existing data, to study the spatial and temporal variations of grassland NPP [3]. Recently, climate-vegetation models have been drawing much attention and have been widely applied internationally [4, 9, 18], having been shown to yield 'reasonable estimates' of global patterns of productivity [4, 9, 14, 16]. The Miami Model [38], Schuur Model [15] and Classification Indices-based Model [22, 24, 25, 27, 28] are examples of these climate-vegetation models. But model inter-comparison under a consistent classification system simultaneously has not been attempted. In particular, the quantitative and spatial descriptions of global grassland classes and their NPP values have not been reported.

The purpose of this section is: 1) to compare NPP estimated from the Miami Model [38], Schuur Model [15], and the Classification Indices-based Model [22,24,25,27,28] with NPP derived from measurements at 37 sites around the world, to evaluate the applicability and reliability of the Classification Indices-based Model; and 2) to simulate the spatial distribution patterns and associated NPP characteristics of global potential grassland under recent past climate scenario using the IOCSG approach and NPP-climate models.

3.3.1. Data acquisition and methods

3.3.1.1. Observed NPP database

A reference data set (n = 113) of grassland NPP field observations with paired climatic variables was compiled for this study. Total observed NPP data were gathered from the Oak Ridge National Laboratories (ORNL) Net Primary Production database (http://www-eosdis.ornl.gov/NPP/npp_home. html), and were the sum of aboveground and belowground NPP following standard methods [9]. The grasslands including the Tundra and alpine steppe, Cold desert, Semi-desert, Steppe, Temperate humid grassland, Warm desert and Savanna, or other intensively managed sites were omitted from this study. Locations of sampling sites were plotted using their associated geographic coordinates, as shown in Figure 11.

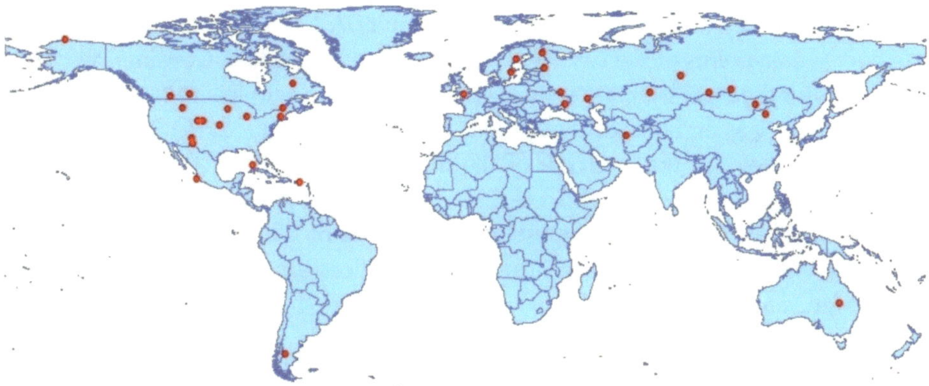

Figure 11. Spatial distribution of grassland NPP observations collected from the ORNL DAAC NPP database (http://www-eosdis.ornl.gov/NPP/npp_home.html).

3.3.1.2. Monthly precipitation and mean temperature grid data

The monthly precipitation and mean temperature grid data, with spatial resolution of 30 arc seconds (i.e., about 1 km), for global land areas excluding Antarctica over 50 years from 1950 to 2000, were generated using the software package ANUSPLIN Version 4.3 (Australian

National University, Canberra, ACT, Australia) [67,68]. We chose this data set because the method that created it has been used in other global studies [69, 70], and it has performed well in comparison with the other multiple interpolation techniques [68, 71].

3.3.1.3. IOCSG approach and NPP-climate model operation

A common approach to extrapolating grassland NPP to the biosphere is to use vegetation maps together with models of plant productivity at a biome level. The IOCSG potential biome maps and its NPP were created and processed by the ArcGIS software (Esri Inc., Redlands, CA, USA) under current climate (1950-2000). In the IOCSG biome (super-class group) classification map, lake water, permanent snow and ice were excluded by using the Moderate Resolution Imaging Spectroradiometer-International Geosphere Biosphere Program (MODIS-IGBP) land-cover classification dataset in 2001, found at http://wist.echo.nasa.gov. The MODIS-IGBP land-cover classification dataset was used to calculate the area of potential grassland biome. The simulated NPP was an ideal potential value, and land-use practices were not taken into account. The total NPP has a unit of Tg C (1 Tg = 10^{12}g) for biome or Pg C (1 Pg = 10^{15}g) for world.

3.3.2. Results

3.3.2.1. Comparison between NPP observations and predictions

Figure 12 plots predicted grassland NPP versus observed NPP values. Data results indicate that the Schuur Model over-predicted grassland NPP. The Miami Model over-predicted the large observations while the Classification Indices-based Model achieved a better balance between over-predicted and under-predicted data points. A 1:1 line fits almost throughout the entire cloud of data points of the Classification Indices-based Model. Considering the slopes and intercepts for the regression of predicted NPP by the three models versus observed NPP, the regression of the Miami Model has a slope (0.77) similar to that (0.70) of the Classification Indices-based Model, but the intercept of the regression line (151.41) of the Classification Indices-based Model is much closer to zero than the regression line (232.52) of the Miami Model. The intercept differences suggested that the Miami Model overestimated the high values of observed NPP, whereas the Classification Indices-based Model slightly underestimated the high values.

Comparison of the predicted with the observed grassland NPP indicated that the CVRMSE value using the Classification Indices-based Model was less than 30%, demonstrating that the prediction suitability of the Classification Indices-based Model is considered fair, while the prediction suitability of the Miami model is considered poor, with the CVRMSE=34.12%. As the CVRMSE value using the Schuur Model is 42.13%, its predictive ability is considered the poorest. The results of linear regression showed that the Classification Indices-based Model explained, on average, a variation of 65.86% (R^2) in observed NPP, while the Miami Model showed the explanations of 59.46% (R^2). With the highest E value, the Classification Indices-based Model ranked first for predicting NPP, indicating that the model can estimate the grassland NPP more precisely (Table 5).

http://dx.doi.org/10.5772/57297

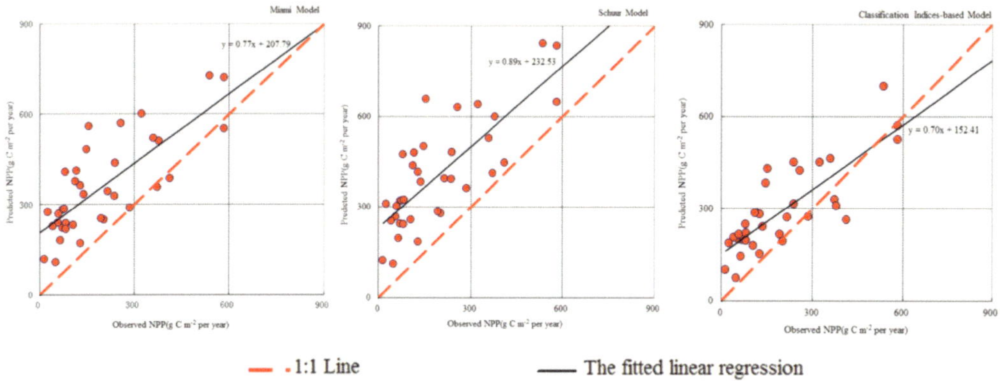

Figure 12. Performance of the Miami, Schuur and Classification Indices-based Models expressed as scatter-plots of predicted versus observed grassland NPP in World.

Model	MBE	CVRMSE (%)	R^2	E
Miami Model	-162.93	34.12	0.5946	-0.11
Schuur Model	-211.46	42.13	0.6116	-0.07
Classification Indices-based Model	-94.92	23.12	0.6586	0.10

Table 5. The MBE, CVRMSE, coefficients of determination (R^2), and model efficiency (E) for the comparison of the predicted with the observed grassland NPP in World.

3.3.2.2. Spatial distribution patterns and characters of grassland biomes over 1950-2000

The spatial distribution of grassland biomes during the period of 1950-2000 is shown in Figure 13a. From Figure 13a, the potential grassland is characterized by a significant distribution pattern in latitudinal and altitudinal directions. From the Equator to the North Pole, grassland biomes were mainly distributed in sequence as follows: (1) Savanna as the largest terrestrial biome, mainly distributed in the east and middle of Africa, Central America and Oceania; (2) Warm desert, mainly distributed in North Africa, North America and the largest part in Asia; (3) Semi-desert and Cold desert, and mixed with Steppe in central Eurasia Continent and southwest of North America, as well as Tundra and alpine steppe in the Tibetan Plateau; (4) Temperate humid grassland, distributed over the entire Eurasian continent, and extending from the east of North America to the west coast with the Pacific Ocean; and (5) Tundra and alpine steppe, mainly distributed in the north of North America, Greenland and the most northerly part of Eurasia. In the Southern Hemisphere, from the Equator to the southernmost edge of Oceania, there are two major zones, including (1) dominated by Savanna, and mixed with a little bit of Steppe, and Warm desert, mainly distributed in the south and middle of Africa, the north and middle of South America, and the south of Southeast Asia and the north of Oceania; (2) dominated by Semi-desert, mainly distributed in southernmost South America, Africa and Oceania. Probably affected by interactions among topography, climate and vegetation, there are more grassland biomes distributed in the east of South American (Figure

13a). Results of statistical analysis indicated that Grassland and Forest covers 55.76% and 44.24% of global land area, respectively (Table 6). Grassland biomes were mainly distributed as follows: Savanna (accounting for 17.19% of global land area); Warm desert (16.29%); Tundra and alpine steppe (6.32%); Semi-desert (6.03%); Steppe (4.83%); Temperate humid grassland (3.86%) and Cold desert (1.25%) (Table 6). These percentages correspond to the actual grassland distribution.

Biome (super-class group)	**Area (million hectares)**	**TNPP(Tg C)**		
		Miami Model	**Schuur Model**	**Classification Indices-based Model**
Tundra and alpine steppe	827.96	1884.82	2269.96	2492.34
Frigid desert	163.57	120.65	23.87	92.18
Semi-desert	789.81	1576.73	1177.32	1365.04
Steppe	633.15	1706.27	4705.38	1371.51
Temperate humid grassland	505.43	1049.52	7515.71	841.33
Warm desert	2134.27	1516.60	3138.55	1111.00
Savanna	2251.42	8616.21	2536.42	8103.96
Forest, including Temperate forest, Sub-tropical forest and Tropical forest	5795.39	38904.59	40147.86	39128.38

Table 6. Total NPP (TNPP) of major global terrestrial biomes estimated by Miami, Schuur and Classification Indices-based Models in recent past (1950- 2000)

Comparisons with the Classification Indices-based Model and Miami Model, Schuur Model estimated substantially higher TNPP, especially for Temperate humid grassland, Steppe and Warm desert (Table 6), but the simulated spatial pattern for global NPP is very similar (Figure 13bcd). The NPP increases from the South and North poles toward the Equator, corresponding to changes in precipitation and temperature. Comparing the NPP maps (Figure 13bcd) with the Table 6 shows highest TNPP of grassland in Savanna, followed by Tundra and alpine grassland and Steppe, respectively. Although Warm desert grassland covers the second greatest area among grassland super-classes, TNPP for Warm desert grassland is still lower due to lower NPP.

In conclusion, under recent past climate conditions, the main parts of global grassland are the Savanna and Tundra and alpine grassland, which together account for 42.15% of total global grassland area and 63.76-68.91% of global grassland TNPP (Table 6).

http://dx.doi.org/10.5772/57297

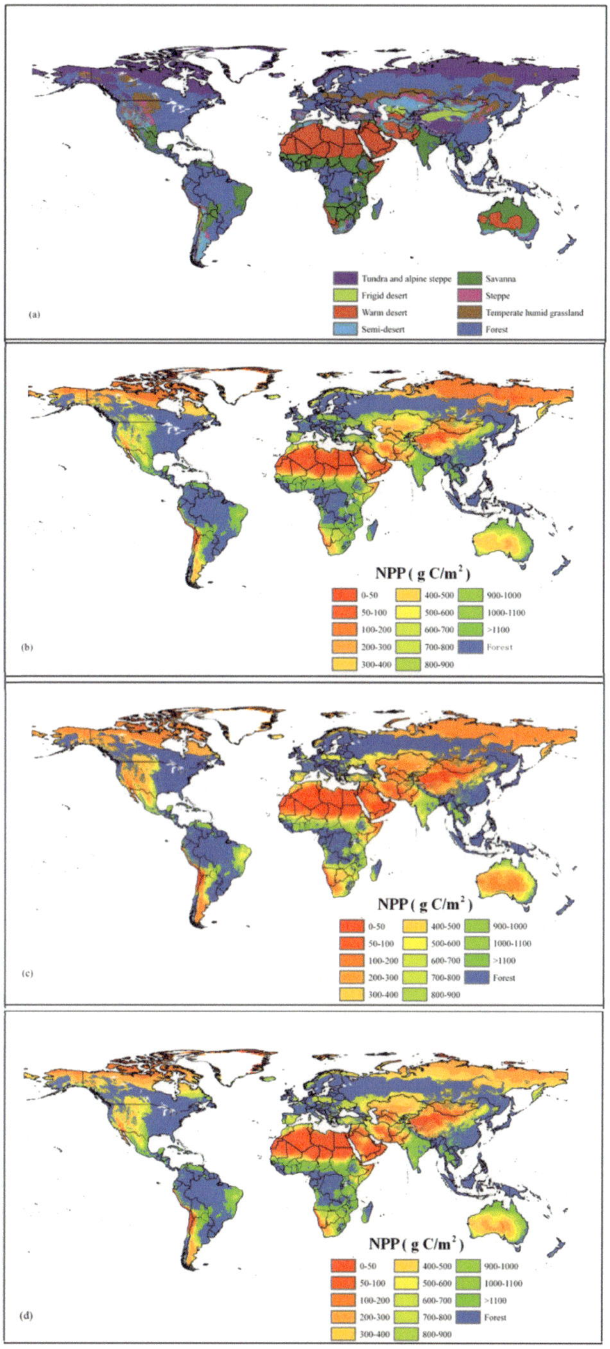

Note: (a) Potential biome (super-class group) classification map; (b, c, d) global pattern of grassland NPP evaluated by the Miami Model, Schuur Model and Classification Indices-based Model.

Figure 13. Global spatial pattern of the potential biome and the grassland NPP for recent past (1950-2000)

3.3.3. Discussion

3.3.3.1. Model-data comparison

GDD, MAP and the moisture index (K value) are used as criteria to determine the class in the IOCSG. The Classification Indices-based Model connected with the IOCSG by using the classification indices as independent variables and considers three meteorological measurements and their interactions in its formulation. The use of these variables allows the Classification Indices-based Model to more accurately predict changes of NPP among grassland classes or super-classes [17, 22, 24, 25, 27, 28]. The model's relative simplicity and ability to make reasonable estimates of the patterns of NPP in global grasslands is attractive. Despite being a newer NPP-climate relationship model, the Classification Indices-based Model provides more accurate conclusions and may also reduce the uncertainty of climate change impacts on grasslands on a global scale. Furthermore, it has the potential to rapidly advance grassland NPP research and application in developing/undeveloped regions or countries that generally lack the detailed and complex data required by other models (i.e., BIOME4).

3.3.3.2. Grassland carbon budgets

The Classification Indices-based Model estimates global TNPP at 54.51 Pg C compared to 55.38 Pg C and 61.52 Pg C using the Miami model and the Schuur Model under recent past climate conditions (Table 6). The Classification Indices-based Model estimate is closer to the estimates of 46 Pg C, 48 Pg C and 55 Pg C by the National Center for Ecological Analysis and Synthesis (NCEAS) model [4], the Carnegie-Ames-Stanford approach (CASA) Biosphere model [72], and the MODIS NPP [73], respectively. This wide range in the reported potential global TNPP could be due to their use of different estimation methods, inconsistent classification systems, and/or different time periods [74]. The potential carbon accumulated in plants was calculated based on simulated NPP. This calculation does not account for the potential impacts from human disturbances, but assumes that the full potential NPP is reached in the grassland. The Classification Indices-based Model estimates global grassland TNPP at 15.38 Pg C compared to 16.47 Pg C and 21.37 Pg C using the Miami model and the Schuur Model under recent past climate conditions (Table 6). Our global-scale NPP estimates can be used to improve understanding of environmental controls on global-scale NPP, help understand how NPP might have been altered by land use [75] and assess the human appropriation of net primary productivity.

4. Grassland in response to climate change

4.1. Introduction

Global climate change will seriously affect terrestrial ecosystems. In global climate change research, modelling is a key method for the study of climate-vegetation, which is an important component in the area of global change research. The current period of climate change and high CO_2 concentrations in recent years has resulted in warm summers and longer growing seasons, to a change in vegetation patterns can be expected. This change would cause changes in vegetation classes and their NPP. For example, the dominance of some cold-season C_3 grasses would decrease, while that of warm season C_4 grasses could be expected to increase. Global warming also causes changes in precipitation patterns and the total amount of precipitation. The far-reaching consequences of climatic phenomena such as La Niña and El Niño receive widespread attention and appear to be occurring more frequently in recent decades. Zhang et al. [26] recently showed that global warming is already affecting the world's rainfall patterns, bringing more precipitation to northern Europe, Canada and northern Russia, but less to sub-Saharan Africa, southern India and Southeast Asia. They explained that human activity is causing alterations to the water cycle, moving more water vapor away from the warmest parts of the planet and pushing it toward the poles. As a result, wet areas are becoming wetter and dry areas are becoming drier. Thus, there has been an increase in the demand by the broader scientific community and policymakers for better projections of regional impacts of future greenhouse-gas-induced climatic change. Global environmental change, signified by 'global warming' and its possible effects on ecosystems have been drawing increasing attention from the scientists and governments of every country and from all circles of the world [9, 13, 76]. Study of responses of terrestrial NPP to climate changes will help us understand the feedback between climate systems and terrestrial ecosystems and the mechanisms of increased NPP in the northern, middle and high latitudes [77]. NPP is not only used for assessing the carbon balance on regional and global scales, it also plays an important role in demonstrating compliance with the Kyoto Protocol on greenhouse gas reduction [11]. So, accurate estimates of NPP are very important not only for scientifically guiding ecosystem management but also for the study of global climate change [53]. Therefore, study of NPP and its response to global change is one of the key focuses for the global scientific community [9, 78].

Climate change has been identified as having far-reaching implications for the world's grasslands [16, 79-83]. Therefore, understanding the sensitivity of grassland to climate change and the effect of these changes on grassland ecosystems is a key issue in global carbon cycling. NPP in grassland is a key variable in our understanding of carbon exchange between the biosphere and atmosphere, both currently and under climate change conditions [3, 4, 13, 84]. The mechanisms and modelling of responses in grassland NPP to global change will enhance understanding of how grassland ecosystems will respond to global change and the projection of grassland NPP to global change. This will enable the development of countermeasures to maximize the grassland NPP under conditions of global change, especially to provide sustainable development of pasturing in arid and semiarid areas, by using these theoretical and technological methods. The IPCC report [85] states that climate change will present challenges for future grassland use by livestock and other grazers, and for the resulting public policy designed to manage grasslands. Given their vast area and diversity of classes, China's grasslands play an important role in both regional and global carbon cycling [33]. In order to effectively manage grassland ecosystems and maintain their sustainability, a deeper understanding of how these ecosystems will respond to growing pressures is needed. Large-scale analysis and modelling is needed to develop a better grasp of the spatial distribution of grasslands, their productivity, and potential variations in response to climate changes [3, 4, 13, 20, 23, 66].

Climate change will have profound effects on the class level, which in turn will influence vegetation types and their NPP. Clearly, this requires improved projections of regional climate and understanding of how vegetation types will respond to climate change. Climate is a major driver of variation in NPP, but a clear understanding of the impact of climate change on NPP is lacking [13]. Not only are grassland researchers concerned with the performance of NPP simulation models under contemporary climate, but also they want to know the behavior of these models under extrapolated future environmental conditions [3, 4, 13]. A growing number of research efforts have demonstrated the importance of climate-vegetation interaction in understanding climate sensitivity and climate change [18]. The study of climate-vegetation interaction is the basis for research on terrestrial ecosystems' responses to global change and mainly comprises two important components: climate vegetation classification and NPP of natural vegetation. From the view of generic relationships among all the vegetation types, the IOCSG can be used to predict climate-linked spatial or temporal succession from an original class to a new class as these climatic factors change. Hence, there is potential to develop future scenarios based on possible changes in vegetation type and its NPP in response to climate change in coming years. The IOCSG is a suitable tool to incorporate the effects of both climate change and vegetation management measures on plant growth activity into NPP estimation at the same resolution as satellite data. The Classification Indices-based Model connects with the IOCSG by using the classification indices as independent variables. The use of these variables allows the Classification Indices-based Model to project changes of NPP among grassland classes or super-classes more accurately [17, 22, 24, 25, 27, 28]. The section on Model Validation and Model Inter-comparison has shown that the Classification Indices-based Model is suitable to predict grassland NPP at regional, Chinese or global scale. In addition, the Classification Indices-based Model presented the closest values to the observed in situ meteorological variables. Furthermore, its relative simplicity and ability to generate reasonable patterns of NPP is attractive. Also, GDD and the moisture index (K value) are used as proxies to represent the water and thermal properties of the growing season, and are effective drivers for modelling NPP. The Classification Indices-based Model not only takes into account dynamical classes [17], it also simulates NPP of corresponding classes. As a new NPP-climate relationship model, the Classification Indices-based Model has the potential to evaluate the possible effects of climate change by improving the accuracy of the NPP prediction and reducing the evaluation uncertainty of the possible effects of climate change [22, 24, 25, 27, 28].

In order to effectively model climate change impacts on global grassland distributions and associated NPP, it is important to understand climate dynamics in the recent past (1950-2000) as well as climatic projections for the future (2001-2050). Hence, in this section we: (1) simulate the spatial distribution patterns and associated NPP characteristics of the world's and China's potential grassland under a future climate scenario using the IOCSG approach and the Classification Indices-based Model; and (2) estimate future trends in response to climate change in the first half of the 21st century by comparing the variation in distribution of NPP with the potential total NPP (TNPP) of grassland between the recent past and the projected future climate scenario. Such findings should improve knowledge about changes in global grassland primary productivity under global warming.

4.2. Data acquisition and methods of analysis

4.2.1. Climate data

Two different global climatic datasets were used in this study. The first was the monthly precipitation and mean temperature grid datasets over 50 years from 1950-2000, described at section of 3.3.1.2. The second dataset was the global monthly precipitation and mean temperature prediction data set (also with 30 arc-second resolution) excluding Antarctica from 2001 to 2050 under the A2a scenario (see below), which was simulated by Australia's Commonwealth Scientific and Industrial Research Organisation (CSIRO) and is available through the website http://www.worldclim.org /futdown.htm [86]. The A2a scenario takes into consideration the following land-use changes: (1) high rate of population growth; (2) slow technological change; and (3) increased energy use, and describes a highly heterogeneous future world in light of regionally-oriented economies. The gridded baseline climate (recent past) and gridded future (2000-2050) A2a climate scenarios were used to estimate changes in mean annual temperature (MAT), growing degree-days (GDD) and mean annual precipitation (MAP).

4.2.2. IOCSG approach and NPP simulation

The IOCSG potential super-class group (biome) maps and its NPP were created and processed by the ArcGIS software (ESRI Inc., Redlands, CA, USA) under the baseline climate (1950-2000) and future climate A2a scenario (2000-2050) on average level. In the IOCSG biome (super-class group) classification map, lake water, permanent snow and ice were excluded by using the Moderate Resolution Imaging Spectroradiometer-International Geosphere Biosphere Program (MODIS-IGBP) land-cover classification dataset in year 2001, found at http://earthdata.nasa.gov. The MODIS-IGBP land-cover classification dataset was used to calculate the area of potential grassland biome. The simulated NPP by the Classification Indices-based Model was an ideal potential value and land-use practices were not taken into account. The recent past map was chosen as a baseline and was used for comparison with projected future climate to estimate trends in changes of class or grassland biome distribution and grassland NPP distribution in response to climate change. Both the IOCSG and the Classification Indices-based Model were applied to analyze the changes in grassland biomes and to measure the change in TNPP of grassland biomes from the recent past (1950-2000) to the future A2a scenario (2001-2050). The total NPP has a unit of Tg C (1 Tg = 10^{12}g) for biome or Pg C (1 Pg = 10^{15}g) for China or the world.

4.3. Modelling global-scale potential grassland changes in spatio-temporal patterns to global climate change

4.3.1. Characteristics of potential grassland over 2001-2050 under A2a scenario

In the A2a scenario, the Savanna, Warm desert, Semi-desert, Tundra and alpine steppe, Steppe, Temperate humid grassland and Cold desert occupy 17.68%, 16.06%, 5.97%, 5.91%, 4.83%, 4.30% and 1.24% of global land area, respectively (Table 7). Table 7 shows the highest TNPP of grassland biomes in the Savanna, followed by Steppe and Semi-desert.

In conclusion, the main grassland biomes found in the world were the Savanna, Steppe and Semi-desert, occupying 50.86% of total global grassland areas and 78.06% of global grassland TNPP under future A2a scenario (Table 7).

Biome (super-class group)	Area (million hectares)	total net primary productivity (TNPP)(Tg C)
Tundra and alpine steppe	774.70	1313.62
Cold desert	162.80	84.76
Semi-desert	782.41	1440.56
Steppe	633.15	1854.16
Temperate humid grassland	563.89	1157.55
Warm desert	2104.30	1235.21
Savanna	2316.12	10197.64
Forest, including Temperate forest, Sub-tropical forest and Tropical forest	5766.10	40389.90

Table 7. TNPP of major global terrestrial biomes estimated by Classification Indices-based Model in future (2001-2050) A2a scenario

4.3.2. Numerical simulations of potential grassland dynamics

Both the IOCSG approach and the Classification Indices-based Model were applied to analyze the succession of classes and super-classes and to measure the change of super-class TNPP from recent past (1950-2000) to future A2a scenario (2001-2050).

4.3.2.1. Change of class and super-class area

The class called Frigid-extrarid frigid desert, alpine desert (IA1), mainly distributed in America and Asia, continuously decreased from recent past to A2a scenario and the rate of decrease per decade was 4.72%. If decreasing rates persist, Frigid-extrarid frigid desert, alpine desert (IA1) could potentially disappear by the second half of this century, while the class called Cold temperate subhumid montane meadow steppe (II D23), mainly distributed in Asia and America, continuously increased from recent past to A2a scenario, by 18.40%.

Climate change also influences the spatial distribution of super-classes. The distribution area for various super-classes in these two time periods shows differing dynamic characteristics from recent past (1950-2000) to future A2a scenario (2001-2050) (Tables 6 and 7). The area of Tundra and alpine steppe will decrease by 53.26 million hectares compared with its size during 1950-2000. Likewise, Warm desert will decrease by 29.97 million hectares. However, trends show that Savanna and Temperate humid grassland will strongly increase, to nearly 123.16 million hectares. Overall, the Forest category will decrease by 29.29 million hectares (Tables 6 and 7).

4.3.2.2. Shift trend of super-classes

The A2a scenario predicts changes in global temperature and rainfall patterns, making wet areas wetter, and dry areas drier (Figure 14), resulting in class conversions from an original class to a new class due to changes in climate conditions. For example, the Frigid perhumid rain tundra, alpine meadow (IF36) class will change to Cold temperate-humid montane meadow (IIE30) with climate warming, or Cold temperate perhumid taiga forest (IIF37) with climate drying. The conversion of classes from recent past to future A2a scenario will result in changes in area and TNPP distribution patterns among biomes (Table 8 and Figure 15a). For example, in the super-class of Tundra and alpine steppe, an area of 656.09 million hectares will be converted to Forest (Legend 18 in Figure 15a and Table 8) in this long-run sequence, which will mainly take place in Asia (accounting for 58.96%) and America (32.21%), while another 182.92 million hectares will be converted to Temperate humid grassland (Legend 15 in Figure 15a and Table 8), a change that will mainly take place in Asia (65.90%). An area of 590.06 million hectares of Forest under recent past climate conditions will be converted into the Temperate humid grassland under A2a scenario (Legend 85 in Figure 15a and Table 8), mainly taking place in Asia (51.14%), the Americas (26.22%) and Europe (22.58%). An area of 389.29 million hectares of Forest will be converted into Savanna (Legend 87 in Figure 15a and Table 8), mainly taking place in the Americas (49.62%) and Africa (35.52%). An area of 253.28 million hectares of Temperate humid grassland under recent past climate conditions will be converted into Steppe under the A2a scenario (Legend 54 in Figure 15a and Table 8), mainly taking place in Asia (48.76%), the Americas (28.89%) and Europe (22.22%). An area of 243.37 million hectares of Steppe will be converted into Semi-desert (Legend 43 in Figure 15a and Table 8), mainly taking place in Asia (49.72%) (Table 8).

An area of 856.54 million hectares of Grassland category will be converted to Forest category and an area of 1089.77 million hectares of Forest category will be converted into Grassland category (Figure 15b. and Table 8). There is a clear increasing trend in the Grassland category. As a whole, areas of Grassland category will increase by 31.76 million hectares, while areas of Forest category will decrease by 29.29 million hectares. The relative area of Grassland category will increase by 0.43% over this period (1950-2050), whereas the relative area of Forest category will decrease by 0.51% over the same period (Tables 6 and 7).

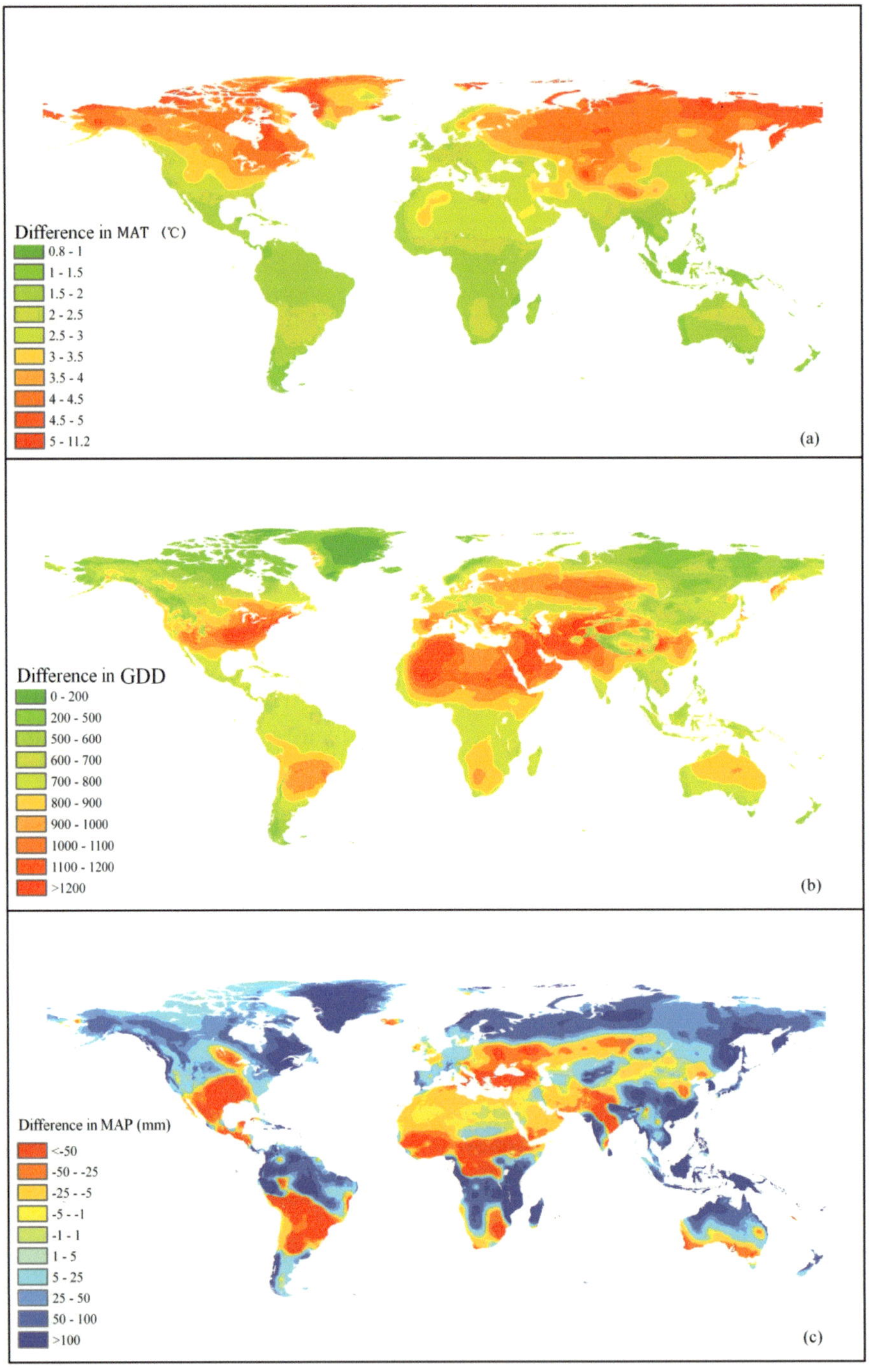

Figure 14. Maps of anomalies in future A2a climatic scenario (relative to the average for 1950-2000) for (a) mean annual temperature (MAT), (b) growing degree-days (GDD) and (c) mean annual precipitation (MAP).

Code	World	Oceania	America	Asia	Africa	Europe
11	774.7	0.06	492.08	305.36	0.01	25.75
12	3.94	-	1.33	2.61	-	-
13	19.89	-	3.20	16.69	-	-
14	13.68	-	1.69	11.98	-	-
15	182.92	-	62.38	120.54	-	-
18	656.09	0.26	211.34	386.85	0.01	57.63
22	108.13	-	12.91	95.21	0.01	-
26	102.80	-	8.26	93.78	0.76	-
32	57.91	-	13.03	44.87	0.01	-
33	502.10	4.51	154.74	299.02	14.40	29.42
34	0.01	-	0.01	-	-	-
36	39.93	0.64	7.80	26.45	5.03	-
37	193.86	43.51	51.26	44.29	38.44	16.35
43	243.37	4.17	65.53	121.01	3.77	48.89
44	70.67	2.27	31.85	28.68	1.94	5.92
47	69.29	6.22	25.44	7.23	13.10	17.30
48	0.20	-	0.07	-	0.13	-
53	29.17	-	2.96	21.38	-	4.83
54	253.28	0.07	73.16	123.50	0.28	56.27
55	290.36	0.11	116.84	158.83	0.01	14.56
58	159.59	0.92	26.57	29.96	0.63	101.51
66	1871.47	168.27	60.84	523.87	1118.23	0.27
67	31.94	25.78	0.21	0.93	5.01	-
76	102.28	20.06	14.40	20.87	46.91	0.03
77	2209.36	427.04	472.40	305.72	993.60	10.61
78	40.66	1.75	0.43	1.90	36.57	-
83	4.18	-	1.93	0.03	0.02	2.20
84	106.24	5.54	18.10	21.51	6.41	54.68
85	590.06	0.18	154.71	301.77	0.18	133.23
87	389.29	7.38	193.16	42.58	138.27	7.90
88	4255.05	84.89	1895.38	1204.96	578.05	491.76

Note: the two digits in the code represents the biome conversion, where the first number depicts the biome under recent past climatic conditions and the second number is under A2a scenario; numbers 1-8 represent Tundra and alpine steppe, cold desert, Semi-desert, Steppe, Temperate humid grassland, Warm desert, Savanna, and Forest, respectively. For example, Code 18 means Tundra and alpine Steppe under recent past climatic conditions would convert into forest in A2a scenario. The number in table is the area of conversion in unit of million hectares.

Table 8. The global biomes conversion matrix from recent past (1950-2000) to future (2001-2050) A2a scenario

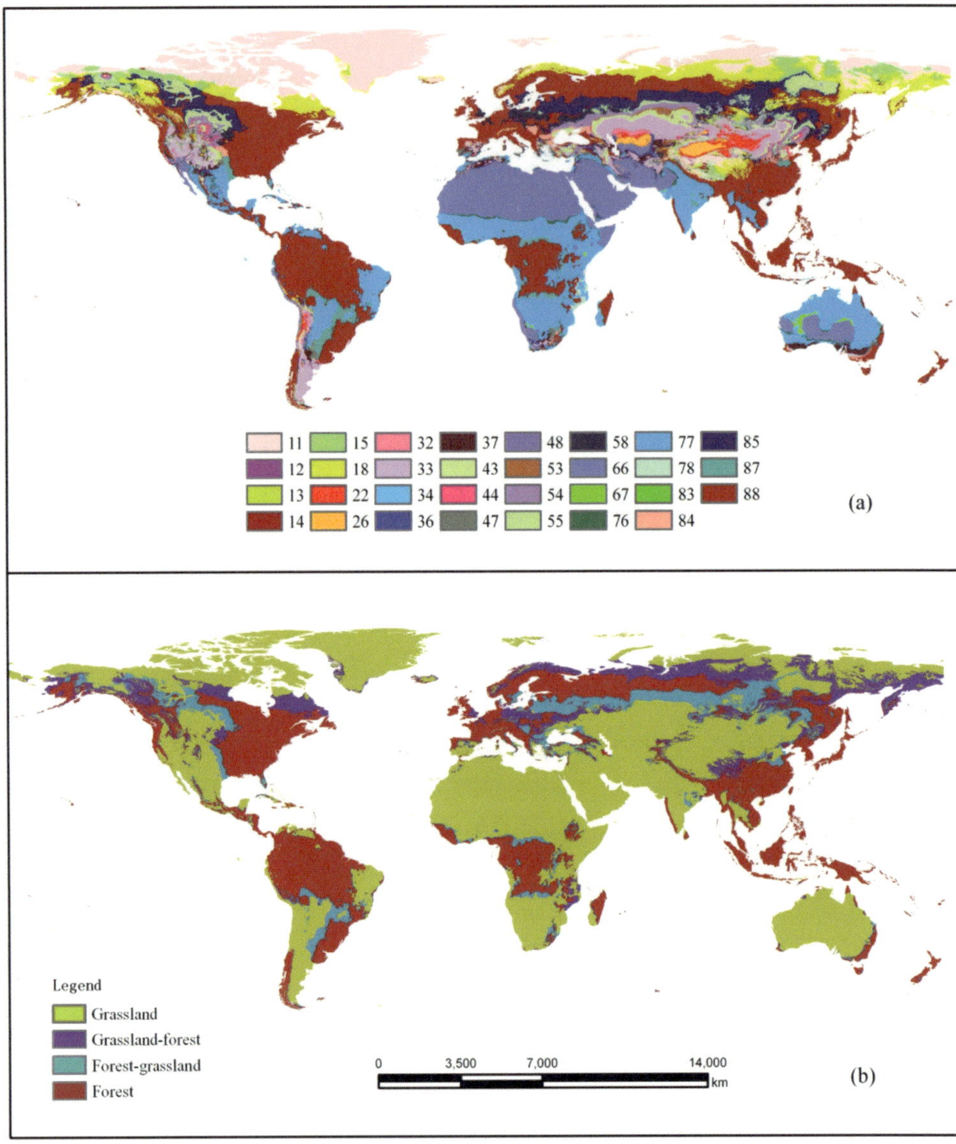

Note: (a) biomes shift. In the legend, the two digits represent the biomes conversion, where the first number depicts the biome under recent past climatic conditions and the second number is under A2a scenario. The numbers 1-8 represent Tundra and alpine steppe, Cold desert, Semi-desert, Steppe, Temperate humid grassland, Warm desert, Savanna, and Forest, respectively. For example, Legend 54 means Temperate humid grassland under the recent past climatic conditions would convert into Steppe under the A2a scenario. (b) Grassland (including Tundra and alpine steppe, Cold desert, Semi-desert, Steppe, Temperate humid grassland, Warm desert, and Savanna) and Forest (including Temperate forest, Sub-tropical forest, and Tropical forest) categories change: Grassland–forest means Grassland category would be converted into the Forest category, Forest-grassland means the Forest category would be converted into the Grassland category; the others are no change.

Figure 15. Spatial distribution dynamics of global potential biomes from recent past to future A2a scenario.

4.3.2.3. Change of TNPP

The Classification Indices-based Model shows NPP either neutral or increasing for most global grid cells during the periods from recent past to A2a scenario (Figure 16).

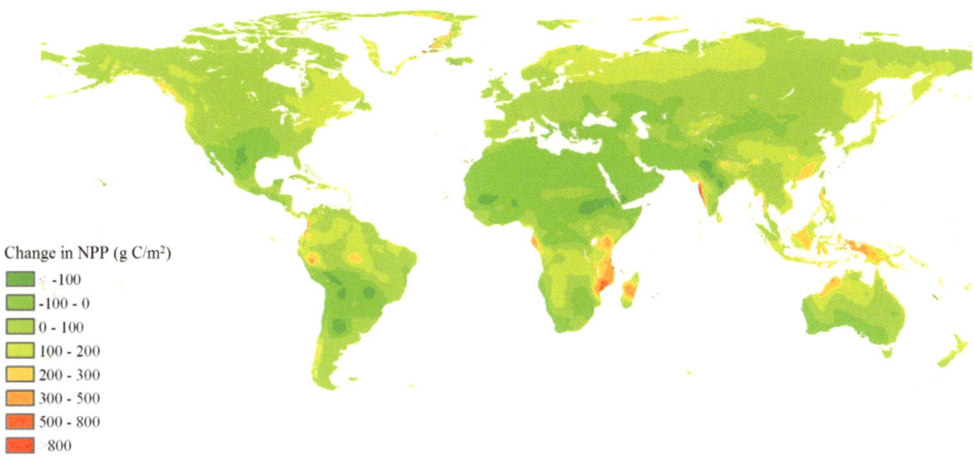

Note: The legends show the differentiated NPP and negative values indicate C loss for NPP.

Figure 16. Trends in global NPP anomalies from 1950 to 2050 computed by the Classification Indices-based Model.

At the class level, the largest increases are in Tropical-humid seasonal rain forest (VIIE 35), Tropical arid tropical desert brush (VIIB14), Tropical-subhumid tropical xerophytic forest (VIID28) and Tropical-semiarid savanna (VIIC 21), as a result of increased MAT and MAP in Africa and the Americas, while TNPP of Frigid perhumid rain tundra, alpine meadow (IF 36), Cold temperate perhumid taiga forest (IIF 37), Sub-tropical perhumid evergreen broad leaved forest (VIF 41), Warm-perhumid deciduous-evergreen broad leaved forest (VF 40), Warm temperate perhumid deciduous broad leaved forest (IVF 39), and Subtropical-subhumid sclerophyllous forest (VID27) decrease sharply. The TNPP of Subtropical-semiarid subtropical brush steppe (VIC20) in Africa shows a marked decrease due to decreased precipitation (Table 9).

At the super-class level, the TNPP of Savanna shows the largest increase, followed by Steppe, Temperate humid grassland, Warm desert and Semi-desert. These increases in TNPP are 2093.68 Tg C, 482.65 Tg C, 316.22 Tg C, 124.21 Tg C and 75.52 Tg C, respectively. The TNPP of Tundra and alpine steppe shows the largest decrease. The simulated TNPP of Tundra and alpine steppe will decrease by 1178.72 Tg C compared with its size during 1950-2000. The decrease of TNPP for Cold desert will be 7.42 Tg C (Table 9).

At the category level, the TNPP of Grassland category will increase 12.40% to 1906.14 Tg C over this period (1950-2050) (Tables 6, 7 and 9).

Biome (super-class group)	Class-ID	Global		Asia		Europe		Africa		America		Oceania	
		Current	A2a	Current	A2a	Current	A2a	Current	A2a	Current	A2a	Current	A2a
Tundra and alpine steppe	IA1	0.02	0.96	0.01	0.23	-	0.07	-	0.05	0.01	0.46	-	0.15
	IB8	5.47	6.82	4.35	6.38	-	-	-	-	1.12	0.44	-	-
	IC15	4.77	5.58	4.04	4.46	-	-	-	-	0.74	1.13	-	-
	ID22	7.10	12.38	6.15	5.95	-	-	-	-	0.95	6.44	-	-
	IE29	38.79	70.51	27.49	34.69	-	-	-	-	11.30	35.83	-	-
	IF36	2436.19	1217.36	1249.53	433.86	215.92	71.70	0.12	0.04	967.04	710.92	3.58	0.84
Frigid desert	IIA2	2.51	1.77	1.81	1.06	-	-	-	-	0.70	0.71	-	-
	IIIA3	23.32	15.78	21.16	12.20	-	-	-	-	2.16	3.58	-	-
	IVA4	66.35	67.21	60.71	59.96	-	-	0.54	0.01	5.10	7.24	-	-
Semi-desert	IIB9	26.13	21.89	19.38	18.90	-	-	-	-	6.75	3.00	-	-
	IIIB10	357.26	276.52	263.64	214.83	1.75	3.90	-	-	91.88	57.78	-	-
	IVB11	470.08	762.99	218.25	400.46	56.49	134.40	35.45	3.53	158.16	224.59	1.73	-
	VB12	511.56	379.16	88.41	104.32	37.32	56.06	98.73	41.67	161.61	151.16	125.50	25.94
Steppe	IIC16	27.93	26.26	22.02	21.05	-	-	-	-	5.91	5.20	-	-
	IIIC17	363.16	312.94	203.41	194.63	33.32	9.36	0.07	-	126.35	108.94	0.01	-
	IVC18	308.63	611.41	85.69	162.77	111.39	284.52	17.09	5.38	84.42	156.78	10.04	1.97
	VC19	271.69	235.16	16.99	60.69	62.59	65.19	52.30	27.77	103.17	53.83	36.64	27.69
	IVD25	400.10	668.39	96.81	105.43	178.71	423.78	23.22	6.62	75.78	119.67	25.59	12.90
Temperate humid grassland	IID23	53.99	101.44	43.96	60.01	-	-	-	-	10.03	41.44	-	-
	IIID24	460.59	573.19	193.04	342.42	90.80	64.68	0.67	0.02	175.48	166.03	0.60	0.05
	IIE30	326.75	482.92	218.82	338.35	0.25	-	-	-	107.68	144.57	-	-
Warm desert	VA5	56.55	52.91	42.65	44.25	-	-	7.83	1.50	6.07	7.17	-	-
	VIA6	351.22	265.31	97.70	109.45	0.25	0.22	96.29	56.62	24.16	31.20	132.82	67.82
	VIIA7	703.24	916.99	212.04	248.00	-	0.03	336.03	420.98	9.29	32.51	145.87	215.47
Savanna	VIB13	1367.24	1337.50	77.23	140.47	19.07	106.51	566.23	342.92	288.82	408.11	415.89	339.49
	VIIB14	3357.22	4848.10	487.68	655.00	-	1.38	1685.12	2316.33	476.02	780.40	708.40	1094.98
	VIC20	714.88	573.17	16.79	75.25	18.77	74.33	372.22	167.42	215.48	204.94	91.63	51.23
	VIIC21	2664.61	3438.87	533.02	629.19	-	0.11	1355.63	1592.66	581.57	973.80	194.40	243.12
Forest, including temperate forest, Sub-tropical forest and tropical forest	VD26	275.02	252.60	65.16	46.27	30.99	70.36	52.70	20.71	99.26	78.71	26.91	36.56
	VID27	768.75	569.52	32.99	136.59	7.98	23.34	438.75	157.32	260.74	229.46	28.30	22.82
	VID28	2991.91	4141.40	668.94	674.43	-	-	1187.64	1629.08	973.51	1643.08	161.82	194.82
	IIIE31	846.52	1313.47	232.00	549.65	433.70	411.61	2.20	0.63	175.74	350.61	2.88	0.96
	IVE32	628.69	860.05	125.12	116.91	257.76	383.06	27.15	8.52	160.99	319.82	57.68	31.74
	VE33	475.38	488.95	211.14	79.74	19.27	63.57	45.45	20.42	176.58	293.46	22.94	31.75
	VIE34	1876.90	1763.09	279.98	419.33	2.34	7.47	785.68	310.50	780.53	993.43	28.38	32.35
	VIIE35	6155.32	7697.00	912.79	1054.92	-	-	2556.00	3506.24	2617.13	3036.96	69.40	98.88
	IIF37	3558.44	2714.89	1447.30	1357.99	658.84	320.88	1.54	0.27	1424.26	1022.19	26.50	13.56
	IIIF38	2575.79	2967.37	466.08	798.74	1064.25	868.80	9.44	4.09	946.25	1233.80	89.76	61.93

Biome (super-class group)	Class-ID	Global		Asia		Europe		Africa		America		Oceania	
		Current	A2a	Current	A2a	Current	A2a	Current	A2a	Current	A2a	Current	A2a
	IVF39	1611.53	1384.49	475.71	356.44	172.41	242.06	44.36	13.16	784.34	634.75	134.71	138.09
	VF40	1094.95	565.79	526.52	298.72	7.25	36.75	73.45	28.14	466.20	168.82	21.53	33.35
	VIF41	2723.58	1889.94	1288.70	1161.66	-	4.23	294.95	194.80	1109.11	511.00	30.82	18.25
	VIIF42	13545.59	13781.35	4329.50	5041.14	-	-	1507.98	1158.16	7621.36	7475.14	86.74	106.92

Note: The class name refers to Figure 7 Explanation.Unit of TNPP is TgC.

Table 9. Comparison of TNPP of biomes and classes estimated by the Classification Indices-based Model under the baseline climate (recent past (1950-2000)) and those under future (2000-2050) climatic scenario (A2a)

4.3.3. Discussion

4.3.3.1. Biome conversions

The changing climate will affect the growth of plants, alter the species composition of plant communities and ultimately change community structure [87]. Under these conditions, the classes will shift [3,5] and result in biome conversions [3,5]. Our analysis indicated that Steppe will experience the largest variation in biome conversions. The result is the same as the conclusion of Liu et al. [88], who confirmed that under climate warming, the temperate steppe in the arid and semiarid regions of northern China will act as a net C source. The Savanna and Temperate humid grassland will have the highest percentage of growth. The Tundra and alpine steppe biome will have the greatest change in total area, followed by Forest (Tables 6, 7 and 9).

4.3.3.2. Trends in grassland NPP in response to climate change

The Classification Indices-based Model estimates global potential grassland TNPP at 16.47 Pg C and global TNPP at 57.67 Pg C under future climatic A2a scenario (Table 7). Climate change is expected to affect the TNPP of grassland. Understanding climate and NPP feedback cycles in grasslands requires knowledge of the changes in spatial patterns on the NPP of grasslands. The dynamic change characteristics of potential biomes (Table 8 and Figure 15) reflect further evidence of increasing global warming since the early 1980s (Intergovernmental Panel on Climate Change, 2007). The Classification Indices-based Model estimated an increase of 1.09 Pg C in TNPP for global potential grassland from recent past to future A2a scenario (Tables 6, 7 and 9). It will bring an increase of 3.91 Pg in annual sequestration of CO_2 and 2.88 Pg of annual contribution to the atmospheric O_2 cycle respectively (1.63 g of CO_2 is absorbed and 1.2 g of O_2 is released to generate 1 g of DM according to photosynthesis stoichiometry; [89]), which will impose a new issue for future grassland research to support sustainable development and to provide relevant knowledge to meet the challenge of climate change [90].

4.4. Spatio-temporal dynamics on the distribution, extent and NPP of potential grassland in response to climate changes in China

4.4.1. Characters of potential grassland over 2001-2050 under A2a scenario

In the A2a scenario, the Tundra and alpine steppe, Cold desert, Semi-desert, Steppe, Temperate humid grassland, Warm desert, and Savanna were projected to occupy 8.9%, 8.8%, 11.1%, 12.0%, 5.9%, 6.2%, and 1.8% of land area of China, respectively (Table 10). Under scenario A2a, the Classification Indices-based Model projected the highest TNPP to be in the Forest category. Table 10 shows that the highest TNPP values of grassland biomes were projected to be in the Steppe, followed by Semi-desert, Tundra and alpine steppe, and Temperate humid grassland, respectively. In conclusion, it was projected that the two main grassland biomes found in China will be in the Steppe and Semi-desert, and they will occupy 42.2% of China's total grassland areas and 58.3% of China's Grassland TNPP in the future climatic A2a scenario (Table 10).

4.4.2. Comparing the area and TNPP of China's major terrestrial biomes of 1950-2000 with those of 2001-2050 under the A2a scenario

Climate change was projected to influence the spatial distribution of grassland biomes, with a combined distribution ranging from the northeastern plain adjacent to Mongolia to the southern Tibetan Plateau, with warm desert showing the greatest discrepancies. The distribution of the area for various grassland biomes in these two time periods was projected to change considerably from the recent past (1950-2000) (Table 4) to the future (2001-2050) in the A2a scenario (Table 10). The area of Tundra and alpine steppe was projected to decrease by 107.9 m ha compared with its size during 1950-2000. The TNPP of Tundra and alpine steppe was projected to change from 250.7 to114.0 Tg C. This is a huge reduction of 54.5% due to global climate change. Likewise, the areas of the Cold desert and Temperate humid grassland were projected to decrease by 40.9 and 0.5 m ha, respectively, while the TNPPs of the Cold desert and Temperate humid grassland were projected to decrease by 1.4 Tg C and 14.1 Tg C, respectively. The Warm desert and Savanna super-class groups were projected to increase the most. The area of Warm desert, Steppe, Semi-desert, and Savanna were projected to increase considerably, by nearly 138.2 m ha (Tables 4 and 10).

4.4.3. Numerical simulations on potential grassland dynamics

4.4.3.1. Changes in class

The changing climate is likely to affect the growth of plants, alter the species composition of plant communities, and ultimately change plant community structure [87]. It is anticipated that the classes will shift [3, 5] and will result in biome conversions. Under the A2a scenario, a new class (sub-tropical-extra-arid sub-tropical desert, VIA6) was projected to appear in conjunction with climate warming and drying, and this would be distributed in the Xinjiang Uygur Autonomous Region. The class called Frigid per-humid rain tundra, alpine meadow (IF 36), mainly distributed in the Tibetan Plateau, is projected to decrease continuously under the A2a scenario, with a rate of decrease per decade of 6.8%. If such rates persist, Frigid per-

Biome(super-class group)	Area(million hectares)	TNPP (Tg C)
		Classification Indices-based Model
Tundra and alpine steppe	83.54	114.02
Cold desert	82.36	34.68
Semi-desert	103.66	163.53
Steppe	112.38	318.95
Temperate humid grassland	55.23	110.58
Warm desert	58.41	11.03
Savanna	16.97	75.49
Forest, including Temperate forest, Sub-tropical forest and Tropical forest	424.97	2455.82

Table 10. TNPP of major China's terrestrial biomes estimated by the Classification Indices-based Model from 2001 to 2050 under A2a scenario

humid rain tundra, alpine meadow (IF 36) could potentially disappear by the end of the century.

4.4.3.2. Trends in super-classes

The A2a scenario projects changes in China's temperature and rainfall patterns, making wet areas wetter and dry areas drier (Figure 17), resulting in class succession from an original class to a new class if one just considers changes in climate conditions. The model outputs suggest that the conversion of classes from the recent past to the A2a scenario would result in changes in distribution patterns of the area and TNPP among biomes. Table 11 showed the projected results of the changes in areas of biome conversions between the recent past and the future climatic A2a scenario. For example, in this long time sequence, an area of 58.0 million hectares of Tundra and alpine steppe would be converted to Forest (Legend 18 in Figure 18a and Table 11), which would take place mainly in the Tibet Autonomous Region (accounting for 40.1% of the change) and Qinghai Province (35.9% of change), while another 28.9 million hectares would change to Temperate humid grassland (Legend 15 in Figure 18a and Table 11), and this change would mainly take place in the Tibet Autonomous Region (78.1%). Likewise, an area of 56.9 million hectares of Cold desert under recent climate conditions would be converted into Warm desert under the A2a scenario (Legend 26 in Figure 18a and Table 11), mainly taking place in the Xinjiang Uygur Autonomous Region (99.1%). An area of 43.7 million hectares of Temperate humid grassland would be converted into Steppe (Legend 54 in Figure 18a and Table 11), mainly taking place in the Inner Mongolia Autonomous Region (42.5%). An area of 36.3 million hectares of Forest under recent climate conditions would be converted into Temperate humid grassland under the A2a scenario (Legend 85 in Figure 18a and Table 11), mainly taking place in both Heilongjiang Province (38.9%) and Inner Mongolia Autonomous Region (29.3%), respectively. There would be an area of 28.5 million hectares of Steppe changing to Semi-desert

(Legend 43 in Figure 18a and Table 11), mainly taking place at the Inner Mongolia Autonomous Region (44.8%).

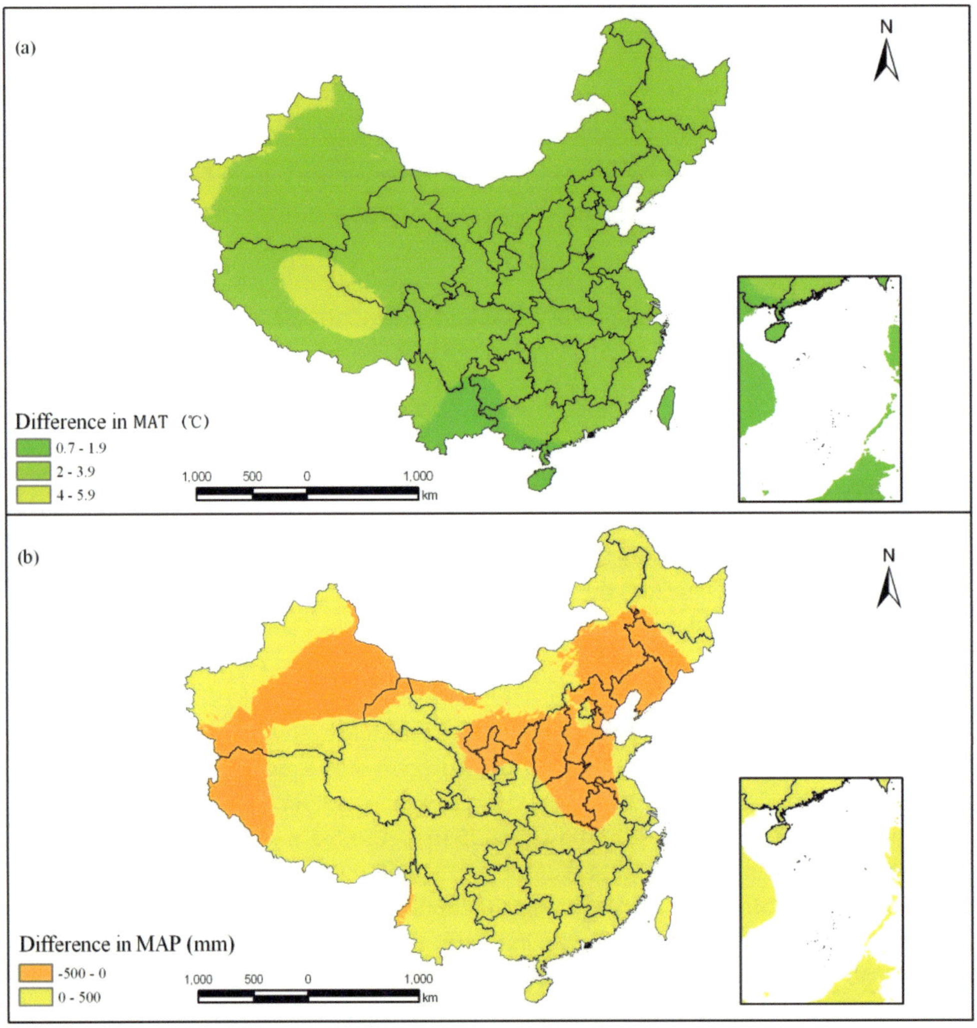

Figure 17. Maps of changes under the future climatic A2a scenario (1950-2000) relative to the average for 1950-2000 for (a) mean annual temperature (MAT) and (b) mean annual precipitation (MAP) in China.

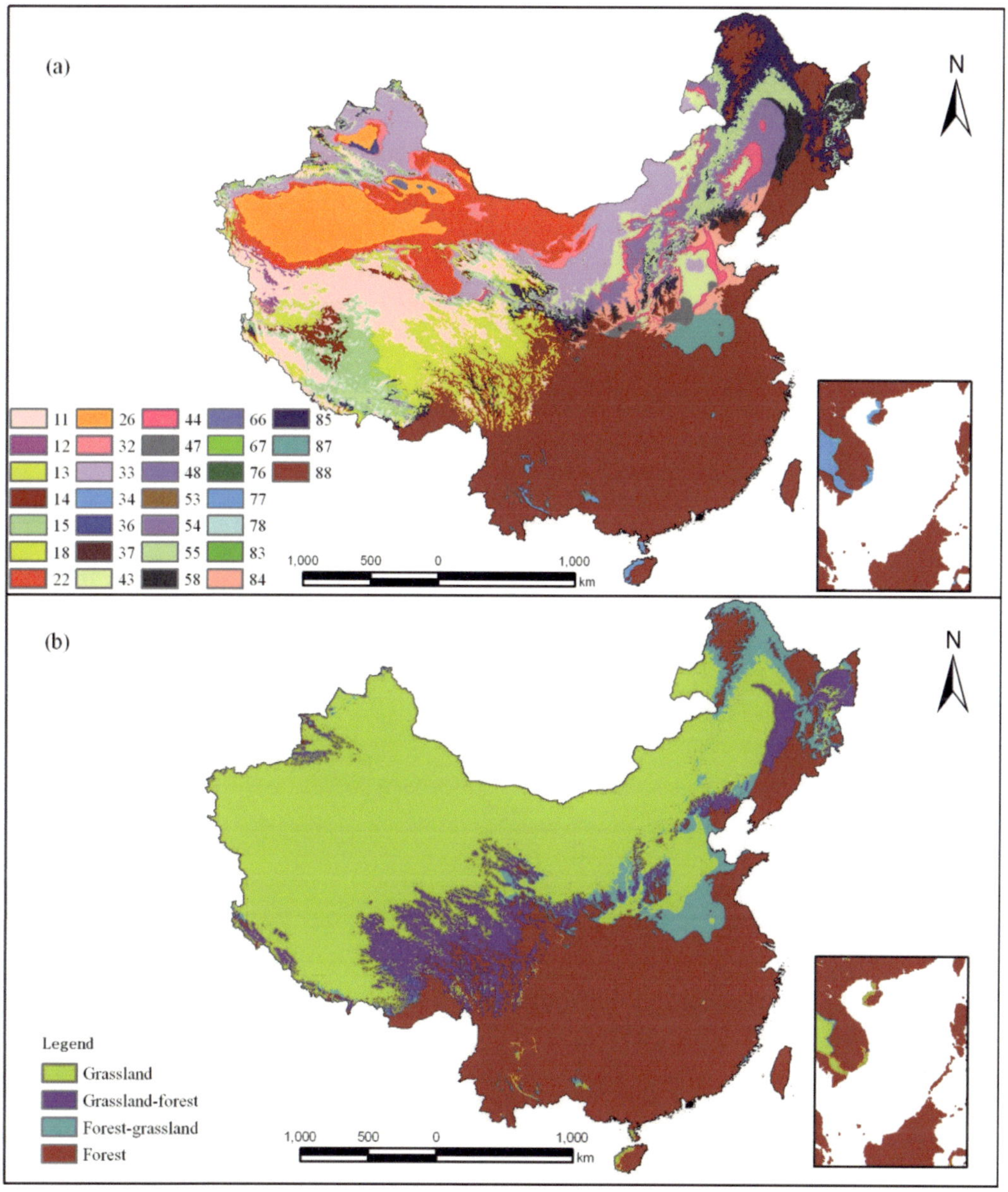

Figure 18. Spatial distribution dynamics of potential biomes in China from the recent past to the future (2001-2050) in the A2a scenario.

Code	China	Inner Mongolia Autonomous Region	Tibet Autonomous Region	Gansu Province	Qinghai Province	Xinjiang Uygur Autonomous Region	Heilongjiang Province
11	83.54	-	39.62	1.29	24.41	18.76	-
12	2.60	-	0.91	0.01	0.03	1.65	-
13	12.95	-	5.66	0.37	1.49	5.44	-
14	8.46	-	5.70	0.32	0.81	1.62	-
15	28.90	-	22.56	0.70	2.31	3.33	-
18	58.02	0.01	23.26	1.67	20.82	2.22	-
22	66.61	18.72	-	11.40	8.41	28.09	-
26	56.94	-	-	0.49	-	56.45	-
32	13.48	3.52	0.30	2.37	0.71	6.57	-
33	62.17	22.95	0.16	4.47	3.98	26.51	-
36	0.81	-	-	-	-	0.81	-
37	0.06	-	-	0.01	-	-	-
43	28.53	12.77	0.07	1.73	0.44	3.35	-
44	18.81	8.33	0.01	0.38	0.48	0.58	0.12
47	4.64	-	-	0.08	-	-	-
53	0.76	-	0.03	0.03	-	0.71	-
54	43.72	18.49	0.55	3.06	1.07	3.34	3.45
55	33.00	12.15	3.86	1.04	2.16	1.59	6.76
58	23.36	0.20	0.72	2.66	0.02	0.07	8.77
66	0.80	-	-	-	-	0.80	-
77	1.25	-	-	-	-	-	-
78	0.09	-	-	-	-	-	-
84	18.26	-	0.03	1.52	-	-	-
85	36.31	10.62	2.95	1.36	2.08	1.32	14.10
87	12.64	-	-	0.13	-	-	-
88	329.27	6.93	14.33	5.50	2.52	0.46	12.02

Note: The symbols are defined in the same way as in Table 8.

Table 11. Projected changes in area(million hectares) of biome conversions from 1950 to 2050

An area of 81.5 million hectares of the Grassland category would be converted to the Forest category and 67.2 million hectares of the Forest category would be converted into the Grassland category (Figure 18b and Tables 4, 10, 11). There is a clear trend for a decrease in the area of the Grassland category by 11.1 million hectares, whereas areas of the Forest category would increase by 11.3 million hectares (Tables 4, 10 and 11). The relative area of the Forest category would increase by 2.7% over this period (1950-2050), whereas the relative area of the Grassland category would decrease by 2.1% over the same period.

4.4.3.3. Change in TNPP

The Classification Indices-based Model shows NPP either neutral or increasing for most Chinese grid cells during the periods from the recent past through the A2a scenario (Figure 19). At the class level, the largest increases would be in Subtropical-humid evergreen broad-leaved forest (VIE 34), Tropical-per-humid rain forest (VIIF 42), Subtropical per-humid evergreen broad-leaved forest (VIF 41), Subtropical-subhumid sclerophyllous forest (VID 27), and Tropical-humid seasonal rain forest (VIIE 35), as a result of increased MAT and MAP in southern China. The TNPP of Warm temperate-semi-arid warm temperate typical steppe (IVC, 18) in the north-east of the Inner Mongolia Plateau and the Plain of Northern China would increase by 71.69 Tg C year^{-1} because of increased precipitation in these regions. Warmer temperatures would be primarily responsible for increases in NPP in the Tibet Autonomous Region, i.e., the TNPP of Warm temperate-arid warm temperate zonal semi-desert (IVB 11), Cold temperate per-humid taiga forest (IIF 37), and Cold temperate-humid montane meadow (IIE 30) would increase, whereas the TNPP of Frigid per-humid rain tundra, alpine meadow (IF 36) would decrease sharply. The TNPP of Cool temperate-arid temperate zonal semi-desert (IIIB 10) in both the middle of the Xinjiang Uygur Autonomous Region and the northern area of the Inner Mongolia Autonomous Region would show a marked decrease due to decreased precipitation (Table 12).

Biome (super-class group)	Class-ID	China		Inner Mongolia Autonomous Region		Tibet Autonomous Region		Gansu Province		Qinghai Province		Xinjiang Uygur Autonomous Region	
		Recent past	A2a	Recent past	A2a	Recent past	A2a	Recent past	A2a	Recent past	A2a	Recent past	A2a
Tundra and alpine steppe	IA1	0.10	2.44	-	-	-	0.24	-	-	-	-	0.10	0.11
	IB8	37.93	57.01	-	-	9.45	30.04	0.14	0.05	0.90	0.13	22.65	26.78
	IC15	31.53	39.20	-	-	11.32	19.66	0.18	0.12	0.66	0.74	14.28	18.69
	ID22	41.24	46.78	-	-	20.36	27.82	0.21	0.34	0.59	2.04	14.62	16.58
	IE29	91.93	106.91	-	-	55.21	77.19	0.33	2.16	1.00	8.34	24.73	19.23
	IF36	2304.27	887.90	0.25	-	1032.10	321.61	3.49	17.82	45.98	420.81	164.80	76.11
Frigid desert	IIA2	18.12	10.75	-	-	-	3.70	0.38	0.12	10.33	0.70	7.41	6.22
	IIIA3	128.53	57.75	44.26	0.01	-	-	36.31	11.83	9.62	27.20	38.33	18.71
	IVA4	214.23	278.30	38.47	110.28	-	-	12.36	52.25	-	-	163.40	115.77
Semi-desert	IIB9	54.70	110.24	0.04	-	2.30	41.86	6.38	5.88	17.71	17.79	28.28	44.70
	IIIB10	691.68	463.69	313.17	271.30	-	0.01	60.47	22.08	19.65	44.18	266.21	124.96

Biome (super-class group)	Class-ID	China		Inner Mongolia Autonomous Region		Tibet Autonomous Region		Gansu Province		Qinghai Province		Xinjiang Uygur Autonomous Region	
		Recent past	A2a	Recent past	A2a	Recent past	A2a	Recent past	A2a	Recent past	A2a	Recent past	A2a
	IVB11	188.23	793.74	38.62	318.08	-	-	8.51	71.95	-	0.30	92.40	276.57
	VB12	2.14	267.67	-	-	-	-	0.17	0.08	-	-	-	0.71
Steppe	IIC16	40.90	110.53	3.26	-	0.68	66.13	3.37	4.21	7.00	11.72	26.59	28.47
	IIIC17	597.23	483.52	414.72	342.19	-	8.80	35.78	17.42	5.90	22.27	31.95	51.72
	IVC18	519.03	1235.92	2.49	279.73	0.31	4.67	4.37	107.37	-	6.05	-	10.31
	VC19	117.75	509.91	-	-	-	0.62	1.56	19.49	-	-	-	-
	IVD25	841.62	849.66	-	6.07	0.72	16.09	61.37	119.16	-	0.58	-	2.12
Temperate humid grassland	IID23	62.18	187.51	14.37	0.05	4.53	138.15	4.05	4.80	9.43	18.23	29.68	26.27
	IIID24	976.28	594.58	397.16	263.99	9.38	59.84	63.01	25.79	10.86	31.26	15.83	30.75
	IIE30	208.57	323.66	75.29	0.17	55.82	233.95	10.61	8.16	24.67	43.97	40.26	37.40
Warm desert	VA5	0.39	109.54	-	-	-	-	-	0.28	-	-	0.39	109.26
	VIA6	-	0.74	-	-		-		-		-		0.74
Savanna	VIB13	1.47	117.36	-	-	-	-	-	1.32	-	-	-	-
	VIIB14	3.85	7.68	-	-	-	-	-	-	-	-	-	-
	VIC20	31.08	568.06	-	-	-	-	0.02	8.37	-	-	-	-
	VIIC21	30.73	61.80	-	-	-	-	-	-	-	-	-	-
Forest, including Temperate forest, Sub-tropical forest and Tropical forest	VD26	583.92	393.29	-	-	-	0.35	2.50	20.88	-	-	-	-
	VID27	219.17	1299.22	-	-	-	-	-	1.80	-	-	-	-
	VIID28	90.26	189.93	-	-	-	-	-	-	-	-	-	-
	IIIE31	1247.72	1240.25	209.53	322.60	24.87	62.07	104.06	36.77	11.71	36.23	6.52	23.37
	IVE32	1118.61	925.41	-	-	1.64	54.73	48.60	47.17	-	-	-	-
	VE33	2050.76	755.60	-	-	-	4.75	-	0.69	-	-	-	-
	VIE34	1911.89	3859.07	-	-	-	0.20	-	-	-	-	-	-
	VIIE35	622.82	1623.86	-	-	-	-	-	-	-	-	-	-
	IIF37	1502.54	1879.78	388.80	81.96	229.04	611.51	96.39	75.86	95.61	546.35	33.71	49.97
	IIIF38	1827.77	1456.20	34.63	105.08	96.09	177.73	56.28	70.14	0.53	31.00	0.18	8.03
	IVF39	1904.99	1027.94	-	-	119.58	71.14	0.82	2.83	-	-	-	-
	VF40	3402.34	1221.15	-	-	99.21	75.22	-	-	-	-	-	-
	VIF41	5166.77	6509.44	-	-	229.74	231.03	-	-	-	-	-	-
	VIIF42	628.69	2177.01	-	-	54.25	203.87	-	-	-	-	-	-

Note: The class name refers to Fig.7. Explanation

Table 12. Comparison of the total net primary productivity (10^{11} g C) of biomes and classes of 1950-2000(recent past) with those of 2000-2050 under A2a scenario

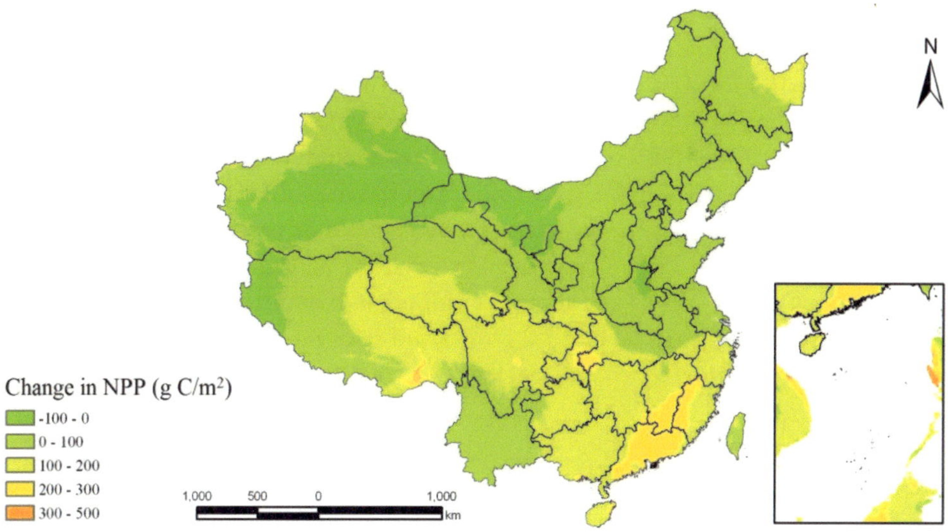

Figure 19. Trend of Changes in net primary productivity (NPP) from the recent past to the future (2001-2050) in the A2a scenario. The changes show the differentiated NPP, and negative values indicate C loss for NPP.

At the grassland biome level, the TNPP of Steppe would show the largest increase, followed by Semi-desert, Savanna and Warm desert. These increases in TNPP are 107.3, 69.9, 68.8 and 11.0 Tg C, respectively. The TNPP of Tundra and alpine steppe would show the largest decrease, followed by Temperate humid grassland and Cold desert. The simulated TNPP of Tundra and alpine steppe would decrease by 136.7 Tg C compared with its size during 1950-2000. The decrease of TNPP for Temperate humid grassland and Cold desert would be 14.1 Tg C and 1.4 Tg C, respectively (Tables 4, 10 and 12).

At the category level, the TNPP of Grassland category would increase by 14.5% to 104.7 Tg C over this period (1950-2050), whereas the TNPP of Forest category would increase by 10.2% to 228.0 Tg C (Tables 4, 10 and 12).

4.4.4. Discussion

4.4.4.1. Grassland carbon budgets

The Classification Indices-based Model projects China's TNPP at 3.28 Pg C (Table 10) under the future climatic A2a scenario and is close to the projections of 3.33 and 3.34 Pg C by the Vegetation Interaction Model (AVIM) [91] and the CEVSA (Carbon Exchange between Vegetation, Soil, and Atmosphere) model [92], both derived by using long-term annual average data (1980-2005). The results could potentially favor China's position in supporting its effort to reduce global warming gas as outlined in the Kyoto Protocol. The model outputs suggest that the potential TNPP of grasslands in China is 0.83 Pg C in the A2a scenario (Table 10).

Climate change is expected to affect the TNPP of grassland. Understanding climate and NPP feedback cycles in grasslands requires knowledge of the changes in spatial patterns on the NPP of grasslands. The dynamic change characteristics of potential biomes (Table 12 and Figure 18) highlights the risk that China's grassland faces with a climate that is 'already hot, dry and variable' [85] under global warming. The Classification Indices-based Model estimated an increase of 104.71Tg C in TNPP for potential grasslands from recent past to the future A2a scenario. The increase in TNPP for grasslands at northern latitudes was mainly driven by increased temperature and, in lower latitudes, by changes in precipitation (Figure 17 and Figure 19). The increase in TNPP for grasslands will bring an increase of 375.5 Tg of annual sequestration of CO_2 and 276.4 Tg of annual contribution to the atmospheric O_2 cycle respectively (1.63g of CO_2 are absorbed and 1.2 g of O_2 are released to generate 1g of dry matter according to photosynthesis stoichiometry) [89].

4.4.4.2. 'Safe' livestock carrying capacity

The goal of sustainable livestock husbandry is to balance the consumption of forage by livestock with the annual production of the grassland, so that future grassland production is not diminished. The concept of a 'safe' livestock carrying capacity was developed to estimate the capacity of the grassland resource to sustainably carry livestock (and other herbivores) in the long term (>30 years) [83]. Predictions of NPP give a good estimate of the annual production of the grasslands, and therefore NPP can be used to balance the food demands of different kinds of livestock, allowing a calculation of the maximum carrying capacity of livestock that can be supported on the grassland. This maximum carrying capacity of the grassland should be used to develop appropriate policies of grazing control to preserve sustainable production on the grassland [16]. The implication of climate change projections for 'safe' livestock carrying capacity remains an open question. Global warming causes changes in temperature and precipitation patterns, resulting in fluctuations in the NPP of grasslands. Several researchers have predicted that increases in temperature will cause a shift in the ratio of C_4 to C_3 grasses, with C_4 increasing [17, 93]. This may decrease carrying capacity, since C_4 grasses are generally considered to be of lower nutritive value to grazing animals than C_3 species [94]. Moreover, increasing temperature and declining precipitation are also considered to decrease dietary crude protein, protein availability, and digestible organic matter [93]. In situations where climates become warmer and drier, grazing livestock are likely to encounter a protein-limited diet. The future declines in forage quality would also produce greater methane production from ruminant livestock [93]. Thus, it is important to assess climate change projections in terms of their implications for future 'safe' livestock carrying capacity and the risk of resource degradation [83]. However, the questions of what the 'safe' livestock carrying capacity is and of how best to integrate a wide range of factors, such as grassland classes, climatic variability and animal nutrition, are unresolved. So further research and development is needed to identify the regional trends for the 'safe' livestock carrying capacity to maintain sustainable resource condition and reduce the risk of resource degradation. This important task remains a challenge for all grassland scientists and practitioners.

Acknowledgements

The research was funded by the National Natural Science Foundation of China (No. 31172250) and the project of the Humanities and Social Sciences Planning Fund of the Chinese Ministry of Education (No. 10YJAZH047). This chapter is dedicated to Wanquan Lin, the father of Huilong Lin, who left this earth on July 3, 2013. He was a source of constant encouragement, wise counsel, and a remarkable example of hard work, integrity and a loving father.

Author details

Huilong Lin*

Address all correspondence to: linhuilong@lzu.edu.cn

State Key Laboratory of Grassland Agro-Ecosystems, College of Pastoral Agriculture Science and Technology, Lanzhou University, Lanzhou City, China

References

[1] Hui DF, Jackson RB. Geographical and interannual variability in biomass partitioning in grassland ecosystems: a synthesis of field data. New Phytologist 2006; 169 85-93.

[2] Scurlock JM, Hall D. The global carbon sink: a grassland perspective. Global Change Biology 1998; 4 229-233.

[3] Scurlock LMO, John K, Olson RJ. Estimating net primary productivity from grassland biomass dynamics measurements. Global Change Biology 2002; 8 736-753.

[4] Del Grosso S, Parton W, Stohlgren T, et al. Global potential net primary production predicted from vegetation class, precipitation, and temperature. Ecology 2008; 89 2117-2126.

[5] Holdridge LR. Determination of world plant formations from simple climatic data. Science 1947; 105 367-368.

[6] Lieth H. Modelling the primary productivity of the world. Nature and Resources 1972; 8 (2) 5-10.

[7] Uchijima Z. An agroclimatic method of estimating net primary productivity of natural vegetation. JARQ 1988; 21 244-250.

[8] Uchijima Z, Seino H. Agroclimatic evaluation of net primary productivity of natural vegetation. I. Chikugo model for evaluating net primary productivity. Journal of Agricultural Meterology 1985; 40 343-352(in Japanese with English summary).

[9] Zaks DPM, Ramankutty N, Barford CC, et al. From Miami to Madison: Investigating the relationship between climate and terrestrial net primary production. Global Biogeochemical Cycles 2007; 21 GB3004, DOI: 10. 1029/2006GB002705.

[10] Zhou GS, Zhang XS. A natural vegetation NPP model. Acta Phytoecologica Sinica 1995; 19(3) 193-200 (In Chinese with English abstract).

[11] Zhou GS, Wang YH, Jiang YL, et al. Estimating biomass and net primary production from forest inventory data: a case study of China's Larix forests. Forest Ecology and Management 2002; 169 149-157.

[12] Zhu ZH. A model for estimating net primary productivity of natural vegetation. Chinese Science Bulletin 1993; 38(22) 1913-1917.

[13] Melillo JM, Mcguire AD, Kicklighter DW, et al. Global climate change and terrestrial net primary production. Nature 1993; 363 234-240.

[14] Adams B, White A, Lenton TM. An analysis of some diverse approaches to modelling terrestrial net primary productivity. Ecological Modelling 2004; 177 353-391.

[15] Schuur EAG. Productivity and global climate revisited: the sensitivity of tropical forest growth to precipitation. Ecology 2003; 84 1165-1170.

[16] Zhang GG, Kang YM, Han GD, et al. Effect of climate change over the past half century on the distribution, extent and NPP of ecosystems of Inner Mongolia. Global Change Biology 2011; 17 377-389.

[17] Ren JZ, Hu ZZ, Zhao J, et al. A Grassland classification system and its application in China. The Rangeland Journal 2008; 30 199-209.

[18] Lobo A, Rebollar JL. Model-based discriminant analysis of Iberian potential vegetation and bio-climatic indices. Physics and Chemistry of the Earth 2010; 35 52-56.

[19] Zhou GS, Wang YH. Global change and climate-vegetation classification, Chinese Science Bulletin 2000; 45 577-584.

[20] Liang TG, Feng QS, Cao JJ, et al. Changes in global potential vegetation distributions from 1911 to 2000 as simulated by the Comprehensive Sequential Classification System approach. Chinese Science Bulletin 2012; 57 1298-1310.

[21] Kaplan JO, Bigelow NH, Prentice IC, et al. Climate change and arctic ecosystems II: Modelling , paleodata-model comparisons, and future projections. Journal of geophysical research 2003; 108 8171.

[22] Lin HL. A New Model of Grassland Net Primary Productivity (NPP) Based on the Integrated Orderly Classification System of Grassland. In: Chen YX, Deng HP, Zhang DG, et al (eds). The Sixth International Conference on Fuzzy Systems and Knowl-

edge Discovery. Tianjing, IEEE Computer Society Conference Publishing Services 2009; (1) 52-56. DOI: 10. 1109/FSKD. 2009. 705

[23] Liang TG, Feng QS, Huang XD, et al. Advance in the Study of Integrated Orderly Classification System of Grassland. Acta prataculture sinica 2011; 20 (3) 79-82(in Chinese with English abstract).

[24] Lin HL, Zhao J, Liang TG, et al. A Classification indices-based model for net primary productivity (NPP) and potential productivity of vegetation in China. International Journal of Biomathematics 2012; 5(3) 1-23. DOI: 10. 1142/S1793524512600091

[25] Lin HL, Zhang YJ. Evaluation of six methods to predict grassland net primary productivity along an altitudinal gradient in the Alxa Rangeland, Western Inner Mongolia, China. Grassland Science 2013; 59 100-110. DOI: 10. 1111/grs. 12019.

[26] Zhang L, Wylie B, Loveland T, et al. Evaluation and comparison of gross primary production estimates for the Northern Great Plains grasslands. Remote Sensing of Environment 2007; 106 173-189.

[27] Lin HL, Feng QS, Liang TG, et al. Modelling global-scale potential grassland changes in spatio-temporal patterns to global climate change. International Journal of Sustainable Development & World Ecology 2013; 20(10) 83-96.

[28] Lin HL, Wang XL, Zhang YJ, et al. Spatiotemporal dynamics on the distribution, extent and NPP of potential grassland in response to climate changes in China. The Rangeland Journal 2013;35(4) 409-425. DOI: 10. 107/RJ12024.

[29] Nash JE, Sutcliffe JV. River flow forecasting through conceptual models. Part 1. A discussion of principles. Journal of Hydrology 1970; 10 282-290.

[30] Lin HL, Zhuang QM, Fu H. Habitat Niche-fitness and Radix Yield Prediction Models for Angelica sinensis Cultivated in the Alpine Area of the Southeastern Region of Gansu Province, China. Plant Production Science 2008; 11(1) 42-58.

[31] Whittaker RH. A criticism of the plant association and climatic climax concepts. Northwest Science 1951; 25 17-31.

[32] Ni J. Net primary productivity in forests of China: scaling-up of national inventory data and comparison with model predictions. Forest Ecology Management 2003; 176 485-495.

[33] Ni J. Forage yield-based carbon storage in grasslands of China. Climatic Change 2004a; 67 237-246.

[34] Ni J. Estimating net primary productivity of grasslands from field biomass measurements in temperate northern China. Plant Ecology 2004b; 174 217-234.

[35] Kang L, Han XG, Zhang ZB, et al. Grassland ecosystems in China: review of current knowledge and research advancement. Philosophical Transactions of the Royal Society B: Biological Sciences 2007; 362 997-1008.

[36] Akiyama T, Kawamura K. Grassland degradation in China: Methods of monitoring, management and restoration. Grassland Science 2007; 53 1-17.

[37] Jiang Y, Kang MY, Liu S, et al. A study on the vegetation in the east side of Helan Mountain. Plant Ecology 2000; 149 119-130.

[38] Lieth H, Box E. Evapotranspiration and primary productivity. In: Memorial Model (Ed Thornthwaite W). Publications in Climatology. New Jersey: Intech; 1972. p37-46.

[39] Zhou GS, Zhang XS. Study on Chinese climate-vegetation relationship. Acta Phytoecologica Sinica 1996; 20 (2) 113-119(In Chinese with English abstract).

[40] Long SP, Garcia Moya E, Imbamba SK, et al. Primary productivity of natural grass ecosystems of the tropics: a reapprisal. Plant and Soil 1989; 115 155-166.

[41] Ren JZ. Research methods of grassland science. China Agricultural Press, Beijing, China: Intech; 1998. (In Chinese).

[42] Ohta S, Uchijima Z, Oshima. Probable effects of CO_2-induced climatic changes on net primary productivity of terrestrial vegetation in East Asia. Ecological Research 1993; 8 199-213.

[43] Fisher JB, Whittaker RJ, Malhi Y. ET come home: potential evapotranspiration in geographical ecology. Global Ecology and Biogeography 2011; 20 1-18.

[44] Fang JY, Song YC, Liu HY, et al. Vegetation-Climate Relationship and Its Application in the Division of Vegetation Zone in China. Acta Botanica Sinica 2002; 44(9) 1105-1122.

[45] Wang P, Sun R, Hua J, et al. Measurements and simulation of forest leaf area index and net primary productivity in Northern China. Journal of Environmental Management 2007; 85 607-615.

[46] Nadler IA, Wein RW. Spatial interpolation of climate normals: test of a new method in the Canadian boreal forest, Agricultural and Forest Meteorology 1998; 92(4) 211-225.

[47] Zhao J, Li F, Fu HY, et al. A DEM-based partition adjustment for the interpolation of annual cumulative temperature in China. In Geoinformatics 2007: Geospatial Information Science (Proceedings of the SPIE) 2007; 6753 6753.

[48] Sun R, Zhu QJ. Distribution and Seasonal Change of Net Primary Productivity in China from April, 1992 to March, 1993. Acta Geographica Sinica 2000; 55 36-45(in Chinese with English abstract).

[49] Feng QS, Liang TG, Huang XD, et al. Characteristics of global potential natural vegetation distribution from 1911 to 2000 based on comprehensive sequential classfication system approach. Grassland Science 2013; 59 87-99. DOI: 10. 1111/grs. 12016

[50] Piao SL, Fang JY, He JS, et al. Spatial distribution of grassland biomass in China. Acta Phytoecologica Sinica 2004; 28(4) 491-498(in Chinese with English abstract).

[51] Ni J. Carbon storage in grassland of China. Journal of Arid Environments 2002; 50 205-218.

[52] Yu DY, Shao HB, Shi PJ, et al. How does the conversion of land cover to urban use affect net primary productivity? A case study in Shenzhen city, China. Agricultural and Forest Meteorology 2009; 149(11) 2054-2060.

[53] Fang JY, Piao SL, Tang ZY, et al. Interannual variability in net primary productivity and precipitation. Science 2001; 293 1723.

[54] Piao SL, Fang JY, Ciais PP, et al. The carbon balance of terrestrial ecosystems in China. Nature 2009; 458 1009-1013.

[55] Haberl H. Human appropriation of net primary production as an environmental indicator: implications for sustainable development. Ambio 1997; 26(3) 143-146.

[56] Haberl H, Krausmann F, Erb KH, et al. Human appropriation of net primary production. Science 2002; 296 1968-1969.

[57] Krausmann F. Land use and industrial modernization: an empirical analysis of human influence on the functioning of ecosystems in Austria 1830-1995. Land Use Policy 2001; 18 17-26.

[58] Kohlheb N, Krausmann F. Land use change, biomass production and HANPP: The case of Hungary 1961-2005. Ecological Economics 2009; 69(2) 292-300.

[59] Haberl H, Plutzar CK. Erb H, et al. Human appropriation of net primary production as determinant of avifauna diversity in Austria. Agriculture, Ecosystems & Environment 2005; 110 119-131.

[60] Haberl H, Gaube V, Delgado RD, et al. Towards an integrated model of socioeconomic biodiversity drivers, pressures and impacts: A feasibility study based on three European long-term socio-ecological research platforms. Ecological Economics 2009; 68(6) 1797-1812.

[61] OmannI, Stocker A, Jäger J. Climate change as a threat to biodiversity: An application of the DPSIR approach. Ecological Economics 2009; 69(1) 24-31. DOI: 10. 1016/j. ecolecon. 01. 003.

[62] Haberl H, Schulz NB, Plutzar C, et al. Human appropriation of net primary production and species diversity in agricultural landscapes. Agricultural Ecosystem Environment 2004; 102(2) 213-218.

[63] Vitousek PM, Ehrlich PR, Ehrlich AH, et al. Human appropriation of the products of photosynthesis. BioScience 1986; 36 363-373.

[64] Anderson JM. The effects of climate change on decomposition processes in grassland and coniferous forests. Ecological Application 1991; 1(3) 326-347.

[65] White R, Murray S, Rohweder M. Pilot analysis of global ecosystems (PAGE): grassland ecosystems. Washington, DC: World Resources Institute: Intech; 2000.

[66] Liang TG, Feng QS, Yu H, et al. Dynamics of natural vegetation on the Tibetan Plateau from past to future using a comprehensive and sequential classification system and remote sensing data. Grassland Science 2012; 58 208-220.

[67] Hutchinson MF. Anusplin Version 4. 3. Centre for Resource and Environmental Studies. Canberra (Australia): The Australian National University 2004.

[68] Hijmans RJ, Cameron SE, Parra JL, et al. Very high resolution interpolated climate surfaces for global land areas. International Journal of Climatology 2005; 25 1965-1978.

[69] New M, Hulme M, Jones P. Representing twentieth-century space-time climate variability. Part I: Development of a 1961-90 mean monthly terrestrial climatology. Journal of Climate 1999; 12 829-856.

[70] New M, Lister D, Hulme M, et al. A high-resolution data set of surface climate over global land areas. Climate Research 2002; 21 1-25.

[71] Jarvis CH, Stuart N. A comparison among strategies for interpolating maximum and minimum daily air temperatures. Part II: the interaction between the number of guiding variables and the type of interpolation method. Journal of Applied Meteorology and Climatology 2001; 40 1075-1084.

[72] Potter CS, Randerson JT, Field CB, et al. Terrestrial ecosystem production: a process model based on global satellite and surface data. Global Biogeochemical 1993; 70(4) 811-841.

[73] Zhang J, Ishii J, Fujikura K. A biogeochemistry approach on the bivalves: REE distribution of white clam in Off Hatsushima cold seepage. Chinese Science Bulletin 1999; 44(supp.) 258-260.

[74] Alexandrov GA, Oikawa T, Esser G. Estimating terrestrial NPP: what the data say and how they may be interpreted? Ecological Modelling 1999; 117(2-3) 361-369.

[75] Feng YM, Lu Q, Wu B, et al. Landuse dynamics of alpine-cold decertified area in the Qinhai-Tibetan Plateau in the last 30 years: a case study in Guinan County, Qinghai Province, China. International Journal of Sustainable Development & World Ecology 2011; 18(4) 357-365.

[76] Ge QS, Dai JH, Zheng JY, et al. Advances in first bloom dates and increased occurrences of yearly second blooms in eastern China since the 1960s: further phonological evidence of climate warming. Ecological Research 2011; 26 713-723.

[77] Piao SL, Fang JY, Chen AP. Seasonal Dynamics of Terrestrial Net Primary Production in Response to Climate Changes in China. Acta Botanica Sinica 2003; 45(3) 269-275.

[78] Twine TE, Kucharik CJ. Climate impacts on net primary productivity trends in natural and managed ecosystems of the central and eastern United States. Agricultural and Forest Meteorology 2009; 149(12) 2143-2161.

[79] Zheng YR, Xie ZX, Robert C, et al. Did climate drive ecosystem change and induce desertification in Otindag sandy land, China over the past 40 years? Journal of Arid Environments 2006; 64 523-541.

[80] Harle KJ, Howden SM, Hunt LP, et al. The potential impact of climate change on the Australian wool industry by 2030. Agricultural Systems 2007; 93 61-89.

[81] Henry BK, McKeon GM, Syktus JI, et al. Climate variability, climate change and land degradation. In: Sivakumar MVK and Ndiang'ui N editors. Climate and Land Degradation. Berlin: Springer-Verlag 2007; 205-221.

[82] Howden SM, Crimp SJ, Stokes CJ. Climate change and its effect on Australian livestock systems. Australian Journal of Experiment Agriculture 2008; 48(7) 780-788.

[83] McKeon GM, Stone GS, Syktus JI, et al. Climate change impacts on northern Australian rangeland livestock carrying capacity: a review of issues. The Rangeland Journal 2009; 31 1-29.

[84] Friedlingstein P, Delire C, Muller JF, et al. The climate induced variation of the continental biosphere: a model simulation of the Last Glacial Maximum. Geophysical Research Letters 1992; 19 897-900.

[85] IPCC. Climate change. 2007: Synthesis Report. Contribution of Working Groups I, II and III to the Fourth Assessment Report of the Intergovernmental Panel on Climate Change. Geneva, Switzerland: Intech; 2007.

[86] Evangelista PH, Kumar S, Stohlgren TJ, et al. Assessing forest vulnerability and the potential distribution of pine beetles under current and future climate scenarios in the Interior West of the US. Forest Ecology and Management 2011; 262 307-316.

[87] Yang HJ, Wu MY, Liu WX, et al. Community structure and composition in response to climate change in a temperate steppe. Global Change Biology 2011; 17 452-465.

[88] Liu WX, Zhang Z, Wan SQ. Predominant role of water in regulating soil and microbial respiration and their responses to climate change in a semiarid grassland. Global Change Biology 2009; 15(1) 184-195.

[89] Dong XB, Yang WK, Ulgiati S, et al. The impact of human activities on natural capital and ecosystem services of natural pastures in North Xinjiang, China. Ecological Modelling 2012; 225 28-39.

[90] Naustdalslid J. Climate change-the challenge of translating scientific knowledge into action. International Journal of Sustainable Development & World Ecology 2011; 18(3) 243-252.

[91] He Y, Dong WJ, Ji JJ, et al. The net primary production simulation of terrestrial ecosystems in China by AVIM. Advance in Earth Science 2005; 20 345-349 (in Chinese with English abstract).

[92] Gao ZQ, liu JY. Simulation study of China's net primary production. Chinese Science Bulletin 2008; 53(3) 434-443.

[93] Craine JM, Elmore AJ, Olson KC, et al. Climate change and cattle nutritional stress. Global Change Biology 2010; 16 2901-2911.

[94] Ehleringer JR, Cerling TE, Dearing MD. Atmospheric CO_2 as a global change driver influencing plant-animal interactions. Integrated and Comparative Physiology 2002; 42 424-430.

Chapter 8

Carbon Sequestration in Central European Forest Ecosystems

Robert Jandl and Andreas Schindlbacher

Additional information is available at the end of the chapter

http://dx.doi.org/10.5772/57264

1. Introduction

The forests of Central Europe play an important role in the retention of atmospheric carbon dioxide. Carbon sequestration is rarely the main objective of forest management and is often an underappreciated by-effect of practical forest management, whereas timber production and protection against natural hazards in mountain regions are the main forest functions. Climate change is expected to alter practical forest management profoundly, by causing changes in productivity in two directions: some regions will be increasingly affected by prolonged drought periods and the growth rate of forests is expected to decline; in mountain regions the growth rate of forests will often increase because climate change will manifest itself by the prolongation of the growing season. Many forests will be affected by an increasing pressure from pests and pathogens, both due to the expansion of the natural habitat, and due to the inadvertent introduction of pathogens in an increasingly globalized market. In the long run climate change will have implications for the tree species distribution. The presently successful Norway spruce will be found in a smaller area because parts of the present habitat will not fulfill the site requirements of this tree species in the future. Such tendencies have far-reaching consequences for the timber industry, so concepts for adaptive forest management need to be formulated. Besides increasing efforts for monitoring forest conditions, development of the opportunities offered by the genetic variability of tree species is an example of pro-active planning. In order to base recommendations for adaptive forest management on a sound basis, climate manipulation experiments are conducted. A network of such experiments enables researchers to make educated guesses on the expected climate change impact on forests. Nevertheless, due to manifold uncertainties with respect to the extent of climate change, its regional realization, the future expectations of society towards forest ecosystems and other imponderabilities, the predictions remain uncertain.

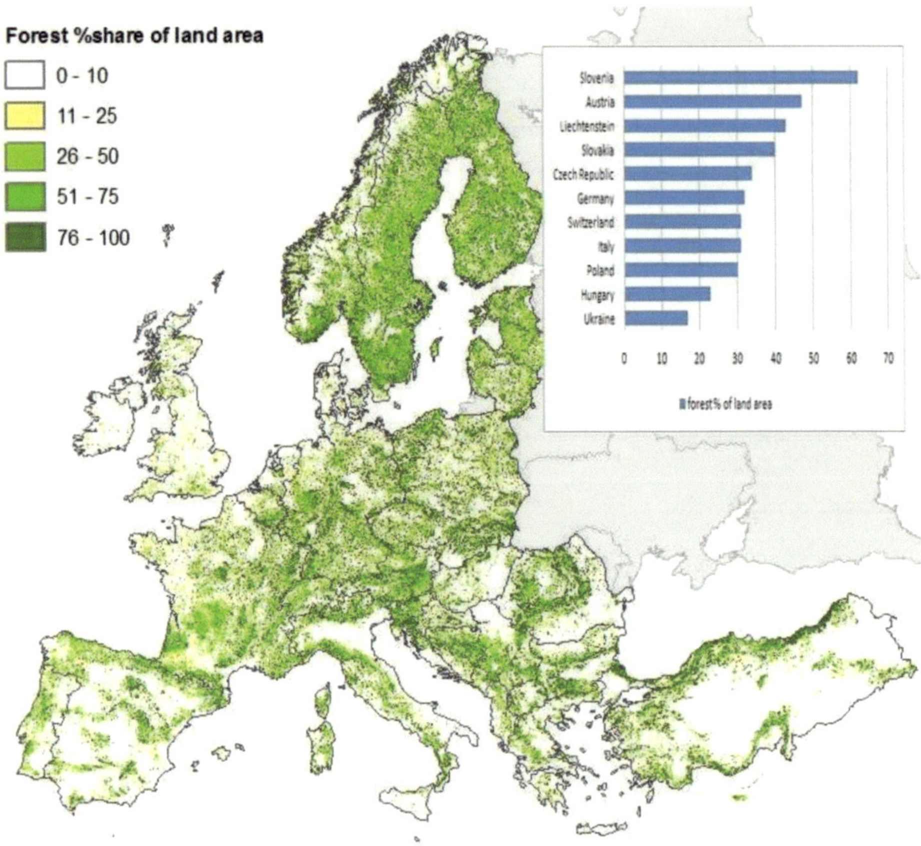

Figure 1. Forest cover of Europe with share of forest in the total land area in Central Europe (inset). European forest cover map provided by the European Forest Institute [2–4].

2. Spatial extent of forests in Central Europe with emphasis on mountain forests

Europe has a total forest area of more than 1 billion (10^9) hectares and has a an average forest cover of ≈ 45% (Figure 1). Forests are unevenly distributed. Densely forested areas are in the Nordic countries and in high elevation ecosystems. Central Europe, here comprising Austria, Czech Republic, Germany, Hungary, Italy, Poland, Slovakia, Slovenia and Ukraine, has a total forest area of more than 52 million (10^6) ha with an average forest cover of 35%. This figure is close to the global average of a forest cover of 31% [1]. The regional distribution of forests is determined by the human population density, the land management options, and by physio-geographic constraints. In low-elevation regions agriculture and settlements are often dominant land covers, whereas in mountainous areas the share of forests increases. A part of the land at high elevation is unproductive because it is situated above the timberline (Figure 2). The economic relevance of agriculture is often inversely reflected in the forest cover. Currently, abandonment of agricultural land allows the expansion of forests, both by encroachment of marginal crop- and grassland, and by concerted afforestation programs.

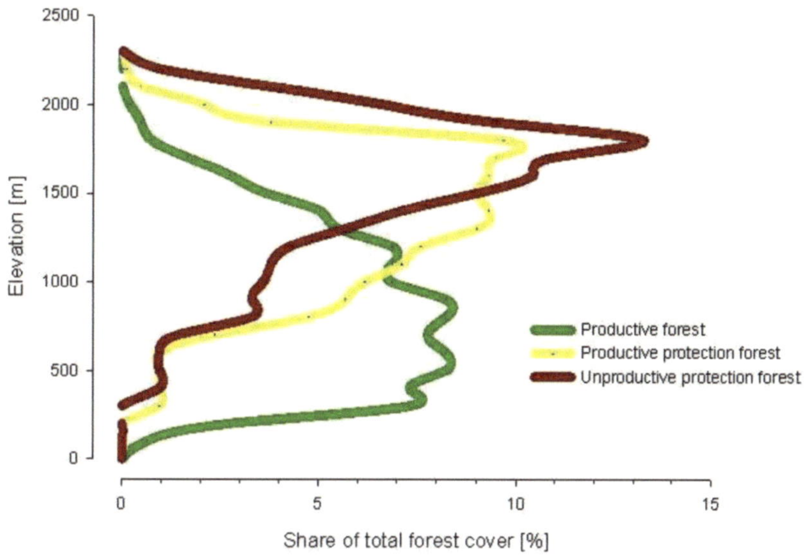

Figure 2. Typical elevational distribution of forests in a mountainous region with productive forests at low elevation and an increasing share of protection forests at high elevation; Data from Austrian National Forest Inventory [5].

The forest cover on managed land increases with altitude and its carbon pools in the aboveground and belowground biomass and in the soil are highly relevant for climate change policies. Although the rate of carbon sequestration in slowly growing forests is low, the maintenance of the carbon pool by minimizing forest disturbances is important. Mountain forests are often well adapted to harsh climate conditions and are, despite natural disturbances, stable elements of mountain ranges. However, when the reporting modalities for national greenhouse gas budgets are confined to actively managed ecosystems, as under the Kyoto Reporting Rules, the role of many mountain forests is not fully recognized because they are not managed [6, 7].

Natural forest dynamics are driven particularly by climate factors. Under increasingly favorable climate conditions, the major constraints to forest growth at high altitudes, such as short growing seasons and low temperatures, are being alleviated and the timberline is moving upwards. Superimposed on this tendency are changes in land use practices. In Central Europe the active management of mountain pastures is declining for economic reasons, and forests are expanding into these areas [8, 9]. The main economic reason is that high-altitude cattle grazing is no longer competitive in an increasingly globalized market for agricultural produce; the political reasons are strong public preferences towards ecosystem conservation in mountain ranges at the cost of practical forestry.

Climate change has positive outcomes for some tree species but may lead to reduction of habitat size or even extinction of other trees [10, 11]. Such changes have far-reaching consequences such as the alteration of the provision of drinking water from forested mountain watersheds, alteration of the protective function of for forests against natural hazards, and the biological and structural diversity of landscape. The impact of these changes is felt in timber production, but also by the local community that is making a living in mountain areas. While rising temperatures benefit the growth of trees, they also facilitate the more rapid growth and expansion of populations of pests and disease-causing organisms [12, 13]. This pressure is further exacerbated by the rapid distribution of such organisms through global market streams. Together, these global change processes increase the pressure on mountain forests. In regions of extensive forestry, outbreaks of pests and diseases have caused major damage to mountain forests: for instance, millions of hectares of lodgepole pine (*Pinus contorta*) in the Canadian Rocky Mountains have been lost to mountain pine beetles (*Dendroctonus ponderosae*), releasing considerable quantities of carbon and putting forest-dependent communities at risk. An often observed phenomenon is bark beetle infestation in the wake of storm damages that has been observed in Central and Northern Europe [14–16]. In regions of intensive forestry, monitoring of pest and pathogen densities is becoming ever more essential and widespread, and silvicultural costs will increase due to the need for international and national alerting and monitoring programs to combat insect and disease outbreaks at an early stage.

In regions where precipitation is increasing, weather conditions will contribute to reduced fire frequency but, at the same time, increased tree cover promotes fire. Regions that are becoming drier and often warmer are likely to have a higher risk of wildfires [17]. Both of these trends require better understanding of the processes involved, and more effective monitoring and rapid response in the case of emerging problems. Nevertheless, many fires are started intentionally or accidentally by people, so that climate change is only one of many factors to be considered. While changes in average temperature and precipitation lead to certain negative impacts, changes likely to result from increased numbers of extreme events are at least as critical. Examples include windstorms and heavy rain and snowfall. Such events will endanger the stability of mountain forests and hence the provision of their ecosystem services.

The role of mountain forests in the mitigation of climate change is rather passive. Their loss would release large amounts of CO_2 into the atmosphere. Their protection, conservation and management are therefore important, so some governments are promoting the establishment of forests in mountain areas as part of their climate change policies or to benefit from carbon credits, sometimes through the United Nations Collaborative Program on Reducing Emissions from Deforestation and Forest Degradation in Developing Countries (REDD). Active mitigation - the absorption of CO_2 in biomass and soils - is slow because the growth of trees and chemical processes at high altitudes is slow. A second field of mitigation is the substitution of fossil fuels with renewable forms of energy. In this context, mountain forests play a modest role since trees generally grow slowly. Nevertheless, some species, which are not of great value for timber, can be regarded as reliable sources of biofuel - as can parts of large trees, such as smaller branches, that are not valuable as timber. Given growing energy demands, all efforts to increase the efficient use of energy are important, so that key needs are to ensure that fuelwood is used more efficiently and that, wherever possible, other renewable energy sources are used.

Adaptation to climate change may be reactive or planned. Reactive adaptation includes actions such as changes in harvesting levels and schedules, the removal of downed trees after storms and the use of the resulting wood, and the development of socioeconomic support programs for communities experiencing negative impacts. Planned adaptation implies the redefinition of forestry goals and practices, taking into consideration the risks and uncertainties associated with climate change [18]. Management plans will have to include uncertainty and the increased probability of extreme events, and be regularly updated. Farmers, communities and forestry companies and agencies alike need to identify plant species or provenances which will grow well and ensure stable and productive forests in the climates of the later decades of the 21st century, contribute to local livelihoods and, in particular, ensure forest cover in areas with an increased risk of erosion, landslides or other natural hazards. For communities, adaptation may include better governance of forest resources, capacity building to monitor and cope with extreme events, increased use of agroforestry practices, and diversification of income sources. At national and global levels, key actions may include the development and implementation of monitoring and reporting systems and tools for vulnerability assessments and adaptation planning, as well as increasing the flexibility of organizational cultures, structures, and forest-related policies before crises arise [19].

3. The carbon pools in forests

The United Nations Framework Convention on Climate Change calls for its Parties to develop, periodically update and report national inventories of anthropogenic emissions by sources and removals by sinks of all greenhouse gases not controlled by the Montreal Protocol, using comparable methodologies. Estimating and reporting anthropogenic emissions and removals to the international community are important to develop policies to control the emissions and removals of greenhouse gases (GHG), which are believed to be the principle causes of the recent anthropogenic climate change. Each country needs to submit annual reports on emissions and removals of GHGs for all relevant sectors of the economy. The estimation and reporting must be complete, accurate, transparent, consistent over time and comparable between countries [7, 20]. For forest ecosystems, five compartments are reported:

- aboveground biomass
- belowground biomass
- dead wood
- litter layer
- soil organic matter

Among the principles of estimating changes in the carbon pool size is that the demonstration should be "scientifically sound". However, the accounting rules are not explicit about the strategy of dealing with uncertainty. A common approach is the calculation of a "central estimate" (mean value) and the uncertainty around it (standard deviation). Another approach is to assess the margins of the uncertainty range [21].

The reporting is done differently in individual countries. The Good Practice Guidance identifies key categories that have a significant influence on a country's greenhouse gas inventory. When land use (agriculture, forestry, other land use 'AFOLU') is a key category countries use methods of higher complexity and with higher data demands (tier levels). Tier 1, the simplest version, applies to countries in which forests and the biomass carbon pool) is not a key category and where no country-specific activity data are available. Tier 2 applies where forests and biomass carbon form a key category and where country-specific estimates of activity data (*e.g.* forest inventories) are available. Tier 3 applies where the "forests remaining forests" and biomass carbon form a key category. It requires detailed national forest inventory data supplemented by dynamic models or allometric equations calibrated to national circumstances that allow for direct calculation of biomass increment [7, 22].

The most reliable data are available for the aboveground biomass due to the highly developed forest inventory systems in Europe [23]. The rotation period of managed forests is variable between ecosystem types, yet well recorded. Therefore, the estimates of the longevity of carbon residing in the aboveground biomass are well supported by data.

The information on the belowground biomass of forest ecosystems is much scarcer for several reasons. The root biomass is economically irrelevant and the required sampling effort is high. There is substantial uncertainty about the representative rooting depth of trees and the arbitrarily chosen reporting depth of 30 to 50 cm may greatly underestimate total belowground biomass. The longevity of the root biomass varies in a wide range. Fine roots turn over quickly, whereas the decomposition of root stumps requires many decades and is affected by numerous site factors [24–26]. Often the belowground biomass is estimated by means of generic equations [27]. Despite the challenges in quantifying the standing stock of roots and their turnover rate, it is of paramount interest to improve this understanding. Roots are important sources of soil carbon. It has even been argued that the intricate network of roots and fungi is the most important source of slowly turning over soil carbon in boreal forests. Hence their role for long-term soil carbon sequestration is large [28].

Dead organic matter, *i.e.*, the residues of fallen logs and decomposing branches and twigs, is also difficult to quantify. The amount of coarse woody debris left on-site after harvesting operations depends on the local demand for biomass on the industrial market, and on the pursued forest management concept.

Litter layer and soil organic matter are difficult to assess because their carbon pool sizes are spatially heterogeneous. There is no unambiguous evidence for either positive or negative impacts of climate change on soil carbon stocks. There are suggestions for enhancing soil C stocks at higher atmospheric CO_2 concentration and reducing soil C stocks when temperatures are rising. The factors of climate change affect soil C, with the effect on soils of CO_2 being indirect through photosynthesis and the effects of weather factors being both direct and indirect. Climate change affects soil carbon pools by affecting each of the processes in the C-cycle: photosynthetic C-assimilation, litter fall, decomposition, surface erosion, and hydrological transport. Due to the relatively large gross exchange of CO_2 between atmosphere and soils and the significant stocks of carbon in soils, relatively small changes in these large but opposing fluxes of CO_2 may have significant impact on climate and soil quality. Therefore, managing these fluxes through proper soil management can help mitigate climate change considerably [29–31].

4. Role of forestry sector in national greenhouse gas budgets

Greenhouse gases comprise several gases but not all of them are equally relevant in forest ecosystems. The dominant gas with the highest relevance for European forest ecosystems is CO_2, whereas nitrogen oxides (*e.g.*, N_2O) and methane play a smaller role. In Central Europe, forests are mostly sinks of carbon dioxide.[1] The reasons are the increase in the forest area at the expense of agricultural land and the discrepancy between high growth rates and low harvesting rates in forests. Land-use change is an ongoing process and there have been successful policy incentives for a more efficient use of forests. Forest products are carrying several elements of a knowledge-based bioeconomy of the European Union. Fibers from woody plants are relevant as raw material for industrial processes and the use of wood biomass as a source of energy is being promoted in order to reduce the dependency of the continent from fossil fuels. However, the entire sector plays a rather marginal role in greenhouse gas emissions. Less than 5% of the total greenhouse gas emissions are compensated by the sector 'land use, land-use change, forestry' (Figure 3). This figure is in reasonable agreement with the estimate of 7 to 12% of CO_2 absorption in terrestrial ecosystems based on modeling results [32]. Nevertheless, the graph shows that forestry is the only sector that is a reliable sink of greenhouse gases. A second observation is that the emissions of greenhouse gases are declining because of the economic crisis in 2007. On a global scale no decline in greenhouse gas emissions was observed. On the contrary, due to the rapid development in previously less developed regions in Asia the emissions remained at high levels and the rate of annual emissions is even accelerating [33]. The top four emitters in 2011 covered 62% of the global emissions: China (28%), United States (16%), EU27 (11%), and India (7%). Among them, emissions have strongly increased in 2010/11 in China and India, and slightly decreased in the United States and Europe [34, 35].

Only recently was the carbon-dioxide sink strength of forests on a global level evaluated as strong and persistent, although the absolute size of the sink and the major contributing regions remained unclear [36, 37]. However, climate change includes elements that in the long run are prone to lead to carbon losses from terrestrial ecosystems [38]. However, currently European forests are a sink of CO_2. For the 25 member states of the European Union (EU-25) and the timespan 1990 to 2005, the long-term carbon forest sink was estimated to be 75 ± 20 gC/m^2/year, with approximately a third of the carbon going into the forest soil and two thirds ending up in unharvested tree biomass [39]. For European forests there are already some indications that the sink strength of forests will not continue at the present rate. The rate of stem growth is decreasing and thus the sink is declining after decades of increase. In addition, the increasing demand for agricultural land and for the establishment of settlements and infrastructures leads to deforestation. The subsequent vegetation cover after forest clearance has a lower carbon density. And finally, an increasing rate of ecosystem disturbances due to extreme climate events and due to biotic hazards exerted by pests and pathogens are leading to the loss of forest stands [40].

The narrative for the greenhouse-gas sink in the forestry sector is given by the example of the Austrian National Inventory Report of the year 2012 [42].

> In 2010 Article 3.3 activities, *i.e.* afforestation, reforestation and deforestation activities, were a net sink in Austria: Net CO_2 removals amounted to 1259 Gg CO_2. Removals from afforestation/reforestation amounted to 2621 Gg CO_2. About 2/3 of these gains were caused by the C stock increases in soil and litter, 1/3 was due to

[1] National Inventory Reports: http://unfccc.int/national_reports/annex_i_ghg_inventories/national_inventories_submissions/items/7383.php; accessed Aug 18, 2013

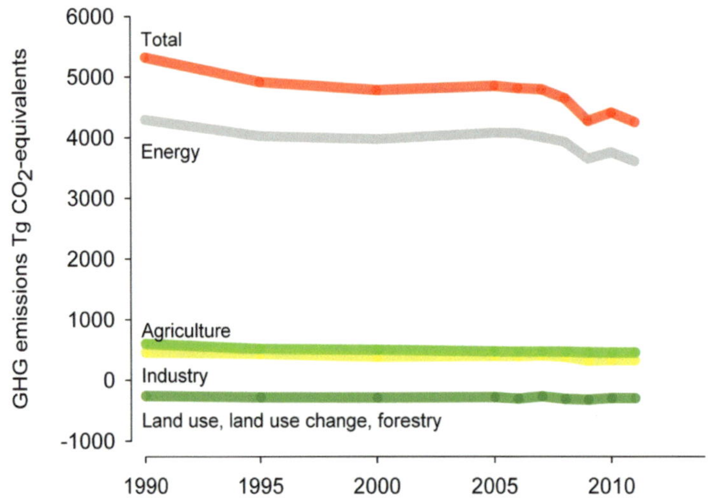

Figure 3. Overview of EU-27 GHG emissions in the main source and sink categories 1990 to 2011 in CO_2- equivalents (Tg); [41]

biomass growth in the afforestation/reforestation areas. Approximately 37% of these removals occurred in afforestation/reforestation areas from grassland, 18% in afforestation/reforestation areas from 'other land', 17% from settlement, 16% from cropland and 12% from wetlands. In the same year, emissions from deforestation amounted to 1362 Gg CO_2. A bit more than 2/3 were due to C stock losses in litter and soil, and 1/3 due to biomass losses in the deforestation areas. Approximately 30% of these losses occurred in deforestation areas to grassland, 28% to settlement, 25% to other land and the rest in deforestation areas to cropland and wetlands.

The net carbon stock changes of forests vary considerably between single years. The annual growth rate and the annual harvest rate can differ significantly year by year due to variations in weather conditions, the market situation for timber, and the intensity of ecosystem disturbances. In the example of Austria, forests have been a strong and reliable CO_2 sink for many years. The reason is that harvesting rates were far below the annual increment. A successful political campaign for the utilization of timber resources and high timber prices gave sufficient incentives for increased harvesting rates. The positive effect on the timber market had the immediate by-effect that the carbon sink strength of forests was reduced. In addition, a series of storm damages of forests unintentionally led to high harvest and salvaging rates. For a single year these disturbances even caused forests to be a source of CO_2 rather than a sink. Figure 4 gives a clear indication that the main inter-annual volatility of the GHG-budget in forests is determined by forests remaining forests. Land-use change, *e.g.*, the conversion of cropland to forest and grassland, to forest plays a marginal role.

5. Effect of land use change on carbon sequestration

The establishment of the agricultural societies starting 60 000 years ago has shaped the European landscape. Forests were cleared to enable the establishment of cropland and

http://dx.doi.org/10.5772/57264

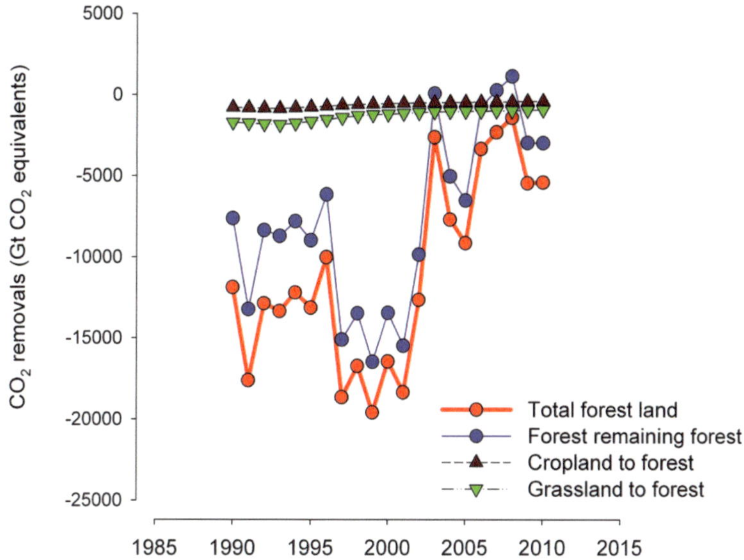

Figure 4. Removals and emissions of carbon dioxide in the forests of Austria according to the IPCC category 'Forest remaining forest', and forest land conversions from 1990 to 2010 (GT) CO_2 equivalents).

pastures. Forests were used both for the wood and non-wood products [43]. Extensive European deforestation started around 1000 BC [44].

Mountain regions play a peculiar role. Their settlements are exposed to natural hazards such as flooding and avalanches. Forests can mitigate these dangers because they affect the hydrology and the ability to stabilize a snow cover. A prerequisite for the protective function is the sustainable maintenance of an optimal stand structure [45, 46]. With awareness of the primary relevance of permanent forest cover in alpine regions, forest management schemes were greatly restricted. Already in the 16th century the exploitation of mountain forests in the Alps was constrained by Forest Acts in order to avoid natural disasters and the rapid degradation of mountain regions. However, mountain regions were often poverty traps because the low productivity of the land impaired economic development and the politcal representation was low. In politically labile times such as in the aftermath of the Napoleonic Wars the legal status of the forests in parts of the Alps was unclear. Quickly, overexploitation of the forests took place as a manifestation of the "tragedy of the commons" [47]. Often farmers were in economically desperate situations and means of increasing the productivity of agricultural lands such as mineral fertilizers were not yet available. The remedy was the deforestation of land at high altitudes to provide pasture land for an increasing number of livestock. The dire consequences followed within a short period of time. As early as 1852 a stringent Forest Act ("Reichsforstgesetz") was implemented and administrative and technical bodies for torrent control were established. Incontestable proof of the relevance of a dense and sustainable forest cover was given by the classic 'Sperbelgraben/Rappengraben' hydrological experiment in Switzerland [30, 48]. Even the establishment of academic forestry curricula in Central Europe was a direct consequence of a series of natural disasters.

Figure 5. The poor condition of mountain forests led to numerous natural disasters. Mourning the death after avalanches in Villgraten, Eastern Tyrol in 1931 (left), catastrophic flooding in the city of Sterzing, Southern Tyrol (center), and soil erosion in Eastern Tyrol (right); [50].

In recent history, the forest area in mountain regions was reduced several times. The Oetztal in the western Austrian province of Tyrol is such an example. Between 1774 and 1880 the forest area was reduced by more than 50%. The political unrest during the Napoleonic Wars led to little control of land use. The precarious income situation for agriculture in the valleys was based on potato, beets, and cabbage as main crops. The land was used at its maximum carrying capacity [49]. The extension of agricultural land in high altitudes offered one of the few options to increase the income in agriculture. The immediate consequence of this overuse of mountain regions and the removal of mountain forests was an increase in the frequency and severity of natural disasters 5. It was even pondered whether permanent settlements in some parts of the Alps should be given up and whether the population should be relocated to safer ground [45].

Today, the reverse process is underway. Mountain areas in the Alps are prospering economically as tourism is growing as a pillar of the economy. They are widely recognized as water towers with an immense impact on the lowlands. In addition, a part of the increase in the Central European forest area is due to the expansion of mountain forests. In particular at high elevation were mountain pastures are abandoned, forests are encroaching on the now unmanaged land [5]. The process has several implications. One is a loss in biodiversity because the forested landscape is becoming more uniform in comparison to the variety of pasture lands with a wide range of management intensities. However, from the perspective of climate mitigation the tendency towards more forests is beneficial because the carbon density in forests is higher than in other types of terrestrial ecosystems [36, 51–53]. It has been concluded that the effect of climate change has an impact on the soil carbon pool size in the longer term. However, the climate signal is much weaker than the effect of land use change and land management [31].

6. Expected climate change impact on tree species and pests and pathogens

There is agreement that in many regions of Europe, and even globally, climatic change will manifest itself in terms of little change in the precipitation regime, but in higher temperatures and longer growing seasons. Precise regionally valid climate predictions are still elusive and the uncertainty of the estimates is expected to remain large [54, 55]. Even apparently simple phenomena such as the length and frequency of drought periods are difficult to quantify and widely different statements on the expected temporal trend of drought periods are presented [56, 57]. However, there is clear evidence that unprecedented rates of tree mortality are happening. Particularly in already dry areas, it is questionable whether forests can re-establish themselves or whether they are regionally going to be replaced by shrublands. Nevertheless, almost all forest types are expected to be affected by drought [17, 58]. The effects of that for carbon sequestration are dramatic. Currently, forests in Europe are a potent sink for carbon dioxide in the atmosphere. However, under increasing stress, the sink strength will be difficult to maintain [39, 40]. Outside of Europe there is already widespread evidence of drought events with massive deleterious effects on the carbon storage in forests [59].

Forest fires are recognized as a substantial future risk for Central European forests. Due to active fire suppression, the incidents of forest fires are still rare and are often caused by human negligence rather than being triggered by climatic phenomena. However, this danger has to been kept on the agenda, particularly because the lengthening of the growing seasons will also extend the fire season. Wildfires not only diminish the tree biomass, they also have long-term adverse effects on the soil carbon pool [60, 61].

Despite the widespread recognition of drought as a threat to forests, currently the greatest damages in Central European forests are caused by storms and bark beetle attacks. There is some evidence that the frequency and severity of storms is increasing [62]. However, the observed signal needs to be interpreted with high caution. The scientific evidence whether or not the strength of storms is increasing is still insufficiently corroborated, although indications are recorded [63, 64]. Furthermore, the predictability of storms in global climate models is confined to advection events whereas locally severe events can be caused by convective processes that are not fully covered in regionalized climate models. A consequence of devastative storm damages is the attack of the broken stems by bark beetles. Due to constraints in the available work force and the difficult accessibility of storm-damaged sites, dead wood can be left on site for long periods. Especially in the first weeks after the disaster the still moist wood presents an ideal breeding ground for *Ips typographus* and other bark beetles. Without effective and often expensive counter-measures, the bark beetle effect can spread from the damaged sites (Figure 6). Foresters in Central Europe often have the chance to salvage damaged forest stands, thereby containing the risk of large-scale damages. In parts of the world where forests are unmanaged or only extensively managed bark beetle infestations can reach dramatic sizes, as shown from the example of forests in Western Canada [15].

An emerging threat is the occurrence of introduced pests and pathogens. This phenomenon is not new. Elm disease, the spread of potato weevils (*Cylas puncticollis*), and the vine fretter (*Phylloxera vastatrix*) are iconic examples. Globalization aggravates the challenge. In globally transported woody packaging material and in transported organic material such as soils, plants and fruits that are vectors of diseases are inadvertently carried to new ecosystems. A recent example of the introduction of new pests is the infestation of trees in northern Austria with the Asian longhorned beetle (*Anoplophora glabripennis*) and the longhorned citrus beetle (*Anoplophora chinensis*) [65, 66]. These species are native to Asia and have been introduced into Europe mainly with packaging wood material and ornamental plants from China or other Asian countries. Under suitable site conditions these endemic species can be persistent and even invade new territories. Favored by climate change the new pests can often use the new habitat as a starting point for a widerrs a poleward shift in the distribution of pests and pathogens by almost 3 km/year has been observed [12].

Figure 6. The bark beetle, *Ips typographus*, is a common pest in secondary spruce forests (left). The infestation extends from single trees to the adjacent previously healthy trees (center), and can ultimately kill the entire forest stand within a few months (right); pictures©: Christian Tomiczek.

From a human perspective, disturbances of forests mostly have a negative connotation. Besides the economic loss, carbon dioxide is also lost from the ecosystem. The slogan 'slow-in/fast-out' has been coined for the quick release of carbon dioxide due to a storm event after decades of carbon accrual in the biomass [67]. However, disturbances are also acknowledged to be a part of the natural ecosystem dynamics [68–71]. Many trees in harsh environments can only germinate and develop because uprooted trees provide site conditions that are far better than in the immediate vicinity of a fallen stem. Moreover, the recent increase in damages in Central Europe can be explained by a combination of climatic factors and forest management strategies. Forests have become more vulnerable: (i) ecologically less stable tree species have been planted for economic reasons, and (ii) the low harvesting rates have led to denser and higher forest stands that are less able to withstand storms [72].

The tree species in Central Europe are currently distributed as shown in Figure 7. The distribution only partly reflects the potential natural vegetation. The proportion of Norway spruce (*Picea abies* Karst.) has been greatly increased by forest managers because of the high productivity of spruce forests. The instability of secondary spruce forests has been known for more than 100 years. Nevertheless, the economic opportunities of a spruce-based timber industry in the Alps have made the higher production risk acceptable [73].

Predictions on the effect of climate change on the tree species composition are highly uncertain because a complex interaction of species-specific parameters and competitive processes within forest stands needs to be considered. In the past the potential natural tree species distribution has served as a useful reference for a target composition of adapted forests for a given set of site parameters [74]. Under the influence of climate change this reference is losing relevance as an indication of the 'non-analogues future'. The present potential natural vegetation is of little relevance when climate change affects the thermal regime and when nitrogen deposition leads to fundamentally altered soil fertility conditions, thereby creating site conditions that have not been encountered before. Niche models are a useful and transparent tool for creating 'species-distribution maps' [75–77]. These maps show a potential distribution and as such an end-member of a long process of a fundamental change in the vegetation. Niche models yield no information on the temporal course of species migration and do not include the adaptive capacity of species. In addition, they are based on information on the currently realized niche of tree species. The fundamental niche may or may not be substantially larger. Hence, their main purpose is constraining the many

Figure 7. Tree species in Central Europe based on compiled data from National Forest Inventories. Green shades represent coniferous forests, while reddish shades are deciduous forests; [79].

future possible tree species composition to a transparent model realization. The challenge for practical foresters is the evaluation of adaptive management no-regret strategies in order to foster a development where forests can fulfill the expected ecosystem services [78].

A comprehensive niche-modeling exercise led to a prediction of the future tree species distribution for Europe [10]. It is predicted that under now realistic climate change scenarios the share of Norway spruce is going to decline. Spruce will still be the dominant tree in the Alps, but the spruce stands at low elevation are not expected to cope with future climate conditions (Figure 8). Beech (*Fagus sylvatica* L.) is expected to benefit in the next decades. However, in the long run its habitat is expected to shrink. The tree species that are expected to expand as a consequence of climate change are pines (*Pinus sp.*) and oaks (*Quercus sp.*). Such a development is highly undesirable for the forestry industry because the most productive tree species will be less abundant whereas tree species with low growth rates or less economic value are on the rise. The benefit of niche models is that practical forestry is already supplied with this information now. Independent scrutiny of the modeling results is given by observations in the southern part of Switzerland, where the proposed species shift according to Figure 8 has been documented in the field [80]. There is still time to develop strategies to deal with climate change effects. Among the emerging fields is the efficient use of the genetic variability of tree species. This strategy is supported by the inherent genetic heterogeneity of many tree species. In particular trees that have a history of migration into different types of ecosystems after the last glaciation have potentially formed several phenotypes that can be efficiently used in forest management [81–83].

The forced migration of a tree species due to climate change is already an extreme manifestation of the change in site conditions. For a substantial amount of time it is likely that trees will adjust their growth rate to the prevailing climate conditions. In Figure 9 the change in productivity is given for different time intervals with the main confounding factor being the pressure from bark beetles (Figure 6). The analysis is based on data from

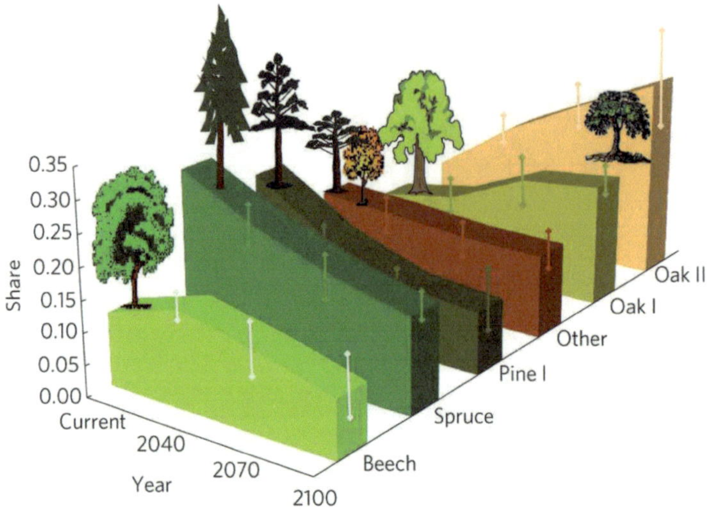

Figure 8. Potential trend of the share of tree species in Europe under the influence of climate change; [10].

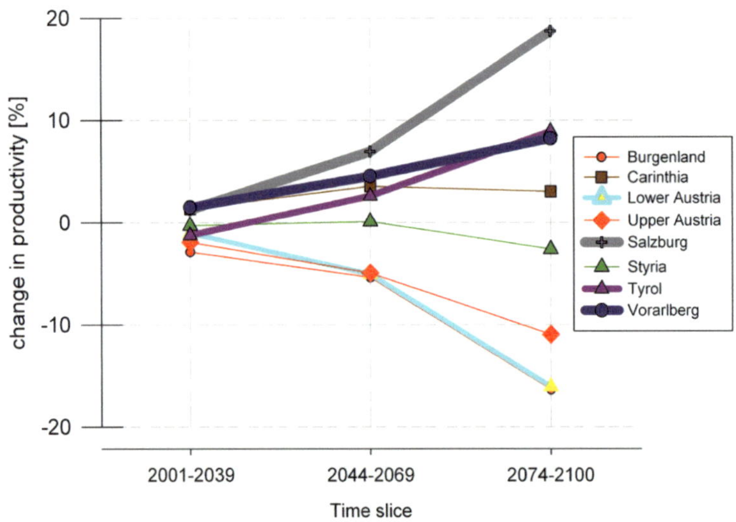

Figure 9. Altered productivity of Austrian forests due to climate change effects, differentiated by provinces.

an earlier experiment [84]. Obviously, climate change does not affect the regions equally. In mountain areas such as Tyrol, Vorarlberg, and Salzburg, the productivity is even going to increase because the main effect of climate change is extension of the growing season. At lower elevations, droughts are going to limit tree growth and the pressure from bark beetles is expected to increase. Therefore, the geographic heterogeneity of landscapes will lead to widely different climate change effects on forest ecosystems.

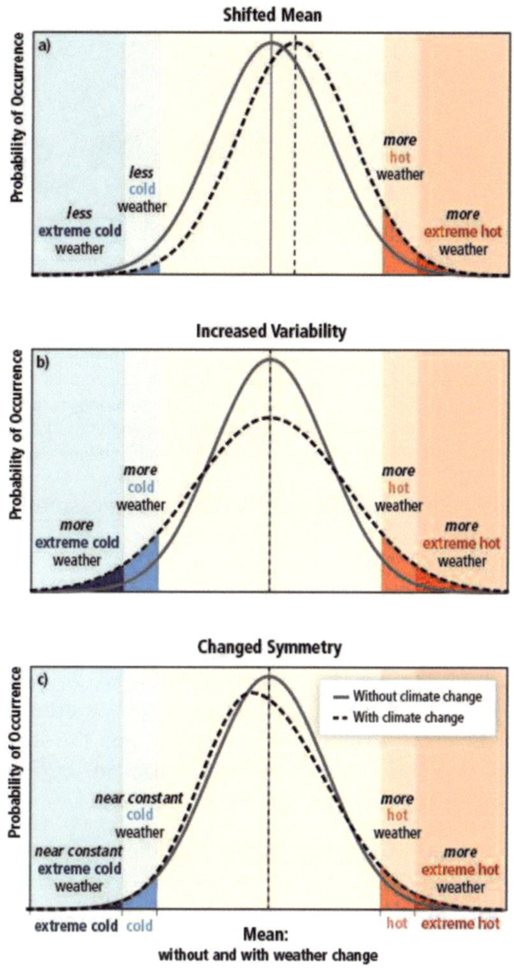

Figure 10. The effect of changes in temperature distribution on extremes. Different changes in temperature distributions between present and future climate and their effects on extreme values of the distributions: (a) effects of a simple shift of the entire distribution toward a warmer climate; (b) effects of an increase in temperature variability with no shift in the mean; (c) effects of an altered shape of the distribution, in this example a change in asymmetry toward the hotter part of the distribution ; [87].

7. Climate manipulation experiments

Climate extremes, such as severe drought, heat waves and periods of heavy rainfall, will affect tree growth greatly. Global climate change is expected to increase both the frequency and intensity of climate extremes and there is an urgent need to understand their ecological consequences, in particular with respect to carbon sequestration [85, 86]. The Special Report on Extreme Events by the IPCC distinguishes three forms of the change [87] (Figure 10).

Field experiments are essential for the assessment of ecosystem responses to changing conditions, especially when the complexity of the ecosystems does not allow the deduction of results based on existing knowledge. A challenge in climate change manipulation

Figure 11. Elements of climate manipulation experiments. Roofs under the tree canopy enable the creation of artificial drought conditions (left), automatic chambers allows the continuous measurement of greenhouse-gas emissions from soils (center), stem diameter growth is a common response variable for the quantification of climate change effects (right).

experiments in forests is their size and longevity. For the assessment of soil processes under conceivable warmer conditions, soil warming experiments have been set up in different forest ecosystems [88–91] (Figure 11).

Another avenue of climate change experiments is the exposure of trees to elevated concentrations of CO^2. Several Free-Air-Carbon dioxide-Enrichment (FACE) experiments have been conducted with young and fast-growing trees [92, 93]. Maximum stimulation of tree growth occurred at concentrations up to 560 ppm unless other limiting factors such as water shortage and insufficient nutrient supply set in. For the long-term C sequestration potential of forest ecosystems the FACE experiments are not entirely relevant because the response to elevated CO_2 is probably transient [94].

A challenge is setting up experiments with a sufficient duration. Funding schemes of scientific experiments often favor short-duration experiments whereas long-term experiments in ecology are often burdened with insufficient funding [30, 95]. Among the climate manipulation experiments the investigation of droughts has been quite successful and is widely applied. One reason is that the installation of roofs covering experimental plots is rather simple (Figure 11). The technique is already long known and has been used in acid rain research since the 1980s [96].

Much more difficult is the assessment of other extreme events in climate manipulation experiments. Early and late frost events are difficult to create under field conditions. A partial remedy is the formation of scientific networks such as CLIMMANI (http.//www.climmani.org) that bring together key researchers within climate change experiments to build coherent interdisciplinary databases. Such climate manipulation experiments can be combined with long-term ecological research installations [97].

Acknowledgements

This chapter is an outcome of several projects, such as the MANFRED Project ("Management strategies to adapt Alpine Space forests to climate change risks projects" (Interreg Alpine Space Programme), Ecosoc Cuvée, sponsored by the Austrian Academy of Sciences, and COST 639.

Author details

Robert Jandl, Andreas Schindlbacher

Forest Research Center (BFW), Vienna, Austria

References

[1] FAO. *Global Forest Resources Assessment 2010 – Main report*, volume 163 of *FAO Forestry Paper*. FAO, Rome, 2010.

[2] P Kempeneers, F Sedano, L Seebach, P Strobl, and J San-Miguel-Ayanz. Data fusion of different spatial resolution remote sensing images applied to forest-type mapping. *IEEE Transactions on Geoscience and Remote Sensing*, 49(12):4977–4986, 2011.

[3] R Päivinen, M Lehikoinen, A Schuck, T Häme, S Väätäinen, P Kennedy, and S Folving. *Combining Earth Observation Data and Forest Statistics*, volume 14 of *EFI Research Report*. European Forest Institute, Joint Research Centre - European Commission, 2001.

[4] A Schuck, J Van Brusselen, R Päivinen, T Häme, P Kennedy, and S Folving. *Compilation of a calibrated European forest map derived from NOAA-AVHRR data.*, volume 13 of *EFI Internal Report*. European Forest Institute, 2002.

[5] Wolfgang Russ. Mehr Wald in Österreich. *BFW-Praxisinformation*, 24:3–5, 2011.

[6] IGBP-TC-WG. The terrestrial carbon cycle: implications for the kyoto protocol. *Science*, 280:1393–1394, 1998.

[7] IPCC. *2006 IPCC Guidelines for National Greenhouse Gas Inventories - Agriculture, Forestry and Other Land Use*, volume 4. National Greenhouse Gas Inventories Programme IGES, Japan, Japan, 2006.

[8] Erich Tasser, Janette Walde, Ulrike Tappeiner, Alexandra Teutsch, and Werner Noggler. Land-use changes and natural reforestation in the eastern central Alps. *Agriculture, Ecosystems and Environment*, 118:115–129, 2007.

[9] Ulrike Tappeiner, Erich Tasser, Georg Leitinger, Alexander Cernusca, and Gottfried Tappeiner. Effects of historical and likely future scenarios of land use on above- and belowground vegetation carbon stocks of an alpine valley. *Ecosystems*, 11:1383–1400, 2008.

[10] Marc Hanewinkel, Dominik A. Cullmann, Mart-Jan Schelhaas, Gert-Jan Nabuurs, and Niklaus E. Zimmermann. Climate change may cause severe loss in the economic value of European forest land. *Nature Climate Change*, 3:203–207, 2012.

[11] Niklaus E. Zimmermann, Robert Jandl, Mark Hanewinkel, Georges Kunstler, Christian Kölling, Patrizia Gasparini, Andrej Breznikar, Eliane S. Meier, Signe Normand, Ulrich Ulmer, Thomas Gschwandtner, Holger Veit, Maria Naumann, Wolfgang Falk, Karl Mellert, Maria Rizzo, Mitja Skudnik, and Achilleas Psomas. *Potential future ranges of tree species in the Alps*, chapter 4. InTech, 2013.

[12] Daniel P. Bebber, Mark A. T. Ramotowski, and Sarah J. Gurr. Crop pests and pathogens move polewards in a warming world. *Nature Climate Change*, 2013.

[13] Holger Griess, Holger Veit, and Ralf Petercord. *Risk assessment for biotic pests under prospective climate conditions*, chapter 5. InTech, 2013.

[14] M. Jonášová, E. Vávrová, and P. Cudlín. Western Carpathian mountain spruce forest after a windthrow: Natural regeneration in cleared and uncleared areas. *Forest Ecology and Management*, 259(6):1127 – 1134, 2010.

[15] W. A. Kurz, C. C. Dymond, G. Stinson, G. J. Rampley, E. T. Neilson, A. L. Carroll, T. Ebata, and L. Safranyik. Mountain pine beetle and forest carbon feedback to climate change. *Nature*, 452:987–990, 2008.

[16] Anders Lindroth, Fredrik Lagergren, Achim Grelle, Leif Klemedtsson, Ola Langvalls, Per Weslien, and Janno Tuulik. Storm can cause a Europe-wide reduction in forest carbon sink. *Global Change Biology*, 15:346–355, 2009.

[17] Craig D. Allen, Alison K. Macalady, Haroun Chenchouni, Dominique Bachelet, Nate McDowell, Michel Vennetier, Thomas Kitzberger, Andreas Rigling, David D. Breshears, E.H. (Ted) Hogg, Patrick Gonzalez, Rod Fensham, Jorge Castro Zhen Zhang, Natalia Demidova, Jong-Hwan Limp, Gillian Allard, Steven W. Running, Akkin Semerci, and Neil Cobb. A global overview of drought and heat-induced tree mortality reveals emerging climate change risks for forests. *Forest Ecology and Management*, 250:660–684, 2010.

[18] Peter Brang, Marc Hanewinkel, Robert Jandl, Andrej Breznikar, and Bernhard Maier. *Managing Alpine Forests in a Changing Climate*, chapter 20. InTech, 2013.

[19] Robert Jandl and Martin Price. Climate change. In Martin F Price, Georg Gratzer, Lalisa Alemayehu Duguma, Thomas Kohler, Daniel Maselli, and Rosalaura Romeo, editors, *Mountain Forests in a Changing World - Realizing Values, addressing challenges*, chapter 8, pages 62–63. FAO, Rome, 2011.

[20] Zoltan Somogyi, Ciara Hayes, Alexandra Freudenschuss, and Rainer Baritz. *Reporting requirements for the estimation of greenhouse gas emissions and removals of soils in the land use sector*, chapter 9, pages 219–238. Wiley-Blackwell, 2011.

[21] Z. Somogyi, A. Bidló, I. Csiha, and G. Illé. Country-level carbon balance of forest soils: a country-specific model based on case studies in Hungary. *European Journal of Forest Research*, 2013.

[22] Robert Jandl and Mats Olsson. *Introduction to carbon in sensitive European ecosystems: from science to land management*, chapter 1, pages 1–12. Wiley-Blackwell, 2011.

[23] Erkki Tomppo, Thomas Gschwantner, Mark Lawrence, and Ronald E McRoberts, editors. *National Forest Inventories - Pathways for common reporting*. Springer, Heidelberg, 2010.

[24] I. Brunner, M. R. Bakke, R. G. Björk, Y. Hirano, M. Lukac, X. Aranda, I. Børja, T. D. Eldhuset, H. S. Helmisaari, C. Jourdan, B. Konôpka, B. C. López, C. Miguel Pérez, H. Persson, and I. Ostonen. Fine-root turnover rates of European forests revisited: an analysis of data from sequential coring and ingrowth cores. *Plant and Soil*, 362:357–372, 2013.

[25] Bohdan Konôpka, Jozef Pajtík, Vladimir Šebeň, and Martin Lukac. Belowground biomass functions and expansion factors in high elevation Norway spruce. *Forestry*, 2010.

[26] Klaus Lorenz. Carbon dynamics and pools in major forest biomes of the world. In Klaus Lorenz and Rattan Lal, editors, *Carbon Sequestration in Forest Ecosystems*, chapter 4. Springer, New York, 2010.

[27] Dimitris Zianis, Petteri Mukkonen, Raisa Mäkipää, and Maurizio Mencuccini. Biomass and stem volume equations for tree species in Europe. *Silva Fennica Monographs*, 4:63p, 2005.

[28] K. E. Clemmensen, A. Bahr, O. Ovaskainen, A. Dahlberg, A. Ekblad, H. Wallander, J. Stenlid, R. D. Finlay, D. A. Wardle, and B. D. Lindahl. Roots and associated fungi drive long-term carbon sequestration in boreal forest. *Science*, 339(6127):1615–1618, 2013.

[29] R. Hiederer and T. Durrant. Evaluation of biosoil demonstration project - preliminary data analysis. Technical Report EUR 24258 EN, Office for Official Publications of the European Communities, Luxembourg, 2010.

[30] Robert Jandl, Mirco Rodeghiero, Cristina Martinez, M. Francesca Cotrufo, Francesca Bampa, Bas van Wesemael, Robert B Harrison, Iraê Amaral Guerrini, Daniel deB Richter Jr., Lindsey Rustad, Klaus Lorenz, Abad Chabbi, and Franco Miglietta. Current status, uncertainty and future needs in soil organic carbon monitoring. *The Science of the Total Environment*, 468-469:376–383, 2014.

[31] René Schils, Peter Kuikman, Jari Liski, Marcel van Oijen, Pete Smith, Jim Webb, Jukka Alm, Zoltan Somogyi, Jan van den Akker, Mike Billett, Bridget Emmett, Chris Evans, Marcus Lindner, Taru Palosuo, Patricia Bellamy, Jukka Alm, Robert Jandl, and Ronald Hiederer. *Review of existing information on the interrelations between soil and climate change*. Alterra, Wageningen, 2008.

[32] Ivan A. Janssens, Annette Freibauer, Philippe Ciais, Pete Smith, Gert-Jan Nabuurs, Gerd Folberth, Bernhard Schlamadinger, Ronald W. A. Hutjes, Reinhart Ceulemans, E.-Detlef Schulze, Riccardo Valentini, and A. Johannes Dolman. Europe's terrestrial biosphere absorbs 7 to 12% of European anthropogenic CO_2 emissions. *Science*, 300:1538–1542, 2003.

[33] Glen P. Peters, Gregg Marland, Corinne Le Quéré, Thomas Boden, Josep G. Canadell, and Michael R. Raupach. Rapid growth in CO_2 emissions after the 2008-2009 global financial crisis. *Nature Climate Change*, 2:2–4, 2011.

[34] Corinne Le Quéré, Michael R. Raupach, Josep G. Canadell, Gregg Marland, Laurent Bopp, Philippe Ciais, Thomas J. Conway, Scott C. Doney, Richard A. Feely, Pru Foster, Pierre Friedlingstein, Kevin Gurney, Richard A. Houghton, Joanna I. House, Chris Huntingford, Peter E. Levy, Mark R. Lomas, Joseph Majkut, Nicolas Metzl, Jean P. Ometto, Glen P. Peters, I. Colin Prentice, James T. Randerson, Steven W. Running, Jorge L. Sarmiento, Ute Schuster, Stephen Sitch, Taro Takahashi, Nicolas Viovy, Guido R. van der Werf, and F. Ian Woodward. Trends in the sources and sinks of carbon dioxide. *Nature Geoscience*, 2:831–836, 2009.

[35] C. Le Quéré, R. Andres, T. Boden, T. Conway, R. Houghton, J. House, G. Marland, G. Peters, G. van der Werf, A. Ahlström, R. Andrew, L. Bopp, J. Canadell, P. Ciais, S. Doney, C. Enright, P. Friedlingstein, C. Huntingford, A. Jain, C. Jourdain, E. Kato, R. Keeling, K. Klein Goldewijk, S. Levis, P. Levy, M. Lomas, B. Poulter, M. Raupach, J. Schwinger, S. Sitch, B. Stocker, N. Viovy, S. Zaehle, and N. Zeng. The global carbon budget 1959-2011. *Earth System Science Data Discussions*, 5:1107–1157, 2012.

[36] Yude Pan, Richard A. Birdsey, Jingyun Fang, Richard Houghton, Pekka E. Kauppi, Werner A. Kurz, Oliver L. Phillips, Anatoly Shvidenko, Simon L. Lewis, Josep G. Canadell, Philippe Ciais, Robert B. Jackson, Stephen Pacala, A. David McGuire, Shilong Piao, Aapo Rautiainen, Stephen Sitch, and Daniel Hayes. A large and persistent carbon sink in the world's forests. *Science*, 333:988–993, 2011.

[37] Robert Jandl. Forests: The carbon conundrum. *IUFRO Spotlight*, 4:1, 2012.

[38] Peter M Cox, Richard A Betts, Chris D Jones, Steven A Spall, and Ian J Totterdell. Acceleration of global warming due to carbon-cycle feedbacks in a coupled climate model. *Nature*, 408:184–187, 2000.

[39] S. Luyssaert, P. Ciais, S. L. Piao, E.-D. Schulze, M. Jung, S. Zaehle, M. J. Schelhaas, M. Reichstein, G. Churkina, D. Papale, G. Abril, C. Beer, J. Grace, D. Loustau, G. Matteucci, F. Magnani, G. J. Nabuurs, H. Verbeeck, M. Sulkava, G. R. Van Der Werf, I. A. Janssens, and members of the CARBOEUROPE-IP SYNTHESIS TEAM. The European carbon balance. Part 3: forests. *Global Change Biology*, 16(5):1429–1450, 2010.

[40] Gert-Jan Nabuurs, Marcus Lindner, Pieter J. Verkerk, Katja Gunia, Paola Deda, Roman Michalak, and Giacomo Grassi. First signs of carbon sink saturation in European forest biomass. *Nature Climate Change*, advance online publication, 2013.

[41] Nicole Mandl. Annual European Union greenhouse gas inventory 1990-2011 and inventory report 2013. Submission to the UNFCCC Secretariat. Technical Report 8, European Environment Agency, Copenhagen, May 27 2013.

[42] Michael Anderl, Alexandra Freudenschuss, Angela Friedrich, Simone Haider, Heide Jobstmann, Traute Köther, Martin Kriech, Verena Kuschel, Christoph Lampert, Katja Pazdernik, Stephan Poupa, Maria Purzner, Melanie Sporer, Barbara Schodl, Gudrun Stranner, Elisabeth Schwaiger, Katrin Seuss, Peter Weiss, Manuela Wieser, and Gerhard Zethner Andreas Zechmeister. Austrias National Inventory Report 2012. Submission under the United Nations Framework Convention on Climate Change and the Kyoto Protocol. Technical Report REP-0381, Umweltbundesamt, Vienna, 2012.

[43] Gerhard Glatzel. Historic land use and its possible implications to recently accelerated tree growth in central Europe. In Timo Karjalainen, Heinrich Spiecker, and Olivier Laroussinie, editors, *Causes and consequences of accelerating tree growth in Europe*, volume 29 of *EFI Proceedings*, pages 65–74. EFI, Joensuu, 1999.

[44] Jed O. Kaplan, Kristen M. Krumhardt, and Niklaus Zimmermann. The prehistoric and preindustrial deforestation of Europe. *Quaternary Science Reviews*, 28(27-28):3016 – 3034, 2009.

[45] Robert Jandl, Andreas Schindlbacher, Silvio Schüler, and Dieter Stöhr. *Wald- und Waldgrenzenforschung in Obergurgl - Vergangenheit und Zukunft*, volume 2 of *Alpine Forschungsstelle Obergurgl*, chapter 5, pages 125–145. Innsbruck University Press, 2012.

[46] Hannes Mayer. *Gebirgswaldbau - Schutzwaldpflege*. Gustav Fischer Verlag, Stuttgart, 1976.

[47] Garrett Hardin. The tragedy of the commons. *Science*, 162:1243–1248, 1968.

[48] C Hegg, BW McArdell, and A Badoux. One hundred years of mountain hydrology in Switzerland by the WSL. *Hydrological Processes*, 20:371–376, 2006.

[49] Walter Moser and Jeannie Peterson. Limits to Obergurgl's growth. *Ambio*, 10(2-3):68–72, 1981.

[50] Dieter Stöhr. Gegen die Destabilisierung der Bergwälder. In *Ecological and economic benefits of mountain forests*, Innsbruck, September 15-18 2002.

[51] T. Eglin, P. Ciais, S. L. Piao, P. Barre, V. Bellassen, P. Cadule, C. Chenu, T. Gasser, C. Koven, M. Reichstein, and P. Smith. Historical and future perspectives of global soil carbon response to climate and land-use changes. *Tellus B*, 62(5):700–718, 2010.

[52] Christopher Poeplau, Axel Don, Lars Vesterdal, Jens Leifeld, Bas Van Wesemael, Jens Schumacher, and Andreas Gensior. Temporal dynamics of soil organic carbon after land-use change in the temperate zone carbon response functions as a model approach. *Global Change Biology*, 17(7):2415–2427, 2011.

[53] Angelika Thuille and Ernst-Detlef Schulze. Carbon dynamics in successional and afforested spruce stands in thuringia and the alps. *Global Change Biology*, 12:325–342, 2006.

[54] Richard H. Moss, Jae A. Edmonds, Kathy A. Hibbard, Martin R. Manning, Steven K. Rose, Detlef P. van Vuuren, Timothy R. Carter, Seita Emori, Mikiko Kainuma, Tom Kram, Gerald A. Meehl, John F. B. Mitchell, Nebojsa Nakicenovic, Keywan Riahi, Steven J. Smith, Ronald J. Stouffer, Allison M. Thomson, John P. Weyant, and Thomas J. Wilbanks. The next generation of scenarios for climate change research and assessment. *Nature*, 463:747–756, 2010.

[55] Retto Knutti and Jan Sedláček. Robustness and uncertainties in the new CMIP5 climate model projections. *Nature Climate Change*, 2012.

[56] Ernst Gebetsroither and Johann Züger. *Drought hazard estimations according to climate change in the Alpine area*, chapter 11. Intech, 2013.

[57] Justin Sheffield, Eric F.Wood, and Michael L. Roderick. Little change in global drought over the past 60 years. *Nature*, 491:435–438, 2012.

[58] Brendan Choat, Steven Jansen, Tim J. Brodribb, Hervé Cochard, Sylvain Delzon, Radika Bhaskar, Sandra J. Bucci, Taylor S. Feild, Sean M. Gleason, Uwe G. Hacke, Anna L. Jacobsen, Frederic Lens, Hafiz Maherali, Jordi Martínez-Vilalta, Stefan Mayr, Maurizio Mencuccini, Patrick J. Mitchell, Andrea Nardini, Jarmila Pittermann, R. Brandon Pratt, John S. Sperry, Mark Westoby, Ian J. Wright, and Amy E. Zanne. Global convergence in the vulnerability of forests to drought. *Nature*, 491:751–756, 2012.

[59] Simon L. Lewis, Paulo M. Brando, Oliver L. Phillips, Geertje M. F. van der Heijden, and Daniel Nepstad. The 2010 amazon drought. *Science*, 331(6017):554, 2011.

[60] Giacomo Certini. Effects of fire on properties of forest soils: a review. *Oecologia*, 143:1–10, 2005.

[61] A. L. Westerling, H. G. Hidalgo, D. R. Cayan, and T. W. Swetnam. Warming and earlier spring increase western U.S. forest wildfire activity. *Science*, 313(5789):940–943, 2006.

[62] Barry Gardiner, Kristina Blennow, Jean-Michel Carnus, Peter Fleischer, Frederik Ingemarson, Guy Landmann, Marcus Lindner, Mariella Marzano, Bruce Nicoll, Christophe Orazio, Jean-Luc Peyron, Marie-Pierre Reviron, Mart-Jan Schelhaas, Andreas Schuck, Michaela Spielmann, and Tilo Usbeck. *Destructive Storms in European Forests: Past and Forthcoming Impacts*. EFI, Joensuu, 2010.

[63] M. G. Donat, D. Renggli, S. Wild, L. V. Alexander, G. C. Leckebusch, and U. Ulbrich. Reanalysis suggests long-term upward trends in European storminess since 1871. *Geophysical Research Letters*, 38:L14703, 2011.

[64] U. Ulbrich, G. C. Leckebusch, and J. G. Pinto. Extra-tropical cyclones in the present and future climate: a review. *Theoretical and Applied Climatology*, 96:117–131, 2009.

[65] Christian Tomiczek and Ute Hoyer-Tomiczek. Der Asiatische Laubholzbockkäfer (*Anoplophora glabripennis*) und der Citrusbockkäfer (*Anoplophora chinensis*) in Europa - ein Situationsbericht. *Forstschutz Aktuell*, 38:2–5, 2007.

[66] Robert Jandl, Silvio Schüler, Andreas Schindlbacher, and Christian Tomiczek. Forests, carbon pool, and timber production. In Rattan Lal, Klaus Lorenz, Reinhard F Hüttl, Bernd Uwe Schneider, and Joachim von Braun, editors, *Ecosystem Services and Carbon Sequestration in the Biosphere*, chapter 6, pages 101–130. Springer Verlag, Dordrecht, 2013.

[67] Christian Körner. Slow in, rapid out - carbon flux studies and Kyoto targets. *Science*, 300:1242–1243, 2003.

[68] Virginia H. Dale, Linda A. Joyce, Steve McNulty, Ronald P. Neilson, Matthew P. Ayres, Michael D. Flannigan, Paul J. Hanson, Lloyd C. Irland, Ariel E. Lugo, Chris J. Peterson,

Daniel Simberloff, Frederick J. Swanson, Brian J. Stocks, and B. Michael Wotton. Climate change and forest disturbances. *BioScicence*, 51(9):723–734, 2001.

[69] Jerry F. Franklin, Thomas A. Spies, Robert Van Pelt, Andrew B. Carey, Dale A. Thornburgh, Dean Rae Berg, David B. Lindenmayer, Mark E. Harmon, William S. Keeton, David C. Shaw, Ken Bible, and Jiquan Chen. Disturbances and structural development of natural forest ecosystems with silvicultural implications, using Douglas-fir forests as an example. *Forest Ecology and Management*, 155:399–423, 2002.

[70] Marc Hanewinkel, Johannes Breidenbach, Till Neeff, and Edgar Kublin. Seventy-seven years of natural disturbances in a mountain forest area - the influence of storm, snow, and insect damage analysed with a long-term time series. *Canadian Journal of Forest Research*, 38(8):2249–2261, 2008.

[71] Rupert Seidl, Paulo M. Fernandes, Teresa F. Fonseca, François Gillet, Anna Maria Jönsson, Katarína Merganicová, Sigrid Netherer, Alexander Arpaci, Jean-Daniel Bontemps, Harald Bugmann, Jose Ramon González-Olabarria, Petra Lasch, Céline Meredieu, Francisco Moreira, Mart-Jan Schelhaas, and Frits Mohren. Modelling natural disturbances in forest ecosystems: a review. *Ecological Modelling*, 222(4):903 – 924, 2011.

[72] Rupert Seidl, Mart-Jan Schelhaas, and Manfred J Lexer. Unraveling the drivers of intensifying forest disturbance regimes in Europe. *Global Change Biology*, 17(9):2842–2852, 2011.

[73] Ernst Assmann. *Waldertragskunde - Organische Produktion, Struktur, Zuwachs und Ertrag von Waldbeständen*. BLV Verlagsgesellschaft, München, 1961.

[74] Hannes Mayer. *Wälder des Ostalpenraumes – Standort, Aufbau und waldbauliche Bedeutung der wichtigsten Waldgesellschaften in den Ostalpen samt Vorland*. G. Fischer, Stuttgart, 1974.

[75] Xavier Morin and Wilfried Thuiller. Comparing niche- and process-based models to reduce prediction uncertainty in species range shifts under climate change. *Ecology*, 90(5):1301–1313, 2009.

[76] W. Thuiller, B. Lafourcade, R. Engler, and M.B. Araújo. Biomod - a platform for ensemble forecasting of species distributions. *Ecography*, 32:369–373, 2009.

[77] Niklaus E. Zimmermann, Nigel G. Yoccoz, Thomas C. Edwards Jr., Eliane S. Meier, Wilfried Thuiller, Antoine Guisan, Dirk R. Schmatz, and Peter B. Pearman. Climatic extremes improve predictions of spatial patterns of tree species. *PNAS*, 106:19723–19728, 2009.

[78] Kristina Blennow, Johannes Persson, Margarida Tomé, and Marc Hanewinkel. Climate change: Believing and seeing implies adapting. *PLoS ONE*, 7:e50182, 11 2012.

[79] D. J. Brus, G. M. Hengeveld, D. J. J. Walvoort, P. W. Goedhart, A. H. Heidema, G. J. Nabuurs, and K. Gunia. Statistical mapping of tree species over europe. *European Journal of Forest Research*, 131:145–157, 2011.

[80] Andreas Rigling, Christof Bigler, Britta Eilmann, Elisabeth Feldmeyer-Christe, Urs Gimmi, Christian Ginzler, Ulrich Graf, Philipp Mayer, Giorgio Vacchiano, Pascale Weber, Thomas Wohlgemuth, Roman Zweifel, and Matthias Dobbertin. Driving factors of a vegetation shift from scots pine to pubescent oak in dry alpine forests. *Global Change Biology*, 19(1):229–240, 2013.

[81] Małgorzata Latałowa and W.O. van der Knaap. Late Quaternary expansion of Norway spruce *Picea abies* (L.) Karst. in Europe according to pollen data. *Quaternary Science Reviews*, 25(21-22):2780 – 2805, 2006.

[82] Stefan Kapeller, Manfred J. Lexer, Thomas Geburek, Johann Hiebl, and Silvio Schueler. Intraspecific variation in climate response of norway spruce in the eastern alpine range: Selecting appropriate provenances for future climate. *Forest Ecology and Management*, 271:46–57, 2012.

[83] T. Wang, A Hamann, Y Yanchuk, GA O'Neill, and SN Aiken. Use of response functions in selecting lodgepole pine populations for future climates. *Global Change Biology*, 12:2404–2416, 2006.

[84] Rupert Seidl, Mart-Jan Schelhaas, Marcus Lindner, and ManfredJ. Lexer. Modelling bark beetle disturbances in a large scale forest scenario model to assess climate change impacts and evaluate adaptive management strategies. *Regional Environmental Change*, 9(2):101–119, 2009.

[85] Markus Reichstein, Michael Bahn, Philippe Ciais, Dorothea Frank, Miguel D. Mahecha, Sonia I. Seneviratne, Jakob Zscheischler, Christian Beer, Nina Buchmann, David C. Frank, Dario Papale, Anja Rammig, Pete Smith, Kirsten Thonicke, Marijn van der Velde, Sara Vicca, ArianeWalz, and Martin Wattenbach. Climate extremes and the carbon cycle. *Nature*, 500:287–295, 2013.

[86] Melinda D. Smith. The ecological role of climate extremes: current understanding and future prospects. *Journal of Ecology*, 99(3):651–655, 2011.

[87] IPCC. *Managing the Risks of Extreme Events and Disasters to Advance Climate Change Adaptation. A Special Report of Working Groups I and II of the Intergovernmental Panel on Climate Change.* Cambridge University Press, Cambridge, UK, and New York, NY, USA, 2012.

[88] Frank Hagedorn, Melissa Martin, Christian Rixen, Silvan Rusch, Peter Bebi, Alois Zürcher, Rolf T. W. Siegwolf, Sonja Wipf, Christophe Escape, Jacques Roy, and Stephan Hättenschwiler. Short-term responses of ecosystem carbon fluxes to experimental soil warming at the Swiss alpine treeline. *Biogeochemistry*, 97:7–19, 2010.

[89] Jerry M. Melillo, Sarah Butler, Jennifer Johnson, Jacqueline Mohan, Paul Steudler, Heidi Lux, Elizabeth Burrows, Francis Bowles, Rose Smith, Lindsay Scott, Chelsea Vario, Troy Hill, Andrew Burton, Yu-Mei Zhouj, and Jim Tang. Soil warming, carbon-nitrogen interactions, and forest carbon budgets. *PNAS*, 108(23):9508–9512, 2011.

[90] Andreas Schindlbacher, Sophie Zechmeister-Boltenstern, Barbara Kitzler, and Robert Jandl. Experimental forest soil warming: response of autotrophic and heterotrophic soil respiration to a short-term 10°C temperature rise. *Plant and Soil*, 303:323–330, 2008.

[91] A. Schindlbacher, S. Wunderlich, W. Borken, B. Kitzler, S. Zechmeister-Boltenstern, and R. Jandl. Soil respiration under climate change: Prolonged summer drought offsets soil warming effects. *Global Change Biology*, 18:2270–2279, 2012.

[92] Martin Lukac, Carlo Calfapietra, and Douglas L Godbold. Production, turnover and mycorrhizal colonization of root systems of three populus species grown under elevated CO_2 (POPFACE). *Global Change Biology*, 9:838–848, 2003.

[93] Richard J. Norby and Donald R. Zak. Ecological lessons from free-air CO_2 enrichment (face) experiments. *Annual Review of Ecology, Evolution, and Systematics*, 42(1):181–203, 2011.

[94] Christian Körner, Roman Asshoff, Olivier Bignucolo, Stephan Hättenschwiler, Sonja G. Keel, Susanna Peláez-Riedl, Steve Pepin, Rolf T. W. Siegwolf, and Gerhard Zotz. Carbon flux and growth in mature deciduous forest trees exposed to elevated CO_2. *Science*, 309:1360–1363, 2005.

[95] Robert Jandl, Stefan Smidt, Andreas Schindlbacher, Michael Englisch, Sophie Zechmeister-Boltenstern, Christian Mikovits, Philipp Schöftner, Friederike Strebl, and Gabriele Fuchs. The carbon and nitrogen biogeochemistry of a montane Norway spruce (*Picea abies* (L.) Karst.) forest: a synthesis of long-term research. *Plant Ecology & Diversity*, 5(1):105–114, 2012.

[96] N.P. Lamersdorf, C. Beier, K. Blanck, M. Bredemeier, T. Cummins, E.P. Farrell, K. Kreutzer, L. Rasmussen, M. Ryan, W. Weis, and X.-J. Xu. Effect of drought experiments using roof installations on acidification/nitrification of soils. *Forest Ecology and Management*, 101:95–109, 1998.

[97] Daniel D Richter Jr., Michael Hofmockel, Mac A. Callaham jr., David S. Powlson, and Pete Smith. Long-term soil experiments: Keys to managing Earth's rapidly changing ecosystems. *Soil Science Society America Journal*, 71(2):266–279, 2007.

Chapter 9

The Role of Simulation Models in Monitoring Soil Organic Carbon Storage and Greenhouse Gas Mitigation Potential in Bioenergy Cropping Systems

Manyowa N. Meki, James R. Kiniry,
Kathrine D. Behrman, Meghan N. Pawlowski and
Susan E. Crow

Additional information is available at the end of the chapter

http://dx.doi.org/10.5772/57177

1. Introduction

There is an increased demand on agricultural systems worldwide to provide food, fiber, and feedstock for the emerging bioenergy industry, raising legitimate concerns about the associated impacts of such intensification on the environment [1, 2]. In the US, the revised National Renewable Fuel Standard program has mandated 30 billion gallons of renewable fuel by 2020 [3]. This aggressive promotion of the bioenergy industry in the US, and elsewhere in the world, is being fueled by several factors: the opportunity to reduce dependence on fossil fuels through renewable energy; the search for energy independence or security; its potential to reduce net greenhouse gas (GHG) emissions and hence provide mitigation options to combat climate change; and the possibility to improve farmers' incomes and rural economies, reduce national budget deficits and trade imbalances [4]. There is thus immense pressure on agricultural systems to provide the much needed feedstocks for this fast emerging industry. Of critical concern, however, is the fact that bioenergy feedstock production depends on finite resources such as land and water, and if not sustainably managed could have huge negative impacts on ecosystem services.

Of the many ecosystem services that could be impacted by the large-scale production of bioenergy feedstocks, soil organic carbon (SOC) plays a pivotal role in maintaining and enhancing the natural soil resource base and the GHG emission balance. According to the [5], periodic checks of SOC should be conducted to ensure that soil quality is not sacrificed for renewable bioenergy. Soil degradation causes carbon (C) loss and release of GHGs as a result of accelerated decomposition from land use change or unsustainable land management practices [6]. To date, there is little or no information that can be used to assess the long-term impacts of bioenergy production on SOC sequestration and storage.

In this chapter we discuss and examine the potential application of process-based simulation models in monitoring SOC and its GHG mitigation potential. The first section highlights the critical functions and benefits of SOC in maintaining soil quality, and hence productivity and sustainability of cropping systems. In addition, we examine the ability of bioenergy cropping systems to sequester SOC and hence mitigate climate change. In the second part, we focus on the need for periodic checks and long-term monitoring of SOC. We point out the challenges of measuring and quantifying SOC at the field level and review selected modeling studies that illustrate how process-based simulation models can be applied to monitor SOC and GHG emissions. A discussion of the importance of life cycle assessment (LCA) in evaluating a bioenergy cropping system's effect on the GHG emission balance or global warming potential (GWP) is presented. Finally, we offer our perspectives on how simulation models could provide the tools to track SOC for the Carbon Credit Markets (CCMs), and briefly examine government policies that are necessary to ensure success.

2. Soil organic matter and soil organic carbon – Functions and benefits

Soils contain C in both organic and inorganic forms. In most soils the majority of C is held as SOC. The term soil organic matter (SOM) is used to describe the organic constituents in the soil, while SOC refers to the C occurring in the SOM [7]. The terms SOM and SOC are often used interchangeably because SOC is the main constituent of SOM, which is typically around 57% C. Besides carbon, SOM also contains other important elements such as oxygen (O, ~40%), nitrogen (N, ~3%), as well as smaller amounts of phosphorus (P), potassium (K), calcium (Ca), magnesium (Mg), sulfur (S) and micronutrients [8].

Soil organic matter is composed of dead plant and animal material at various stages of decomposition. These materials are mainly plant residues and leaf litter, animal wastes and remains, plant roots and exudates, and soil biota, e.g., microbes and earthworms. Soil biota are continually breaking down the SOM through physical and biochemical reactions, releasing C and nutrients back to the soil, and GHGs such as carbon dioxide (CO_2), nitrous oxide (N_2O) and methane (CH_4) to the atmosphere. Soils also contain inorganic C in the form of biochar and carbonate minerals, such as calcite, dolomite and gypsum.

Soil organic carbon plays a very important function in the growth of plants through maintenance as well as improvement of many soil properties. These functions of SOC can be classified into three broad categories: biological, physical, and chemical. According to [9], these categories are not static entities; instead, a change in one property will likely affect the other soil properties as well. An overview of the principal functions and benefits of SOM in soils is given in Figure 1.

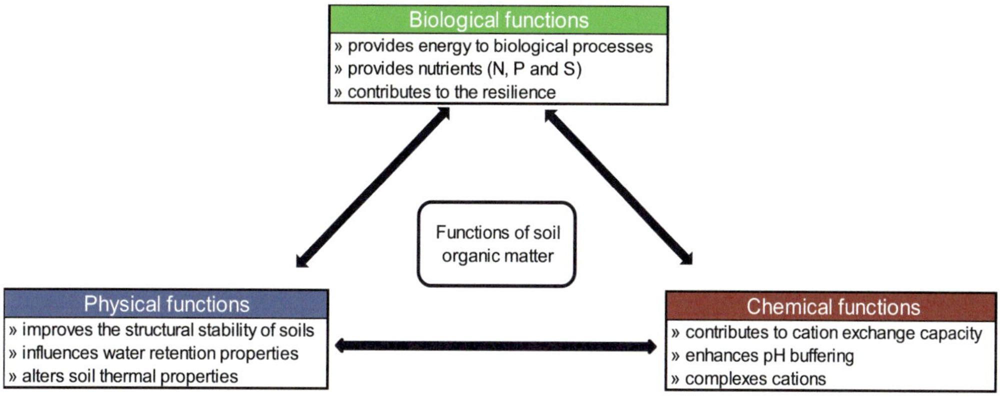

Figure 1. The biological, physical, and chemical functions of soil organic matter (SOM). Source: adapted from [10].

Biological functions – Organic matter is a source of energy for many soil biota. Soil fauna play an important role in the initial breakdown of complex and large pieces of organic matter, making it easier for soil microorganisms to release C and plant nutrients during the process of decomposition. While microorganisms only make up less than 5% of SOM [11], they are the key driver of important processes such as decomposition, nutrient cycling, N fixation, and symbiotic plant relationships. These processes have direct and indirect effects on plant nutrient availability. It is estimated that 90-95% of the soil N, and about 2% of available P and S in natural ecosystems, come from organic matter.

Chemical functions – Humus is finely divided stable organic matter. Due to the large exposed surface area, humus contributes 30-70% of the cation exchange sites that adsorb plant nutrients that are eventually taken up by plants. Humus chelates micronutrients (usually iron, zinc, copper, or manganese) into soluble compounds, hence increasing their general mobility in soils and availability to plants. Humus also plays an important role in buffering the soil against any changes in acidity, salinity, and damage by pesticides. The exchange sites are also believed to help in cleaning up contaminated water by temporarily adsorbing heavy metal pollutants such as lead, cadmium, copper and zinc.

Physical functions – Plant roots, microbially-derived polysaccharides, and fungal hyphae cement soil particles into aggregates, thereby improving soil aeration, rate of water infiltration, and hence water holding capacity [12]. Build-up of organic matter lowers the soil bulk density, which allows the soil to be tilled with less horsepower, and improves plant rooting. Organic matter left on the soil surface, such as plant residues, also enhances rainfall capture, which reduces sediment and nutrient losses by runoff. Furthermore, organic matter shades the soil, which reduces excessive evapotranspiration and soil temperature in the summer, while keeping the soil warmer in winter.

A review of the literature shows that the productivity and sustainability of cropping systems are positively related to SOC content [13]. The ability to increase SOC in agricultural ecosystems is of interest both for sequestering atmospheric CO_2 and for restoring organic-matter pools important to soil health [14]. The challenge is to identify soil management practices that promote SOC sequestration while ensuring productivity and profitability for farmers.

2.1. Bioenergy crop yields, impacts on soil organic carbon storage and greenhouse gas mitigation potential

Although most current bioenergy crop production is from annual row crops such as corn (*Zea mays* L.) and soybean [*Glycine max* (L.) Merr.], we do not discuss these here because they offer few or no environmental benefits over current agricultural practises [16]. Instead, we focus on second-generation bioenergy crops (Table 1) as identified by the US Department of Energy (DOE) [15, 16]: (1) the highly productive perennial warm-season grasses (WSGs) as emerging biofuel feedstocks for cellulosic ethanol production [e.g. switchgrass (*Panicum virgatum L.*), miscanthus (*Miscanthus giganteus*), Napiergrass (elephantgrass) (*Pennisetum purpureum* Schum.), and energy cane (*Saccharum officinarum* L.)]; and (2) the short-rotation woody crops (SRWCs) [e.g. hybrid poplar (*Populus* spp.) and willow (*Salix* spp.)].

In the US, average yields of selected WSGs ranged from 4.0 to 14.0 Mg ha^{-1} yr^{-1} [15]. More recent results indicate the potential to reach 20.0 Mg ha^{-1} yr^{-1} in several locations. Energy cane and Napiergrass can produce yields from 12.2 to 32.4 Mg ha^{-1} yr^{-1}. However, these species are very sensitive to frost and are limited to the southern regions of the US. Short rotation woody crop yields range from 10.0 to 17.0 Mg ha^{-1} yr^{-1}. This is 2-3 times the yields normally achieved by traditional forest management. Observed positive attributes of these bioenergy crops include: being perennials, they require less tillage than conventional row crops (tillage is costly and is detrimental to maintenance of SOC stocks); their deep and extensive rooting patterns effectively exploit and extract water and soil nutrients deep-down the soil profile; they require little or no external nutrient inputs due to nutrient retention and efficient nutrient cycling between growing years or harvests; most are adapted to low-quality land (nutrient-depleted, compacted, poorly drained, and eroded); and they are tolerant to a host of abiotic stresses, such as drought, heat, cold, low pH, metal toxicity and salinity. The perennial grasses switchgrass and miscanthus have the additional advantage of the C_4 pathway of photosynthesis, which is associated with high water use efficiency (WUE).

To date, there are a few empirical studies of SOC sequestration under both WSGs and SRWCs cropping systems. Therefore, the following discussion is based on data from studies carried out mainly in the US and Europe. More data is needed to provide a comprehensive understanding for these cropping systems.

2.1.1. Impacts of Warm Season Grasses

Growing WSGs offers promise to displace fossil fuels and reduce net CO_2 emissions through SOC sequestration [17]. When compared to row crops, the greater SOC storage under WSGs is attributed to greater aboveground biomass and root biomass concentration [18, 19]. Switchgrass produces about 6.7 Mg ha^{-1} yr^{-1} of total root biomass and half of this is found below the 30-cm soil depth [19], while about 10% is found below the 60-cm soil depth. Warm-season grass root systems, which often extend to nearly 3 m in the soil profile, enhance translocation

Common name	Scientific name	Observed yields (Mg ha^{-1})
Perennial Warm Season Grasses (WSGs)		
Switchgrass	*Panicum virgatum*	4 to 14
Miscanthus	*Miscanthus giganteus*	
Napiergrass (elephantgrass) (ratoon)	*Pennisetum purpureum*	12 to 32
Energy cane (ratoon)	*Saccharum spp.*	
Short rotation woody crops (SRWCs)		
Poplars, cottonwoods & hybrids	*Populus spp.*	
Willows	*Salix spp.*	
Eucalypts	*Eucalyptus spp.*	
Sycamore	*Platanus occidentalis*	10 to 17
Silver maple	*Acer saccharinum*	
Sweetgum	*Liquidambar styraciflw*	
Black locust	*Robinia pseudoacacia*	

Source: Modified from [15].

Table 1. Model bioenergy crops identified by the US Department of Energy (DOE).

of SOC to deeper layers, thereby reducing decomposition of SOC and promoting long-term C storage [19]. Data from a regional study across 42 paired sites in Minnesota, North Dakota and South Dakota showed that switchgrass stored about 2.0 Mg ha^{-1} of SOC in the 0- to 5-cm depth, 7.7 Mg ha^{-1} in the 30- to 60-cm depth, and 4.4 Mg ha^{-1} in the 60- to 90-cm depth [20]. The fraction of SOC stored in deeper (>30 cm) soil profiles is critical for long-term SOC sequestration because this fraction has longer residence times and slower turnover due to reduced microbial processes and fluctuations in soil water content and temperature [21]

In central and northwest Europe, annual increases of SOC for the conversion of croplands to miscanthus differed widely among studies and among sites within one study, ranging from 6.9 to 7.7 Mg C ha^{-1} yr^{-1} [22 - 24]. Data from soil samples of miscanthus fields and adjacent reference croplands at four different locations showed SOC change rates ranging from 2.6 to 2.8 Mg C ha^{-1} yr^{-1} [22], while [23] found change rates of 0.8 to 2.2 Mg C ha^{-1} yr^{-1}. Since all the above studies were conducted in central and northwest Europe, this scatter emphasizes the impact of site-specific factors on SOC sequestration. Carbon sequestration rates in miscanthus fields were determined at 0.1-0.2 Mg C $ha^{-1}yr^{-1}$ [25], while a comparative study of C sequestration in miscanthus and willow in Poland reported sequestration rates of 0.6 and 0.3 Mg C ha^{-1} yr^{-1}, respectively [26]. Observations by [27] show that the establishment of miscanthus on grasslands may lead to SOC losses and that the co-benefit of SOC sequestration mainly occurs under miscanthus established on former croplands.

The amount of SOC sequestered by WSGs is a function of site-specific factors; soil texture, management practices, initial SOC levels, and climate [28, 29]. The sequestration rate can be greater in soils with initial low SOC levels than in those that are about to reach their C saturation level. On average, perennial WSGs can sequester between 0 and 3.0 Mg C ha^{-1} yr^{-1} in the 0- to 5-cm soil depth [28]. In a clay loam in Iowa, switchgrass stored 0.8 Mg C ha^{-1} yr^{-1} in plant and litter biomass, which was higher than that in cool-season grass and row crops [30]. The SOC content under switchgrass was 45% more than under row crops planted in a sandy loam in Alabama [31]. A comparative evaluation of SOC sequestration in croplands, and under 5 to 8-yr-old WSGs across 10 sites in Indiana, showed only moderate increases in SOC storage: WSGs - 22.4 g C kg^{-1}, compared to 19.8 g C kg^{-1}under croplands [32]. A review of 115 studies of grassland SOC sequestration from 17 countries including Australia, Brazil, Canada, New Zealand, the UK, and the US reported SOC sequestration rates ranging from 0.1 to 3.0 Mg C ha^{-1} yr^{-1} [33]. Site-specific rates depended on biome type and climate.

Carbon sequestration rates during early grassland establishment on disturbed lands ranged from 0.3 to 0.5 Mg C ha^{-1} yr^{-1} [33, 34]. For the short term, perennial grasses may offer better economic returns due to the quicker establishment and the annual harvests. Data on SOC sequestration under Napiergrass and energy cane is scarce. One study on Napiergrass in the southeast US showed an average total soil C increase of 3.2 Mg ha^{-1} after four years of cultivation [35].

While growing WSGs increases SOC concentration by 0.3 to 0.5 Mg ha^{-1} yr^{-1} [33, 34], crop residue removal (e.g., corn stover) reduces SOC sequestration by 1 to 1.5 Mg ha^{-1} yr^{-1} in the 0- to 30-cm soil depth [12]. Growing WSGs thus increases SOC concentration while providing alternative feedstocks for cellulosic ethanol production, unlike crop residue removal, which results in large losses of SOC. The increase in SOC with WSGs not only plays a major role in mitigating GHG emissions, it also enhances ecosystem functions.

2.1.2. Impacts of Short Rotation Woody Crops

Information on SOC storage has not been widely reported for SRWCs. The limited data available so far comes from some few studies in the US, Canada and Europe. Two US studies were established in Minnesota to specifically address the issue of SOC sequestration under SRWCs [36, 37]. In the study by [36], soil C content was monitored in poplar plantations established on previously tilled agricultural prairie land, and compared to adjacent control grass and arable fields, while in the study by [37], poplar plantations were established on grass and arable lands. In both studies, there was a net loss of SOC over the first 6–12 years of initial establishment of the plantations. The SOC loss was largely from the surface 30 cm of soil, suggesting that it was due to enhanced decomposition. While [37] observed no increase in SOC under the poplar plantations, [36] reported an average rate of SOC increase of 1.6 Mg C ha^{-1} yr^{-1}. Other research also indicates that SRWCs may result in high erosion and runoff rates during the first year(s) of establishment.

As with perennial WSGs, SRWCs have a huge potential to increase SOC due to the abundant above- and belowground biomass input. For example, one study by [38] showed that willow trees produce finer root biomass than corn (5.8 Mg ha^{-1} versus 0.9 Mg ha^{-1} for corn). On average,

SRWCs can store between 0 and 1.6 Mg C ha^{-1} yr^{-1} in the 0- to 100-cm soil depth [39]. On relatively fertile soils in Canada, willow trees stored more SOC than either corn or switchgrass after 4 yr of establishment [40]. A determination of C sequestration in forests and willow cultivations showed sequestration rates at about 0.2 Mg C $ha^{-1}yr^{-1}$ [25]. In the US Midwest region, gains in SOC under SRWCs were greater than in croplands when the initial SOC concentration was low [41].

In the UK, estimates of SOC sequestration rates under SRWCs were based on willow and poplar plantations that were coppiced for varying time periods at three sites [42]. The average annual increase in SOC under SRWCs was estimated at between 1.2% and 2.2% [43]. In a study of three mixed poplar, aspen and willow plantation sites across Germany, substantial increases in SOC (0.1 to 0.6 Mg C ha^{-1} yr^{-1}) in the upper surface soil were observed after seven to nine years [44]. For much of Europe, research investigating potential SOC sequestration under SRWCs is still in its infancy.

According to [41], SRWCs should be grown in marginal and degraded lands where there will be greater accumulation of SOC rather than in croplands or natural forests. However, [45] argue that perennial WSGs may be more suitable for C sequestration since SRWCs take time for canopy closure, making soil more prone to SOC loss. If a management strategy increases the above- and belowground biomass, then a net gain in SOC sequestration can result.

2.1.3. Greenhouse gas mitigation potential

In contrast to row crops, evidence shows that biofuels from perennial WSGs and SRWCs have a positive GHG mitigation potential. Carbon sequestration associated with bioenergy-based cropping systems could potentially offset 1 to 2 Pg C yr^{-1} globally [46]. Three GHGs: CO_2, N_2O and CH_4 contribute directly to the GHG emission balance of cropping systems [47]. Nitrous oxide is produced in soils mostly through nitrification and denitrification. The majority of agricultural soils are minor sinks or sources of CH_4 [48]. Carbon dioxide emissions come from diesel fuel used in farm operations and during the manufacture of agricultural machinery, seed, and agrochemicals. Accounting for the GHG emission balance of bioenergy cropping systems entails evaluating their net impact on all emissions associated directly and indirectly to bioenergy feedstock production [1].

In the US, 60 million (M) ha of land is potentially available for conversion to bioenergy plantations. The conversion to bioenergy crops could mitigate C emissions at a rate 5.4 Mg C ha^{-1} yr^{-1}. On 60 M ha, that would represent 324 Tg C yr^{-1}, which is equivalent to a 20% reduction from current fossil-fuel CO_2 emissions [39]. In one study it was estimated that SRWCs sequester soil C at an average rate of 0.7 Mg C ha^{-1} yr^{-1} (Table 2). In addition, SRWCs could also generate substantial reductions in fertilizer and fuel use, reducing CO_2 and N_2O emissions, for a net GHG mitigation potential of 1.1 Mg C ha^{-1} yr^{-1}. Accordingly, as much as 40 M ha of highly eroded, degraded or mined lands in the US could be planted with SRWCs with limited negative impact on the production of key food and fiber crops [49].

GHG category	Switch to short rotation woody crops
Number of observations	**35**
	Mg C ha^{-1} yr^{-1}
Soil carbon	**0.7** (-2.1 to 3.6)
Nitrous oxide	**0.2**
Methane	**no data**
Process and upstream emissions	**0.2**
Sum of GHGs	**1.2** (-1.6 to 4.0)
Maximum U.S. applicable area, M ha	**40**

Positive numbers depict removal of greenhouse gases from atmosphere or prevented emissions. Source: [49].

Table 2. Carbon sequestration and greenhouse gas (GHG) mitigation potential in short rotation woody crops (SRWCs) – based on 35 field-based observations.

Fluxes of N_2O and CH_4 of SRWCs remain unknown. Uncertainties regarding N_2O emissions from nitrification and denitrification complicate the estimation of a system's GHG mitigation potential or GWP [1]. However, estimates of CH_4 consumption by soils are small, generally found to have little impact on net GWP [48]. One study across Europe noted much lower N_2O emissions in forested versus agricultural lands [50]. A review of the potential of SOC sequestration and CO_2 offset by SRWCs in the US [51] reported C sequestration rates ranging from 0 to 1.6 Mg C ha^{-1} yr^{-1}, and a GHG mitigation potential of 5.4 Mg C ha^{-1} yr^{-1}. In contrast, little difference was found in the N_2O emissions of annual crops versus those of poplar plantations [52]. Other researchers have estimated the bioenergy displacement of fossil fuels from SRWCs being as much as 4.9 to 5.5 Mg C ha^{-1} yr^{-1} [49]. In the majority of studies, N_2O emissions are the main source or driver of the cropping system's GWP. The assumption is that C sequestration by bioenergy crops of around 0.25 Mg C ha^{-1} yr^{-1} makes bioenergy neutral in terms of the GHG emission balance [53]. Quantifying these factors is necessary to determine the net effect of bioenergy cropping systems on soil C sequestration and GHG emissions.

2.2. Monitoring soil organic carbon storage and greenhouse gas fluxes

Soil organic carbon is the single most important indicator of soil quality [54, 55]. In addition to the importance of SOC as an indicator of soil quality, accurate measurements of soil C are also crucial in the assessment of the GHG emission balance of bioenergy systems. In general, methods for the determination of soil C involve direct and indirect approaches. Direct methods employ field and laboratory measurements of soil C stocks or CO_2 fluxes over time, while indirect methods combine data collected in direct soil C measurements and process-based simulation models to predict soil C changes temporally and spatially.

2.2.1. Field monitoring

The International Climate Change Treaty (Kyoto Protocol) recognized removal of CO_2 from the atmosphere by plants as a valid approach to mitigating climate change [56] and identified the need to conduct long-term monitoring of C stocks under various land uses. While the methods for analyzing the SOC concentration of a given soil sample are well established and easily carried out with high precision and negligible analytical error [57], field documentation of SOC changes faces many challenges because of the heterogeneity of soils, environmental conditions, land use history, sampling methods and analytical errors [58]. Here we highlight some of the challenges.

Variability in soil C estimates can be due to a number of factors or conditions: uncertainties in bulk density measurements as impacted by above and below-ground biomass inputs; rock and stone contents; compaction; sediment loss; and deposition by wind and water erosion. These conditions can drastically affect the accuracy of soil C sequestration estimates, implying that a relatively large number of soil samples will be required to obtain more reliable estimates. For example, [59] suggest that 20 to 30 soil samples (250 cm^3) will be needed for estimates of mean SOC within ±10% at the 95% confidence level, while [60] estimate that more than 100 samples (2.54 cm cores) would be needed to detect a 2–3% change in soil C under switchgrass.

The small annual incremental changes of soil C are difficult if not impossible to detect by present analytical methods. According to [61], noticeable soil C changes require a longer time period (e.g., at least 20 years). Statistically significant differences in SOC were observed after 5 to 10 yr [34, 62 - 64]. Coleman et al. [41] collected soil samples from 27 study sites across north central US to compare the soil C of short rotation poplar plantations to adjacent agricultural crops and woodlots. Soil organic carbon ranged from 20 to more than 160 Mg ha^{-1} across the sampled sites. Woodlots consistently contained greater SOC than the other crops, especially at depth. Surprisingly, [41] observed little difference between paired poplar and switchgrass. Furthermore, there was no evidence of changes in poplar SOC relative to adjacent agricultural soils when considered for stand ages up to 12 years.

A review of the different approaches to measurement and monitoring of SOC stocks shows that the main challenge is in designing an efficient, cost-effective sampling and SOC stock estimation system [65]. Current field sampling methods rely on a set of measurements that are extrapolated to represent a given geographic area. However, such intensive sampling is time consuming and costly. Furthermore, quantifying SOC changes at national or regional scales requires even higher sampling densities [66]. Conant et al. [65] are however optimistic that alternative measurement techniques, such as the infrared probe used by [67] or gamma-ray spectroscopy [68], could allow field-wide measurements of SOC stocks, even though the technologies currently require extensive site-specific instrument calibration. In addition, eddy covariance methods have improved greatly over the last two decades, and if coupled with appropriate and thoroughly tested terrestrial biosphere models, could be an alternative method to identify portions of the net ecosystem exchange associated with SOC [69]. Until remotely sensed measurements are calibrated to subsurface information, remotely sensed SOC inventories should only be considered as minimum estimates [70].

Given the many benefits and functions of SOC, a number of soil monitoring networks (SMN) that target soil C and other related data have been implemented or are under development in a number of countries throughout the world. For example, the European SMN – ENVASSO (ENVironmental ASsessment of Soil for mOnitoring; [71] has partners from 25 countries and a total of 33 334 soil C monitoring sites. The density of the European networks is high, with a median coverage of 300 km^2 for each monitoring site [58]. Although most SMNs are designed primarily for documenting soil C stocks and related data, their objectives do vary from country to country. For a more detailed review of the concepts, design and objectives of some of the SMNs across the world, see [72].

In the US there has been considerable investment in soil C surveys in accordance with the Energy Independence and Security Act (EISA) [73]. Two such programs are (1) the U.S. Geological Survey (USGS) carbon sequestration program (www.usgs.gov/climate_landuse/carbon_seq/). Two USGS national-scale assessments (subsurface and terrestrial) covering all 50 states and primary ecosystems are currently under way, and will provide the most comprehensive accounting of C storage potential in the US. The second program (2) is the USDA Natural Resources Conservation Service (NRCS) Rapid Assessment of US Soil Carbon (RaCA) (http://soils.usda.gov/survey/raca/). In general, the two programs were initiated to develop statistically reliable quantitative estimates of amounts and distribution of C stocks for US soils under various lands, differing agricultural management; to provide data to support model simulations of soil C change related to land use change, agricultural management, conservation practices, and climate change; and to provide a scientifically and statistically defensible inventory of soil C sequestration and GHG emissions for the US.

In Canada, a group of energy industries formed the GEMCo (Greenhouse Gas Emissions Management Consortium) in recognition of the potential of soil C sequestration as a possible GHG offset mechanism, together with the need to develop soil C measuring and verification protocols for C trading [74, 75]. The GEMCo, in partnership with Canadian soil conservation associations and government agencies, went on to initiate work aimed at monitoring and documenting soil C sequestration at the field level [69].

Some national [76, 77] and global [78, 79, 80] scale assessments of soil C stocks have been made. The World Bank-Global Environmental Fund's sustainable development project is supporting similar soil C monitoring initiatives in developing countries, e.g., Mexico [69]. In the US, the USDA NRCS soil databases; (1) State Soil Geographic (STATSGO) [81], provides generalized soil data for the entire US, and (2) the Soil Survey Geographic (SSURGO) provides detailed soil information for selected US counties [82]. Both these databases can be used to estimate SOC stocks [69]. Although at a lower resolution, the FAO's Soil Map of the World [79, 83] provides global estimates of SOC stocks.

Globally, a major issue with SMNs or any other soil C monitoring program is that there is no consensus on soil C measurement, monitoring, and verification methods. Post et al. [69] stress that there is an urgent need to develop robust, science-based, flexible, and agreed-upon practical protocols for monitoring and verifying temporal and spatial changes in soil C.

2.2.2. *Application of simulation models*

Computer simulation models can complement and extend the applicability of information collected in field trials [84] and can be applied to obtain realistic assessments of bioenergy feedstock production effects on SOC storage and the GHG emission balance. Combining measurement of SOC with models provides more flexibility and reliability because the model can be continually updated with new field sample data, and will be able to account for both C sequestration and GHG mitigation potential through LCA.

Although several soil C and GHGs emission models have been developed for conventional agricultural and forest systems, most of them have not been fully parameterized and effectively tested for application on WSGs and SRWCs. Detailed documentation on the main SOC models, their main features and comparative performance is widely available in the literature [85, 86, 87, 88, 89]. We focus on the applicability of a select number of process-based models that have been, or are currently being applied and hence could be adopted or adapted to monitor SOC sequestration and the GHG emission balance of bioenergy crops.

The amount of SOC in a given soil can increase or decrease depending on numerous factors, including climate, vegetation type, nutrient availability, disturbance, land use, and management practice. Most common models that describe this dynamic behavior or turnover of SOC follow the approach by [90], who proposed that soil C dynamics can be represented by two pools comprising one C pool and one residue pool, and can be described as follows:

$$dC_s/dt = hC_i - kC_s \tag{1}$$

where C_s is soil carbon (Mg ha^{-1}), t is time (yr), C_i is carbon inputs from plant residues and roots (Mg ha^{-1}), h is a humification factor and k is the apparent soil decomposition rate (yr^{-1}). This model, while simple, is adequate for hypothesis development and testing, and for field-level data interpretation [91]. However, as more information on SOC dynamics became available, new multi-compartment models accounting for variations in turnover rates for different SOC pools were developed [92]. For example, the most widely applied soil C simulation models, CENTURY [93] and RothC [94], divide SOM into different pools with varying turnover rates (Table 3). This partitioning is supposed to reflect the average biochemical and physical properties of SOM in the pools (e.g., metabolic, structural and recalcitrant) or the accessibility of the SOM to soil microbes or catalytic enzymes, or constraints imposed on decomposition by environmental conditions [95].

In the CENTURY model, SOC is divided into the active, slow and passive pools, with mean residence times (MRTs, i.e., the time required for half of the SOM to decompose) of 1.5, 25, and about 1000 yr, respectively. The organic inputs are separated into metabolic (readily decomposable; MRT of 0.1–1 yr) and structural (difficult to decompose; MRT of 1–5 yr) pools as a function of the lignin:N ratio [96]. A more detailed discussion of the SOC pools is given in [95]. It is, however, important to note that the C pool compartments are only conceptual, and have not been verified experimentally [96]. Despite this drawback, compartmental models have proved to be reasonably good in simulating changes in SOC stocks.

Although the CENTURY model was originally developed for grasslands [97], versions of the model have successfully been developed to simulate crops [98] and forest systems [99]. Among the nine SOC models that were evaluated against long-term datasets at a NATO Advanced Research Workshop held at Rothamsted, UK, in 1995, the CENTURY model had the best potential for adaptation to bioenergy cropping systems because of its integrated plant-soil approach and the availability of specific forestry subroutines [88]. A similar comparative study of six simulation models; RothC, CENTURY, EPIC [100], DNDC (DeNitrification DeComposition) [101], ECOSYS [102] and SOCRATES (Soil Organic Carbon Reserves And Transformations in agro-EcoSystems) [103] was conducted by [87]. Based on the model simulated outputs, the SOCTRATES model was best able to describe temporal (soil C changes observed in long-term experiments) and spatial SOC dynamics, and hence could be applied to scale-up soil C sequestration at the regional scale [87]. In another study, the model successfully predicted SOC change at eighteen long-term crop, pasture and forestry trials from North America, Europe and Australia [104]. A major advantage of the SOCRATES model is that it requires minimal data inputs, and is specifically designed to examine the impact of land use and land use change on SOC storage.

Despite sharing several basic concepts with the CENTURY model, the RothC model's applicability in monitoring SOC stocks in bioenergy cropping systems is rather limited due to the fact that it is solely concerned with soil processes, and does not contain a sub-model for plant production. In addition, it does not calculate annual returns of plant C to the soil from above-ground yields [88].The model has however been successful applied to describe SOC evolution under several long-term experiments, across a variety of land use and climate types [105].

Residue type	Century	RothC	Residence time (yr)	C:N	Compounds
Litter	Metabolic	DPM	0.1–0.5	10–25	Simple sugars
					Amino acids
					Starch
	Structural		2–5	100–200	Polysaccharides
SOM	Active	BIO	1–2	15–30	Living biomass
		DPM			POM
					Polysaccharides
	Slow	RPM	15–100	10–25	Lignified tissues
					Waxes
					Polyphenols
	Passive	HUM	500–5000	7–10	Humic substances
		IOM			Clay: OM complexes
					Biochar

DPM, decomposable plant material; BIO, microbial biomass; RPM, resistant plant material; HUM, humified organic matter; IOM, inert organic matter; POM, particulate organic matter; OM, organic matter. Source: [95].

Table 3. The pools of SOM defined according to their mean residence times (MRTs) and corresponding compound classes.

Due to the general applicability of the CENTURY model, a number of process-based simulation models such as EPIC, CropSyst [106] and DAYCENT [107], either adopted or adapted the SOC modeling algorithms from the CENTURY model. EPIC has since evolved into a comprehensive agro-ecosystem model with a SOC module capable of simulating SOC dynamics in a wide range of plant species, including crops, native grasses and trees [100]. Recently, [108] successfully applied the EPIC model to simulate the production and environmental effects of perennial biomass feedstocks grown on marginal lands. The comparative assessment of six alternative cropping systems over 20 years showed that once well established, perennial herbaceous grasses have a direct GHG emissions mitigation capacity (-2.3±0.1 Mg C ha^{-1} yr^{-1}) which is higher than that of conventional crops. Meki et al. [84] applied the sister model, ALMANAC [109], to evaluate sustainable energy sorghum biomass harvest thresholds and tillage effects on SOC and bulk density. Model results showed that 75% of energy sorghum biomass can be removed from a continuous no till system without any detrimental effects on SOC storage. This level of biomass removal, however, significantly increased soil bulk density, a critical indicator of soil compaction.

By integrating outputs from the REAP model [111] with environmental impact estimates derived from the EPIC model, [110] were able to evaluate the environmental and economic impacts of increased ethanol production from switchgrass. Highlights from this study indicated that commercial-scale switchgrass production could be associated with reduced erosion and reduced nutrient losses, but the projected increases in N fertilizer application and the associated N_2O emissions could offset soil C sequestration benefits and result in substantial increases in total GHGs emissions from the agricultural sector.

In many European countries, soil C and GHG simulation relies on the Yasso07 soil C model [112]. This model describes litter decomposition and soil C cycle based on the chemical quality of organic matter and the climatic conditions. Total soil C stocks are estimated for the top-most 1 m soil layer. Since Yasso07 is a reasonably new model [113], not many studies on the model's applicability to GHG gas inventory purposes are available. The strategy when developing the Yasso07 model was to achieve a soil C model that had fewer input data requirements, thus ensuring wider applicability [114]. A recent study in southern Finland by [112] showed the Yasso07 model accurately predicting the long-term soil C stocks of a coniferous forest soil (67.0 Mg ha^{-1}), which accumulated at an average rate of 0.12 ± 0.06 Mg ha^{-1} yr^{-1}. Previously, [115] evaluated the performance of the Yasso07 model in predicting the soil C stock changes in afforestation and deforestation sites. Overall, there was a good agreement between simulated and measured soil C stocks at most sites.

Garten and Wullschleger [60] used a simple, parameterized two-compartment model for SOC dynamics to predict the potential for soil C sequestration under switchgrass in the southeastern US. Model simulations indicated a measurable and verifiable recovery of soil C (~12% increase) on degraded lands through one decade of switchgrass production. A sensitivity analysis of the model indicated that switchgrass potential to sequester C depends on initial soil C inventories, prevailing climate, soil type, and site management.

The Terrestrial Ecosystem Model (TEM) is a process-based model that describes C and N dynamics of plants and soils for terrestrial ecosystems (http://ecosystems.mbl.edu/tem/). The

TEM uses spatially referenced information on climate, elevation, soils, vegetation and water availability as well as soil- and vegetation-specific parameters to make monthly estimates of C and N fluxes and pool sizes of terrestrial ecosystems. Conversion of agricultural lands to bioenergy crops consequently affects the ecosystem C balance. Qin et al. [116] applied the TEM model to evaluate the impacts of this change on the C balance, assuming several land use change scenarios from corn, soybean, and wheat to bioenergy crops of switchgrass and miscanthus. Overall, SOC increased significantly when land use changed from corn, soybean or wheat, to either switchgrass or miscanthus, reaching an average of ~100 Mg C ha^{-1} in the two bioenergy crops.

Relatively simple statistical models can also be useful in estimating SOC losses associated with land conversion. Anderson–Teixeira, et al. [117] used reduced ANOVA and the forced- and free-intercept regression statistical models to estimate and compare SOC losses associated with land conversion and SOC sequestration rates under corn, sugarcane, switchgrass, miscanthus and mixed native grass. Their predicted SOC accumulation rates under switchgrass, miscanthus and mixed native grass ranged from 0.1- 1.0 Mg C ha^{-1} yr^{-1} in the top 30 cm, while SOC losses under corn and sugarcane ranged from 3.0–8.0 Mg C ha^{-1} yr^{-1}. These results are consistent with data from numerous field studies reported elsewhere in the literature [20, 28, 31, 40].

Life-cycle assessments are commonly used to determine whether a biomass biofuel system results in a net reduction in GHG emissions or an improved energy source. In LCAs, all input and output data in all phases of the product's life cycle including biomass production, feedstock storage, feedstock transportation, biofuel production, biofuel transportation and final use are required. Because field measurements of bioenergy LCA components are difficult to carry out, simulation models have been applied effectively to obtain realistic estimates of the bioenergy crop's GWP or GHG mitigation potential.

Standard and useful tools for LCA are multidimensional spreadsheet models such as the GREET [118] and GHGenius [119] software models, which were specifically designed to address full LCA of biofuels. These spreadsheet models have advantages in that they are user-friendly, publicly available, straightforward, and relatively transparent. In a study commissioned by the US Department of Energy, Wang [120] applied the GREET model to characterize the soil C sequestration and GHG mitigation potential for the bioenergy crops switchgrass, poplars and willows. Model estimates indicated that there is a 70-85% reduction in GHGs emission when agricultural lands are converted to grasslands and forests.

DAYCENT is the daily time-step version of the CENTURY biogeochemical model [121]. DAYCENT simulates fluxes of C and N among the atmosphere, vegetation and soil [107]. Adler et al. [122] routinely applied the DAYCENT model to assess and compare GHG fluxes and biomass yields for corn, soybean, alfalfa, hybrid poplar, reed canarygrass, and switchgrass as bioenergy crops in Pennsylvania, US. All simulated cropping systems provided net GHG sinks compared with the fossil fuel life cycle, even in the long term when there were no further increases in soil C sequestration due to SOC storage reaching a new steady state. Switchgrass and hybrid poplar (Populus spp.) provided the largest net GHG sinks, (> 0.5 Mg C ha^{-1} yr^{-1}) for biomass conversion to ethanol, and > 1.1 Mg C ha^{-1} yr^{-1} for biomass gasification for electricity generation. Compared with the LCA of fossil fuels, switchgrass-derived biofuels reduced GHG emissions by more than 115%. Similar results were reported by [123], who also applied the DAYCENT model to calculate the net GHG fluxes, when converting from cotton and unmanaged grasslands to a switchgrass system in the southern U.S. Long-term simulations predicted a decrease in GHG emissions (0.3–1.0 Mg C ha^{-1} yr^{-1}) for conversions from cotton to switchgrass at N application rates of 0–135 kg N ha^{-1}. Conversely, conversion from unmanaged grasses to switchgrass resulted in annual increases of net GHG emissions (0.1–0.4 Mg C ha^{-1} yr^{-1}) for switchgrass at no and low (45 kg N ha^{-1}) fertilization rates. Using the DAYCENT model, Davis et al. [124] estimated the effects of replacing corn ethanol with switchgrass and miscanthus on GHG emissions in Central US. Overall, there was a 29 to 473% reduction in GHG emissions even after accounting for emissions associated with indirect land-use change. Conversion from a high-input annual crop to a low-input perennial crop for biofuel production can thus transition the central US from a net source to a net sink for GHGs. A simulation study with the DAYCENT model showed that cultivation of energy cane on former pasture on Spodosol soils in the southeast US has the potential for high biomass yield and a positive GHG emission balance [35].

The USGS General Ensemble Biogeochemical Modeling System (GEMS) [125] was developed to integrate well-established ecosystem biogeochemical models, such as CENTURY, with various spatial databases for the simulation of biogeochemical cycles over large areas. GEMS accounts for the impacts of atmospheric and climatic changes (e.g., changes in precipitation and temperature, CO_2 enrichment, N deposition), land use change (e.g., biofuel crops, tillage, forest logging, reforestation, urbanization, fire fuel treatment), natural disturbances and extreme events (land fire, insect outbreak, drought, flooding, hurricane) on ecosystem C balances and GHG emissions.

Even with all the uncertainties associated with LCAs, most suggest that WSGs and SRWCs have higher net energy balances and positive GHG emission balances than conventional crops [122, 126, 127]. Sarker [128] concludes that the use of computational models to quantify GHG fluxes in LCAs of bioenergy crops gives a more accurate representation of the system than using constant IPCC [129] emission factors. The USDA-ARS GRACEnet (Greenhouse gas Reduction through Agricultural Carbon Enhancement network: www.ars.usda.gov/research/GRACEnet) research initiative compares various GHG mitigation strategies based on field studies in several agroecoregions of the US. It is hoped that results from this environmental research program will be important to inform public policy on valuing the benefits of bioenergy crop production, and in addition, provide the much needed data for model GHG emission calibration and testing.

2.3. Carbon Credit Markets and Enabling Policies

It has been demonstrated conclusively that bioenergy crops can sequester soil C and mitigate GHG emissions. Accreditation of bioenergy production to CCMs will provide the much needed incentives for land managers and bioenergy feedstock producers to expand production, both for GHG mitigation benefits and the resulting improvements to soil health. Pioneered by the European Union's (EU) Emissions Trading Schemes (ETSs) launched in 2005, several CCMs are now operating across the world - see [130]. For agriculture, the schemes or programs allow producers and landowners to earn income by storing C in their soil through GHG capture projects. Soil C credits are then traded on CCMs, much like any other agricultural commodity. Large companies and other entities purchase C credits on the CCMs to offset their own GHG emissions into the atmosphere.

Carbon credits are allocated to an enterprise or project based on verifiable GHG emission reductions. For bioenergy feedstock producers to participate in CCMs there is need for low-cost, simple, but yet reliable methods for verifying soil C change in bioenergy cropping systems under different soil types, climates and management practices. As discussed earlier, field monitoring of soil C to document soil C changes is not only cost prohibitive but is generally not feasible due to the spatial variability in C stocks, and for most systems, the small annual incremental changes are not detectable by present analytical methods. According to [131], quantifying soil C may be even more costly than C stored in above-ground biomass, considering that soil C cannot be observed in the ways that biomass can. It is apparent that fully parameterized process-based models could be adopted or adapted to provide the much needed tools to track soil C sequestration and GHG emissions for the CCMs. As pointed out by [132], a major reason for developing soil C or agroecosystem models is to be able to use them to predict SOC changes under climate-soil management combinations lacking in soil C measurements.

Models such as Century, RothC, EPIC, SOCRATES, and more recently, C-Farm [92] can be used to assess and verify C sequestration rates and GHG mitigation potential for the CCMs. For most models, applicability is often limited because they require numerous data inputs. More user-friendly tools with minimum calibration requirements are needed. The Comet 2.0 tool (http://www.cometvr.colostate.edu/) is an easy to use Carbon Management Tool and Greenhouse Gas Accounting system for farm-ranch-orchard operations. Growers in the US can apply the Comet tool to obtain farm level mean estimates and uncertainty of C sequestration and GHG emissions from annual crops, hay, pasture and range, perennial woody crops (orchards, vineyards), agroforestry practices, and fossil fuel usage. The tool links a large set of databases containing information on soils, climate and management practices to dynamically run the CENTURY model as well as empirical models for soil N_2O emissions and CO_2 from fuel usage for field operations. Similarly, C-FARM is a user-friendly cropping systems model that allows calculating rates of C storage for the soil profile on a layer-by-layer basis. For simplicity, the model is built around a robust single-pool C model and can provide estimates of short- and long-term on-farm C storage rates. According to [92], C-FARM is a useful tool for growers, consultants, and state and federal agencies that need to develop C storage estimates for local conditions.

The energy crisis of 1973, and more recent concerns about global climate change and energy security have led to significant policy support for bioenergy throughout the world. The United States, European Union (EU), Australia, Canada and Switzerland spent at least $11 billion on biofuel subsidies in 2006 [133]. Here we briefly examine policies that could stimulate increased C sequestration by promoting bioenergy feedstocks production. Some suggest that, policy changes must focus on finding new avenues to divert public support to producers and landowners in the form of stewardship or green payments as well as making progress in encouraging the success of private C trading systems [28].

Most producers are reluctant to take their land out of conventional crop production into bioenergy crops for a long period of time. Governments should introduce policies that include programs that help farmers make the transition to growing bioenergy crops. Although only partially implemented, the USDA Biomass Crop Assistance Program (BCAP) is meant to provide cost-share payments for establishing dedicated energy crops and annual payments to cover the foregone income from alternative uses of the land [28]. Such payments give producers the opportunity to preserve their land by increasing the SOC pool, while at the same time producing a crop that helps to meet future energy needs.

It is however important to note that while subsidies do create incentives for a transition to biofuels, they may also encourage increased fuel consumption by lowering the price of blended fuel, thus mitigating the benefits of land conversion to biofuels through increased GHG emissions. Many environmental economists recognize that a tax or fee on CO_2 emission from fossil fuel sources is the most efficient system to reduce emissions and spread the burden equitably across all sources: industrial and personal [134]. A CO_2 tax is currently under discussion in a number of countries, including the US and EU. In general, the tax would be applied to all fossil fuels before combustion according to the C content of the respective fuel.

Although CCMs keep developing throughout the world, their sustained future existence is uncertain as they face many challenges: the global economic crisis and its impacts on trading markets; failure by governments to pass cap-and-trade legislation, especially the US, one of the world's largest GHG emitters; great uncertainty over future investments in some developing countries as CCM investors do not know where demand will come from; and finally, the international community's inability to agree on a post-2012 Kyoto framework in Copenhagen (COP15) and Cancun (COP16), which has seriously damaged the confidence of the private sector in the long-term viability of existing CCMs [135].

3. Concluding remarks

The potential for bioenergy landscapes under WSGs and SRWCs to sequester soil C and mitigate GHG emissions is indisputable. The functions and benefits of SOC in the maintenance of soil quality, productivity and sustainability of cropping systems are well documented. It is imperative to monitor the SOC integrity of bioenergy crops through the establishment of stringent SOC measurement and quantification methodologies. Available information suggests the best approach would be to combine field measurements with process-based simulation modeling. While models can capture complex C dynamics as impacted by site-specific factors or conditions, field measurements are crucial for the verification of model outputs. Besides promoting participation in CCMs by producers, accurate monitoring of SOC stocks will also allow assessment of whether a given bioenergy cropping system meets government-stipulated life cycle GHG emission reductions. Whether or not countries implement policies that stimulate increased C sequestration through bioenergy cropping, the importance of enhancing the soil resource base should be increasingly highlighted.

Acknowledgements

Preparation of this book chapter was supported by Texas A&M AgriLife Research and the USDA-ARS, Grassland, Soil and Water Research Laboratory, Temple, Texas, through Specific Cooperative Agreement: 58-6206-1-053, and partially funded by the U.S. Navy, Office of Naval Research. USDA is an equal opportunity provider and employer.

Author details

Manyowa N. Meki[1*], James R. Kiniry[2], Kathrine D. Behrman[2], Meghan N. Pawlowski[3] and Susan E. Crow[3]

*Address all correspondence to: nmeki@brc.tamus.edu, jim.kiniry@ars.usda.gov

1 Texas A&M AgriLife Blackland Research and Extension Center, Temple, US.

2 USDA, Agricultural Research Service, Grassland Soil and Water Research Laboratory, Temple, US.

3 Dept. of Natural Resources and Environmental Management, University of Hawaii at Manoa, Honolulu, US.

References

[1] Meki NM., Kemanian AR., Potter SR., Blumenthal JM., Williams JR., Gerik, TJ. Cropping systems effects on sorghum grain and biomass yield, soil organic carbon and global warming potential in central and south Texas. Journal Agricultural Systems 2013;117: 19-29.

[2] Miller SA., Landis AE., Theis TL. Environmental tradeoffs of biobased production. Environmental Science and Technology 2007;41:5176–5182.

[3] US EPA (Environmental Protection Agency). Renewable Fuel Standard Program: (RFS2) Regulatory Impact Analysis EPA-420-R-10-006, February 2010. http://www.epa.gov/otaq/renewablefuels/index.htm (accessed 15 August 2013).

[4] Meki NM., Kiniry RJ. A Dynamic Tool. International Innovation: The Global Forecast, October 2013. Research Media, UK, p118-120, ISSN 2051-8544. 2013.

[5] USDA NRCS (USDA Natural Resources Conservation Service). Crop residue removal for biomass Production: Effects on soils and recommendations. Soil Quality – Agronomy Technical Note No. 19. 2006.

[6] Lal, R. Managing soils and ecosystems for mitigating anthropogenic carbon emissions and advancing global food security. BioScience 2010;60: 708-721.

[7] Milne E. Soil organic carbon. Encyclopedia of the Earth. CAMEL. 2012

[8] Broadbent FE. The soil organic fraction. Advances in Agronomy 1953;5: 153-183.

[9] Krull ES., Skjemstad JO., Baldock JA. Functions of soil organic matter and the effect on soil properties. In: Grains Research and Development Corporation. p129. http://www.grdc.com.au/uploads/documents/cso000291.pdf. 2004. (accessed 19 August 2013).

[10] GRDC (Grains Research and Development Corporation). Carbon farming (factsheet), Adelaide and Mount Lofty Ranges. Natural Resources Management Board. Government of South Australia. http:// www.grdc.com.au/uploads/ documents/GRDC_ CarbonFarming_4pp.pdf. 2009. (accessed 20 August 2013).

[11] Overstreet FL., DeJong-Higes J. The importance of soil organic matter in cropping systems of the Northern Great Plains. University of Minnesota. Extension. http://www.extension.umn.edu/distribution/cropsystems/M1273.html. 2013. (accessed 20 August 2013).

[12] Blanco-Canqui H., Lal R. Corn stover removal for expanded uses reduces soil fertility and structural stability. Soil Science Society American Journal 2009;73: 418–426.

[13] Wilhelm WW., Johnson JMF., Hatfield JL., Voorhees WB., Linden DR. Crop and soil productivity response to corn residue removal: A literature review. Agronomy Journal 2004;96: 1–17.

[14] Hooker BA., Morris TF., Peters R., Cardon ZG. Long-term effects of tillage and corn stalk return on soil carbon dynamics. Soil Science Society of America Journal 2005;69: 188–196.

[15] Hohenstein GW., Wright LL. Biomass energy production in the United States: An overview. Biomass and Bioenergy 1994;6: 161-113.

[16] Tolbert VR., Schiller A. Environmental enhancement using short-rotation woody crops and perennial grasses as alternatives to traditional agricultural crops. Proceedings of the Conference, Environmental Enhancement Through Agriculture. Tufts University, Medford. Massachusetts. USA. 1996.

[17] Robertson GP., Dale VH., Doering OC., Hamburg SP., Melillo JM., et al. Sustainable biofuels redux. Science 2008;322: 49-50.

[18] Brown RA., Rosenberg NJ., Hays CJ., Easterling WE., Mearns LO. Production and environmental effects of switchgrass and traditional crops under current and green-

house gas-altered climate in the Central United States: A Simulation study. Agriculture, Ecosystems and Environment 2000;78: 31–47.

[19] Frank AB., Berdahl JD., Hanson JD., Liebig MA., Johnson HA. Biomass and carbon partitioning in switchgrass. Crop Science 2004; 44: 1391–1396.

[20] Liebig MA., Johnson HA., Hanson JD., Frank AB. Soil carbon under switchgrass stands and cultivated cropland. Biomass Bioenergy 2005;28: 347–354.

[21] Blanco-Canqui H. Energy crops and their implications on soil and environment. Agronomy Journal 2010;102: 403-419.

[22] Kahle P., Boelcke B., Zacharias S. Effects of *Miscanthus x giganteus* cultivation on chemical and physical soil properties. Journal of Plant Nutrition and Soil Science 1999;162: 27–32.

[23] Hansen EM., Christensen BT., Jensen LS., Kristensen K. Carbon sequestration in soil beneath long-term Miscanthus plantations as determined by C-13 abundance. Biomass and Bioenergy 2004;26: 97–105.

[24] Felten D., Emmerling C. Accumulation of Miscanthus-derived carbon in soils in relation to soil depth and duration of land use under commercial farming conditions. Journal of Plant Nutrition and Soil Science 2012;175: 661–670.

[25] Bradley RI., King JA. 2004. A review of farm management techniques that have implications for carbon sequestration – validating an indicator. OECD Expert Meet. Farm Management Indicators and the Environment. 8-12 March, 2004, Palmerston North. http://webdomino1.oecd.org/comnet/agr/farmind.nsf/22afaebba539ba74c1256a3b004d5175/b3b8d25f219f4ae3c1256bd5004874f1/$FILE/Bradley1.pdf (accessed 25 August 2013).

[26] Borzêcka-Walker M., Faber A., Borek R. Evaluation of carbon sequestration in energetic crops (*Miscanthus* and coppice willow). International Agrophysics 2008;22: 185-190.

[27] Don A., Osborne B., Hastings A. et al. Land-use change to bioenergy production in Europe: implications for the greenhouse gas balance and soil carbon. Global Change Biology Bioenergy 2012;4: 372–391.

[28] Lemus R., Lal R. Bioenergy crops and carbon sequestration. Critical Reviews in Plant Sciences 2005;24: 1–21.

[29] Blanco-Canqui H., Lal R., Lemus R. Soil aggregate properties and organic carbon for switchgrass and traditional agricultural systems in the southeastern United States. Soil Science 2005;170: 998–1012.

[30] Tufekcioglu A., Raich JW., Isenhart TM., Schultz RC. Biomass, carbon and nitrogen dynamics of multi-species riparian buffers within an agricultural watershed in Iowa, US. Agroforestry Systems 2003;57: 187–198.

[31] Ma Z., Wood CW., Bransby DJ. Soil management impacts on soil carbon sequestration by switchgrass. Biomass Bioenergy 2000;18: 469–477.

[32] Omonode RA., Vyn TJ. Vertical distribution of soil organic carbon and nitrogen under warm-season native grasses relative to cropland in west-central Indiana. Agriculture, Ecosystems and Environment 2006;117: 159–170.

[33] Conant RT., Paustian K., Elliott ET. Grassland management and conversion into grassland: effects on soil carbon. Ecological Applications 2001;11: 343–355.

[34] Post WM., Kwon, KC. Soil carbon sequestration and land-use change: processes and potential. Global Change Biology 2000;6: 317–327.

[35] Duval BD., Anderson-Teixeira KJ., Davis SC., Keogh C., Long SP., Parton WJ., DeLucia EH. Predicting greenhouse gas emissions and soil carbon from changing pasture to an energy crop. PLOS ONE 2013;8(8):e72019. doi:10.1371/journal.pone.0072019. (accessed 27 August 2013).

[36] Hansen EA. Soil carbon sequestration beneath hybrid poplar plantations in the north-central United States. Biomass and Bioenergy 1993;5: 431-436.

[37] Grigal DF., Berguson WE. Soil carbon changes associated with short-rotation systems. Biomass and Bioenergy 198;14: 371-377.

[38] Bonin C., Lal R. Physical properties of an Alfisol under biofuel crops in Ohio. J. of Technology Innovations in Renewable Energy 2012;1: 1-13.

[39] Sartori F., Lal R., Ebinger MH., Miller RO., Parrish DJ. Potential soil carbon sequestration and CO2 off set by dedicated energy crops in the US. Critical Reviews Plant Science 2006;25: 441–472.

[40] Zan CS., Fyles JW., Girouard P., Samson RA. Carbon sequestration in perennial bioenergy, annual corn and uncultivated systems in southern Quebec. Agriculture, Ecosystems and Environment 2001;86: 135–144.

[41] Coleman MD., Isebrands JG., Tolsted DN., Tolbert VR. Comparing soil carbon of short rotation poplar plantations with agricultural crops and woodlots in north central United States. Environmental Management 2004;3: 229–308.

[42] Matthews R. Modelling of carbon and energy budgets of wood fuel coppice systems. Biomass Bioenergy 2001;21: 1-19.

[43] Smith P., Powlson DS., Glendining M., Smith J. Potential for carbon sequestration in European soils: preliminary estimates for five scenarios using results from long-term experiments. Global Change Biology 1997;3: 67-79.

[44] Jug A., Makeschin M., Rehfuess KE., Hofmann-Schielle C. Short-rotation plantations of balsam poplars, aspen and willows on former arable land in the Federal Republic of Germany. III. Soil ecological effects. Forest Ecology and Management 1999;121: 85-99.

[45] Harmon ME., Ferrell W., Franlkin JF. Effects of carbon storage of conversion of old-growth forests to young forests. Science 1990;247: 699–702.

[46] Cannell MGR. Carbon sequestration and biomass energy offset: Theoretical, potential, and achievable capacities globally, in Europe and the UK. Biomass and Bioenergy 2003; 24: 97–116.

[47] Bronson KF., Mosier AR. Nitrous oxide emissions and methane consumption in wheat and corn cropped systems. In: Harper LA., Mosier AR., Duxbury JM., Rolston DE. (eds.) Agriculture Ecosystem Effects on Trace Gases and Global Climate Change. ASA-CSSA-SSSA, Madison, WI. 1993. p133–144.

[48] Robertson GP., Paul EA., Harwood RR. Greenhouse gases in intensive agriculture: contributions of individual gases to radiative forcing of the atmosphere. Science 2000;289 : 1922–1925.

[49] Tuskan GA., Walsh ME. Short-rotation woody crop systems, atmospheric carbon dioxide and carbon management: A US case study. Forestry Chronicle 2001;77: 259–64.

[50] Machefert SE., Dise NB., Goulding KT., Whitehead PG. Nitrous oxide emission from a range of land uses across Europe. Hydrological Earth System Science 2002;6: 325–337.

[51] Sartori F., Lal R., Ebinger, HE., Parrish DJ. Potential Soil Carbon Sequestration and CO2 Offset by Dedicated Energy Crops in the USA. Critical Reviews in Plant Sciences 2006;25: 441-472.

[52] Scheer C., Wassmann R., Kienzler K., Ibragimov N., et al. Methane and nitrous oxide fluxes in annual and perennial land-use systems of the irrigated areas in the Aral Sea Basin. Global Change Biology 2008;14: 2454–2468.

[53] Volk TA., Verwijst T., Tharakan PJ., Abrahamson LP., White EH. Growing fuel: a sustainability assessment of willow biomass crops. Frontiers in Ecology and the Environment 2004;2(8): 411- 418.

[54] Reeves DW. The role of soil organic matter in maintaining soil quality in continuous cropping systems. Soil Tillage Research 1997;43: 131–167.

[55] Karlen DL., Varvel GE., Johnson JM., Baker JM., Osborne SL., Novak JM., Adler PR., Roth GW., Birrell SJ. Monitoring soil quality to assess the sustainability of harvesting corn stover. Agronomy Journal 2011;103: 288–295.

[56] Marland G., Schlamadinger B. The Kyoto Protocol could make a difference for the optimal forest-based CO_2 mitigation strategy: some results from GORCAM. Environmental Science and Policy 1999;2: 111-124.

[57] Nelson DW., Sommers LE. Total carbon, organic carbon, and organic matter. In: Sparks DL. (ed.) Methods of soil analysis. Part 3: Chemical methods. Madison, WI: Soil Science Society of America. 1996.

[58] Saby NPA., Bellamy PH., Morvan X., Arrouays D., Jones RJA., Verheijen FGA., Kibblewhite MG., Verdoot AY., Üveges JB., Freudenschuß A., Simota C. Will European soilmonitoring networks be able to detect changes in topsoil organic carbon? Global Change Biology 2008;14: 1-11

[59] Mollitor AV., Leaf AL., Morris LA. Forest Soil Variability on Northeastern Flood Plains. Soil Science Society American Journal 1980;44: 617–620.

[60] Garten CT., Wullscheleger SD. Soil carbon dynamics beneath switchgrass as indicated by stable isotype analyses. Journal of Environmental Quality 2000;29: 645–653.

[61] Rasmussen PE., Albrecht S.L., Smiley RW. 1998. Soil C and N changes under tillage and cropping systems in semi-arid Pacific Northwest agriculture. Soil Tillage Research 1998;47: 197-205.

[62] Izaurralde RC., Nyborg M., Solberg ED., Janzen HH., Arshad MA., Malhi SS., Molina-Ayala M. Carbon storage in eroded soils after five years of reclamation techniques. In: Lal R., Kimble J., Follett R., Stewart BA. (eds.) Soil Processes and the Carbon Cycle. Advanced Soil Science CRC Press, Inc., Boca Raton, FL. 1998. p369–386.

[63] Smith P., Powlson D., Glendining M. 1996, 'Establishing a European GCTE Soil Matter Network (SOMNET). In; Powlson DS., Smith P., Smith JU. (eds.) Evaluation of soil organic matter models using existing long-term datasets. NATO ASI Series, Vol. 38, Springer-Verlag, Berlin. 1996. p81–97.

[64] Paul EA., Paustian K., Elliot ET., Cole CV. (eds.) Soil Organic Matter in Temperate Agroecosystems: Long-Term Experiments in North America, CRC Press, Inc., Boca Raton, FL. 1997.

[65] Conant RT., Ogle SM., Paul EA., Paustian K. Measuring and monitoring soil organic carbon stocks in agricultural lands for climate mitigation. Frontiers in Ecology 2010;9: 169–73.

[66] Makipaa R, Hakkinen M., Muukkonen P., Peltoniemi M. The costs of monitoring changes in forest soil carbon stocks. Boreal Environmental Research 2008;13: 120–30.

[67] Christy CD. Real-time measurement of soil attributes using on-the-go near infrared reflectance spectroscopy. Computers and Electronics in Agriculture 2008;61: 10–19.

[68] Wielopolski LG., Hendrey KH., Johnsen KH., et al. Non-destructive system for analyzing carbon in soil. Soil Science Society American Journal 2008;72: 1269–77.

[69] Post WM., Izaurralde RC., Mann, LK., Bliss, N. Monitoring and verifying changes of organic carbon in soil. Climatic Change 2001; 51: 73–99, 2001.

[70] Waltman WS., Olson C., West L., Moore A., Thompson J. Preparing a soil organic carbon inventory for the United States using soil surveys and site measurements: Why carbon stocks at depth are important. 19th World Congress of Soil Science, Soil Solutions for a Changing World 1–6 August 2010, Brisbane, Australia. 2010.

[71] Kibblewhite M., Jones RJA., Baritz R., Huber S., Arrouays D., Micheli E., Dufour MJD. ENVASSO. Environmental assessment of soil for monitoring. In: EC desertification meeting, Brussels. 2005. http://eusoils.jrc.ec.europa.eu/projects/envasso/ (accessed on 11 September 2013)

[72] Wesemael Bas van, Paustian K., Andrén O., Cerri EPC., et al. How can soil monitoring networks be used to improve predictions of organic carbon pool dynamics and CO_2 fluxes in agricultural soils. Plant Soil 2011;338: 247–259.

[73] EISA (Energy Independence and Security Act) http://www.afdc.energy.gov/laws/eisa. 2007. (accessed 11 September 2013).

[74] Ellert, BH., Janzen HH., McConkey B. A Method for measuring soil carbon change on the Canadian Prairies. In: Lal R., Kimble J., Follett R. (eds.) An International Workshop on Assessment Methods for Soil C Pools. Lewis Publishers. CRC Press, Inc., Boca Raton, FL. 2000.

[75] Izaurralde RC., McGill WB., Bryden A., Graham S., Ward M., Dickey P. Scientific challenges in developing a plan to predict and verify carbon storage in Canadian Prairie soils. In: Lal R., Kimble J., Follett R. (eds.) An International Workshop on Assessment Methods for Soil C Pools. Lewis Publishers. CRC Press, Inc., Boca Raton, FL. 1998. p433–446.

[76] Kern JS., Turner DP., Dodson RF. Spatial patterns in soil organic carbon pool size in the Northwestern United States. In: Lal R., Follett R., Stewart BA. (eds.) Soil Processes and the Carbon Cycle. Advanced Soil Science. CRC Press, Inc., Boca Raton, Fla., 1998. p29-44.

[77] Tarnocai C., Ballard M. Organic carbon in Canadian soils. In: Lal R., Kimble J., Levine E. (eds.) Soil Processes and Greenhouse Effect. USDA, Soil Conservation Service, National Soil Survey Center, Lincoln, NE, USA. 1994. p31–45.

[78] Post WM., Pastor J., Zinke PJ., Stangenberger AG. Global soil nitrogen. Nature 1985;317: 613–616.

[79] Eswaran H., Van den Berg E., Reich P., Kimble J. Global Soil Carbon Resources. In: Lal R., Kimble J., Levine E., Stewart BA. (eds.), Advances in Soil Science: Soils and Global Change, Lewis Publishers, CRC Press. Boca Raton, FL. 1995. p. 27–43.

[80] Batjes NH. 1996. Total carbon and nitrogen in the soils of the world. European Journal Soil Science 1996;47: 151–163.

[81] NRCS (Natural Resources Conservation Service) 1994, State Soil Geographic (STATSGO) Data Base - Data Use Information. US Department of Agriculture. http://soils.usda.gov/survey/geography/ssurgo/description_statsgo2.html. (accessed 9 September 2013).

[82] NRCS (Natural Resources Conservation Service). 1995 Soil Survey Geographic (SSURGO) Data Base – Data Use Information. US Department of Agriculture. http://

soils.usda.gov/survey/geography/ssurgo/description.html. (accessed 9 September 2013).

[83] Sombroek WG., Nachtergaele FO., Hebel A. Amounts, dynamics and sequestration of carbon in tropical and subtropical soils. Ambio 1993;22: 417–426.

[84] Meki NM., Snider JL., Kiniry JR., Raper RL., Rocateli AC. Energy sorghum biomass harvest thresholds and tillage effects on soil organic carbon and bulk density. Industrial Crops and Products 2013;43: 172-182.

[85] McGill WB. Review and classification of ten soil organic matter models. In: Powlson DS., Smith P., Smith JU. (eds.) Evaluation of Soil Organic Matter Models Using Existing Long-Term Datasets. Springer-Verlag, Heidelberg. 1996. p111–132.

[86] Paustian K., Levine E., Post WM., Ryzhova IR. 1997, The use of models to integrate information and understanding of soil C at the regional scale. Geoderma 1997;79: 227–260.

[87] Izaurralde RC., Hauren-Kozyra KL., Jans DC., McGill WB., Grant RF., Hiley JC. Soil C dynamics: Measurement, simulation and site-to-region scale-up. In: Lal R., Kimble J. Follett R. (eds.) An International Workshop on Assessment Methods for Soil C Pools. Lewis Publishers. CRC Press, Inc., Boca Raton, FL. 2000.

[88] Grogan P., Matthews R. Review of the potential for soil carbon sequestration under bioenergy crops in the UK. Report for UK Dept. of Environment, Food, and Rural Affairs. Contract NF 0418. Institute of Water and Environment. Cranfield University, Silsoe. Bedfordshire MK45 4DT. UK. 2001.

[89] Nair SS., Kang S., Zhang X., Miguez F., Izaurralde, CR., Post W., Dietze MC., Lynd L., Wullschleger SD. Bioenergy crop models: descriptions, data requirements, and future challenges. GCB Bioenergy 2012;4(6): 620-633.

[90] Hénin S., Dupuis, et.M. Essai de bilan de la matière organique du sol. Annals Agronomy 1945;15: 17–29.

[91] Huggins DR., Clapp CE., Allmaras RR., Lamb JA., Laysee MF. Carbon dynamics in corn-soybean sequences as estimated from natural Carbon-13 abundance. Soil Science Society America Journal 1998;62: 195–203.

[92] Kemanian AR., Stöckle CO. C-farm: a simple model to evaluate the carbon balance of soil profiles. European Journal Agronomy 2010;32: 22–29.

[93] Parton WJ. The CENTURY model. In: Powlson DS., Smith P., Smith JU. (eds.) Evaluation of soil organic matter models using existing long-term datasets. Springer-Verlag, Berlin, Germany. 1996. p283–291.

[94] Coleman DC., Jenkinson DS. RothC-26.3 – A model for the turnover of carbon in soil. In: Evaluation of Soil Organic Matter Models using Existing Long-Term Datasets. In:

Powlson DS., Smith P., Smith JU. (eds.) NATO ASI Series I. Springer, Berlin; 1996. p237–246.

[95] Dungait AJ., Hopkins DW., Gregory AS., Whitmore AP. Soil organic matter turnover is governed by accessibility not recalcitrance. Global Change Biology 2012;18: 1781–1796.

[96] Six J., Jastrow JD. Organic matter turnover. Encyclopedia of Soil Science. 2002. p936-942.

[97] Parton WJ., Schimel DS., Cole CV., Ojima DS. Analysis of factors controlling soil organic matter levels in Great Plains grasslands. Soil Science Society America Journal 1987;51: 1173-1179.

[98] Parton WJ., Rasmussen PE. Long-term effects of crop management in wheat/fallow: II. CENTURY model simulations. Soil Science Society America Journal 1994;58: 530-536.

[99] Johnson K., Scatena FN., Pan Y. Short- and long-term responses of total soil organic carbon to harvesting in a northern hardwood forest. Forest Ecology and Management 2010;259: 1262–1267.

[100] Izaurralde RC., Williams JR., McGill WB., Rosenberg NJ., Jakas MC. Simulating soil C dynamics with EPIC: Model description and testing against long-term data. Ecological Modeling 2006;192: 362-384.

[101] Li C., Frolking S., Harriss R. Modeling carbon biogeochemistry in agricultural soils. Global Biogeochemical Cycles 1994;8: 237–254.

[102] Grant RF., Juma NG., McGill WB. Simulation of carbon and nitrogen transformations in soil: mineralization. Soil Biology and Biochemistry 1993;25: 1317–1329.

[103] Grace PR., Ladd JN. SOCRATES v2.00 User Manual. Cooperative Research Centre for Soil and Land Management, PMB 2 Glen Osmond 5064, South Australia. 1995.

[104] Grace PR.,, Ladd JN., Robertson GP., Gage SH. SOCRATES - A simple model for predicting long-term changes in soil organic carbon in terrestrial ecosystems. Soil Biology and Biochemistry 2006;38: 1172–1176.

[105] Kelly RH., Parton WJ., Crocker GJ., Grace PR., Klir J., Korschens M., Poulton, PR., Richter DD. Simulating trends in soil organic carbon in long-term experiments using the century model. Geoderma 1997;81: 75-90.

[106] Stockle CO., Donatelli M., Nelson R. CropSyst, a cropping systems simulation model. European Journal of Agronomy 2003;18: 289–307.

[107] Del Grosso SJ., Mosier AR., Parton WJ., Ojima DS. DAYCENT model analysis of past and contemporary soil N_2O and net greenhouse gas flux for major crops in the US. Soil Tillage Research 2005;83: 9-24.

[108] Gelfand I., Sahajpal R., Zhang X., Izaurralde RC., Gross KL., Robertson GP. Sustainable bioenergy production from marginal lands in the US Midwest. 2013 doi:10.1038/nature11811.

[109] Kiniry JR., Williams, JR., Gassman, PW., Debaeke P. A general, process-oriented model for two competing plant species. Transactions ASAE 1992;35: 801–810.

[110] Marshall L., Sugg Z. Fields of Fuel: Market and environmental implications of switching to grass for US transport. World Resources Institute, WRI Policy Note, Energy: Biofuels. 2010.

[111] Johansson R., Peters M., House R. Regional Environment and Agriculture Programming model (REAP). USDA- Economic Research Service. Technical Bulletin No. (TB-1916) pp118. March 2007. www.ers.usda.gov/publications/TB1916. 2007. (accessed 21 August 2013).

[112] Rantakari M., Lehtonena A., Linkosalo T., Tuomi M., Tamminen P., et al. The Yasso07 soil carbon model – Testing against repeated soil carbon inventory. Forest Ecology and Management 2012;286: 137–147.

[113] Tuomi M., Thum T., Järvinen H., Fronzek S., Berg B., Harmon M., Trofymow JA., Sevanto S., Liski J. Leaf litter decomposition – Estimates of global variability based on Yasso07 model. Ecological Modeling 2009;220: 3362–3371.

[114] Liski J., Palosuo T., Peltoniemi M., Sievänen R. Carbon and decomposition model Yasso for forest soils. Ecological Modeling 2005;189: 168–182.

[115] Karhu K., Wall A., Vanhala P., Liski J., Esala M., Regina K. Effects of afforestation and deforestation on boreal soil carbon stocks – comparison of measured C stocks with Yasso07 model results. Geoderma 2011;164: 33–45.

[116] Qin Z., Zhuang Q., Chen M. Impacts of land use change due to biofuel crops on carbon balance, bioenergy production, and agricultural yield, in the conterminous United States. GCB Bioenergy 2012;4: 277–288.

[117] Anderson-Teixeira KJ., Davis SC., Masters MA., DeLucia EH. Changes in soil organic carbon under biofuel crops. GCB Bioenergy 2009;1: 75–96.

[118] Wang, M. Fuel choices for fuel-cell vehicles: Well-to-Wheels energy and emission impacts. Journal of Power Sources 2002;112: 307-321.

[119] Stanciulescu V., Fleming JS. Life Cycle Assessment of Transportation Fuels and GHGenius. EIC Climate Change Technology Conference. 2006. p1-11. Ottawa, ON. Canada.

[120] Wang M. GREET 1.5 — Transportation Fuel Cycles Model Volume 1: Methodology, Use, and Results, ANL/ESD-39, Center for Transportation Research, Argonne National Laboratory, Argonne, Ill. August 1999. IL. US. 1999.

[121] Parton WJ., Hartman MD., Ojima DS., Schimel DS. DAYCENT: Itsland surface sub-model: description and testing. Global Planetary Changes 1998;19: 35-48.

[122] Adler PR., Del Grosso SJ., Parton WJ. Life cycle assessment of net greenhouse gas flux for bioenergy cropping systems. Ecological Applications 2007;17: 675–691.

[123] Chamberlain JF., Miller S., Frederick J. Using DAYCENT to quantify on-farm GHG emissions and N dynamics of land use conversion to N-managed switchgrass in the Southern U.S. Agriculture, Ecosystems and Environment 2011;141: 332-341.

[124] Davis SC., Parton WJ., Del Grosso SJ., Keough C., Marx E., AdlerPR., DeLucia EH. Impact of second-generation biofuel agriculture on greenhouse-gas emissions in the corn-growing regions of the US. Frontiers in Ecology and the Environment 2011. doi: 10.1890/110003.

[125] Liu S., Tan Z., Chen M., Liu J., Wein A., Li Z., Huang J., Oeding J., Young CJ., Verma SB., Suyker AE., Faulkner S., McCarty GW. The General Ensemble Biogeochemical Modeling System (GEMS) and its applications to agricultural systems in the United States. In: Liebig MA., Franzluebbers AJ., Follett RF. (eds.) Managing agricultural greenhouse gases-coordinated agricultural research through GRACEnet to address our changing climate: San Diego, Calif., Academic Press. 2012. p309-323.

[126] Boehmel C., Lewandowsk I., Claupein W. Comparing annual and perennial energy cropping systems with different management intensities. Agricultural Systems 2008;96: 224-236.

[127] Tilman D, Hill J., Lehman, C. Carbon-negative biofuels from low-input high-diversity grassland biomass. Science 2006;314: 1598-600.

[128] Sarkar S., Miller SA., Frederick JR., Chamberlain JF. 2011. Modeling nitrogen loss from switchgrass agricultural systems. Biomass and Bioenergy 2011;35: 4381-4389.

[129] IPCC (Intergovernmental Panel on Climate Change). Greenhouse Gas Inventory Reference Manual. Intergovernmental Panel on Climate Change. Cambridge University Press, Cambridge, UK. 1997.

[130] Perdan S. Azapagic A. Carbon trading: Current schemes and future developments. Energy Policy 2011;39: 6040–6054.

[131] Antle JM., Capalbo SM., Mooney M., Elliott ET., Paustian KH. 2003. Spatial heterogeneity, contract design, and the efficiency of carbon sequestration policies for agriculture. Journal of Environmental Economics and Management 2003;46(2): 231–50.

[132] He X., Izaurralde RC., Vanotti MB., Williams JR., Thomson AM. Simulating long-term and residual effects of nitrogen fertilization on corn yields, soil carbon sequestration, and soil nitrogen dynamics. Journal of Environmental Quality 2006;35: 1608-1619.

[133] GSI (Global Subsidies Initiative). Biofuels at What Cost: Government support for ethanol and biodiesel in selected OECD Countries. International Institute for Sustainable Development, Geneva, Switzerland. 2007.

[134] Schlesinger, W.H. Carbon trading. Science 2006;314: 1217.

[135] IETA (International Emissions Trading Association):www.ieta.org. (accessed 15 August 2013).

Chapter 10

Pre-Injection Phase: Site Selection and Characterization

B. Llamas, M. Arribas, E. Hernandez and
L.F. Mazadiego

Additional information is available at the end of the chapter

http://dx.doi.org/10.5772/57405

1. Introduction

Depleted oil and gas fields, deep saline aquifers and non-mineable coal beds fulfil the special requirements for CO_2 storage and have become global references to develop this industry [2]. Depleted oil or gas fields have been well researched due to the associated industrial and economic value of these emplacements. While deep saline aquifers are among the most promising emplacements, since theoretically these structures offer the highest value in terms of capacity [2],[3], the risk associated with the exploration of these potential sites is greater than that for already investigated depleted oil or gas fields. In such cases, it is likely that multicriteria algorithms can facilitate evaluation to find the best option under consideration, so the results of this process will help the decision-maker to decrease the risk associated with the exploration of these potential emplacements.

The site selection phase comprises the identification, characterization and selection of emplacements that could be suitable for CCS among a list of candidates [4]. This phase is generally completed by the definition of the qualification criteria and the provision of the evidence concerning the reliable functioning of the emplacement according to these criteria. The selection of a suitable site also depends on the scale of the assessment. In this regard, every scale will be related to a different resolution and detail of information. It is possible to differentiate at least three levels, namely, basin level (identifying and quantifying large potential storage areas), regional level (increased level of detail, identifying areas to prospect) and local level (very detailed structures, pre-engineering site selection).

While the main mechanisms for CO_2 storage have been identified, and a series of criteria for site assessment and selection have been developed [5],[6],[7], a multicriteria algorithm to quantify and rank the potential areas under consideration has, to our knowledge, never been applied in an absolute mode, meaning that any alternative will be compared against a pattern.

Thus, the objective of this study was to develop a methodology based on multicriteria algorithms for assessing the best emplacement for CCS within a range of alternatives.

In general, most of the areas which could be suitable for storing CO_2 are not well explored geologically. As a result, further exploration of the subsurface must be carried out, which implies higher cost and risks. In order to reduce the risk of failure, it is necessary to define a previous phase, which should be based on (1) data collection, and (2) the definition of a criterion and multicriteria decision tool.

The collection of data is independent of each country or region. It involves the study and processing of existing data, such as from geophysical surveys along with existing wells (shallow and deep) developed for oil and gas exploration, water resources and exploration, mining activities (exploration and extraction), exploration for nuclear waste deposits and underground natural gas storage activities. The data from these industrial activities can be complemented by other academic evaluations, including postgraduate theses and/or peer reviewed scientific articles.

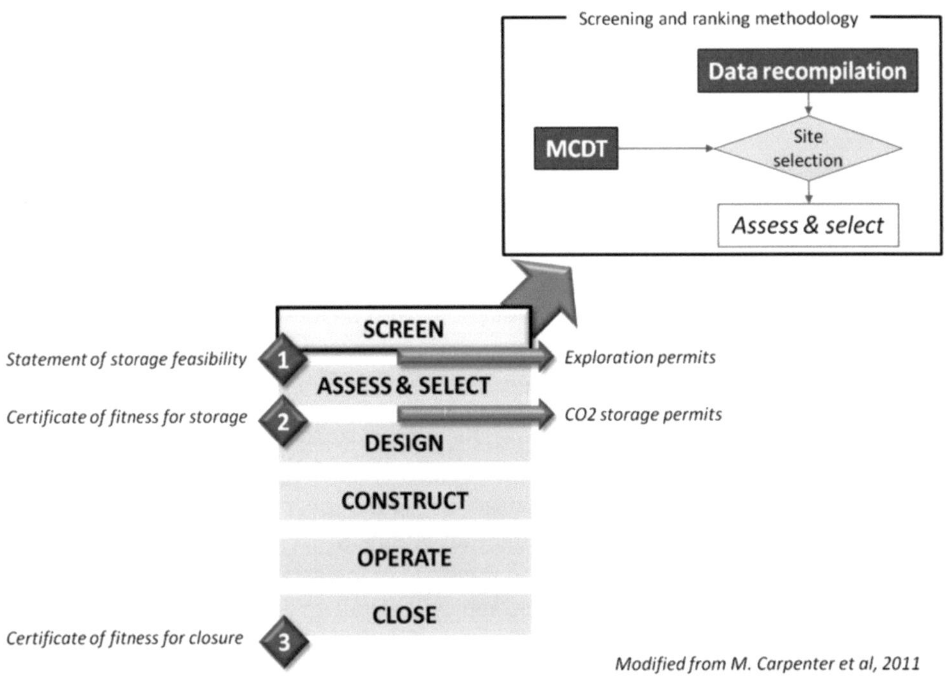

Figure 1. Work flow proposed for basin screening (Definition phase)

Figure 1 represents a proposed work flow [4]. The screening phase could be differentiated by the Data Recompilation task and the Multicriteria Decision Tool. It is integrated as a preliminary phase, and it is connected with a second phase called *characterization phase,* which corresponds to site maturation and testing.

http://dx.doi.org/10.5772/57405

2. Site criteria definition

It is not only necessary to evaluate specific sites with a technical point of view. Sometimes problems can involve economic aspects or social acceptance, which could make a CO_2 storage site much more difficult (and costly). Therefore, it is necessary to establish a high level differentiation between technical and socioeconomic criteria (later we call this differentiation the 1st criteria level of the Analytical Hierarchy Process).

There is no standardization in this aspect, so the selection of the criteria should be as careful as possible, and should include all aspects that can make an area suitable or not. The criteria proposed in this chapter are based on the direct experience of the research group involved in this publication and on the evaluation of several publications and projects that focus on site selection methodology [10],[4].

Even though the criteria may be as described in the next sub-chapters, it is convenient to consider another type of classification, which is based on time scale and the possibility of modifying each criterion along that scale. Moreover, we can describe:

1. Geological, geothermal and hydrodynamic criteria are considered fixed, because they do not change, except in geological scale.

2. Knowledge of the basin is a criterion variable, because information can be increased over time.

3. Economic, political and social criteria are often variable because they change in a short time period.

2.1. Technical criteria

These criteria relate to scientific aspects or parameters to provide confidence in the findings about the subsurface structure. Deep subsurface exploration implies higher risk because the exploration techniques available are expensive and the probability of success is not high. In order to reduce the technical risk, it is necessary to define those criteria relevant for considering every critical issue.

Tectonics. This parameter comprises aspects like the structural definition of the trap at a basin and local level. The former considers whether the sedimentary basin is convergent or divergent, as well as its neotectonic activities. The latter relates to the type of the structure and trap, whether it is an anticline, syncline or seal fault [10]. Furthermore, the geo-mechanic evaluation should be taken into account in order to evaluate the maximum CO_2 injection pressure in the storage formation. This injectivity criterion defines the maximum capacity per unit of time.

Geology. This criterion evaluates the storage and the cap-rock formations. The common formations considered suitable for storing CO_2 are sandstones and limestones. These geological formations tend to have high porosity. While sandstone has a primary porosity, which is much more homogeneous, limestone presents secondary porosity, which is created by processes of diagenesis (e.g., leeching of minerals or the generation of a fracture system). Other parameters to be defined in storage formations are permeability, and thickness of the formation. The plasticity, porosity and thickness of the cap-rock formations should also be evaluated and considered. In relation to the plasticity of the cap-rock formation, it is more desirable to have a ductile rather than a brittle behavior.

Hydrogeology. This parameter should consider both the dynamics and the fluid quality in the reservoir formation. Hydrodynamic criteria describe the natural dynamic flow system and hence the potential for hydrodynamic trapping within the basin under assessment. Shallow, short flow systems therefore do not meet the geological requirements for maintaining supercritical CO_2, in terms of depth, pressure and temperature, and do not have sufficient residence time to immobilize the injected CO_2 by one of the other trapping mechanisms, such as residual trapping, solution trapping or mineral trapping. Flow rate is controlled by the driving forces of the fluid, including its buoyancy and hydraulic gradient, and by the permeability and porosity characteristics of the reservoir rock through which the fluid is moving.

An additional sub-criterion that should be considered is the quality of the fluid under consideration. In this regard, it is mandatory to consider the principle of sustainability, whereby present actions cannot compromise future generations. Fluid quality is measured in Total Dissolved Solids (TDS), which define drinkable water (TDS < 3000 ppm). Thus, when drinkable water is present, the aquifer is not suitable for CO_2 storage. On the other hand, TDS higher than 10,000 ppm should be considered suitable for CO_2 storage operations, as the quality of this water makes it not suitable for any other activity.

Subsurface conditions of the CO_2. The most efficient way to store CO_2 underground is to store it under supercritical conditions, [2]. This special state of the CO_2 provides similar densities to a liquid (i.e., increasing the capacity per volumetric unit), while the viscosity is similar to a gas (i.e., increasing the capacity per unit of time). Supercritical conditions can be reached in CO_2 geology storage when the depth of the storage formation is higher than 800 m and the geothermal gradient is low (below 25 ∘C/km) [5,[6].

Capacity. The current methods for estimating CO_2 storage potential and capacity are based on widely accepted assumptions about geological trapping mechanisms, storage media and operating timeframes reviewed previously by other authors [2][6]. Using the concept of resources and reserves, the CSLF Task Force proposed a Techno-Economic Resource-Reserve Pyramid for CO_2 Storage Capacity, [11]. The various capacities are nested within the resource-reserves pyramid, and defined as Theoretical, Effective, Practical and Matched Storage Capacity.

This parameter is considered a technical criterion, but it is also relevant for economic reasons: it is necessary to define the CO_2 emitter in order to relate CO_2 emissions and capacity.

Other geological formations. CO_2 storage implies a long time period. Considering that the whole life cycle of the project includes characterization (pre-injection), injection, closure and post-closure as the most general phases, in addition to the storage itself, the project can last up to 100 years. Moreover, if there are any future modifications to the original conditions, such as mining activities or the development of oil fields, the confinement mechanisms may not be guaranteed. For this reason, it is necessary to consider every shallow geological formation that could be of potential use in the future. This will fulfill the sustainability principle previously described.

2.2. Socioeconomic criteria

These criteria include both economic aspects and parameters related to the social acceptance of the emplacement and its activity.

The quantity and quality of the geological data, although considered in many cases as an economic aspect, is one of the main criteria, as this information is used to determine and quantify the technical criteria referred to in the above text.

The more information is available, the fewer characterization methods will have to be applied. In this case, the characterization program (geophysics and wells) will be less expensive, and the risk of failure of the geological exploration will be reduced.

CO_2 sources. This criterion relates to the location details of the major stationary sources, and the distance between them and the areas of interest. An additional parameter to be considered is the flue gas quality (CO_2 quality). Certain gases (NO_X or SO_X) can increase the interaction between the storage formation and the fluid injected [12]. These chemical interactions between injected gases and the storage or caprock formation materials can create precipitates or dissolutions, which can modify storage specifications of the area or structure.

Regional location. Whether an area or structure is onshore or offshore has an important economic consideration, since generally it is likely to be cheaper and technically easier to implement a CO_2 injection site onshore rather than offshore. On the other hand, public perception and land use issues may dictate that offshore sites are preferential for many CO_2 storage projects.

Maturity of the Area. This criterion considers those aspects which define, at a local level, the location of the areas under consideration. It includes key aspects such as the climate, existing infrastructure that may be affected by the geological exploration or CO_2 storage, and infrastructure that is required to develop the exploration program and engineering activities to develop the emplacement.

Areas of interest: population, environmental and cultural resources. This criterion refers to aspects that can affect acceptance of the emplacement by the community. Protest against the storage activity will be very demanding and obtaining legal approvals and permits may be delayed. For this reason, some special areas including major cities and those areas protected by the Natura 2000 network should not be considered as optimal areas for CO_2 storage. Furthermore, other areas with relevant monuments or archaeological sites should be taken into account in the model for assessing a potential emplacement.

3. Multicriteria decision making (MCDM)

Multiple criteria decision making (MCDM) is a methodology developed for making decisions in the presence of multiple, usually conflicting, criteria. Evaluation methods and multicriteria decisions include the selection of a set of feasible alternatives, the simultaneous optimization of several objective functions, and a decision-making process and evaluation procedures that must be rational and consistent. The application of a mathematical model of decision-making will help to find the best solution, establishing the mechanisms to facilitate the management of information generated by the various disciplines of knowledge.

Those problems in which decision alternatives are finite are called Discrete Multicriteria Decision problems. Such problems are most common in reality and this case scenario will be applied in solving the problem of *site selection for storing* CO_2. Discrete MCDM is used to assess and decide on issues that by nature or design support a finite number of alternative solutions. Recently, Multicriteria Decision Analysis has been applied to hierarchy policy incentives for CCS [15] or to assess the role of CCS [16].

Assessment methods and criteria decision include selection among a set of feasible alternatives, optimization with various objective functions simultaneously, a decision-maker and rational and consistent procedures for assessment. Its principles are derived from matrix theory, graph theory, organizational theory, measurement theory, theory of collective decisions, operations research and economics.

The main evaluation methods are: linear weighting (scoring), multi-attribute utility (MAUT), overcoming relationships and hierarchical analysis (AHP).

Some of the advantages of AHP over other methods of Multicriteria Decision are:

1. It has a mathematical basis.
2. It enables breaking down and analyzing a problem in parts.
3. It allows measuring quantitative and qualitative criteria using a common scale.
4. It includes participation of different people or groups of interest to build consensus.
5. It enables checking the consistency index and making corrections, if applicable.
6. It generates a synthesis and provides the ability to perform sensitivity analyses.
7. It is easy to use and allows the solution to be complemented with mathematical optimization methods.

3.1. Analytical Hierarchy Process (AHP)

AHP is one of the most extensively used and powerful MCDM. Nowadays it is used by many companies in solving various multicriteria problems, ranking these in the following categories: selection, prioritization and assessment, provision of resources against a standard assessment, management and quality management and strategic planning. For example, AHP has been applied in the analysis of location, resource allocation, outsourcing, evaluation, manufacturing, marketing, supplier selection, finance, energy, education and risk analysis, [17]. This widespread use shows the suitability of AHP in solving various types of business decision-making problems.

http://dx.doi.org/10.5772/57405

The AHP overcomes the problems with a scoring approach by structuring complexity as a hierarchy and by deriving ratio scale measures through pairwise relative comparisons. Pairwise comparisons are basic to the AHP method. Hence, when comparing a pair of criteria, sub-criteria or alternatives, a ratio of relative importance can be established. The pairwise comparison process can be performed using words, numbers, or graphical bars.

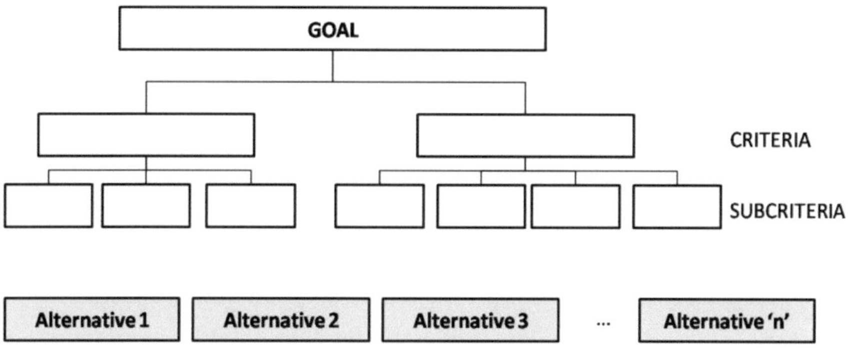

Figure 2. AHP Components: Four steps to build a hierarchy or network structure.

Once the model is built, pairwise comparisons are made with all individual elements (criteria, sub-criteria and alternatives). This process allows giving numerical values to the judgments provided by people, which is also able to measure how each element contributes to each level of the hierarchy. Furthermore, the process is based on a well-defined structure consisting of arrays, and the ability of the eigenvalues to generate values or to approximate weights of each criterion [18], [19], [20]. The problem of finding a nonzero solution to this set of equations is very common in engineering and physics and is known as an *eigenvalue problem*.

In order to carry out these comparisons, the AHP uses a fundamental scale of numbers that have proven absolute in practice and that have been experimentally validated for physical problems and decisions. This scale assigns mathematical values with respect to quantitative or qualitative attributes, homogenizing each valuable criterion.

Figure 4 illustrates the process followed for every criterion. As an example, Original Fluid Quality should be evaluated considering the Water Quality for different uses (agricultural sector, human use) and the European Directive for CO_2 storage; it is possible to establish different mathematical values for each measurable criterion.

3.2. Construction of the decision tree

As a major conclusion, a decision tree has been proposed. This model considers all the criteria described above, and they have been classified.

Weight assessment has been defined considering the AHP method: each level of the criteria and sub-criteria has been compared, using a comparison matrix, which should be constructed considering the consistency principle (it should fulfill the transitivity and reciprocity rules).

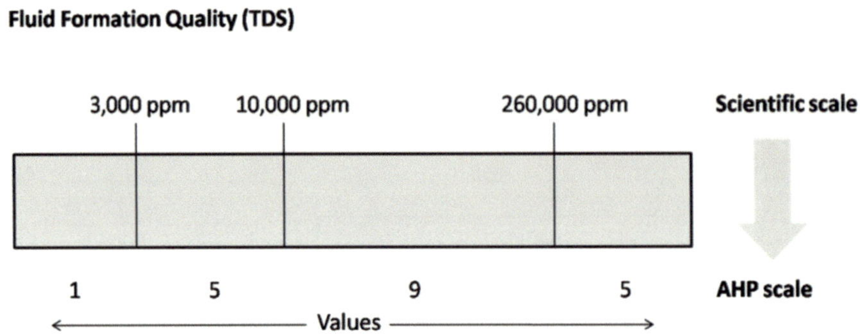

Figure 3. Scientific scale and translation into an AHP scale. This translation facilitates the evaluation of individual criteria (quantitative or qualitative) using a homogeneous scale.

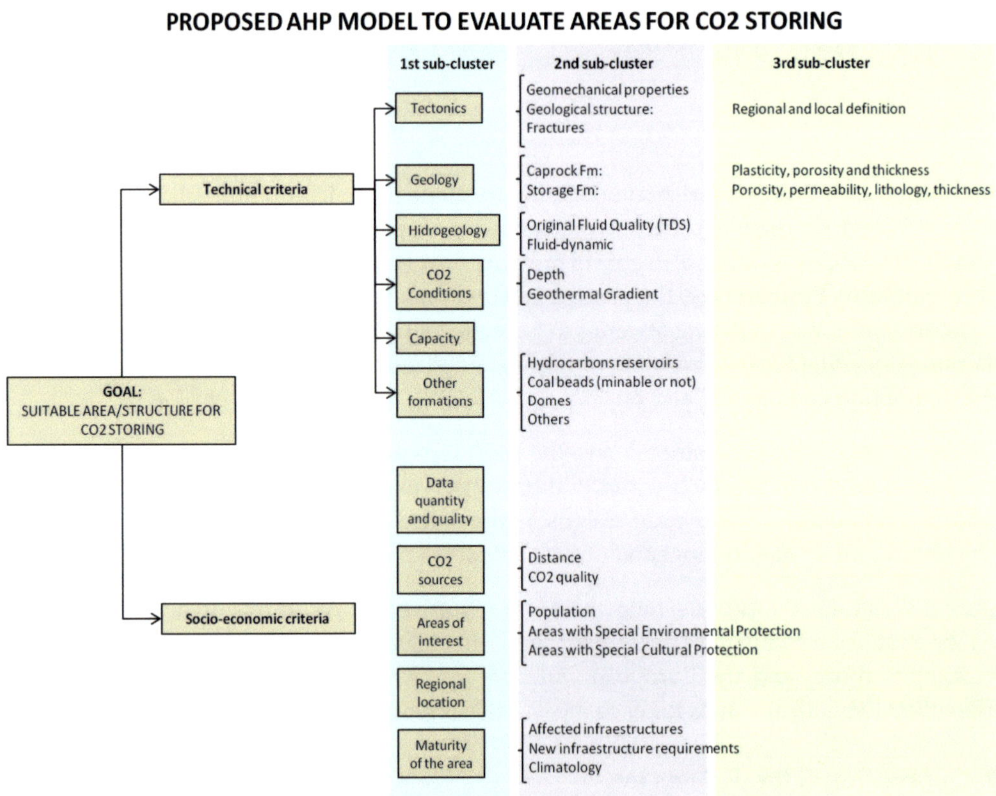

Figure 4. Analytical Hierarchy Process: criterion tree proposed to identify the best site for storing CO_2.

In order to "recover" or find the vector of weights, [w1, w2, w3,..., wn] given to these ratios, the matrix product of matrix A with the vector w can be calculated and considered in an equation, which is described as the eigenvalue matrix equation. The problem of obtaining a

http://dx.doi.org/10.5772/57405

nonzero solution to this set of equations is very common in engineering and physics and is known as an eigenvalue problem.

$$[A]\cdot[W]=n\cdot[W]$$

Where *[A]* is the pairwise comparison matrix – where *n* Is the dimension – and *[W]* is the weight matrix (eigenvalues) for every criterion.

Site (S_n) assessment is evaluated using the formula:

$$S_n=\sum_{i=1}^{i=n} W_i\cdot V_i$$

Where W_i are the weights of each criterion, and V_i are the values assigned for the specific conditions of each site.

AHP as an absolute mode

AHP is a multicriteria methodology which has been developed for use in two different ways: relative and absolute mode. In the first case, all the alternatives are compared between each other, but no more than seven alternatives are recommended for evaluation at the same time.

There are two reasons to justify this limitation: (1) consistency principle and (2) neurons. Pairwise comparisons errors increase due to inconsistent judgments. It is possible to distribute this inconsistency among all the alternatives under evaluation. If the number of alternatives/elements is low, the priorities will be less affected by this inconsistency. The neuronal explanation has its limits in the brain's ability to identify simultaneous events: the more criteria exist for pairwise comparison, the greater the risk of inconsistent judgments will be.

For this study, we consider the AHP algorithm in absolute mode. It requires a standard with which to compare alternatives. The process leads to absolute preservation in the rank of the alternatives no matter how many are introduced. In this case, it is possible to define a standard considering the best values for each criterion (see Table 1and Table 2).

Tectonic, structural				**Hydrogeology**		
Geo mechanical	Lateral continuity	Fractures		TDS	Hydro-dynamic	
Stable domain	Anticline	Weakly fractured. Few faults		> 10.000	Regional	
Storage				**Caprock**		
Porosity (%)	Permeability	Thickness (m)	Lithology	Plasticity	Thickness (m)	Porosity
>25	> 1 D	>100	Sandstone	Ductile	> 100	< 5
CO2 conditions		**Capacity (Mt CO2)**	**Other formations**			
Deep (m)	**Temperature**		**Oil or gas**	**Coal beds**	**Massive saline**	**Other**
>1.000 years	900~2.000	> 150	Yes, huge volume	Yes; deep 200~800m	Beds	Exploitation permits

Table 1. Technical criteria and their best values for CO_2 storage alternatives.

CO_2 source								Maturity	
Quality of the information	Distance (km)	CO_2 Quality	Population (km)	Environmental resources (km)	Cultural resources (km)	Location	Climatology	Affected Infrastructure	New infrastructure
Detail data (GIS) based on deep data (wells and seismic)	< 25	Impurity content less than 1%	>50	>20	>20	Onshore	Moderate	All	None

Table 2. Socio-economical criteria and its best values for CO_2 storage alternative.

4. CO2SITE ASSESS: Informatics tool

As described before, site selection is based on several criteria, values and weights. Even though the methodology proposed in this chapter allows selecting the best option in a quantitative and objective way, it is necessary to consider other points of view of the problem. Indeed, CO_2 storage is a controversial way to remove CO_2 from the atmosphere and even though it has been described a safe and affordable, there are many stakeholders who consider it unreliable.

For those reason, and to manage the huge amount of technical information and different weight definitions of each criterion, a specific program has been developed: CO2SITE ASSESS. This software has been developed in VISUAL BASIC© (easy in terms of programming and speed in obtaining results, robust integration with Data Base, allowing operations in read/write formats). It includes the AHP algorithm (weights and values), so its interaction with the end-user is easy. Many of the technical and socioeconomic parameters can be represented in a Geographical Information System (GIS), so the CO2SITEASSESS results also generate a file which allows representing the results and *CO_2 site storage assessment*.

It is possible to differentiate two different databases: the first one comprises the CO_2 emitters and its data (location, CO_2 emission, primary energy, date of commissioning, and others), whereas the second database includes the CO_2 storage location. Data to be included in this form should be the technical and socioeconomic criteria previously described, and the tool can compare the alternatives using the AHP algorithm and decision tree described in this chapter.

The results classify each area into five levels: optimal, good, normal, poor and very poor. These values will help decision makers to evaluate which areas are the best considered and if it is reliable to go to the next stage.

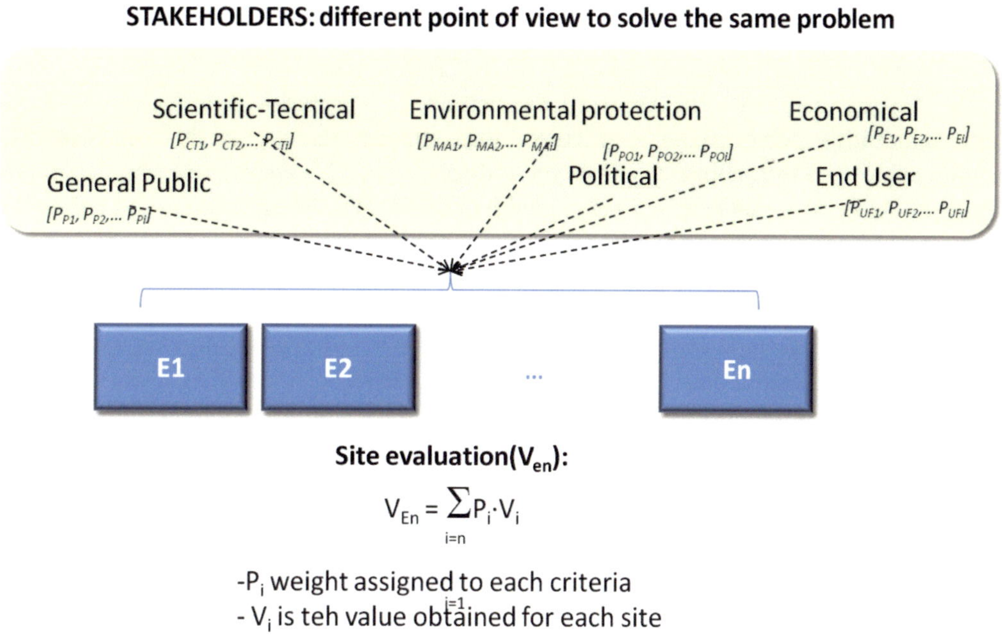

Figure 5. CO2 storage site selection. A complex issue with different stakeholders

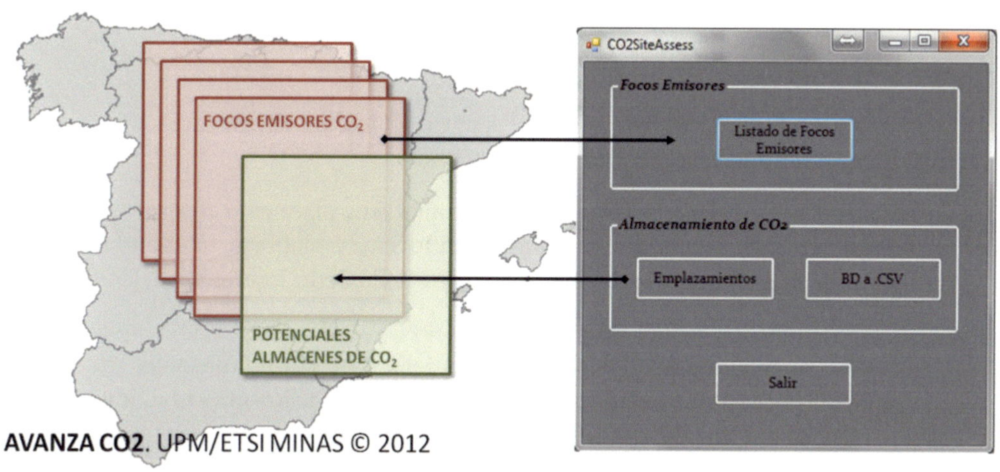

Figure 6. CO2SITEASSES is a program to evaluate multiple criteria and to obtain georeferenced information related to potential CO_2 store areas.

This tool is useful to compare many structures – even if there are many alternatives – and since the algorithm implemented in its code is based on AHP absolute mode, it is possible to compare more than seven alternatives.

5. Site selection: Evaluation of different options in southern Spain

According to the Description of Work of the AVANZA CO2 project (national project supported by the Ministry of Industry and Tourism), the methodology proposed in this chapter has been applied to study specific geological structures. The company that supports this study (SACYR) considers bio-CCS technology as an option to decrease the CO_2 emissions from its biomass plants located in the Guadalquivir basin (South of Spain). This study was carried out by the Technical University of Madrid, in collaboration with Gessal.

Guadalquivir basin: an area of potential interest

The Guadalquivir basin in southern Spain is an ENE–WSW elongated foreland basin developed during the Neogene and Quaternary between the external zones of the Betic Cordillera to the south and Sierra Morena (Iberian Massif) to the north [24], which respectively forms its active and passive margins. The external zones of the Betic Foldchain are made up of Mesozoic and Cenozoic sediments that include thick calcareous and evaporitic formations, as well as siliciclastic units.

Sediments of the basin can be divided into two main stratigraphic: the lower, which includes materials deposited prior to the collision and embodies a long sedimentary process ranging from Cambrian to Permian, culminating in a strong tectonic collision known as the Hercynian orogeny; and the upper, comprising materials of the foreland basin itself, which is known as the alpine stage and begins with an intensive erosional period (Hercynian discordance), and a new subsequent sedimentary period that spreads from Upper Permian to Quaternary. The latter constitutes the proper filling of the foreland basin. It can be divided into five depositional sequences (relatively consistent set of strata, genetically related and whose roof and wall are discontinuity or continuity sequences).

It is possible to individualize two main sub-stages, which are disconnected by the alpine tectonic stage, of Burdigalian age. These sub-stages are:

1. From Permian to Lower Miocene, sedimentation takes place over a passive or Atlantic type margin, which differentiates - from North to South- platform, talus and deep water facies. These paleogeographic realms are known as Prebetic, Intermediate Unit, Subbetic and Flysch.
2. The Alpine orogeny reached its main deformation phase during the lower Miocene. We interpret this phase under a classical deformation model –intra continent subduction type, taking place under NW-SE and E-W compressional vectors.

In the same way as many other alpine forelands, compressive deformation seems to have been established by following a classic model of piggy back or progressive tectonic propagation, from the early active Southern System's Front to the Northern Passive Margin.

The selection of this area was made according to the following information:

1. The interest in the Guadalquivir Basin as a potential area for storing CO_2 has been described in national and European Projects.

2. Technical conditions: Some structures were described in the Geocapacity Project, and others were proposed under the AVANZA CO2 Project.

3. Economic conditions: The specific area of interest is also defined due the interest of the company SACYR. This company has some power stations which use biomass as primary energy near this location. The interest of the company was evaluated in this area to develop the bio-CCS concept.

Structure	Associated wells
A	Fuensanta 1
B	R. Guadalquivir H-1
C	Nueva Carteya-1
D	R. Guadalquivir K-1
Córdoba E	Baeza 1, 2, 3 and Bailen 1
Córdoba S	Guadalquivir N-1

Figure 7. Detail of selected structures, and figures taken from previous studies by the IGME (Geocapacity project).

According to previous stratigraphic, petrological and petrophysical data obtained from exploration wells, it is possible to define preferred targets; both caprock and storage formations. The data include seismic reflection and refraction profiles, well logs, gravity and field observations.

Finally, structural definition was done based on the interpretation of the geophysical data in each area. These interpretations allow us to define specific structures and define the CO_2 capacity of each structure. Well interpretation was used to identify storage and caprock properties.

5.1. Sites evaluation: Application of the AHP model to this area

Some of the structures considered were not evaluated for different conditions (shallow storage formations, lack of data or low thickness of the storage formations). These conditions should not eliminate these structures – indeed they have been included in the CO2SITEASSESS data base – but social or economic aspects will cause them to be considered the worst areas to develop CO_2 storage.

For instance, A, B and C are the alternatives that have been considered. Even if the capacity calculated for each area is not enough for an industrial scale project, it could be considered for a pilot project or demonstration of the bio-CCS technology in Spain.

Moreover, CO2SITEASSESS was used in another region (Duero Basin, also in Spain), where the assessment of this structures are much better than in the Guadalquivir basin.

ID	BASIN		VALUE	NOTE
1	Duero	El Gredal-Utrillas	**1**	**POOR**
2	Duero	El Gredal-Bunt	7.26	GOOD
3	Guadalquivir	Fuensanta de Martos	**1**	**POOR**
4	Guadalquivir	Guadalquivir H1	6.33	NORMAL
5	Guadalquivir	Nueva Carteya	6.54	NORMAL

Table 3. CO2SITEASSESS results.

All the areas defined in the present chapter have been previously defined by several hydrocarbon explorations: geophysical surveys and wells were considered: more than 10 wells were evaluated and hundreds of seismic studies were evaluated. Indeed, this region has active natural gas reservoirs – in two different turbidite systems: the Arenas del Guadiana Fm. in the Poseidón Gas field, in the Gulf of Cadiz, and the Arenas del Guadalquivir Fm., which produces from several small fields that are located onshore, in the Guadalquivir basin.

Moreover, outcrops were analyzed to properly evaluate mineral composition, hydrogeology and maturity conditions.

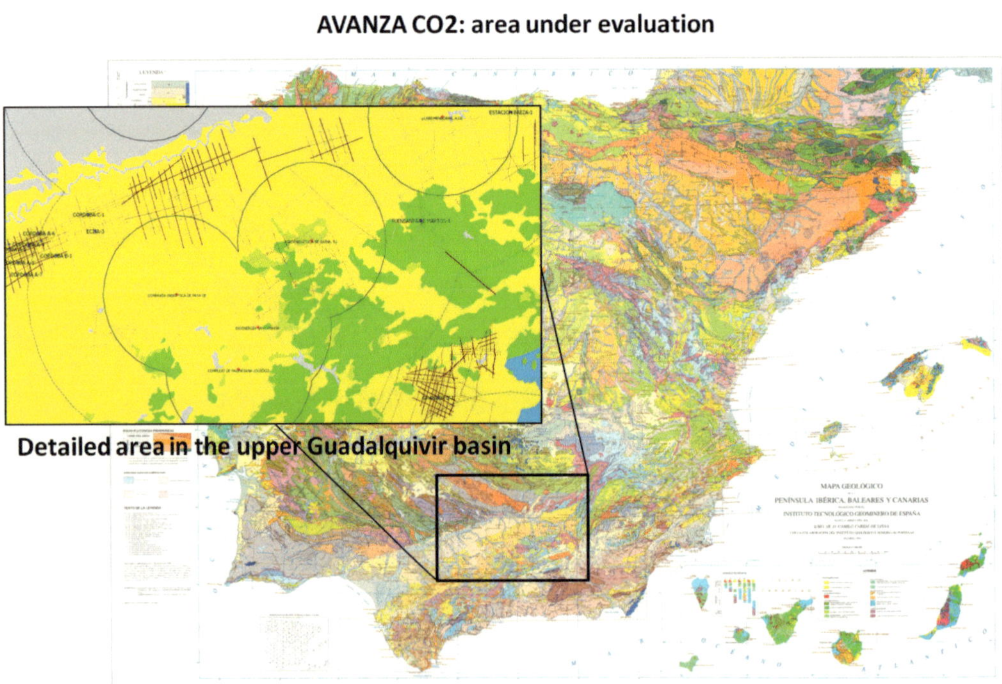

Figure 8. Spanish geological map and detailed area under evaluation. Source: IGME and AVANZA CO2 project.

http://dx.doi.org/10.5772/57405

FUENSANTA (Alternative A): Located in the basin's internal prebetic area (intermediate unit). As reservoir this study evaluated a carbonate rock belonging to Dogger – Lias, whereas a marls is considered as the caprock.

The anticlinal trap, from the data collected, is estimated to have a total area of 15 km^2 of Structure A. The roof of the structure would be located at 1081 m.

Existence of water with low salt content has been confirmed (7 gr/l). Structure-Trap Anticline, elongated in EW direction, preserved under Subbetic materials in its western part, limited by a front thrust in its margin N, N-verging. structural closure to the "spill point" is around 400 milliseconds, which may involve around 700-800 meters.

A

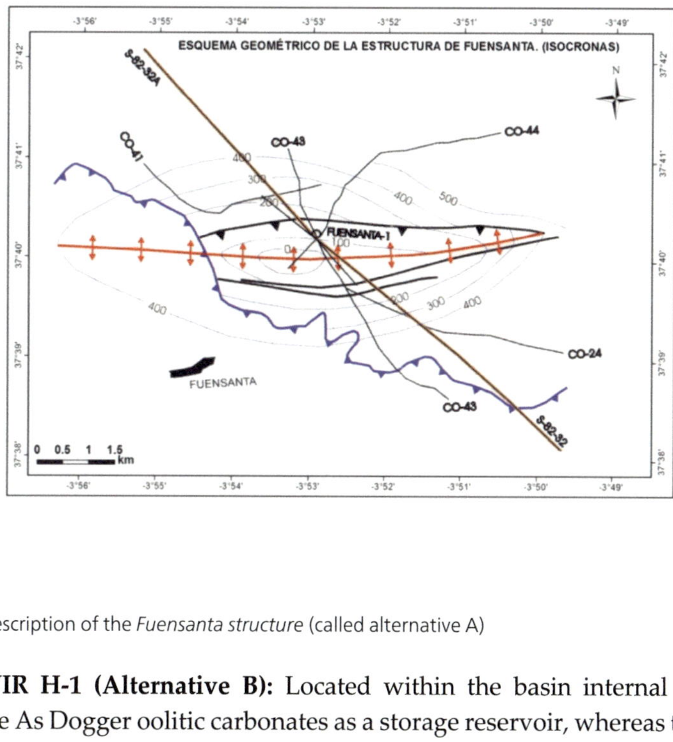

Figure 9. Detailed description of the *Fuensanta structure* (called alternative A)

GUADALQUIVIR H-1 (Alternative B): Located within the basin internal prebetic. It is possible to define As Dogger oolitic carbonates as a storage reservoir, whereas the Malm marl may be considered as its caprock.

The trap is defined as a folding anticlinal; considering the data collected it is estimated to have a total area of 26 km^2. The roof of the store formation would be located at 1668 m. However, the target (corresponding to the Oolític Dogger *Jabalcuz* formation) has low porosity, between 2.25 and 4%.

For instance, it is possible to define a potential reservoir and a caprock.

	Level / Criteria	I	II	III	IV	V
	(math) VALUE	1	3	5	7	9
Tectonic, structural	Geo mechanical		Unstable domain		Stable domain	
	Lateral continuity	Convergent Basins. Volcanic Activity	Convergent Basins. Neo tectonics		Divergent Basins.	
			Syncline		Horizontal or sub horizontal	Anticline
	Fractures		Strongly fractured. Many faults		Weakly fractured. Few faults	
Storage	Porosity (%)	< 10 (5.0)		10–25		>25
	Permeability	< 1 mD		1–100 mD (10-10^2)		> 1 D
	Thickness (m)		< 10	10–100		>100 (175 m.)
	Lithology		Other		Carbonates	Sandstone
Caprock	Plasticity	Fragile		Intermediate		Ductile
	Thickness (m)	< 10		Active region 10–100	Stable region 10–100	> 100 (173 m.)
	Porosity	> 10		10 – 5		< 5
Hydrogeology	TDS	< 3 000		3 000–10 000 >260 000		> 10 000
	Hydro-dynamic	Local < 100 years		Area 100–1000 years		Regional >1 000 years
	Deep (m)	< 600		600–900	> 2 000	900–2 000
	Temperature		Warm basin	Mild temperate basin	Cold basin	
	Capacity (Mt CO_2)	< 10	10–50	50–100	100–150	> 150
Otras formaciones	Oil or gas		NO	Yes, small volume	Yes. Medium volume	Yes, huge volume
	Coal beds		NO	Yes. No methane presence	Yes; deep > 800m	Yes; deep 200–800m
	Massive saline		NO	Domes	Beds	
	Other	Exploration		Evidence. Exploration or research permits		Explotation permits

Figure 10. Evaluation of alternative A, using the AHP model described above.

B

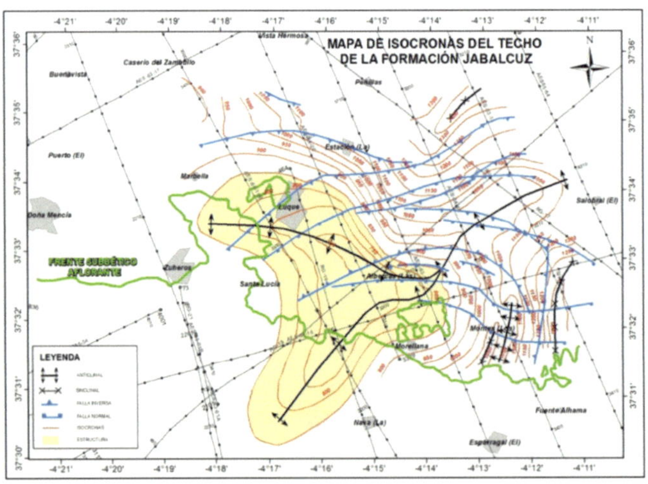

Figure 11. Detailed description of the *Guadalquivir H-1 structure* (called alternative B)

	Criteria / Level	I	II	III	IV	V
	(math) VALUE	1	3	5	7	9
Tectonic, structural	Geo mechanical		Unstable domain		Stable domain	
	Lateral continuity	Convergent Basins. Volcanic Activity	Convergent Basins. Neo tectonics		Divergent Basins.	
			Syncline		Horizontal or sub horizontal	Anticline
	Fractures		Strongly fractured. Many faults		Weakly fractured. Few faults	
Storage	Porosity (%)	< 10 (4)		10–25		>25
	Permeability	< 1 mD		1–100 mD		> 1 D
	Thickness (m)		< 10	10–100 (84)		>100
	Lithology		Other		Carbonates	Sandstone
Caprock	Plasticity	Fragile		Intermediate	marls	Ductile
	Thickness (m)	< 10		Active region 10–100	Stable region 10–100	> 100 (140)
	Porosity	> 10		10 – 5		< 5
Hydrogeology	TDS	< 3 000		3 000–10 000 >260 000		> 10 000 (no documentado)
	Hydro-dynamic	Local < 100 years		Area 100–1000 years		Regional >1 000 years
	Deep (m)	< 600		600–900	> 2 000	900–2 000
	Temperature		Warm basin	Mild temperate basin	Cold basin	
	Capacity ($Mt CO_2$)	< 10	10–50	50–100	100–150	> 150
Otras formaciones	Oil or gas		NO	Yes, small volume	Yes. Medium volume	Yes, huge volume
	Coal beds		NO	Yes. No methane presence	Yes: deep > 800m	Yes: deep 200–800m
	Massive saline		NO	Domes	Beds	
	Other	Exploration		Evidence. Exploration or research permits		Explotation permits

Figure 12. Evaluation of alternative B, using the AHP model described above.

NUEVA CARTEYA - 1. Within the pre-betic terminal basin, the carbonate reservoir rock (Dogger oolitic) was evaluated. Malm marls are considered to be caprock.

The trap is an anticline, with a total area of 30 km^2 and estimated thickness of 160 m. The roof would be located at a depth of 1240 m.

Although a storage formation has been defined, it is necessary to study whether this structure is closed.

C

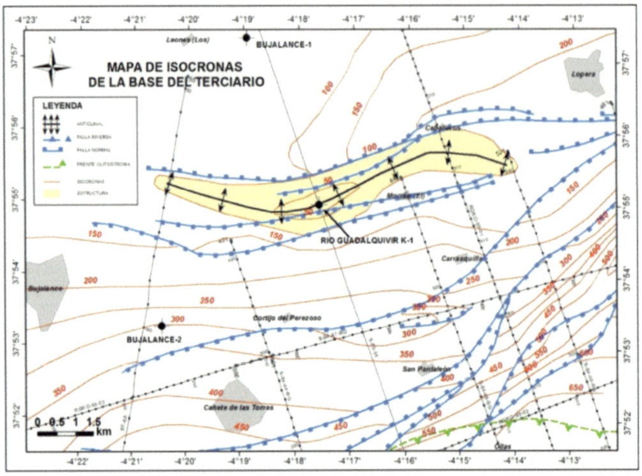

Figure 13. Detailed description of the *Guadalquivir H-1 structure* (called alternative C)

	Level / Criteria	I	II	III	IV	V
	(math) VALUE	1	3	5	7	9
Tectonic, structural	Geo mechanical		Unstable domain		Stable domain	
	Lateral continuity	Convergent Basins. Volcanic Activity	Convergent Basins. Neo tectonics		Divergent Basins.	
			Syncline		Horizontal or sub horizontal	Anticline
	Fractures		Strongly fractured. Many faults		Weakly fractured. Few faults	
Storage	Porosity (%)	< 10 (3.0)		10″25		>25
	Permeability	< 1 mD		1″100 mD		> 1 D
	Thickness (m)		< 10	10″100		>100 (160)
	Lithology		Other		Carbonates	Sandstone
Caprock	Plasticity	Fragile		Intermediate		Ductile
	Thickness (m)	< 10		Active region 10″100 (50)	Stable region 10″100	> 100
	Porosity	> 10		10 ″ 5		< 5
Hydrogeology	TDS	< 3 000		3 000″10 000 >260 000		> 10 000 190 gr/L
	Hydro-dynamic	Local < 100 years		Area 100″1000 years		Regional >1 000 years
	Deep (m)	< 600		600″900	> 2 000	900″2 000 (1220 a 1380 m)
	Temperature		Warm basin	Mild temperate basin	Cold basin	
	Capacity ($Mt\ CO_2$)	< 10	10″50	50″100	100″150	> 150
Otras formaciones	Oil or gas		NO	Yes, small volume	Yes. Medium volume	Yes, huge volume
	Coal beds		NO	Yes. No methane presence	Yes; deep > 800m	Yes; deep 200″800m
	Massive saline		NO	Domes	Beds	
	Other	Exploration		Evidence. Exploration or research permits		Explotation permits

Figure 14. Evaluation of alternative C, using the AHP model described above.

http://dx.doi.org/10.5772/57405

5.2. Socioeconomic values

The area under evaluation contains different features in the same region. For instance, while the socioeconomic parameters can be considered similar for all of them, other parameters are different slightly different (i.e., distance from CO_2 sources, storage area and the nearest town). It was possible to estimate the values marked in the Figure 15.

	Criteria / Class	I	II	III	IV	V
	(math) VALUE	1	3	5	7	9
	Quality of the information	No data. It isn't possible to make any geological interpretation	Few data. It's possible to make an interpretation based on adjacent regions	Detail data and enough deep. General or shallow data.	Digital regional data (GIS).	Detail data (GIS) based on deep data (well and seismic data)
	Distance (km)	> 250	250~100	100~50	50~25	< 25
	CO_2 Quality	Impurity content up to 2%		Impurity content between 1-2%		Impurity content less than 1%
	Population (km)	< 10	10~25 Unfavorable topography	10~25 Favorable topography	25~50	>50
	Environmental resources (km)	0		10~20		>20
	Cultural resources (km)	0		10~20		>20
	Location		Offshore (deep)	Offshore (shallow)	Onshore	
Maturity	Climatology		Extreme		Warm	
Maturity	Affected Infrastructure		None	Little		All
Maturity	New infrastruture		All	Little		None

Figure 15. Evaluation of socioeconomic parameters

The area was explored during the twentieth century, so there is enough information to build a GIS and to define some of the structures (conceptual or static model). As shown in Figure 8, there are industrial CO_2 sources a short distance away and the quality of the flue gases is sufficient. Some of the emitters are biomass power stations, but close to this region it might be possible to identify a larger emitter (Puente Nuevo power plant, close to Córdoba).

There are towns and cities close to each structure, but the topography can be considered favorable, and there are no environmentally protected areas close to the structures under evaluation.

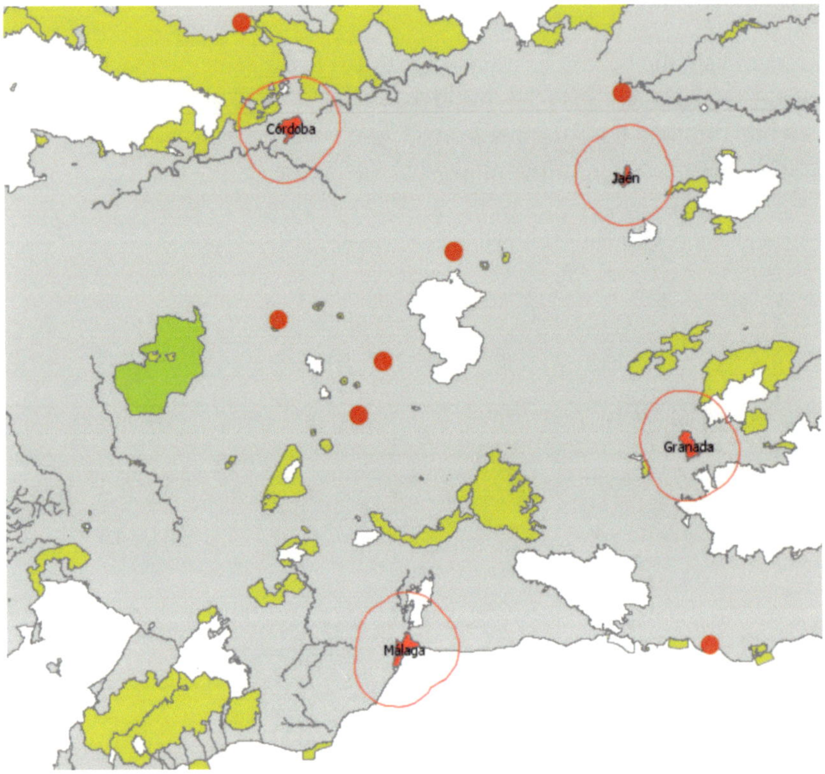

Figure 16. Protected areas (environmental conditions: white and green areas), and towns (red dots) in the area of interest.

6. Conclusions

Site selection for storing CO_2 is a complex issue, especially when deep saline aquifers are under assessment. These geological structures used to be poorly characterized and the risk of unsuccessful geology exploration is high. For this reason, the Multicriteria Decision Tool can be used to evaluate related technical and socioeconomic data on different alternatives under consideration. In addition, there are different stakeholders with different points of view, so the decision maker needs to take these viewpoints into account.

AHP is the proposed multicriteria algorithm. It selects the best area in an objective way. Therefore, its use decreases the risk associated with the site selection phase, and it will easily show the strengths and weaknesses of the information or characteristics of the alternatives under study. Furthermore, it can help increase social acceptance by stakeholders.

An innovative program (CO2SITEASSESS) was been developed and validated, using some defined areas (at basin and regional scale) and structures (local scale). This software also allows obtaining georeferenced data; and the combination of both uses (georeferenced data and AHP

http://dx.doi.org/10.5772/57405

algorithm) has never before been applied to select areas to store CO_2. This combination has some advantages:

1. The results are obtained using a decision tree and multicriteria algorithm (Analytical Hierarchy Processes).
2. The results are objective – all the alternatives are compared with a defined standard (absolute measurements).
3. The results can be represented in a GIS, so the data can be referenced on a map, helping to make decisions.
4. This software easily allows different evaluations – considering different stakeholders with different points of view.
5. It saves a lot of time in decision-making, and generates a range of information useful for taking decisions.

In addition, some of the criteria can change during the pre-injection phase (i.e., data available, other formations of interest, etc.) and this tool can be useful to consider at which stage each alternative is during a period of time (pre-injection: selection, characterization, static or dynamic model and engineering).

The results obtained in the High Guadalquivir basin suggest that this area is not suitable for CO_2 storage on an industrial scale, but some of the structures considered in this chapter could be useful for pilot scale, especially if bio-CCS technology is applied.

Nevertheless, the CO2SITEASSES methodology has been demonstrated as robust to identify the best alternative under evaluation, and it reduces the inherent risk associated with geological explorations.

The AHP is applied in this study in an absolute mode, so it allows the assessment of limitless alternatives. For instance, this method and software can be useful as a standard in different regions (i.e., Spain or Europe).

Another version of the CO2SITEASSESS will be developed in the near future to relate site selection and a program characterization of each alternative. This characterization program will consider the three characterization phases: outcrop, geophysics and wells.

Author details

B. Llamas[1*], M. Arribas[1], E. Hernandez[2] and L.F. Mazadiego[1]

*Address all correspondence to: bernardo.llamas@upm.es

1 Technical University of Madrid. Engineering School of Mines, Madrid, Spain

2 GESSAL, Madrid, Spain

References

[1] Torvanger, Asbjørn, Kristin Rypdal and Steffen Kallbekken, 2005. Geological CO_2 Storage as a Climate Change Mitigation Option. *Mitigation and Adaptation Strategies for Global Change*, 10 (4), 693-715.

[2] Benson, S, Cook P, Metz B. (Ed.), Davidson O. (Ed.), de Coninck H. (Ed.), Loos M. (Ed.), Meyer L. (Ed.) *IPCC Special Report: Carbon Dioxide Capture and Storage*. Cambridge University Press, United Kingdom. 2005.

[3] Bradshaw J., Bachu S., Bonijoly D., Burruss R., Holloway S., Christensen N.P. and Mathiassen O.M., CO_2 storage capacity estimation: Issues and development of standards. *8th International Conference on Greenhouse Gas Control Technology*. Trondheim, Norway, 2006.

[4] M. Carpenter, K. Kvien, J. Aarnes. The CO2QUALSTORE guideline for selection, characterisation and qualification of sites and projects for geological storage of CO_2. *International Journal of Greenhouse Gas Control*, 2011, 5, 942–951.

[5] Bachu S., Sequestration of carbon dioxide in geological media: criteria approach for site selection. *Energy Conv. Manage*, 2000, 41, 953-970.

[6] Bachu S., Screening and ranking of sedimentary basins for sequestration of CO_2 in geological media in response to climate change. *Environmental Geology*, 2003, 44(3), 277-289.

[7] Bachu S., Screening and selection criteria, and characterization for CO_2 storage. In: Marotovaler M (Ed.) *Developments and Innovation in Carbon Dioxide (CO_2) Capture and Storage Technology, vol. 2*, Woodhead Energy Series 16. Woodhead Publishing Ltd, United Kingdom, 2010, 27-56.

[8] Benson S., Hepple R., Apps J., Tsang, C-F, Lippmann M., *Lessons learned from natural and industrial analogues for storage of carbon dioxide in deep geological formations*, Lawrence Berkeley National Laboratory, United States, 2002.

[9] European Parliament: *Directive 2009/31/EC on the geological storage of carbon dioxide*. Official Journal of the European Union L140/114. 2009, Brussels.

[10] Vangkilde-Pedersen, T. et al (2009): *D26 WP4 report: capacity standards and site selection criteria*. GEOCAPACITY Project (6th Framework Programme). 2009

[11] Bachu S., Bonijoly D., Bradshaw J., Burruss R., Christensen N.P., Holloway S. and Mathiassen O.M. CO_2 storage capacity estimation: methodology and gaps, *International Journal of Greenhouse Gas Control*, 2007, 1, 304-443.

[12] De Visser E., Hendriks C., de Koeijer G., Liljemark S., Barrio M., Austegard A., Brown A., *D 3.1.3 DYNAMIS CO_2 quality recommendations*. DYNAMIS Project (6th Framework Programme). The Netherlands, 2007, 16-35.

[13] Carlsson C., and Fullér R., Fuzzy multiple criteria decision making: Recent developments, *Fuzzy Sets and Systems*, 1996, 78, 139-153.

[14] Yang, J., Xu, L., *Introduction to Multi-criteria Decision-making and the Evidential Reasoning Approach*, University of Manchester Institute of Science and Technology, 2001, 106, 1-21.

[15] Christoph von Stechow, Jim Watson, Barbara Praetorius, Policy incentives for carbon capture and storage technologies in Europe: A qualitative multi-criteria analysis. *Global Environmental Change*, 2011, 21, 346–357.

[16] Shackley S., McLachlan C, Trade-offs in assessing different energy futures: a regional multi-criteria assessment of the role of carbon dioxide capture and storage, *Environmental Science and Policy*, 2006, 9, 376-391.

[17] Islam R., Abdullah N.A., Management decision-making by the analytic hierarchy process: a proposed modification for large-scale problems, *Journal for International Business and Entrepreneurship Development*, 2006, 3, 18-40.

[18] Saaty T.L., *The Analytic Hierarchy Process: Planning, Priority Setting, Resource Allocation*, McGraw Hill International, New York, United States, 1980.

[19] Saaty T.L., Axiomatic Foundations of the Analytic Hierarchy Process, *Management Science*, 1986, 32, 841-855.

[20] Saaty T.L., Vargas L.G. *Decision Making in Economic, Social and Technological Environments*, RWS Publications, Pittsburgh, United States, 2006.

[21] Saaty, T.L., *Decision Making for Leaders*, RWS Publications, Pittsburgh, United States, 2001.

[22] Saaty T.L. *Fundamentals of Decision Making and Priority Theory with the Analytic Hierarchy Process*, RWS Publications, Pittsburgh, United States, 2006, 45-197.

[23] Saaty, T.L., Decision making with the analytic hierarchy process, *Int. J. Services Sciences*, 2008, 1, 83-98.

[24] Sanz de Galdeano, C. & Vera, J.A. (1992). Stratigraphic record and palaeogeographical context of the Neo-gene basins in the Betic Cordillera, Spain. *Basin Research*, 4: 21-36.

[25] C-W Hsu, et al. Site selection for carbon dioxide geological storage using analytic network process, *Separation and Purification Technology*, 2011, doi:10.1016/j.seppur.2011.08.019.

Chapter 11

Numerical Simulation of CO_2 Sequestration in Large Saline Aquifers

Zheming Zhang and Ramesh K. Agarwal

Additional information is available at the end of the chapter

http://dx.doi.org/10.5772/57065

1. Introduction

With heightened concerns over CO_2 emissions from pulverized-coal power plants, there has been major emphasis in recent years on the development of safe and economical geological carbon sequestration (GCS) technology. Although it is one of the most promising technologies to address globalwarming due to anthropogenic CO_2 emissions, the detailed mechanisms of GCS are not well understood. As a result, there remain many uncertainties in determining the sequestration capacity of the formation/reservoir and the safety of sequestered CO_2 due to leakage. These uncertainties arise due to lack of information about the detailed interior geometry of the formation and the heterogeneity in its geological properties, such as permeability and porosity, which influence the sequestration capacity and plume migration. Furthermore, the sequestration efficiency is highly dependent on the injection strategy, which includes injection rate, injection pressure, type of injection well employed and its orientation etc. The goal of GCS is to maximize the sequestration capacity and minimize the plume migration by optimizing the GCS operation before proceeding with its large-scale deployment. In this chapter, numerical simulations of GCS are conducted using the US Department of Energy (DOE) multi-phase flow solver TOUGH2 (Transport of Unsaturated Groundwater and Heat). A multi-objective optimization code based on genetic algorithm is also developed to optimize the GCS operation for a given geological formation. It is described in Chapter titled," Optimization of CO2 Sequestration Saline Aquifers". Most of the studies are conducted for sequestration in a saline formation (aquifer). Large-scale GCS studies are conducted for the Mt.Simon, Frio and Utsira saline formations, for which some experimental data and computations performed by other investigators are available. These simulation studies provide important insights as to the key sources of uncertainties that can influence the accuracy of simulations.

2. Saline aquifer geological carbon sequestration (SAGCS)

Studies of GCS have suggested that various geological structures can serve as potential CO_2 storage sites. The major geological carbon sinks include the following structures: (1) conventional hydrocarbon reservoirs, (2) un-minable coal seams, (3) mature oil/gas reservoirs, and (4) deep saline formations. Among these candidates, this chapter focuses on carbon sequestration in saline aquifers considering the following facts.

- Concentrated locations of major sources of CO_2 (such as power plants) are close to existing saline aquifers.
- Geological survey has confirmed vast geological distribution of deep saline formations possibly suitable for GCS in the US and Canada.
- Preliminary estimates have suggested large storage capacity of the existing deep saline formations. The US DOE estimates an aggregate storage capacity of approximately 2102~20043 billion metric tons of CO_2 for SAGCS in US, which accounts for 80~90 percent of US overall GCS potential [1].
- Since most of the saline formations are located deep underground, i.e., at least 800 m below sea level, they provide great potential for secure long-term sequestration.
- Many surveys, research projects and commercial projects have already been conducted for SAGCS, making it attractive for further research and technical contributions.

A huge geographic distribution of deep saline aquifers in North America has been identified by the DOE. The DOE-estimated storage capacity for SAGCS takes into account more than 80 percent of the overall storage capacity of all possible GCS sites. In Table 1, the low-end capacity of 2102 billion metric ton of CO_2 is estimated under the condition that ineffective storage may occur due to improper and non-optimized sequestration approaches. On the other hand, the high-end capacity of 20043 billion metric ton of CO_2 is estimated under the conditions that most effective and optimal storage takes place. It can be seen that the high-end estimated capacity is over nine times the low-end estimated capacity. This large difference means it is important to deploy optimized reservoir engineering techniques for effective utilization of storage potential and successful GCS practice.

3. The approach for numerical simulations of GCS

The large spatial extent (of the order of kilometers) and time duration (centuries) for CO_2 plume migration after injection makes the study of SAGCS very difficult at these large spatial and temporal macro-scales by using laboratory scale experiments, which can be conducted only on relatively small spatial and temporal scales varying from nanometers to a few meters and from nanoseconds to a few days/months. Conducting field tests in large-scale formations before the actual deployment takes place can be very expensive. However, numerical simulations using computation fluid dynamics (CFD) technology can be employed at industrial

RCSPs	Low-end capacity (billion metric tons)	High-end capacity (billion metric tons)
BSCSP (Big Sky Caron Sequestration Partnership)	98	1237
MGCS (Midwest Geological Sequestration Consortium)	11	158
MRCSP (Midwest Regional Carbon Sequestration Partnership)	95	123
PCOR (Plains CO2 Reduction Partnership)	174	511
SECARB (Southeast Regional Carbon Sequestration Partnership)	1376	14089
SWP (Southwest Regional Partnership on Carbon Sequestration)	266	2801
WESTCARB (West Coast Regional Carbon Sequestration Partnership)	82	1124
Total	2102	20043

Table 1. Estimation of saline aquifer storage capacity of GCS for different regional carbon sequestration partnerships (RCSPs) [1]

scale to determine the fate of injected CO_2 in a reservoir. With the development over the past four decades, CFD technology has now become mature and has been widely and successfully applied to various engineering problems. With the proper modeling of the storage formation and ground water transportation, CFD is capable of providing accurate enough analysis for quick estimation of reservoir performance at considerably lower cost. The governing equations of mass/energy transportation and numerical representations of the formation properties have been well explained in the TOUGH2 User's Guide [2],[3].

In a complex simulation like that of SAGCS, it is impractical to integrate all geophysical and geochemical effects into a single model while retaining acceptable computational efficiency. Therefore, careful examination of physical phenomenon of interest in SAGCS is essential to determine simplifications in modeling of features of interest [4],[5]. Another important benefit of numerical simulations is that one can investigate the effect of various injection parameters such as injection rate, injection duration, and injection well orientation and displacement on CO_2 storage efficiency and plume migration in a given reservoir. The advantage of numerical simulations makes it possible to perform optimization studies of these injection parameters to achieve the highest possible storage efficiency and minimum plume migration, as described in Chapter titled," Optimization of CO2 Sequestration Saline Aquifers". Such an optimization capability can be very beneficial in successful cost effective implementation of SAGCS on industrial scale. We have employed genetic algorithm in conjunction with numerical simulator TOUGH2 for optimization of SAGCS practice. Some salient features of the genetic algorithm and its integration with TOUGH2 are described in Chapter titled," Optimization of CO2 Sequestration Saline Aquifers".

4. Simulation code validation using analytical and benchmark solutions

TOUGH2 is available as source files written in Fortran77. TOUGH2 does not provide graphical user interface (GUI) of any kind. All its input files and output results are in ASCII format. TOUGH2 has very high computing efficiency when executing large-scale simulations and it is very convenient for users to make modifications to the source code if needed. However, both the problem setup and result analysis capability in TOUGH2, such as mesh generation and contour map visualization, are not as comprehensive or straightforward as available in some newer commercial multiphase flow field simulators. For any complex problem, the modeling process tends to be tedious and error-prone. To address such deficiency, a third-party GUI for TOUGH2 named PetraSim was also installed on the same machine with TOUGH2. PetraSim preserves the original TOUGH2 binary files to execute simulations, while providing a smooth interface to a user-friendly computing environment. However, PetraSim in its original form lacks the compatibility for integrating a new optimization module and some recently developed equation-of-state modules.

Previous validation simulations on benchmark problems have shown that the simulation results obtained by TOUGH2 and PetraSim are identical [6],[7]. For our code validation purpose, we employed three widely used benchmark problems by GCS researchers worldwide. Simulations were conducted by both TOUGH2 and PetraSim. These three benchmark problems were first defined in the Workshop on Numerical Models for Carbon Dioxide Storage in Geological Formations at the University of Stuttgart, Germany [8],[9],[10],[11],[12]. We study the benchmark problem #1 and #3 using PetraSim, while benchmark problem #2 is simulated using the original version of TOUGH2 because of the limit on the availability of the needed equation-of-state module in PetraSim. When simulation is performed using TOUGH2, necessary post-processing programs such as Tecplot are employed for visualization and analysis.

4.1. Simulation of in situ CO_2 migration and comparison with analytical solution

As a first step towards code validation, simulations for CO_2 plume migration in an ideal simplified reservoir are performed. The analytical solution for this case is available [13], [14], [15] and is obtained as:

$$\frac{b_{ag}(r,t)}{B} = \frac{1}{\mu_b - \mu_{ag}}\left[\sqrt{\frac{\mu_b \mu_{ag} V(t)}{B\pi\phi r^2}} - \mu_{ag}\right] \tag{1}$$

where t is the time elapsed since the inception of injection, r is the distance (radius) from the injection well, $b_{ag}(r,t)$ is the plume thickness as a function of r and t, B is the total thickness of the reservoir, φ is the porosity of the reservoir, μ is the dynamic viscosity, subscripts ag and b stand for the injected CO_2 and brine respectively, and $V(t)=\int Qdt$ is the volume of the injected CO_2 within time t. For a horizontal reservoir, setting $b_{ag}(r,t) = 0$ yields Eq. (2), which gives a quick evaluation of the maximum CO_2 plume migration as:

$$R_{\max} = \sqrt{\frac{\mu_b V(t)}{\mu_{ag} B\pi\phi}} \tag{2}$$

In numerical simulations, a hypothetical deep saline reservoir of thickness of 100 m is assumed. A cylindrical computational domain is considered as shown in Figure 1. Generic hydro-geological properties are used. CO_2 injection rate is set at 1 kg per year for ten years. Detailed model parameters used in the simulations are summarized in Table 1. CO_2 plume migration in each year is computed by the simulation and compared with the analytical solution given by Eq. (2).

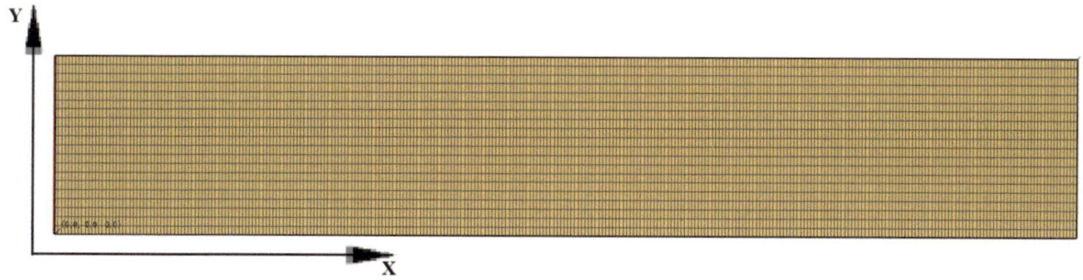

Figure 1. Computational domain and mesh of a generic cylindrical aquifer

Geometry	100 m in thickness; 3000 m in radius
Permeability	Isotropic, 100 mDarcy
Porosity	0.3
Temperature	20°C
Pressure	6.4 MPa
CO_2 Density	789.96 kg/m^3
CO_2 Viscosity	0.0000712905 Pa•s
Brine Density	1029.69 kg/m^3
Brine Viscosity	0.001488427 Pa•s
Relative Permeability	Linear
Brine Residual Saturation	0
CO_2 Residual Saturation	0
Capillary Pressure	None
Injection Rate	1 kg/s
Boundary Condition	Open boundary
Domain Discretization	300 × 20

Table 2. Geometry parameters and hydro-geological properties of the generic saline aquifer

The simulation time is ten years and the CO_2 migration within the aquifer is computed for each of the ten years. In Figure 2, the CO_2 plume after 1, 4, 7 and 10 years since the inception of injection is shown.

As seen in Figure 2, the injected CO_2 migrates upwards very rapidly and then prominently migrates underneath the caprock. A typical plume shape is already identifiable after one year

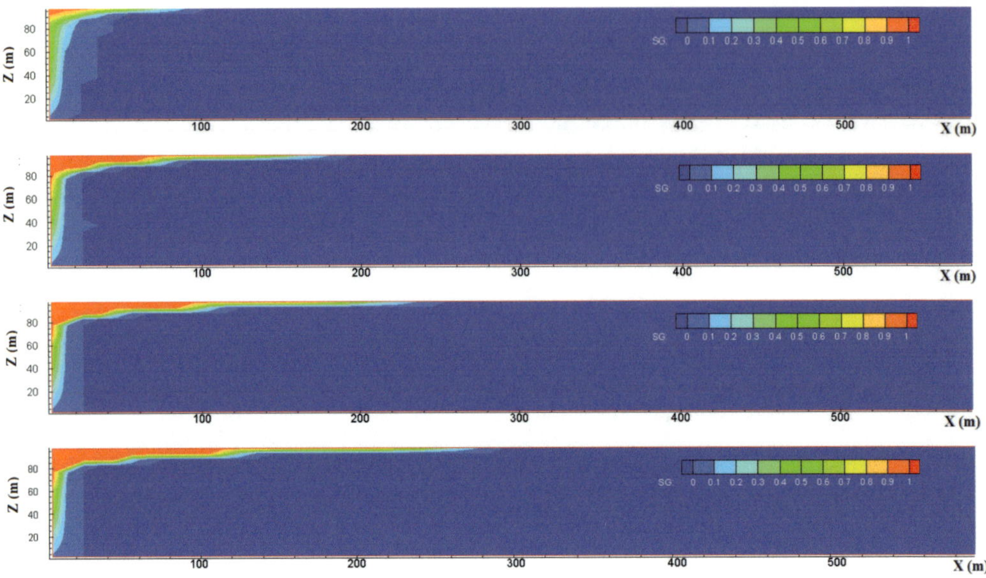

Figure 2. CO_2 plume after 1, 4, 7 and 10 years after injection

of injection, when the farthest CO_2 migration reaches about 100 m from the injection well. In the following nine years, in situ CO_2 keeps migrating outwards and spreads to 300 m after the tenth year of injection. Physically, such a large radial migration of in situ CO_2 is caused by gravity separation. The horizontal extent of the plume can be analytically calculated by utilizing Eq. (2). Taking the necessary values of reservoir and fluid properties from Table 2, the horizontal extent of the plume predicted by Eq. (2) for the first 10 years since injection is summarized in Table 3. The horizontal extent of the plume given by TOUGH2 simulations is also summarized in Table 3 for comparison with the analytical solution.

Year	Maximum Migration based on Numerical Simulation (A)	Maximum Migration based on Analytical Solution (B)	Deviation ((A-B)/B)
1	100.75 m	95.58 m	0.054090814
2	140.49 m	135.17 m	0.039357846
3	168.37 m	165.55 m	0.017034129
4	191.34 m	191.16 m	0.00094162
5	211.27 m	213.72 m	-0.011463597
6	229.23 m	234.12 m	-0.020886725
7	235.34 m	252.88 m	-0.069360962
8	260.81 m	270.34 m	-0.035251905
9	275.70 m	286.74 m	-0.038501779
10	289.36 m	302.25 m	-0.042646816

Table 3. Maximum CO_2 migration underneath the caprock given by the analytical solution and TOUGH2 simulation

As shown in Table 3, TOUGH2 simulations successfully predict the maximum CO_2 plume migration underneath the caprock with excellent agreement with the analytical solution given by Bachu, Nordbotten and Celia [13],[14],[15]. The insignificant difference between the numerical and analytical solutions can be explained by the fact that CO_2 dissolution is accounted for in TOUGH2 while it is neglected in the derivation of the analytical solution. Since CO_2 dissolution is governed by the contact area between CO_2 and the ambient brine, the rate of CO_2 dissolution into brine should gradually increase over time as larger contact area becomes available as the CO_2 plume spreads. Nevertheless, Table 3 validates TOUGH2 as an accurate simulation tool for predicting the migration of in situ CO_2.

A schematic of the CO_2 plume flow is shown in Figure 3. Although based on a simplified analytical model, Figure 3 shows in situ CO_2 migration due to the combined pressure-driven Darcy flow and the buoyancy-drive CO_2 transport.With the injection well located on the left side of Figure 3, the CO_2 plume can be identified as consisting of two distinct regions. The first region is a smaller region on the left, adjacent to the injection well, marked as region (1) in Figure 3. In this region, CO_2 is distributed uniformly through the entire period of the injection interval. This implies strong hydrodynamic force caused by the pressure difference between the pressurized injection well and the unaffected aquifer. Within this region, lateral pressure gradient dominates the movement of CO_2 and Darcy flow occurs, causing CO_2 to migrate more radially through the aquifer. The second region is marked as region (2) in Figure 3, where the CO_2 plume fully develops. In this region, buoyancy due to the density difference between CO_2 and brine becomes dominant and drives the upward movement of CO_2 along with lateral migration. In this region, the vertical movement of CO_2 becomes dominant and results in plume flow. Since it is a phenomenon of fundamental concern in SAGCS, understanding the development of plume flow is critical for the success of SAGCS. The size of the two regions in Figure 3 can vary depending on the properties of the actual aquifer, but under most conditions, region (2) becomes dominant in size, which influences the safety and efficiency of SAGCS operations. Therefore, every effort should be made either to increase the size of region (1) or decrease the size of region (2) for successful and desirable implementation of SAGCS.

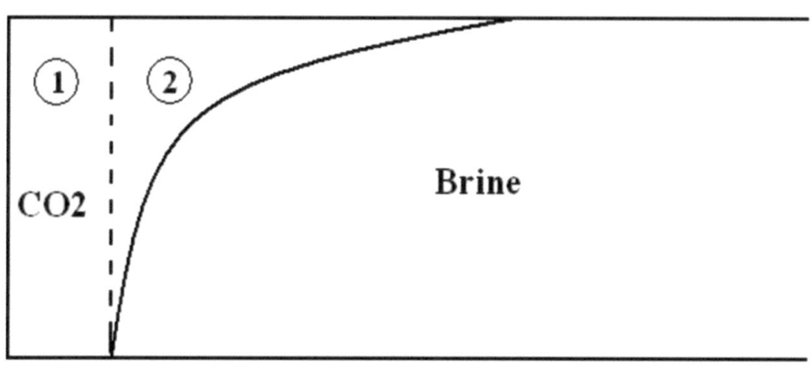

Figure 3. Schematic of the shape of in situ CO_2 plume

4.2. Simulation of benchmark problem #1 – CO_2 plume evolution and leakage through an abandoned well

A three-layered formation is modeled for the first benchmark problem [9]. CO_2 is injected into the deeper aquifer, shown schematically in Figure 4. It spreads in the aquifer and then rises up to a shallower aquifer upon reaching a leaky well. Quantification of the leakage rate, which depends on CO_2 plume evolution and the pressure buildup in the aquifer, is the main objective of this benchmark simulation. Figure 4 shows the schematic of the problem description by providing a cross-section of the formation.

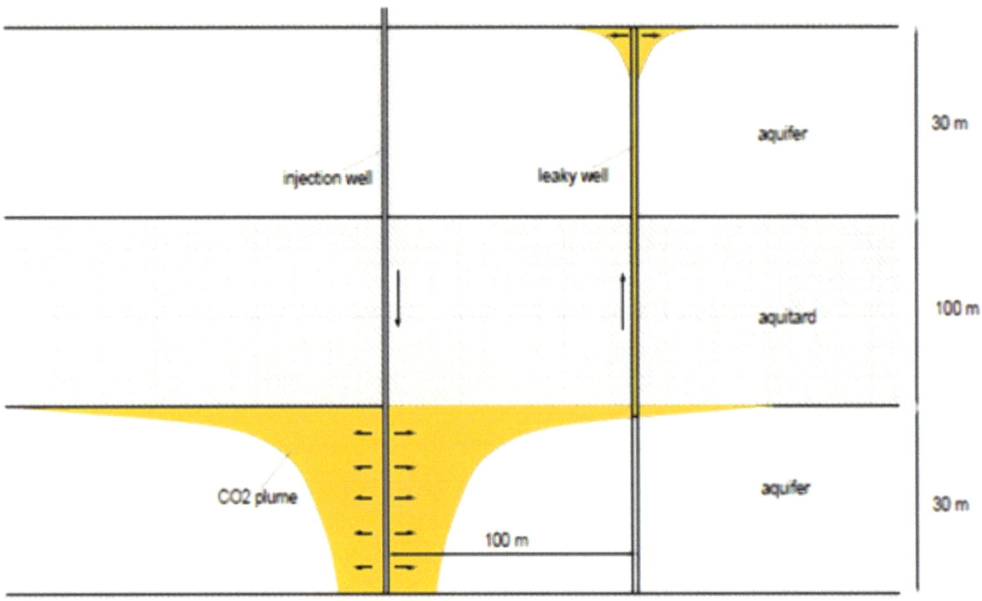

Figure 4. Schematic of benchmark problem #1 (cross-sectional view) [9]

The three layers in Figure 4 are identified as one aquitard layer and two (saline) aquifer layers. The lower aquifer layer is assumed to be 3000m below the ground surface. Typical saline aquifer conditions and hydrogeological properties, such as temperature, salinity, permeability, are assigned to the aquifer layers. The aquitard is assumed to be impermeable to both saline and CO_2, so it is considered an ideal geological seal to flow transportation. An "abandoned well" fully penetrating the three layers is located 100 m away from the CO_2 injection well. It can be either a crack in the formation or a physical abandoned well, which served as a pathway for upward CO_2 migration. Supercritical CO_2 is injected only into the lower aquifer through the injection well. Being less dense than brine, injected supercritical CO_2 gradually migrates to the ceiling of the lower aquifer and forms a plume. The formation and migration of the plume depends on the aquifer's geometric and hydrogeological properties. Table 4 summarizes the geometric properties of the aquifer in benchmark problem #1. It should be noted that the actual geometry of the injection well and abandoned well is circular, with a radius of 0.15

http://dx.doi.org/10.5772/57065

m. Since the use of an unstructured grid is not supported by PetraSim, an approximation to the circular geometry is made. Maintaining an identical cross-sectional area, the original circular injection well and the leaky well are replaced by wells of square cross-section with dimension of 0.266 m × 0.266 m. Such an approximation is acceptable since the details of the flow pattern inside the wells are not critical in achieving the objective of this simulation.

Domain Dimension	1000 m × 1000 m × 160 m
Aquifer Depth	2840 m ~ 3000 m
Aquifer Thickness	30 m
Aquitard Thickness	100 m
Distance between Injection Well and Leaky Well	100 m
Injection & Leaky Well Geometry	0.266 m × 0.266 m

Table 4. Geometry parameters for benchmark problem #1

Figure 5 shows the simulation domain and the structured mesh inside the domain.

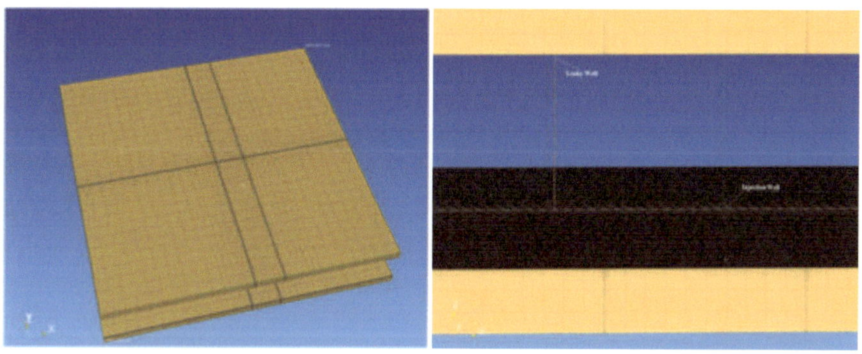

Figure 5. Entire computational domain (left) and the zoomed-in-view (right) for benchmark problem #1

To accurately model the small dimensions of the wells and accurately capture the CO_2 leakage rate, the mesh is highly refined in the neighborhood of the injection and leakage wells, as can be seen in the zoomed-in-view in Figure 5. Since high CO_2 concentration is expected at the ceiling of the lower aquifer due to gravity segregation, the mesh in this part of the lower aquifer is also refined. The mesh in the upper aquifer is not refined since it does not affect the accuracy of simulations and results in less computational efforts. The upper aquifer is uniformly discretized with vertical discretization length of 10 m, since it is assumed that the leakage amount is small and the leakage plume shape is not of great interest. By establishing the simulation domain and the mesh in this manner, reasonably accurate results are obtained while keeping the computational effort relatively low. The hydrogeological properties of the simulation domain are summarized in Table 5.

Aquifer Permeability	20 mDarcy
Leaky Path Permeability	1 Darcy
Porosity	0.15
Residual Brine Saturation	0.2
Residual CO_2 Saturation	0.05
Relative Permeability	Linear ($k_r = S$)
Capillary Pressure	Brooks-Corey
Entry Pressure	1.0×10^5 Pa
Brooks-Corey Parameter	2.0

Table 5. Hydrogeological parameters for benchmark problem #1

Other simulation parameters such as initial conditions and boundary conditions are summarized in Table 6.

Thermal Condition	**Isothermal**
Initial Condition (Temperature)	Geothermal gradient: 0.03 K/m, Initial value at 800 m: 34°C
Initial Condition (Pressure)	Pressure gradient: 1045 Pa/m, 30.86 MPa at 3000 m depth
Boundary Conditions	Fixed-state on lateral boundaries No mass flow on top and bottom boundaries
Initial CO_2 Mass Fraction	$X_{CO2} = 0$
Initial Salt Mass Fraction	$X_{sm} = 0.20$
Injection Rate	8.87 kg/s
Simulation End Time	1000 days

Table 6. Simulation parameters for benchmark problem #1

With the properties and parameters summarized above, the numerical model of benchmark problem #1 is setup in PetraSim. A pre-injection simulation is carried out first with no injection of CO_2 to achieve equilibrium condition under gravity. The equilibrium state is then implemented as an initial condition for the subsequent simulation with CO_2 injection. The equilibrium simulation is critical to provide the simulation with CO_2 injection with realistic initial conditions; this is a prerequisite procedure for all the simulations reported in this chapter. For this benchmark problem, it takes about five minutes of CPU time on the workstation for the simulation to finish. The leakage flux, pressure perturbation, and CO_2 saturation distribution throughout the aquifer after 80 days of CO_2 injection are examined and compared with the simulations of other investigators [12]. The leakage flux is a non-dimensional quantity defined as the ratio of CO_2 leakage rate to CO_2 injection rate. Detailed comparisons using various simulation codes are shown in Figure 6 and are summarized in Table 7.

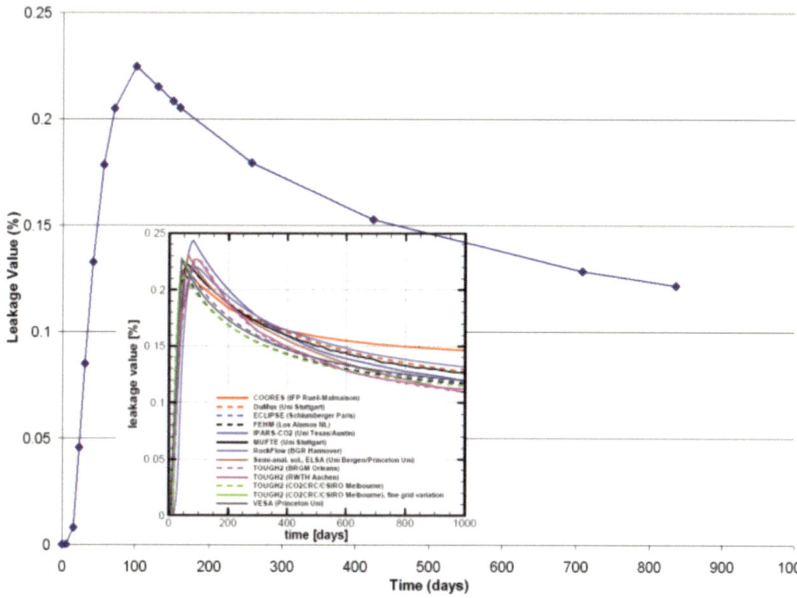

Figure 6. CO_2 leakage flux value obtained with WUSTL-TOUGH2 and other simulation codes

In Figure 6, our result (WUSTL-TOUGH2) is shown by the large graph, while comparisons with other simulation codes are shown in the inner box.

	Max. Leakage	Time at Max. Leakage	Leakage at 1000th Day
TOUGH2 (WUSTL)	0.225 %	100 days	0.115 %
TOUGH2 (BRGM)	0.226 %	93 days	0.110 %

Table 7. Simulation results and comparisons with other codes for benchmark problem #1

As additional comparisons, the pressure perturbation and CO_2 saturation distribution after 80 days of injection are also computed and compared with those from the MUFTE numerical solver [12]. Excellent agreement is obtained, as shown in Figure 7 and Figure 8.

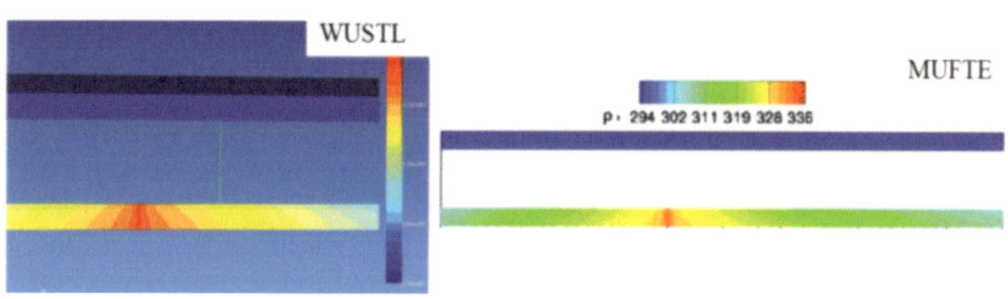

Figure 7. Pressure perturbation within the aquifer after 80 days of injection (left: WUSTL-TOUGH2; right: MUFTE)

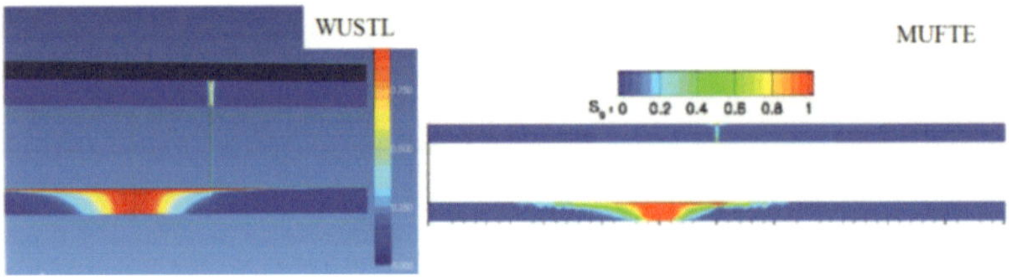

Figure 8. CO_2 distribution within the aquifer after 80 days of injection (left: WUSTL-TOUGH2; right: MUFTE)

As seen in Figure 8, the CO_2 plume is nicely captured in the simulation.

The simulation of benchmark problem #1 is very instructive. Three conclusions can be made: 1) Small variations among the results from different numerical simulators with different users are un-avoidable. Such variations are expected because some parameters are intentionally left unspecified. 2) Our results are in satisfactory agreement with the results of other investigators. 3) The most important CO_2 behavior under reservoir condition, i.e., the plume flow, is well captured and understood by the simulations. This simulation and others for benchmark problems #2 and #3 not only validate our numerical solver but also provide insights needed for the optimization studies reported in Chapter titled," Optimization of CO2 Sequestration Saline Aquifers".

4.3. Simulation of benchmark problem #2 – enhanced CH_4 recovery in combination with CO_2 sequestration in depleted gas reservoirs

For decades, the oil and gas industry has been using a reservoir engineering technique to increase the oil/gas production from mature reservoirs, known as enhanced oil/gas recovery (EOR/EGR). As the original formation fluid (oil or natural gas) gets extracted, pressure in the reservoir gradually decreases. Such de-pressurization process makes it increasingly difficult to maintain the desired production rate. The reservoir needs to be re-pressurized to mitigate the drop of oil/gas production. One of the means to do this is to inject CO_2 into the mature reservoir. With void space being occupied by the injected CO_2, remaining oil/gas is pushed out of the reservoir. Meanwhile, the depleted reservoir becomes an ideal carbon sink for long-term storage. EOR/EGR with CO_2 sequestration, also known as CSEOR/CSEGR, has been frequently used by the industry due to its strong economic merits.

In benchmark problem #2, a five-spot pattern domainis considered for modeling. The five-spot pattern is a common configuration for oil/gas production. A schematic of the reservoir is shown in Figure 3.

Natural gas is produced at the four upper corners of the reservoir, while CO_2 is injected in the middle of the domain at the bottom-most part. This is a direct result of CO_2 being heavier than CH_4 under the reservoir conditions. Injection of CO_2 at the bottom avoids gas mixing and creates better sweep efficiency. The main goal of this benchmark simulation is to identify the gas recovery factor, defined as the ratio of enhanced CH_4 production to the original remaining CH_4 amount until the shutdown of the production well. Additionally, the time until production

http://dx.doi.org/10.5772/57065

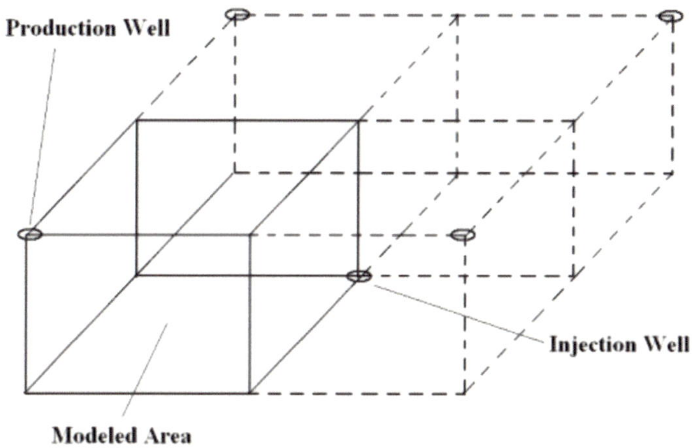

Figure 9. Schematic of the 3D five-spot pattern for benchmark problem #2

shut-down, which is defined as the moment when the production contains up to 20% of CO_2 by mass, needs to be determined.

Due to symmetry, only a quarter of the domain is modeled, as shown in Figure 10 as the volume bounded by the solid lines. The dimension of the modeled domain is 201.19 m ×201.19 m with thickness of 45.72 m. Due to the relatively strong diffusion, the discretization length has a strong influence on the gas mixing [10].It is therefore strictly specified as 4.572 m for both vertical and horizontal direction.Figure 10 shows the CFD model and its mesh in the quarter computational domain of the five-spot reservoir.

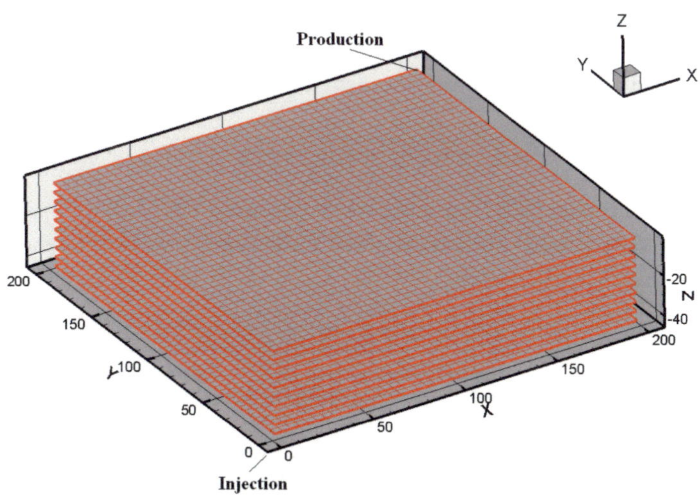

Figure 10. CFD model and its mesh in the quarter of the 5-spot domain for benchmark problem #2

The hydrogeological properties assigned to the model are summarized in Table 8.

Permeability	Horizontal: 50 mDarcy, Vertical: 5 mDarcy
Porosity	0.23
Residual Brine Saturation	0
Relative Permeability	Liquid: immobile, Gas: linear ($k_r = S$)
Capillary Pressure	None

Table 8. Hydrogeological properties of the domain for benchmark problem #2

Initial conditions and boundary conditions and some other parameters of the domain are given in Table 9.

Thermal Condition	**Isothermal**
Initial Condition (Temperature)	66.7 °C
Initial Condition (Pressure)	3.55 MPa
Boundary Conditions	No mass flow at all boundaries; Constant pressure at CH_4 production well
Initial CO_2 Mass Fraction	$X_{CO2} = 0$
Initial CH_4 Mass Fraction	$X_{sm} = 1$
Injection Rate	0.1 kg/s until shut-down

Table 9. Simulation parameters for benchmark problem #2

The termination of the simulation depends solely on the mass fraction of CO_2 in the reservoir. It takes about 30 minutes of CPU time to run 2,000 days of simulation before major CO_2 contamination occurs. The recovery factor, production shut-down time, pressure and CO_2 saturation distribution in the reservoir are investigated and compared with the results of other investigators.

In Figure 11, our results using TOUGH2 are shown as the large graph, while results of simulations from other investigators are shown in the inner box. Table 10 provides comparisons for recovery factor and production well shut-down time with other investigators' simulations [12].

	Recovery Factor	**Production Well Shut-down Time**
TOUGH2 (WUSTL)	61.4%	2063 days
TOUGH2 (CO_2/CRC)	58%	1987 days
MUFTE (U. Stuttgart)	53%	1894 days
IPARS (U. Texas)	55%	1891 days

Table 10. Comparisons of recovery factor and production shut-down time

http://dx.doi.org/10.5772/57065

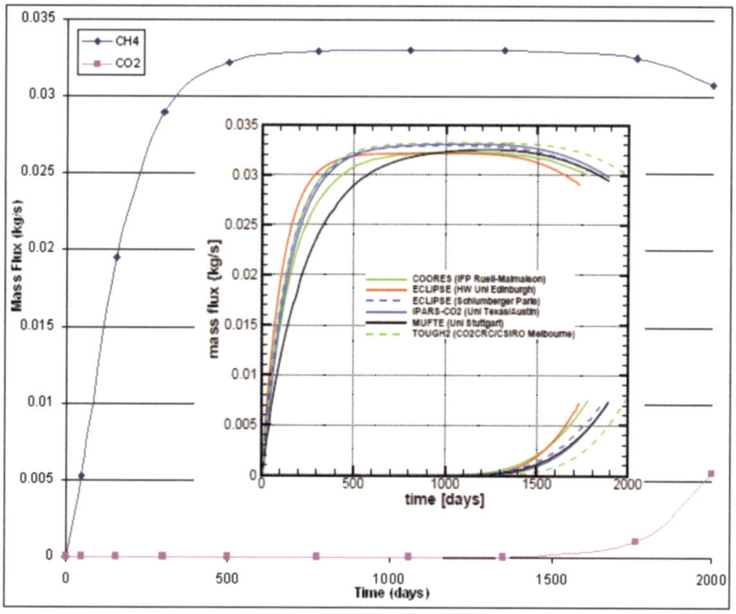

Figure 11. History of enhanced CH_4 recovery for benchmark problem #2

To visualize how the displacement process of CH_4 by CO_2 works, the density and CO_2 mass fraction profiles at production shut-down are examined in Figure 12.

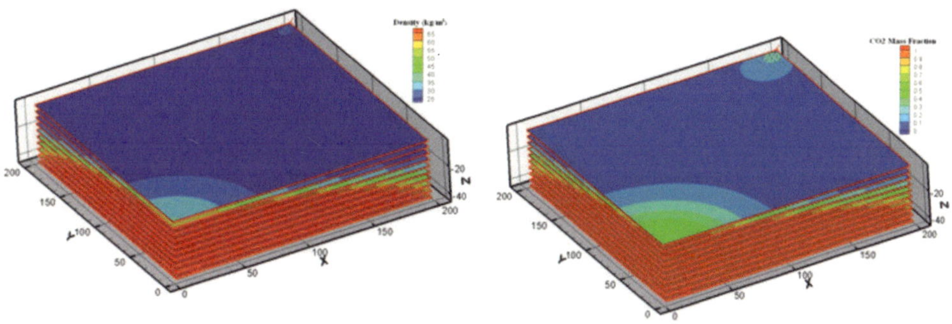

Figure 12. (a) Density profile (b) CO_2 mass fraction profile at production shut-down for benchmark problem #2

Figure 11, Table 10 and Figure 12 lead to the following four conclusions: 1) Small variations in the results among different simulations with different users are unavoidable. Such variations are expected because some parameters are intentionally left unspecified. 2) Our results are in satisfactory agreement with those of other investigators. 3) It can be seen that the injected CO_2 migrates from the near lower corner to the far upper corner in a semi-spherical fashion. Unlike SAGCS, in situ CO_2 tends to sink to the bottom of the reservoir. This indicates strong gravity segregation caused by the density difference. 4) Production gas contamination caused by upward movement of CO_2 occurs at the production well despite the gravity segregation. This is due to the strong convective flow near the production well and mass diffusion.

4.4. Simulation of benchmark problem #3 – CO_2 injection in a heterogeneous geological formation

Accurate estimation of in situ CO_2 dissolution into the ambient brine is another important aspect of SAGCS simulation, since CO_2 becomes securely sequestrated once dissolved. Overestimation of CO_2 dissolution could lead to underestimating the potential leakage; on the other hand, underestimation of CO_2 dissolution would result in inefficient utilization of the formation's storage potential. In the meantime, it is instructive to model a realistic geological reservoir with heterogeneous hydrogeological properties for more realistic estimation of CO_2 storage capacity. In benchmark problem #3, part of the Johanson formation off the Norwegian coast is modeled for SAGCS [11]. The Johanson formation is a highly heterogeneous formation, especially in its porosity and permeability, as shown in Figure 13. CO_2 is injected in the middle of the modeled formation at 50 m from the bottom. The injection lasts for 25 years before it is shut down, and the total simulation time is 50 years. The goal of this benchmark study is to identify the amount of dissolved CO_2, the amount of CO_2 still in gaseous phase, and how these amounts evolve with respect to time. This study is very instructive to understand the dissolution process of injected CO_2 under reservoir conditions.

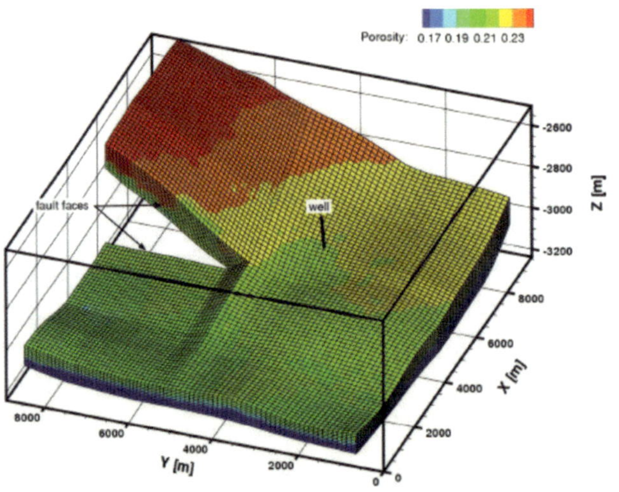

Figure 13. Johanson formation's porosity heterogeneity for benchmark problem #3[11]

The dimension of the modeled portion of the Johanson formation is 9600 m × 8900 m with varying thickness between 50 m to 150 m. The injection well is located at (x = 5440 m, y = 3300 m) over the bottom 50 m of the formation. The coordinates of vertices of 54756 hexahedral cells in Cartesian system have been provided for geometry construction [11]. Figure 14 shows the final CFD model of the Johanson formation.

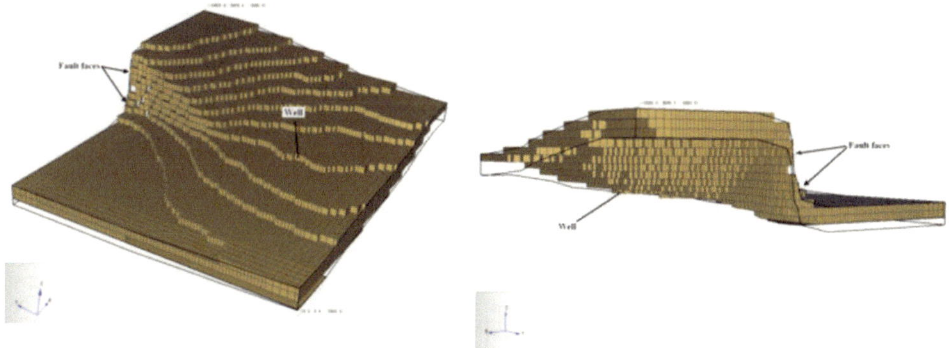

Figure 14. (a) Front view and (b) Rear view of the modeled Johanson formation

The porosity and permeability of the modeled Johanson formation areshown in Figure 15.

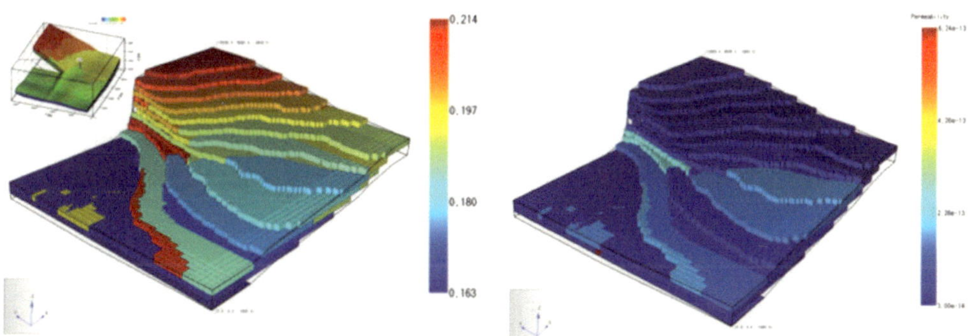

Figure 15. (a) Porosity and (b) Permeability of the modeled Johanson formation

Hydrogeological properties of modeled Johanson formation are summarized in Table 11.

Permeability	Varies
Porosity	Varies
Residual Brine Saturation	0.2
Residual CO_2 Saturation	0.05
Relative Permeability	Brooks-Corey
Capillary Pressure	Brooks-Corey
Entry Pressure	1.0×10^4 Pa
Brooks-Corey Parameter	2.0

Table 11. Hydrogeological properties of the modeled Johanson formation

Initial conditions, boundary conditions and other parameters of the modeled Johanson formation are summarized in Table 12.

Thermal Condition	Isothermal
Initial Temperature	0.03 °C/m; 100 °C at 3000 m depth
Initial Pressure	1075 Pa/m, 30.86 MPa at 3000 m depth
Boundary Conditions	Fixed-state on lateral boundaries No mass flow on fault, top and bottom boundaries
Initial CO_2 Mass Fraction	$X_{CO2} = 0$
Initial Salt Mass Fraction	$X_{sm} = 0.1$
Injection Rate	15 kg/s (for 25 years), 0 kg/s thereafter
Discretization	Number of computational grids: 18804 Non-uniform for *x*-, *y*-, *z*-directions

Table 12. Simulation parameters for the modeled Johanson formation

Both gaseous and aqueous CO_2 accumulations after 50 years are considered as benchmark criteria for making comparisons with the simulations of other investigators. In Figure 16 our results using TOUGH2 are shown as the large graph, while results from other simulations are shown in the inner box.

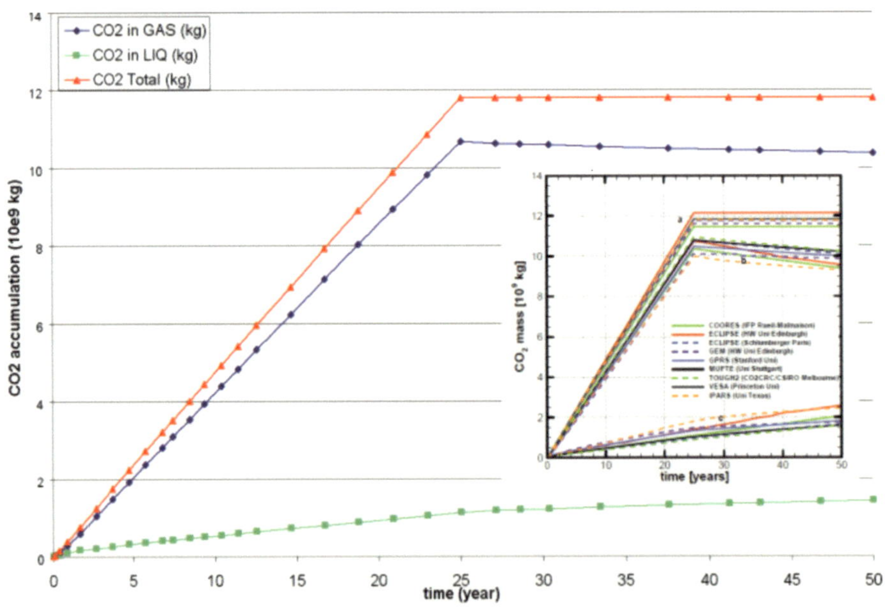

Figure 16. Gaseous and aqueous CO_2 accumulations for 50 years

Table 13 provides additional quantitative comparisons [12].

	Gashouse CO_2 at 50th Year	Dissolved CO_2 at 50th Year
TOUGH2 (WUSTL)	87.9%	12.1%
TOUGH2 (CO_2/CRC)	86.5%	13.5%
IPARS (U. Texas)	79.1%	20.9%

Table 13. Comparisons of gaseous and aqueous CO_2 accumulations after 50 years

A comparison of the CO_2 migration after 50 years is given in Figure 17.

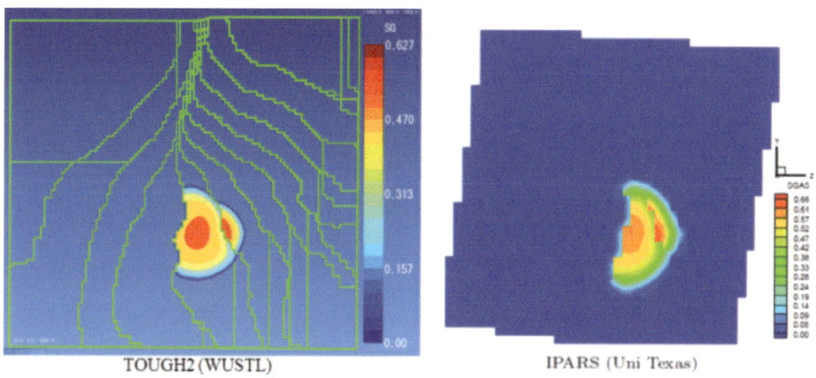

Figure 17. CO_2 saturation in the formation after 50 years, plan view

Figure 16, Table 13, and Figure 17 lead to four conclusions: 1) Small variations in the results among different simulations with different users are unavoidable. Such variations are expected because some parameters are intentionally left unspecified. 2) Our results are in satisfactory agreement with the results of other investigators. 3) CO_2 dissolution into the ambient pore-water is a process that takes place very slowly. 4) The greater slope of aqueous CO_2 during the first 25 years (when injection continues) implies enhanced carbon dissolution due to convection during the first 25 years.

4.5. Conclusions of benchmark problems

The benchmark simulations presented in this chapter demonstrate that the TOUGH2 numerical simulator is capable of producing accurate and consistent results for various types of problems related to GCS. These simulations allow us to conduct simulations of large-scale SAGCS in identified saline formations with confidence, and proceed towards the development of a numerical optimization module for TOUGH2 and perform optimization designs of innovative reservoir engineering techniques for enhanced SAGCS safety and storage efficiency, as reported in Chapter titled," Optimization of CO2 Sequestration Saline Aquifers".

5. Simulations of GCS in identified large scale saline aquifers

Accurate large-scale simulations of existing (completed/continuing) SAGCS projects for identified known aquifers are crucial to create confidence in the future deployment of SAGCS projects. Although detailed history-matching simulations of existing SAGCS projects are challenging due to various uncertainties, e.g., in the reservoir topography and hydrogeology, the simulations can still provide informative insights into several aspects of SAGCS, such as the variance in multiphase flow properties, integrity of the geological seals, and the mechanism of CO_2 trapping. Such insights are essential for better understanding of the nature of SAGCS and its best practices for deployment. Detailed history-matching simulations have always been an important part in the SAGCS research activity. In the following sections, the results of SAGCS simulations for three large-scale identified aquifers are described.

5.1. SAGCS simulation for Mt.Simon formation

Located in the Illinois basin, the Mt.Simon sandstone formationis a huge saline aquifer that covers most of Illinois, southwestern Indiana, southern Ohio and western Kentucky. The estimated storage capacity of the Mt.Simon formation ranges from 27000 to 109000 million tons of CO_2[16],[17]. The Midwest Geological Sequestration Consortium (MGSC) is the regional consortium conducting studies of the possibility of large-scale GCS throughout the Illinois basin. ADM-Decatur GCS Project and FutureGen 2.0 Project are the two best-known SAGCS projects being currently carried out in theMt.Simon formation.

The depth of the Mt.Simon formation varies significantly throughout its coverage [18],[19]. In the southern part, it reaches as deep as 4300 m below mean sea level (MSL) while it increases to 80 m below MSL in the north. Consequently, a south-north geological slope of approximately 8 m/km has been estimated. The thickness of the Mt.Simon formation also changes significantly.A maximum thickness of 800 m in the north has been measured while it diminishes to zero further to the south. Other than the variance in topography, analysis of rock samples has suggested strong anisotropy in the formation's hydrogeological properties, with porosity ranging from 0.062 to 0.2 and permeability ranging from 5 mDarcyto 1000 mDarcy. Low-permeable Eau Claire shale, which sits above the Mt.Simon formation, serves as the caprock. Except for some small regions near the Mississippi River, Eau Claire shale is very thick (more than 90 m) throughout most of the Illinois basin. The security of SAGCS over Mt.Simon formation is therefore assured by the continuous coverage of Eau Claire shale. A Precambrian granite formation stretches beneath the Mt.Simon saline aquifer.

A recent geological survey has suggested an area in the center of the Mt.Simon formation to be the core injection area – an area in which future storage sites are likely to be located. This core injection area is indicated as the area compressed by the white boundary in Figure 12, along with the elevation information of the Mt.Simon formation. As can be seen from Figure 12, both the ADM and FutureGen 2.0 projects are located in the core injection area.

Mt. Simon sandstone is a typical stratified saline formation. According to the geological survey, strong anisotropy in porosity, permeability and capillary pressure exists through the entire

http://dx.doi.org/10.5772/57065

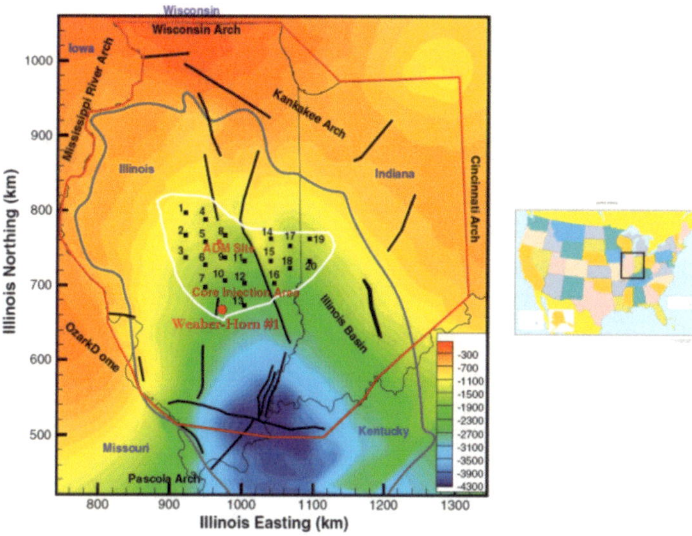

Figure 18. Core injection area and elevation of Mt.Simon sandstone

depth of the formation. Based on variance of porosity, the Mt. Simon formation can be distinguished as four subunits, namely an upper unit with sandstone and shale tidally influenced and deposited, a middle unit with relatively clean sandstone, an Arkosic unit with highly porous and permeable sandstone, and a lower unit with decreased porosity and permeability. The high porosity and permeability of the Arkosic unit makes it an ideal candidate for the injection site. When modeling, these four subunits of Mt.Simon are further divided into 24 layers, each of which has a layer-averaged porosity and permeability value [16],[19]. With the consideration of data availability, a candidate site for future sequestration project, the Weaver-Horn #1 well (WH #1 well is shown as the red dot in Figure 18), has been chosen for our simulation study. The detailed well log of WH #1 is summarized in Table 14[20]. It is desired to model the anisotropy of hydrogeological properties as accurately as possible to capture its effect on in situ CO_2 transport. It should be noted that the lower unit of the Mt.Simon formation is not considered in the model due to its absence near WH #1 well. Both Eau Claire shale and Precambrian granite are modeled as impermeable formations.

A cylindrical model of the Mt.Simon formation is constructed. For thermal condition,the model uses calculated values with a thermal gradient of 9.2°C/km. The reservoir pressure is assumed to be hydrostatic pressure with a gradient of about 10.8 MPa/km from the ground surface. Salinity is assumed to increase with the depth, starting from 235 mg/L at 450 m below ground surface with a gradient of 12.8 mg/L per meter in depth. A north-south geological slope of 8 m/km is also considered in the model. A "no-flux" boundary condition is applied at the top and bottom of the model, representing the impermeable upper and lower bounding formations. A "fixed-state" boundary condition is imposed at the lateral boundary to represent an essentially "open" reservoir. The permeability and porosity of the 24 sublayers can be seen in Figure 19.

Sub-Unit	Sublayer	Layer Depth (m)	Mean Porosity	Mean Permeability (mDarcy)	Characteristic Capillary Pressure (bar)
	1	2140 – 2150	0.061	5	0.37
	2	2150 – 2182	0.109	300	0.06
	3	2182 – 2197	0.074	10	0.28
Upper Unit	4	2197 – 2203	0.083	3.6	0.4875
	5	2203 – 2230	0.195	110	0.1
	6	2230 – 2232	0.071	1.1	0.8
	7	2232 – 2280	0.13	210	0.083
	8	2280 – 2322	0.083	5.4	0.4125
	9	2322 – 2331	0.24	150	0.0875
	10	2331 – 2340	0.088	8	0.35
	11	2340 – 2350	0.156	800	0.095
	12	2350 – 2370	0.25	80	0.125
	13	2370 – 2378	0.163	900	0.095
Middle Unit	14	2378 – 2385	0.195	105	0.1007
	15	2385 – 2399	0.163	800	0.05
	16	2399 – 2406	0.136	72	0.1167
	17	2406 – 2412	0.156	700	0.05
	18	2412 – 2424	0.129	160	0.09
	19	2424 – 2430	0.161	850	0.05
	20	2430 – 2462	0.128	60	0.15
	21	2462 – 2500	0.202	1000	0.05
Arkosic Unit	22	2500 – 2502	0.14	190	0.09
	23	2502 – 2537	0.151	1000	0.04

Table 14. Porosity, permeability and characteristic capillary pressure of the 24 layers of Mt.Simon at injection site WH #1 [20]

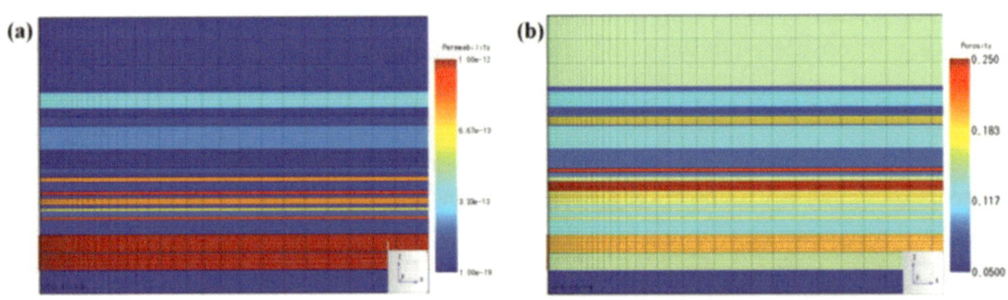

Figure 19. (a) Permeability, (b) porosity, and computational mesh of the 24 sublayers of the Mt.Simon formation model at WH #1 well

Due to the relatively high porosity and permeability, CO_2 injection is assigned at the bottom Arkosic unit (bottom three sublayers). The injection rate is assigned to be 5 million tons per

http://dx.doi.org/10.5772/57065

year and injection lasts for 50 years. CO_2 footprint at 5th year, 25th year and 50th year since the beginning of injection is examined.

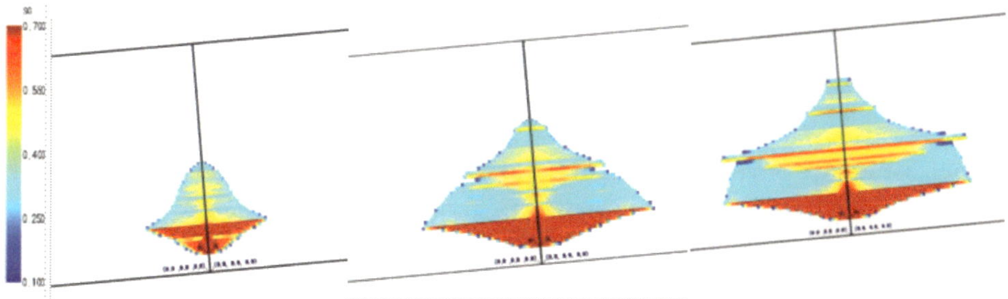

Figure 20. Saturation of gaseous CO_2 at (a) 5th (b) 25th and (c) 50th year of injection

As seen in Figure 20, the CO_2 plume evolves with a complex spatial pattern during the 50 years of injection. Within the Arkosic unit where the injector is located, extensive lateral migration with relatively higher concentration of gaseous CO_2 is observed. In the overlying sublayers, however, a strong secondary sealing effect that retards the vertical migration of gaseous CO_2 is observed, as can be seen from the pyramid-shaped sub-plume. Detailed analysis of secondary sealing effect is done as follows. As shown in Figure 20, the injected CO_2 migrates laterally away from the injector within the highly permeable Arkosic unit in the first 5 years after the start of injection. Simultaneously, buoyancy also leads to upward movement of CO_2 until it encounters the immediately overlying low-permeability sublayer (sublayer #20). The low permeability of sublayer #20 directly results in higher capillary pressure experienced by mobile CO_2, and thus a stronger vertical pressure gradient is required for mobile CO_2 to penetrate sublayer #20. When the capillary pressure is greater than the phase pressure of CO_2, sublayer #20 appears to be "impermeable" to the underlying CO_2 plume. Consequently, gaseous CO_2 accumulates under this layer and continues spreading out laterally, finally reaching a maximum extent of approximately 3000 m. Meanwhile, the increased CO_2 column under sublayer #20 brings up its phase pressure. Once the phase pressure of CO_2 exceeds the entry pressure of sublayer #20, mobile CO_2 breaks the capillary barrier of its overlying layer and starts to penetrate it. Such accumulation-penetration-breakthrough behavior of gaseous CO_2 occurs each time the upward migrating CO_2 encounters an overlying sublayer with lower permeability. Because the high capillary entry pressure of the overlying layer temporarily prevents CO_2 from migrating upwards,this phenomenon is identified as the secondary sealing effect. As can be seen from Figure 20, this effect is a very effective means to retard the upward migration of in situ CO_2. Its contribution makes gaseous CO_2 barely reach the Eau Clair shale even after 50 years of injection.

5.2. SAGCS simulation of Frio formation

The SAGCS pilot project for the Frio deep saline formation near the Mexican Gulf Coast is the subject of study in this simulation. The Frio project has two characteristics that make it attractive for numerical study. First, it is a completed pilot project with detailed field data available; secondly, hysteresis information of relative permeability and capillary pressure has been obtained from the core sample of the Frio saline formation. The hysteresis effect is an important factor in obtaining accurate estimation of CO_2 migration and dissolution for full-term SAGCS simulation. A full-term simulation refers to a simulation that investigates the fate of in situ CO_2 through the entire life cycle of a SAGCS project, which consists of both injection and post-injection periods.

The Frio SAGCS pilot project was conducted at the South Liberty oil field operated by Texas American Resources in Dayton, Texas. Starting from October 4, 2004, 1600 tons of CO_2 was injected into the Frio formation about 1500 m below the ground surface within 10 days. The Frio formation consists of brine-bearing sandstone with high permeability beneath the Gulf Coast. It is a relatively thin sandstone layer only 23 m thick. A steep geological slope of 16° from south to north has been identified for Frio formation [21]. The Frio pilot project employed one injection well and one observation well about 33 m to its north. Other than the conventional pre-injection geological surveys, laboratory analysis of core samples suggested the hysteresis behavior of relative permeability and capillary pressure in the Frio formation. The hysteresis has been considered in our simulations.

The hydrogeological parameters of the modeled Frio formation are summarized in Table 15.

Permeability	Isotropic, 1 mDarcy
Porosity	0.28
Residual Brine Saturation	0.15
Residual CO_2 Saturation	0.2
Relative Permeability	van Genuchten-Mualem with hysteresis
Capillary Pressure	van Genuchten-Mualem with hysteresis
Initial Conditions	$P = 15.2$ MPa, $T = 59$°C
Initial CO_2 Mass Fraction	$X_{CO2} = 0$
Initial Salt Mass Fraction	$X_{sm} = 0.093$

Table 15. Geometry and hydrogeological parameters for Frio formation

Characteristics of capillary pressure and relative permeability have been obtained from mercury-injection laboratory experiments on core samples from Frio formation, given in Figure 21. Hysteresis in both capillary pressure and relative permeability can be clearly observed. Drainage (of pore-water) curves are marked red and imbibition (of pore-water) curves are marked blue. When multiple drainage-imbibition cycles occur, different imbibition curves represent different orders of drainage-imbibition cycles. The primary imbibition curve, i.e., when brine imbibition occurs for the first time, is depicted as a bold solid curve. Since only one drainage-imbibition cycle takes place when continuous CO_2 injection is imposed before it is permanently shut down, only the primary imbibition curve needs to be considered in the model.

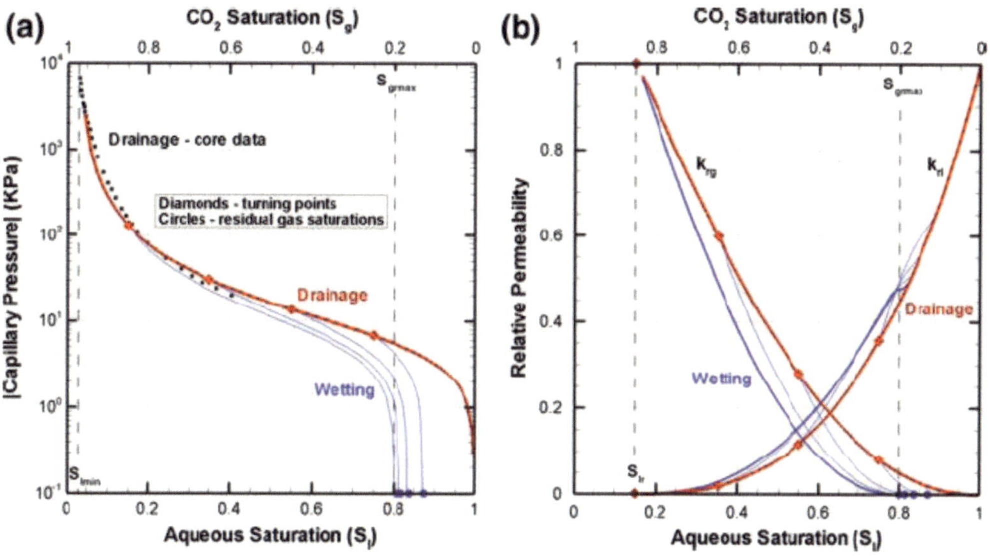

Figure 21. Capillary pressure and relative permeability characteristics of Frio formation[22],[23]

Following Doughty et al.'s work [22], a rectangular portion of Frio formation with dimension of 2500 m northwest-southwest, 800 m northeast-southeast, and 23 m thickness is modeled, as shown in Figure 22. The injection well is located at a point with coordinate (x=560 m, y=800 m) from the lower left corner of the computational domain. Although the formation is 23 m thick, injection takes place only over the first 8 m from the caprock. An observation well is located 33 m to the north. Because flow transport is most intense near the injection and observation wells, it is evolved in a computational domain of dimension 30 m × 30 m whose mesh is refined for accurate capture of the flowpattern. The injection and observation well locations, well depth, computational mesh, and north-south slope of the numerical model are all shown in Figure 22.

The simulation time is set at 10 days to match the actual duration of injection. It takes approximately 12 hours of CPU time for the simulation to finish. The profiles of gaseous phase CO_2 at the end of injection in the vertical cross-section containing both injection and observation wells are shown in Figure 23. Doughty et al.'s result [22] is also shown in Figure 23 for comparison.

Additionally, Figure 24 shows the CO_2 saturation profiles at the injection and observation wells obtained by our simulation. They are compared with those given by Doughty et al. and the reservoir saturation tool (RST) logs [22]. The RST well logs are actual measurements in the field during the pilot project.

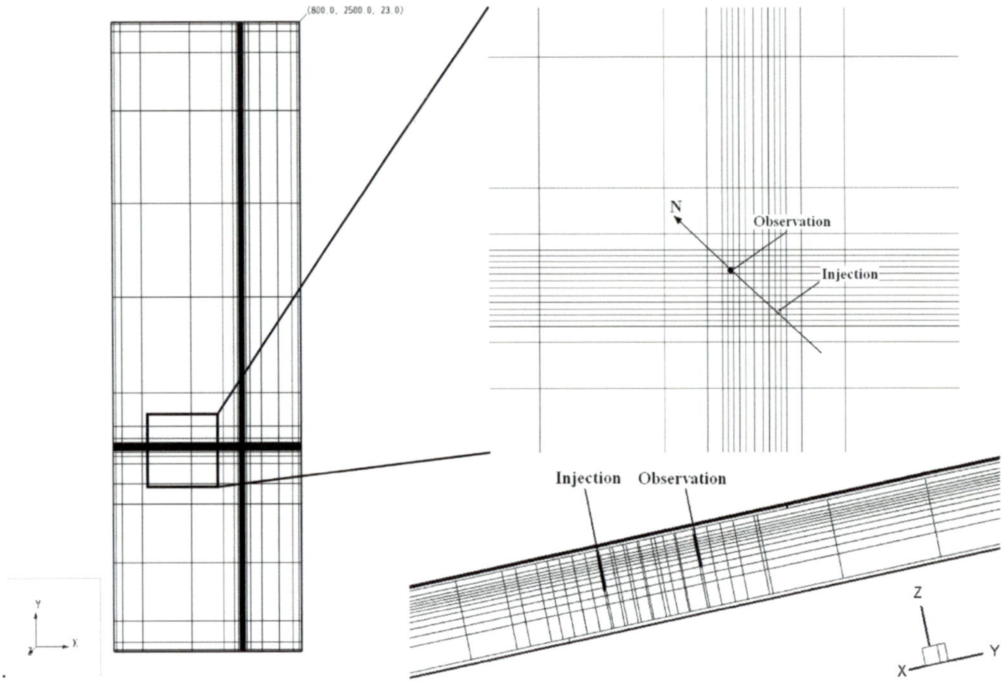

Figure 22. Model geometry and mesh in a portion of Frio formation and zoomed-in side view of the injection and observation wells

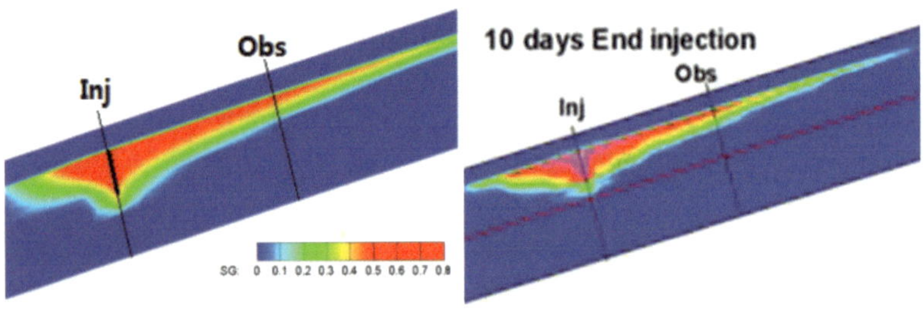

Figure 23. CO_2 footprint on 10th day when injection stops (comparison with Doughty et al.'s work[22])

As seen from Figure 23, the highly asymmetric CO_2 plume suggests a strong tendency of movement upwards in the direction of the geological slope. Unlike the case of Mt. Simon SAGCS, the CO_2 plume in the Frio project is shaped like an inverted pyramid, which implies the lack of secondary sealing effect. Both the asymmetric migration and inverted pyramid-shaped plume indicate strong evidence of the dominant role of gravity segregation in determining the in situ CO_2 migration. Considering the relatively short-term injection (10 days) for the Frio SAGCS project, this implies that in situ CO_2 migrates mostly convectively. Furthermore, it demonstrates that the poor permeability caprock layer above the injection serves quite well as the CO_2 barrier.

http://dx.doi.org/10.5772/57065

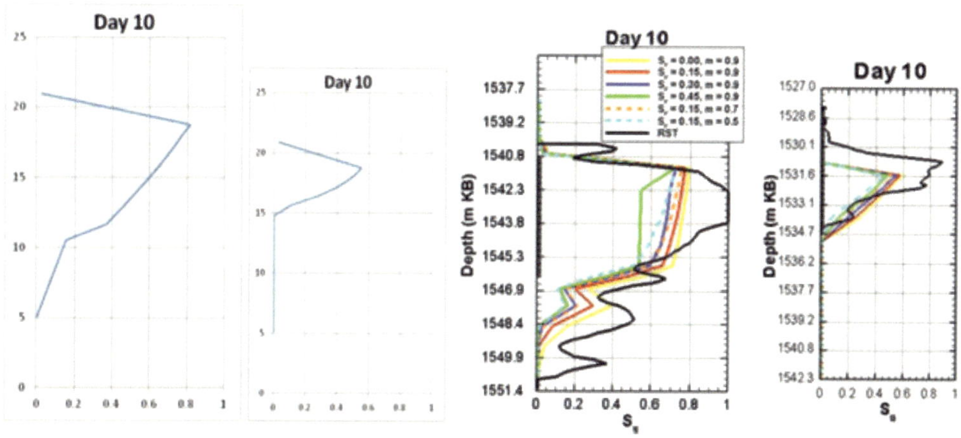

Figure 24. CO_2 saturation profiles given by the numerical simulations and the RST logs

Comparing our results with those from Doughty et al. [22], the following conclusions can be drawn: (1) Overall our results are in good agreement with those of Doughty et al. for the plume shape, tendency of plume migration induced by the slope, distance of plume migration, and gaseous CO_2 saturation, etc. (2) Discrepancy still exists at the detailed simulation level. The results show that in our simulations, CO_2 saturation at the injection well reaches a maximum of 0.8 by the 10th day of injection. Although being consistent with Doughty et al.'s work, it differs from the field data. Results from the RST measurement suggest a CO_2 saturation value of 1.0, i.e., dry-out of brine, occurs adjacent to the injection well. The occurrence of brine dry-out is fairly common near the injection well due to the strong pressure gradient. However, the absence of brine dry-out in both our and Doughty et al.'s simulations can be explained by the designated brine residual saturation value. In our TOUGH2 simulations, a brine residual saturation value of 0.15 is pre-assigned to the entire computational grid including the grid near the injection wells. Since residual saturation describes the minimum saturation value of a certain phase being displaced, it means that at least 15% of the pore space will remain occupied by brine regardless of the pressure gradient. A direct result is the capped CO_2 saturation value at 0.85 and the absence of brine dry-out. (3) Our simulation shows quicker decrease in gas saturation during the injection interval. In Doughty et al.'s work, the gas saturation only drops from 0.8 to 0.65 for the upper 5 m of injection depth. In contrast, it drops from 0.8 to 0.4 in our simulation. This implies stronger buoyancy in our simulation, and thus results in a steeper inclined CO_2-brine interface. This also explains the slight overshoot in the plume migration to the north in our simulation.

5.3. SAGCS simulation for Utsira formation

The Sleipner project near the Norwegian coast on the North Sea is probably the most prestigious, important and successful SAGCS demonstration so far. It has the most complete topographic description, industrial-scale injection amount, and long-term monitoring data. Nevertheless, great uncertainties still exist for accurate reservoir-scale simulation of the

Sleipner SAGCS project. Simulation studies of this project can provide helpful insights in understanding the transport behavior of in situ CO_2 and the reservoir performance.

Starting from 1996, the Sleipner field in the North Sea has been the host of the world's first commercial SAGCS project. CO_2 is captured from the gas mixture produced from a nearby deeper natural gas reservoir. To date, approximately 1 million tons of supercritical CO_2 has been sequestered annually. Utsira saline aquifer is the target formation for permanent carbon sequestration for the Sleipner SAGCS project. The Utsira formation is located at a depth of 800 m – 1100 m from the seabed with thickness of about 200 m – 250 m. The injection site is located at the southern portion of the Utsira formation. A 250 m – 330 m thick shale layer known as the Nordland Formation serves as the caprock, and core testing has suggested its potential for bearing a CO_2 column of at least 100 m but perhaps up to 400 m (depending on the in situ conditions). It is estimated that the Utsira formation has permeability of about 1– 8 mDarcy, porosity of about 0.35 – 0.4, and temperature of about 34°C – 37°C. It is also estimated that the reservoir bears hydrostatic pressure from its overburden formations. Similar to the Mt.Simon formation, the Utsira formation is also highly stratified, consisting of sublayers with high-permeability sandstone and low-permeability shale. Therefore, it is expected that the secondary sealing effect will occur. Figure 25 shows a 2-D seismic image taken in 2008 revealing the CO_2 plume in the Utsira formation. Multiple layers can be distinctly identified from the seismic image.

Two numerical models have been constructed to study the Sleipner SAGCS project. The first model is a generalized axisymmetric layered model for estimating the ballpark migration of in situ CO_2. The purpose of this simulation is to determine the secondary sealing effect and gain an overview of the plume migration within the Utsira formation. The second model describers a 48 km^2 area of detailed topmost sandstone layer (marked as Layer #9 in Figure 26). Layer #9 is of particular interest regarding the safety of the sequestration project, as it is the layer within which the highest concentration of gaseous CO_2 exists and the most significant plume migration occurs. Detailed topography of Layer #9 is shown in model #2, making it a complicated 3D model. The 3D Layer #9 model is introduced to investigate the effect of actual topography on in situ CO_2 migration, while avoiding intensive computational effort associated with full 3D modeling and simulation of the entire Utsira formation.

5.3.1. Model #1 – Generalized stratified model of the Utsira formation

A pre-injection geological survey unveiled the layered structure of the Utsira formation. The majority of the formation can be identified as an eight-layered structure, but one extra layer needs to be added to the structure near the injection site due to the existence of a sand wedge. Therefore, a cylindrical domain with nine alternating sandstone and shale layers is constructed as shown in Figure 26. According to the seismic survey, it is assumed that all four shale layers have identical thickness of 5 m, four shallower sandstone layers have identical thickness of 25 m, and the bottom sandstone layer has a thickness of 60 m. This adds up to a total 180 m thickness for the modeled Utsira formation. The lateral radius of the generalized cylindrical model reaches 100 km, which is about the same as the actual extent of the southern part of the Utsira formation. According to Audigane et al. [25], all sandstone layers have identical and

http://dx.doi.org/10.5772/57065

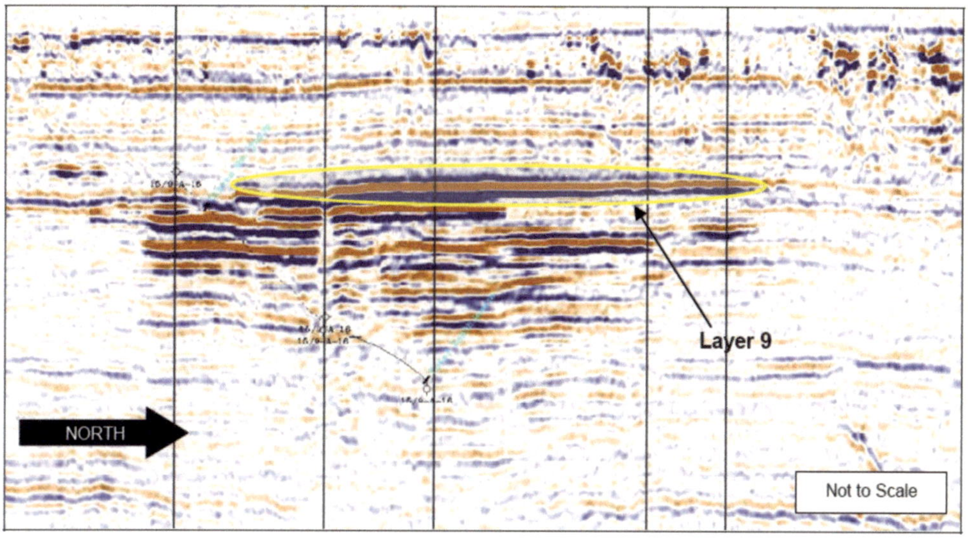

Figure 25. eismic image of Utsira formation after 9 years of injection, S-N cross-section [24]

isotropic hydrogeological properties, as do the shale layers. Figure 26 shows the layered structure and computational mesh of the modeled Utsira formation as well as the location of CO_2 injection. The permeability and porosity are 3 mDarcy and 0.42 for the sand layer and 10 mDarcy and 0.1025 for the shale layer. A temperature of 37°C and pressure of 11 MPa are applied as the pre-injection conditions. van Genuchten-Mualem functions are used to describe both relative permeability and capillary pressure. CO_2 injection of 30 kg/s is assigned as a point source at the middle of the bottom-most sand layer.

Figure 26. Computational mesh and layered structure of the generalized 9-layered model of the Utsira formation

The simulation time is set at 15 years and the CO_2 plume profile is examined for each year. Figure 27 shows the cross-sectional view of gaseous CO_2 in the reservoir for ten consecutive years since the start of injection.

The results shown in Figure 27 provide evidence of strong secondary sealing effect for migration of in situ CO_2. Similar to the case of Mt. Simon SAGCS, the injected CO_2 first migrates upwards driven by buoyancy until it reaches the first shale layer. Due to the low permeability and high capillary entry pressure, CO_2 is confined by this shale layer and is forced to migrate radially. Simultaneously, CO_2 concentration builds up beneath the shale layer and finally breaks through the capillary barrier upon sustaining sufficient CO_2 column height. The accumulation-penetration-breakthrough takes place each time the CO_2 plume encounters a new shale layer and forms an inverted pyramid shaped sub-plume, as documented clearly by the first and second year plume shapes in Figure 27. Due to the secondary sealing effect, in situ CO_2 has very limited contact with the caprock of the Utsira formation by the third year of injection.

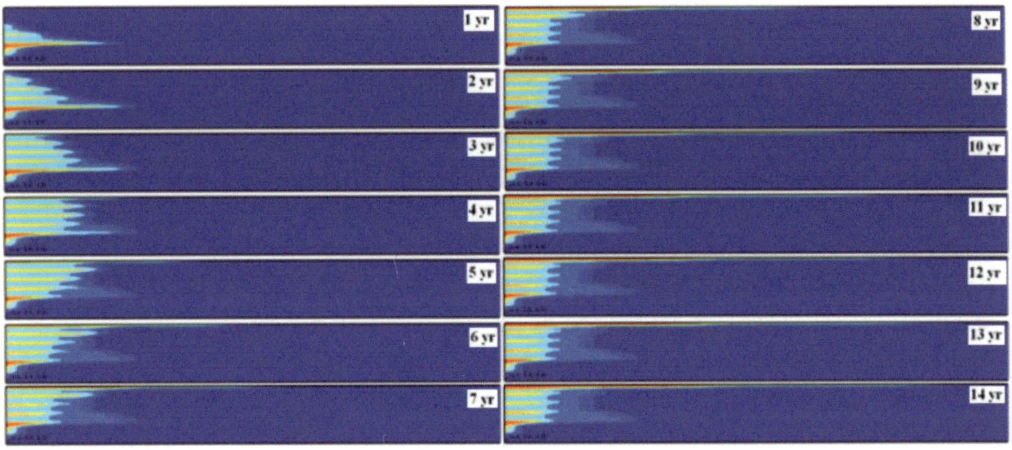

Figure 27. In situ CO_2 distribution for 15 years of injection

Additionally, ten-year CO_2 flux analysis has been made for the topmost sandstone layer (Layer #9) since it is critical to identify the accumulation of CO_2 underneath the caprock. As shown in Figure 30, excellent agreement between our simulation and the seismic amplitudes analysis [26] is observed, suggesting the overall accuracy of our model despite some discrepancy at the detailed level. The flux analysis shown in Figure 28 also implies that the accumulation rate of CO_2 in the topmost sandstone layer tends to increase until it becomes stabilized. This fact can be explained by mechanism of secondary sealing effect.

5.3.2. Model #2 – Detailed 3D model of the Utsira Layer #9 formation

In situ CO_2 possesses strong potential to migrate upward due to buoyancy, and thus accumulates under the caprock unless capillary barrier is compromised. Previous experience has demonstrated that the accumulation of CO_2 under the caprock occurs in a relatively short period compared to the entire time span of SAGCS projects, and it is a major concern for storage security. Therefore, it is critical for a SAGCS project to identify the accumulation of CO_2 and its tendency for migration underneath the caprock. With such information available, precautionary treatments can be deployed to avoid potential leakage. The Utsira formation near the injection site has been identified as a nine-layer structure as shown in Figure 29. The topmost sandstone layer, Layer #9, is of most interest since it has the

http://dx.doi.org/10.5772/57065

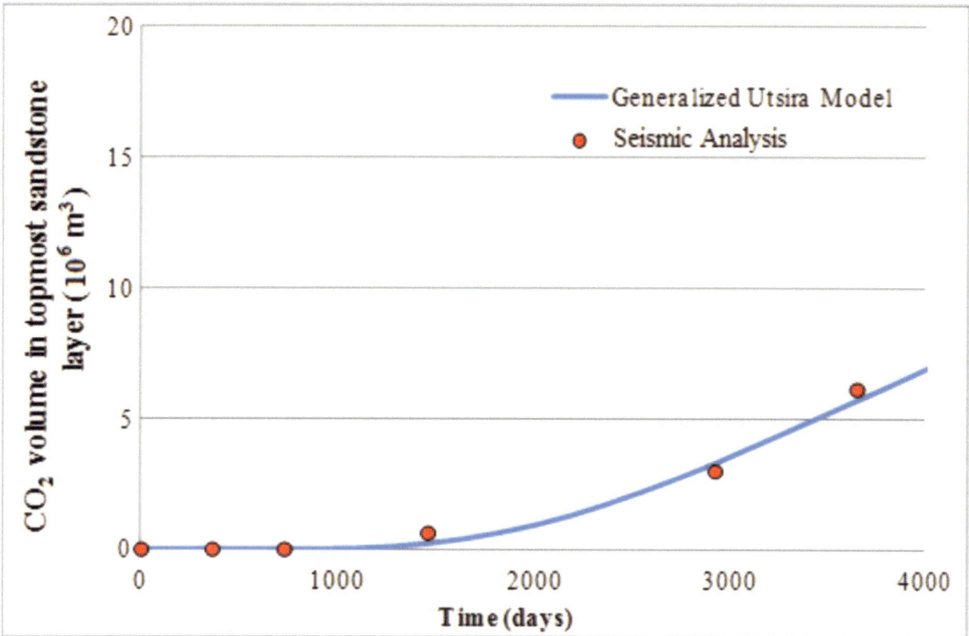

Figure 28. Gaseous CO_2 accumulation in the topmost sandstone layer

highest concentration of gaseous CO_2 and has direct contact with the overlying caprock formation. Seismic surveys have shown striking growth of CO_2 accumulation in Layer #9 between 1999 and 2006 as shown in Figure 29 [26].

The black dot in Figure 29 marks the location of the injection well, which is roughly 200 m under Layer #9. Two distinct local CO_2 accumulations appeared after about three years of injection (recall that injection began in 1996), indicating CO_2 began to accumulate under the caprock. However, CO_2 migration in Layer #9 was not symmetric due to the topography of the caprock. The northward migration of initially impacted CO_2, seen as the "body" of the plume in Figure 29, implies a local topographic dome; a prominent north-tending migration, seen as the "finger" of the plume in Figure 30. It implies the spill of local structurally trapped CO_2 along a north-tending topographic ridge. CO_2 migration along the north-tending ridge was rather fast at about 1 m/day between 2001 and 2004 [24].

In order to examine the plume's evolution within the topmost layer more closely, a 3D model of Utsira Layer #9 is created with detailed topography. It should be noted that only Layer #9, not the entire depth, is modeled because of the following considerations. To ensure the accurate capture of topographic effect on plume shaping, a computational domain with considerable fine mesh resolution has to be modeled based on geological survey data. The computational effort and thus the feasibility of highly detailed model of the entire Utsira formation is very intensive and time consuming. Secondly, CO_2 has to breakthrough several layers of relatively low-permeability shale prior to reaching the topmost layer. While it is difficult to quantify the breakthrough of gaseous CO_2, the quantification of CO_2 feeding into the topmost layer (Layer

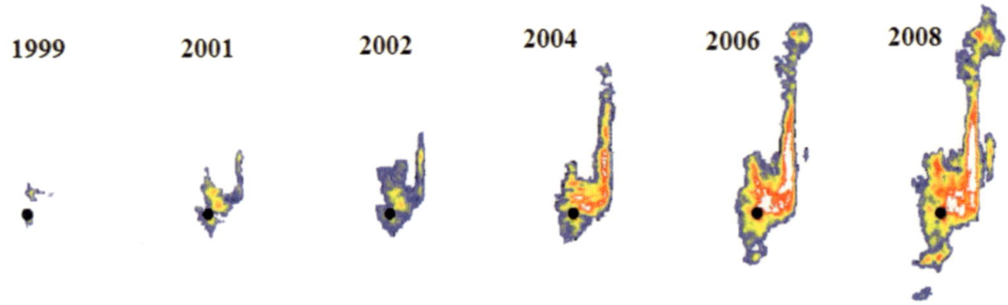

Figure 29. Amplitude maps of Layer #9 from 1999 to 2008, from Singh et al. [26]

#9) is rather reliable. Therefore, a model of only the topmost layer (Layer #9) could provide an ideal platform to investigate the effect of various parameters such as topography on the shaping of the CO_2 plume, and also could be used for optimization purpose to achieve high efficiency sequestration while maintaining an affordable computational effort and cost.

A reservoir model with dimension of 1600 m × 4900 m with varying thickness was constructed. It covers the portion of the Utsira formation where the plume, shown in Figure 31, is located. As mentioned earlier, the topography of this portion of the Utsira formation is accurately modeled based on seismic geological survey data (provided by Zhu and Lu of the University of Indiana [27]) with 50 m × 50 m mesh resolution. Because only Layer #9 is modeled, the thickness of the computational domain varies from 3.5 m to 26.3 m with an average thickness of 11.3 m. However, to accurately capture the accumulation and upward and lateral movement of CO_2, 37 layers are used along the thickness. The topmost layer and the bottom two layers are designated to represent the low permeability shale, while the 34 layers in the middle are assigned the properties of mudstone. In the 3D Layer #9 model, permeability anisotropy is considered with west-east permeability of 2 mDarcy, north-south permeability of 10 mDarcy, and vertical permeability of 200 mDarcy. A 3D overview of the Layer #9 model is shown in Figure 24.

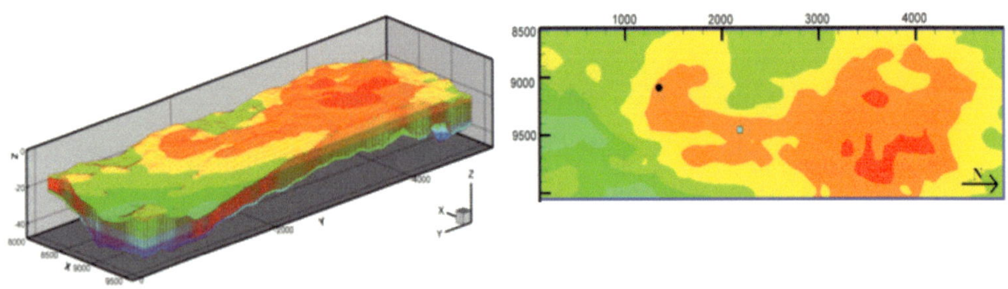

Figure 30. 3D overview and plan view of the 3D Layer #9 model of Utsira indicating feeder locations (black dot: main feeder; cyan square: secondary feeder)

Table 16 summarizes the hydrogeological properties of the Layer #9 model.

Temperature	33 °C
Pressure	8.6 MPa
Total Utsira Formation Area	26100 km^2
Total Utsira Formation Thickness	50 m ~ 300 m
Layer#9 Area	1600 m x 4900 m
Layer#9 Thickness	3.5 m ~ 26.3 m
Shale Permeability	Horizontal: 0.001 mDarcy, Vertical: 0.0001 mDarcy
Mudstone Permeability	W-E: 2 mDarcy, N-S: 10 mDarcy, Vertical: 200 mDarcy
Utsira Porosity (shale/mudstone)	35.7 %
Residual CO_2 Saturation	0.02
Residual Brine Saturation	0.11
Relative Permeability Type	Corey/van Genuchten-Muller
Capillary Pressure	None
Porewater Salinity	3.3 %
North-south Geological Slope	8.2 m/km, 5.8 m/km
CO_2 Feeder Location	Main feeder: W-E: 516 m, N-S: 1210 m, bottom mudstone Secondary feeder: W-E: 925 m, N-S: 2250 m, bottom mudstone

Table 16. Hydrogeological properties of the Utsira Layer #9 model

It should be noted that in the 3D Layer #9 model, the source of CO_2 is identified as "feeder" but not "injector" to emphasize that CO_2 is supplied from the lower aquifer through leakage pathways rather than by direct injection. Since the actual CO_2 injector is located at about 200 m under Layer #9, information on the injection rate recorded at the injector is not applicable for the CO_2 feeders in the Layer #9 model. To determine the CO_2 feeding rate to Layer #9, seismic surveys of CO_2 distribution are used to obtain its volume under in situ conditions, and then converted to mass flow rate. Information of CO_2 accumulative mass provided by Zhu and Lu [28] is summarized inTable 17.

Year	Accumulated Mass (kg)	Yearly Feeding Mass (kg)	Feeding Rate (kg/s)
1999	0.00	0.00	0.00
2000	1.82×10^7	1.82×10^7	0.577
2001	5.52×10^7	3.70×10^7	1.17
2002	9.49×10^7	3.97×10^7	1.26
2003	1.45×10^8	5.01×10^7	1.59
2004	2.13×10^8	6.80×10^7	2.16
2005	3.07×10^8	9.40×10^7	2.98
2006	4.34×10^8	1.27×10^8	4.03
2007	6.03×10^8	1.69×10^8	5.36
2008	8.20×10^8	2.17×10^8	6.88

Table 17. Accumulated CO_2 mass in Layer #9, 1999-2008

Table 17 gives the CO_2 accumulation in Layer #9 from 1999 to 2008. It can be seen that CO_2 feeding rate to Layer #9 keeps on increasing for the recorded nine years. Recalling the analysis of secondary sealing effect given for the previous cases, it is the pressure gradient between the gaseous CO_2 phase pressure at lower aquifer and the capillary pressure of the overlying shale layer that determines the breakthrough of CO_2 and its flow rate. When breakthrough first occurs, the pressure gradient just breaks the equilibrium state, resulting in relatively low breakthrough mass flux to Layer #9. However, as more CO_2 accumulates, the pressure gradient gradually increases and leads to increasing breakthrough mass flux, as depicted in Table 17. A nine-year average feeding rate of about 2.89 kg/s can be obtained from Table 17. Both the nine-year average value and the values in Table 17 have been considered in the simulations.

The significant north-tending plume finger is rather perplexing for regular pressure-gradient driven Darcy flow. Analysis suggests three possible explanations for the cause of the prominent north-tending CO_2 fingering along the ridge: (1) significantly higher permeability applied to the ridge; (2) existence of north-south geological slope which enhances the buoyancy-drive migration along the ridge; and (3) existence of a secondary (or even multiple) CO_2 pathways under the ridge. The hypothesis of significantly higher permeability at the ridge can be easily ruled out since no such evidence is obtained from the geological survey. It is still under debate whether geological slope should be considered when analyzing the CO_2 migration in the Utsira formation. Chadwick and Noy's work [24] suggested two possible values of geological slope based on the seismic images of the cross-section of the Utsira formation, which are 8.2 m/km and 5.8 m/km.

The simulation time is set at nine years, which corresponds to the injection period of 1999~2008. CO_2 plume migration at the topmost layer is examined for each year. Considering all the uncertainties mentioned above, a total of nine simulations are performed until good history-matching is obtained, as shown in Figure 31.

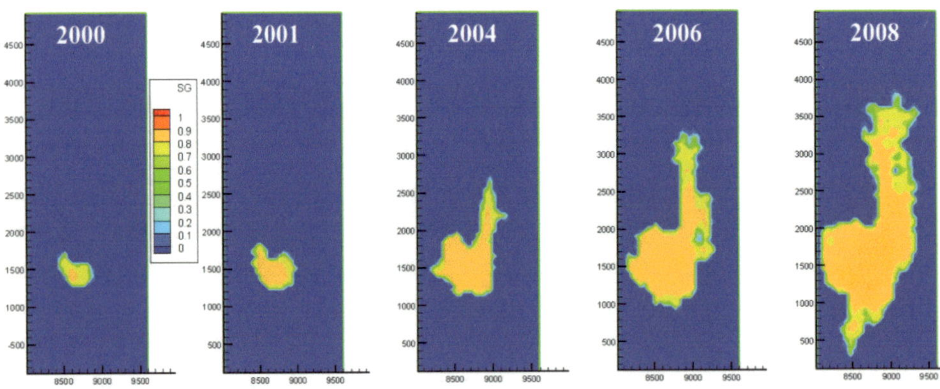

Figure 31. Simulated CO_2 migration in Layer #9, 2000 ~ 2008

Both the 2D generalized Utsira formation model and the 3D detailed Utsira Layer #9 model have generated satisfactory simulation results, as seen by history-matching. In summary, five major conclusions can be reached, as follows. First, the simulations show that the permeability anisotropy should be accurately modeled. Vertical-to-horizontal anisotropy close to 10:1 is needed to accurately capture the upward migration of CO_2. Horizontal anisotropy of 2:10 is needed to capture the northern spill of CO_2 into the north-tending ridge. Second, a secondary feeder is likely to exist directly under the north-tending ridge to generate sufficient plume migration along the ridge. This suggests multiple pathways for CO_2 breakthrough from the lower aquifer structure. Third, the fact that the injection gas is a CO_2-methane mixture is very important in modeling since the presence of 2% methane enhances the buoyancy. Fourth, it is critical that time-dependent feeding of CO_2 be modeled. This is consistent with the behavior of CO_2 path flow breaking the capillary pressure barrier, as noted before for the secondary-sealing effect in the case of the Mt.Simon formation. And finally, simulation results suggest strong mobility of gaseous CO_2 under the caprock (shale) without major leakage, implying that the caprock serves quite well as the non-permeable CO_2 barrier while exerting little resistance for the lateral flow of CO_2 underneath.

The simulation studies of the three actual large-scale identified deep saline aquifers conclude this chapter. These studies have provided important insights and best practices for obtaining accurate simulations of GCS using TOUGH2. In Chapter titled," Optimization of CO2 Sequestration Saline Aquifers", innovative reservoir techniques and their optimization, for more efficient and secured SAGCS operations, are discussed.

6. Concluding remarks

In this chapter, some key factors relevant to saline aquifer geological carbon sequestration (SAGCS) have been investigated. First, numerical simulations have been performed for completed/ongoing SAGCS projects in three large scale identified saline formations using the DOE numerical simulator TOUGH2. Before performing these studies, TOUGH2 was validated against the available analytical solutions and the benchmark numerical test cases. The three case studies of SAGCS in large-scale saline formations have provided important insights into the reservoir performance and sequestration uncertainties.

Acknowledgements

Financial support for this work was provided by the Consortium for Clean Coal Utilization (CCCU) at Washington University in St. Louis, MO, USA.

Author details

Zheming Zhang and Ramesh K. Agarwal

Department of Mechanical Engineering and Materials Science, Washington University in St. Louis, St. Louis, USA

References

[1] US Department of Energy. The 2012 United States Carbon Utilization and Storage Atlas – Fourth Edition; 2008.http://www.netl.doe.gov/technologies/carbon_seq/refshelf/atlasIV/ (accessed 24 May 2013).

[2] Pruess K. TOUGH2: A General Numerical Simulator for Multiphase Fluid and Heat Flow, LawrenceBerkeley Laboratory Report LBL-29400, California: Berkeley; 1999.

[3] Pruess K., Oldenburg C., and Moridis G. TOUGH2 User's Guide, Version 2.0 (revised), Lawrence Berkeley Laboratory Report LBL-43134, California, Berkeley; 2011.

[4] Celia M.A. and Nordbotten J.M. Practical Modeling Approaches for Geological Storage of Carbon Dioxide. Underground Water 2009; 47(5) 627-638.

[5] Celia M.A. and Nordbotten J.M. How Simple Can We Make Models for CO_2 Injection, Migration, and Leakage? Energy Procedia 2001; 4 3857-3864.

[6] PetraSim User Manual, Thunderhead Engineering website, http://www.thunderheadeng.com/downloads/petrasim/PetraSim-4-manual.pdf(accessed 24 May 2013).

[7] PetraSim Example Manual, Thunderhead Engineering website, http://www.thunderheadeng.com/downloads/petrasim/4/examples/PetraSimExamples.pdf(accessed 24 May 2013).

[8] Workshop on Numerical Models for Carbon Dioxide Storage in Geological Formations, University of Stuttgart website, http://www.iws.uni-stuttgart.de/CO2-workshop/(accessed 24 May 2013).

[9] Ebigbo A., Nordbotten J.M., and Class H. Numerical Investigations of CO_2 Sequestration in Geological Formations: Problem-Oriented Benchmarks, Problem 1, CO_2Plume Evolution and Leakage through an Abandoned Well. R&D-Program of GEOTECHNOLOGIEN website, http://www.hydrosys.uni-stuttgart.de/co2-workshop/Problem1.pdf(accessed 24 May 2013).

[10] Ebigbo A., Nordbotten J.M., and Class H. Numerical Investigations of CO_2 Sequestration in Geological Formations: Problem-Oriented Benchmarks, Problem 2, Enhanced CH4 Recovery in Combination with CO_2Storage in Depleted Gas Reservoirs. R&D-Program of GEOTECHNOLOGIEN website, http://www.hydrosys.uni-stuttgart.de/co2-workshop/Problem2.pdf(accessed 24 May 2013).

[11] Ebigbo A., Nordbotten J.M., Class H. and Eigestad G. Numerical Investigations of CO_2 Sequestration in Geological Formations: Problem-Oriented Benchmarks, Problem 3, Estimation of the CO_2 Storage Capacity of a Geological Formation. R&D-Program of GEOTECHNOLOGIEN website, http://www.hydrosys.uni-stuttgart.de/co2-workshop/Problem3.pdf (accessed 24 May 2013).

[12] Class H., Ebigbo A., Rainer H., Dahle H.K., Nordbotten J.M., Celia M.A., Audigane P., Darcis M., Ennis-King J., and Fan Y. A Benchmark Study on Problems Related to CO_2 Storage in Geologic Formations. Computational Geosciences 2009; 13409-434.

[13] Bachu S., Nordbotten J.M., and Celia M.A. Evaluation of the Spread of Acid-Gas Plumes Injected in Deep Saline Aquifers in Western Canada as an Analogue for CO_2 Injection into Continental Sedimentary Basins.In: 7thInternational Conference on Greenhouse Gas Control Technologies, Vancouver, Canada, September 2004.

[14] Celia M.A., Bachu S., Nordbotten J.M., Gasda S., and Dahle H. Quantitative Estimation of CO2 Leakage from Geological Storage: Analytical Models, Numerical Models, and Data Needs. In: 7thInternational Conference on Greenhouse Gas Control Technologies, Vancouver, Canada, September 2004.

[15] Nordbotten J.M., Celia M.A., and Bachu S. Injection and Storage of CO_2 in Deep Saline Aquifers: Analytical Solution for CO_2 Plume Evolution during Injection. Transport in Porous Media 2005; 58(3) 339-360.

[16] Leetaru H.E., Frailey S.M., Damico J., Finley R., McBride J.H., and Morse D.G. Developing a Geological Model for the Phase III (ADM) Saline Sequestration Validation Site. In: 7th Annual Conference on Carbon Capture and Sequestration, Pittsburgh, US,May 2008.

[17] Barnes D.A., Bacon D.H., and Kelley S.R. Geological Sequestration of Carbon Dioxide in the Cambrian Mount Simon Sandstone: Regional Storage Capacity, Site Characterization, and Large Scale Injection Feasibility, Michigan Basin. Environmental Geosciences 2009; 16(3) 163-183.

[18] Midwest Regional Carbon Sequestration Partnership, CO_2 Injection Test in the Cambrian-Age Mt. Simon Formation, Battelle Final Report, April 29, 2011.http://addap.tk/userdata/phase_II_reports/report_only/east_bend_report_final-mainreport.pdf (accessed 24 May 2013).

[19] Zhou Q., Birkholzer J.T., Mehnert E., Lin Y., and Zhang K. Modeling Basin- and Plume-Scale Processes of CO_2 Storage for Full-Scale Deployment. Ground Water 2010; 48(4) 494–514.

[20] Zhang Z. and Agarwal R. Numerical Simulation of Geological Carbon Sequestration in Saline Aquifers - three case studies. In: 12th Annual Conference on Carbon Capture and Sequestration, May 13-16 2013, Pittsburgh, PA, USA.

[21] Hovorka S.D., Benson S.M., Doughty C., Freifeld B.M., Sakurai S., Daley T.M., Kharaka Y.K., Holtz M.H., Trautz R.C., Nance H.S., Meyer L.R., and Knauss K.G. Measuring Permanence of CO_2 Storage in Saline Formations—the Frio Experiment. Environmental Geosciences 2006; 13(2) 105–121.

[22] Doughty C., Freifeld B.M., and Trautz R.C. Site Characterization for CO_2 Geologic Storage and Vice Versa: The Frio Brine Pilot, Texas, USA as A Case Study. Environmental Geology 2008; 54 1635–1656.

[23] Bachu S. and Bennion B. Effects of In-situ Conditions on Relative Permeability Characteristics of CO_2-Brine Systems. Environmental Geology 2007; 54 1635–1656.

[24] Chadwick R.A. and Noy D.J. History- Matching Flow Simulations and Time-Lapse Seismic Data from the Sleipner CO_2 Plume. In: Proceedings of the 7th Petroleum Geology Conference, 2010; 7 1171–1182

[25] Audigane P., Gaus I., Czernichowski-Lauriol I., Pruess K., and Xu T. Two-Dimensional Reactive Transport Modeling of CO_2 Injection in a Saline Aquifer at the Sleipner Site. American Journal of Science 2007; 307 974-1008.

[26] Singh V., Cavanagh A., Hansen H., Nazarian B., Iding M., and Ringrose P. Reservoir Modeling of CO_2 Plume Behavior Calibrated Against Monitoring Data from Sleipner, Norway. Society of Petroleum Engineers 2010; 134891-PP.

[27] Zhu C. CO_2-Water-Rock Interaction in Geological Carbon Sequestration, Seminar presentation at Washington University in St. Louis, 2011.

[28] Zhu C. and Lu P. Personal Communication, Department of Geological Sciences, University of Indiana, 2012.

Chapter 12

Seismic Reflectivity in Carbon Dioxide Accumulations: A Review

Claudia L. Ravazzoli and Julián L. Gómez

Additional information is available at the end of the chapter

http://dx.doi.org/10.5772/57087

Introduction

It is widely recognized that the continuous increase in the concentration of carbon dioxide (CO_2) in our atmosphere is a major cause of global climate change [1]. Consequently, in accordance with the objectives of the Kyoto Protocol, many carbon dioxide capture and storage projects are being developed worldwide to reduce and stabilize the emission of this greenhouse gas into the atmosphere. The geological storage of CO_2 in many cases becomes a feasible option to accomplish this goal, giving rise to the science of *carbon sequestration*, a challenging task for governments, scientists and engineers [2].

The main targets for geological disposal of CO_2 are depleted hydrocarbon reservoirs and deep saline aquifers. While the latter are more numerous, their characterization requires detailed studies because they are typically not explored for prospecting. The geological sequestration of carbon dioxide requires careful prior study and subsequent monitoring to prevent this gas from leaking to the surface. In general, the significant contrast between the physical properties of natural reservoir fluids and those of carbon dioxide allows the utilization of time lapse seismic methods to monitor the evolution of the injected CO_2 volume.

While it is accepted that time lapse surveys are able to monitor the presence or absence of CO_2 within a geological formation, their ability to quantify the saturation, state and volume of this fluid within the reservoir still needs research efforts from the geophysical community. This makes it necessary to search for reliable seismic indicators. Theoretical and numerical modeling provides, in this sense, a tool to explore useful correlations between the seismic response and relevant parameters of the CO_2 repository.

The significance of the *seismic reflectivity* has long been recognized by many authors, resulting in the publication of numerous works for different constitutive models and applications. In close connection with this, the analysis of seismic amplitude variation with angle, hereafter called *AVA* is frequently used in reservoir geophysics to obtain information about the rocks and pore fluids. The behavior of the reservoir reflectivity for different overall CO_2 saturations, distribution types, layer thicknesses, frequencies and CO_2 physical states controls the amplitude of seismic wave reflections and strongly conditions the detectability of the injected CO_2 volume, as reported by [3],[4] and [5] among others.

With this idea, using rock physics models, in this chapter we focus on the theoretical analysis of the properties and variations of the seismic reflectivity and related parameters in a porous rock partially saturated with CO_2 and brine. In our tests we model the seismic effects of different CO_2 accumulations formed below impermeable seals.

We calibrate our models using information on the Utsira formation, a shallow saline aquifer in the Sleipner field (offshore Norway). Millions of tonnes of CO_2 have been separated since October 1996 from natural gas and re-injected into the aquifer, consisting of a highly porous and unconsolidated sandstone, with several thin intra-reservoir shale layers. These shale intervals act as temporary seals causing accumulations of CO_2 beneath them [6],[7], whose saturations increase with time. A 50-100 m thick high velocity layer overlies the sand, forming the reservoir caprock, known as the Nørland formation. It is mainly composed of shales, siltstones and mudstones [8]. Another high-velocity shale layer of about 5 m thick is located about 10 m below the top of the sand.

As pointed out by [9] the *topmost* CO_2 layer is of special interest for monitoring the Sleipner injection site, since its growth reveals the total upward flux and its changes over time. Thus, the analysis of the reflectivity associated with this kind of CO_2 accumulation is an important topic in CO_2 sequestration problems. The present book chapter focuses on this subject and is structured in three main sections considering different aspects of the problem.

First, in Section 1 we present a description of the models and rock physics tools used in the subsequent sections. Next, in Section 2 we model the compressional P-wave reflection coefficient at the interface between a caprock and a permeable porous layer partially saturated by a mixture of brine and CO_2 under liquid, supercritical and gaseous conditions. For this analysis we consider a simple model consisting of two halfspaces, combining a fluid substitution procedure and a Gassmann-Hill formulation to take into account variable spatial distributions of the fluids. We perform a sensitivity analysis of the standard AVA coefficients (intercept, gradient and curvature) in the near offset range, to investigate their correlations with CO_2 saturation and its thermodynamic state within the geologic formation. Also, we analyze the effect of modeling the CO_2 by means of simple and complex equations of state.

The two-halfspace representation considered in Section 2 is valid when the accumulation thicknesses are larger than the involved seismic wavelengths. When this assumption is not valid, the interference between the multiple waves reflected at the different boundaries give rise to strong frequency dependence, *reflectivity dispersion* and *tuning* effects which must be taken into account. To do this, in Section 3 we compute the *generalized* P-wave reflection coefficients by means of a reflectivity method. This leads us to the study of frequency dependent AVA and *amplitude vs. frequency AVF*, which are topics of great interest due to the increasing use of spectral decomposition techniques on seismic data. This also allows us to analyze the pattern of maxima in the reflectivity amplitude and their associated *peak frequencies*, which are strongly related to the thickness of the CO_2 layer.

Seismic monitoring of geological CO_2 injection sites is mostly based on the seismic reflections coming from high saturation CO_2 accumulations. This is due to their large seismic amplitudes and good signal to noise ratio. However, low-saturation zones with dispersed CO_2, or *saturation transition zones* may have an important role in the propagation of waves within the reservoir, giving rise to amplitude and phase changes of the seismic signals. Transition zones have been studied by different authors in the geophysical literature considering linear velocity trends with depth and constant density [10], [11]. Therefore, using the parameters of the Sleipner field, in Section 3.3 we model the reflectivity response of a CO_2 transition layer, defined by a given linear vertical CO_2 saturation profile, which results in a non-linear velocity trend with depth.

In general, emphasis is placed on establishing correlations between the seismic reflectivity and related attributes (intercept, gradient, curvature, peak frequencies) and the overall CO_2 saturation, its physical state and thickness of the accumulation. These results are intended to help in the understanding of the expectable variations in a seismic time lapse study. They can be extended to other CO_2 repositories with proper calibration of the rock and fluid properties.

Many of the results and procedures presented in this chapter are a revision and extension of those published by us in [4], [12], [13], [14] and [15].

1. Theoretical framework and assumptions

In this section we summarize the modeling tools used later for the applications. First we explain the procedure for the calculation of the elastic properties of the partially saturated rocks for variable CO_2 saturation. Next we describe the laws for the computation of the physical properties of the CO_2 and brine for the different temperature and pressure states. Finally, we outline the reflectivity method for the computation of the generalized seismic reflectivity of the layered model

1.1. Elastic properties of CO_2 bearing rocks

To begin with the description of the model we consider that after the injection, a volume of carbon dioxide occupies part of the pore volume of a geologic reservoir which at the pre-injection state was fully saturated with brine. For simplicity we assume that CO_2 displaces, without dissolution, part of the *in-situ* brine giving rise to a two-phase fluid saturation. From now on the corresponding saturations of free CO_2 and brine are denoted as S_g and S_{br}, respectively, so that $S_{br} + S_g = 1$.

For the study of partially saturated rocks it is necessary to compute the bulk density and elastic coefficients (bulk and shear modulus) of the fluid saturated medium. The bulk density ρ is given by

$$\rho = (1 - \phi)\rho_s + \phi\,(S_{br}\,\rho_{br} + S_g\,\rho_g)\,, \tag{1}$$

where ϕ is the rock porosity, ρ_s is the mineral grain density and ρ_g, ρ_{br} are CO_2 and brine densities, respectively.

The formulation and solution of the energy and amplitude splitting problem when a monochromatic plane compressional wave strikes obliquely at a plane interface between two porous saturated media has been discussed by various authors, such as [16], [17] and [18]. In those papers the mechanical behavior of the porous rock was described using the classic constitutive relations and equations of motion of Biot [19],[20]. In the following sections our aim is to study the reflection of elastic waves and related parameters, for frequencies f

within the common seismic range $f \leqslant 120Hz$. When a low frequency seismic wave propagates through a fluid saturated rock, due to fluid viscosity the solid and the fluid move in phase, so the medium behaves as an *effective* medium.

One of the most significant models to estimate the effective bulk modulus and the seismic velocity of a porous fluid saturated rock is to use Gassmann's relations [21] which can be written in the form:

$$K^G = K_s \left(\frac{K_m + Q}{K_s + Q} \right), \quad \text{with } Q = \frac{K_f (K_s - K_m)}{\phi (K_s - K_f)}. \tag{2}$$

In this equation the mechanical behavior of the fluid saturated porous medium is assumed to be elastic and isotropic. The coefficients K_m, K_s and K_f are the bulk modulus of the dry matrix, the mineral grains and the pore fluid, respectively. The physical properties of the two-phase brine-CO_2 fluid are computed using an *effective fluid* whose density is given by their weighted average and its compressibility by the isostress Reuss average of individual fluid bulk moduli [22]:

$$\frac{1}{K_f} = \frac{S_{br}}{K_{br}} + \frac{S_g}{K_g}, \tag{3}$$

where K_{br} and K_g are the bulk moduli of brine and CO_2. In these equations, it is assumed that the mixture of CO_2 and brine at the pore scale can be treated as a viscous single phase effective fluid. However, as pointed out in [23], this approach is strictly valid only when the pore fluids are uniformly mixed at very small scales so that the different wave-induced pressure increments in each fluid have time to diffuse and equilibrate during a seismic period. This critical distance is the so-called *diffusion length*. The presence of fluid heterogeneities over scales greater than this length (in which wave induced pore pressure gradients cannot equilibrate quickly), gives rise to a *patchy saturation*. It has been shown that for low frequencies such as those used in seismic exploration the effective modulus of a rock with patches of brine and CO_2 of arbitrary geometry is given by [24],[23]

$$\frac{1}{(K^P + \frac{4}{3}\mu)} = \frac{S_{br}}{K_{br}^G + \frac{4}{3}\mu} + \frac{S_g}{K_g^G + \frac{4}{3}\mu}; \tag{4}$$

where K_{br}^G and K_g^G are the Gassmann's moduli (2) of the rock fully saturated with brine and CO_2. The coefficient μ denotes the shear modulus of the rock, equal to that of the rock matrix, since the rigidity of a rock does not change due to the saturant fluid.

The existence of heterogeneous fluid distribution may give rise to mesoscopic wave attenuation and velocity dispersion phenomena in the seismic frequency range. There are many studies about numerical modeling of these effects, e.g. [25],[26],[27], showing that they are strongly dependent on the shapes and characteristic lengths of the patches. Given that these parameters are rarely known, to avoid dealing with this uncertainty, in our analysis we assume elastic layers where no attenuation-dispersion phenomena, associated with the patchy fluid distribution take place. However, these effects can be included in our formulation using appropriate viscoelastic models in the frequency domain. Because of the classic *correspondence principle*, this implies the replacement of the real constant elastic moduli by complex frequency dependent moduli [20].

The patchy (Hill) and uniform (Gassmann) fluid distributions are respectively upper and lower bounds for the seismic wave velocities [23]. Taking into account the uncertainties in the knowledge of the *in-situ* CO_2 - brine distribution, in our models we assume that for each saturation it is reasonable to compute the average between both velocities.

It is important to mention the assumption of no chemical interactions between the pore fluids and the frame, which allows us to employ the fluid substitution procedure to consider that the pore space is saturated by brine and CO_2 in variable proportions. Also, since the amount of CO_2 dissolved in brine is a negligible fraction [28],[29], we do not take this effect into account. Nevertheless, when CO_2 is injected in oil reservoirs this effect can not be neglected and should be included in the physical models, as shown in [4] and [28].

As pointed out in [8], no systematic variations in fluid pressure are observed in the Sleipner field. Thus, variations of rock elastic properties with effective pressure (related to the difference between confining and pore pressure), are not taken into account in the model. However, using appropriate effective pressure laws these effects can be included in the computations, as explained in [18].

1.2. Physical properties of carbon dioxide and brine

Depending on the *in-situ* pressure and temperature conditions the CO_2 can exist in the subsurface in different phases. We recall that the *critical point* for CO_2 occurs at a temperature $T_c = 31.1$ °C and a pressure $P_c = 7.39$ MPa. For temperatures higher than T_c and pressures higher than P_c the carbon dioxide is in a *supercritical* state, where it is compressible like a gas but with the density of a liquid. This characteristic of CO_2 is particularly relevant for its underground storage since supercritical CO_2 can fill the available pore volume with minimum buoyancy effects [30]. Temperatures and pressures near the critical point commonly occur in applications involving CO_2, such as enhanced oil recovery techniques and sequestration projects [31]. However, as pointed out by [30], the depth at which CO_2 supercritical conditions are present is highly variable and strongly dependent on surface temperature and geothermal gradients, even within a single basin. In addition, the pressure regime of the basin (normal or abnormal), is also very important and is related to its geologic history, existence of sealing faults, permeability barriers and the occurrence of overpressure generation mechanisms [32].

For the examples, the density and bulk moduli of brine for given *in-situ* temperature and pressure conditions are computed using the semi-empirical relations proposed by Batzle and Wang (BW) [29]. For the computations we consider a typical brine salinity of 50000 ppm. The corresponding properties of CO_2 can be computed using some of the many equations of state (EoS) developed for real gases such as the two-parameter van der Waals (vW) equation [35] and the Peng and Robinson (PR) equation [34], or some more specific one such as Duan, Moller and Weare (DMW) [33], which involves fifteen parameters. To our knowledge, there is no full agreement in the geophysical literature about which EoS is the most appropriate to represent CO_2 properties under the thermodynamic conditions found in geologic reservoirs. Therefore, we use different models to analyze the effect of the EoS in the seismic magnitudes of interest. With that purpose we selected three different temperature

Physical state EoS	CO_2 DMW	CO_2 PR	CO_2 vW	Brine BW
Gaseous				
$T = 40°$ C	$K_g = 0.0049$	$K_g = 0.0089$	$K_g = 0.0050$	$K_{br} = 2.5986$ GPa
$P = 6$ MPa	$\rho_g = 149.8$	$\rho_g = 153.2$	$\rho_g = 142.1$	$\rho_{br} = 1028.7$ kg/m^3
Supercritical				
$T = 36°$ C	$K_g = 0.037245$	$K_g = 0.026120$	$K_g = 0.01999$	$K_{br} = 2.6234$ GPa
$P = 10$ MPa	$\rho_g = 637.6$	$\rho_g = 706.4$	$\rho_g = 490.7$	$\rho_{br} = 1030.4$ kg/m^3
Liquid				
$T = 20°$ C	$K_g = 0.0931$	$K_g = 0.1376$	$K_g = 0.0464$	$K_{br} = 2.5009$ GPa
$P = 10$ MPa	$\rho_g = 853.5$	$\rho_g = 833.1$	$\rho_g = 565.2$	$\rho_{br} = 1036.07$ kg/m^3

Table 1. Physical properties of CO_2 and brine under different thermodynamic conditions. The density and bulk moduli of CO_2 are obtained using three equations of state: DMW [33], PR [34] and vW [35]. The CO_2 bulk moduli estimations are corrected for adiabatic conditions. The properties of brine are computed using BW laws [29] for a salinity of 50000 ppm.

and pressure (T, P) values: (40°C, 6 MPa) corresponding to gaseous state, (36°C, 10 MPa) corresponding to supercritical state and (20°C, 10 MPa) corresponding to liquid state. The resulting estimations are shown in Table 1 and will be used in the following sections. The temperature and pressure of the supercritical state are in agreement with the conditions reported in [8] for Utsira sandstone.

It must be pointed out that the different CO_2 bulk moduli estimations in Table 1, to be used for the computations, were corrected to represent adiabatic conditions. The reason is that the fluid compression due to the passage of a wave is fast, so that the process is not isothermal [29], [37].

1.3. Calibration of the rock physics model

To calibrate our elastic model for the Utsira sandstone, we used the measured bulk density and the compressional and shear wave velocities in the preinjection state (i.e. under full brine saturation), given in [8] and [38]: ρ_b = 2050 kg/m^3, V_p =2050 m/s and V_s = 640 m/s. Using the parameters given for the pore water ρ_{br} = 1040 kg/m^3, K_{br} = 2.305 GPa and the average sandstone porosity $\phi = 0.37$, the mineral grain density results in ρ_s = 2643 kg/m^3. We also performed a Gassmann-inverse calculation to obtain: K_m = 2.569 GPa and μ = 0.84 GPa. For the shale layers we obtained the elastic parameters using the reported values: V_p = 2270 m/s, V_s = 850 m/s and ρ_b = 2100 kg/m^3.

1.4. The generalized reflectivity of a layered medium

The computation of the reflection and transmission coefficients of elastic waves propagating through layered media is a subject of interest in different fields such as seismology, seismic prospecting and underwater acoustics. The first numerical procedures date back to the pioneering works of Thomson [39] and Haskell [40], who proposed matrix methods that transfer stresses and displacements through successive layers. Since then, different approaches and applications have been presented.

In this work we perform an implementation of the *reflectivity method* [41], [42], which is based on the continuity of particle displacements and stress components at the interfaces between sets of plane layers embedded between two halfspaces. This method builds up the reflection and transmission matrices iteratively by starting at the top of a lower bounding halfspace and adding one layer per iteration until the total stack response is constructed. This procedure is intuitively simple and exact [43], taking into account all internal reverberations. The recursion algorithm is

$$\begin{aligned}\mathcal{R}_{j-1} &= R^d_{j-1} + T^u_{j-1}\bar{\mathcal{R}}_j(I - R^u_{j-1}\bar{\mathcal{R}}_j)^{-1}T^d_{j-1}, \\ \mathcal{T}_{j-1} &= \bar{\mathcal{T}}_j(I - R^u_{j-1}\bar{\mathcal{R}}_j)^{-1}T^d_{j-1}; \\ \bar{\mathcal{R}}_j &= E_j\mathcal{R}_jE_j, \\ \bar{\mathcal{T}}_j &= \mathcal{T}_jE_j,\end{aligned} \tag{5}$$

were $\mathcal{R}_j$ and $\mathcal{T}_j$ are the total-reflection and total-transmission matrices of interface j. $R^{u,d}_j$ and $T^{u,d}_j$ are the upward and downward layer reflection and transmission matrices respectively. These matrices depend on the bulk density, P- and SV-wave seismic velocities and ray angles from interface j. E_j is the layer phase-shift diagonal matrix and $(I - R^u_{j-1}\bar{\mathcal{R}}_j)^{-1}$ is known as the reverberation operator [44], with I being the identity matrix.

The recursion starts at the base of the layering $j = n$ with $\mathcal{R}_n = R^d_n = 0$ and $\mathcal{T}_n = T^d_n = I$ and continues to $j = 1$. When the top of the layering is reached, the generalized P-wave reflection coefficient for scaled displacements is obtained from $R_{pp}(f, \theta) = \mathcal{R}_{0[1,1]}$. This coefficient is a complex function of frequency f and incidence angle θ.

For a complete description of the matrices involved in this scheme the reader is referred to [44].

2. AVA analysis at the top of a CO_2 accumulation

In this section we model the seismic compressional wave reflection coefficient and related AVA parameters at the interface between Utsira sandstone, partially saturated by mixtures of CO_2 and brine, overlain by a shale caprock, with the properties described in Section 1.3.

Following the classic approach, for the present applications both media are considered as elastic halfspaces, an assumption valid for layers of thicknesses larger than the wavelengths of the incident waves. Under these conditions the reflection coefficient are real and frequency independent. In this particular configuration, our reflectivity method reproduces the classic results of Zoeppritz [45] for two halfspaces.

In the near offset domain $\theta < 60°$ or below the critical angle, we can assume that the R_{pp} reflection coefficient as a function of the incidence angle θ can be approximated using the expression of Shuey [46]:

$$R_{pp}(\theta) \approx A + B\ sin^2\theta + C\ \left(tan^2\theta - sin^2\theta\right); \tag{6}$$

where A, B and C are known as the *AVA coefficients*. A is called the *intercept*, B the *gradient* and C the *curvature* [47],[48]. The intercept is equal to the normal incidence reflection coefficient and is controlled by the contrast in acoustic impedance between both media. The gradient is related to the contrast in density and in compressional P and shear SV wave

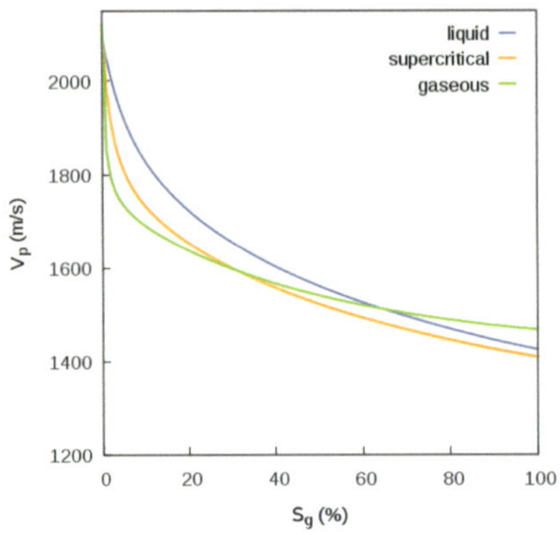

Figure 1. Compressional elastic wave velocity for variable CO_2 saturation S_g obtained using Gassmann-Hill average under different thermodynamic conditions.

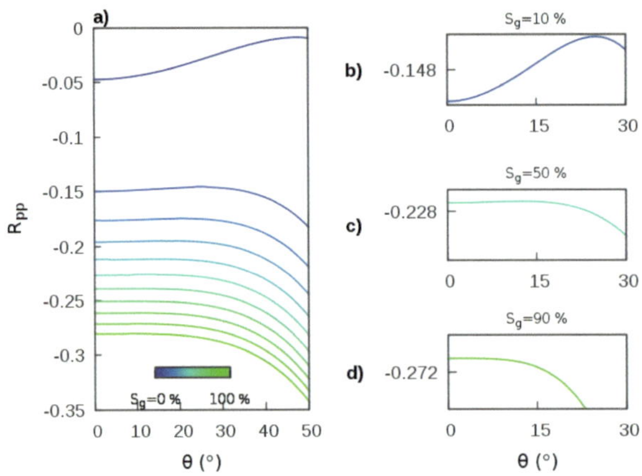

Figure 2. P-wave reflection coefficient vs. ray angle for different fixed partial CO_2 saturations under supercritical conditions.

velocities [49]. The curvature parameter is important at far offsets and near critical angles [44]. Equation (6) is used in this section to carry out a parametric analysis of the AVA coefficients or to study their sensitivity by implementing a standard fitting procedure on the results obtained for $R_{pp}(\theta)$.

In Figure 1 we plot the compressional elastic wave velocity for the partially saturated CO_2-brine Utsira sandstone in the low frequency range. The parameters of the CO_2 at the liquid, supercritical and gaseous states correspond to the PR EoS and are listed in Table 1, along with those of the brine. As mentioned in Section 1, our velocity model for the sandstone corresponds to the Gassmann-Hill average for each saturation state. It is important to note the decreasing tendency of the wave velocity for increasing CO_2 saturation, which becomes less marked as saturation increases, a feature that will also be observed in the AVA coefficients. The velocity model for the supercritical CO_2 is in close agreement with that presented in [8] and [38] for the Utsira sandstone. According to this model, the seismic velocity can discriminate the liquid from the other two physical states, particularly for saturations lower than about 40%.

Next, in Figure 2 a)-d) we show the compressional reflection coefficient versus incidence angle from 0° to 50°, for fixed values of CO_2 saturation in the range 0%–100%. The temperature is 36°C and pressure 10 MPa, corresponding to the supercritical conditions given in Table 1, for the PR equation. As expected, an abrupt change in the reflectivity from the pre-injection to the post-injection state is observed, due to the high contrast between the physical properties of CO_2 and brine in this state. As expected, the CO_2 saturation decreases the impedance of the lower medium with respect to that of the upper one, resulting in a negative reflection coefficient for normal incidence, becoming more negative for higher angles and higher saturations. These curves are consistent with those published in [50]. The curves corresponding to liquid and gaseous states show very similar behavior, and consequently they are not shown. This means that the discrimination of the physical state from the AVA behavior may be difficult. This subject will be analyzed in more detail in terms of the AVA coefficients. According to the well known AVA classification [48], [51], the curves for the

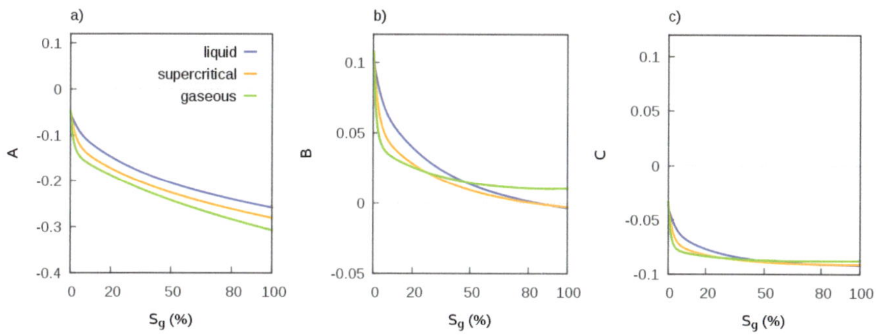

Figure 3. Behavior of AVA parameters intercept A, gradient B and curvature C versus CO_2 saturation in different physical states.

preinjection case and for $S_g = 10\%$ (in Figure 2 b) correspond to the AVA class IV. The set of AVA curves for higher saturations shows a very flat tendency for near angles (i.e. with very low gradients B), which can be observed in more detail in Figures 2 c) and d). Its classification will be discussed later in Figure 4.

To analyze the sensitivity of the AVA parameters we plot in Figure 3 the predicted variations of A, B and C with increasing saturation for the liquid, supercritical and gaseous conditions. We observe that the intercept A is negative, decreasing and monotone for the three physical states under consideration, being a sensitive attribute throughout the saturation range. The gradient B shows a trend similar to that of the intercept except for the gaseous state, in which stabilization is observed. This can be useful to distinguish this particular state from the other two.

The curvature attribute C, is the least sensitive of the three, both to physical state and saturation level. We only note appreciable changes for liquid CO_2 and for saturations in the range from 0 to about 30%. However, the correct determination of such changes may be strongly limited by the seismic resolution. In connection with this, Brown et al. [3] pointed out that AVA variations on the order of 5% are seismically detectable. Taking that numerical threshold into account we can state that, for this model, variations in the parameter C do not contain much information with respect to variations in the CO_2 content. Consequently, an AVO time-lapse analysis can be based solely on the coefficients A and B, where A is the most sensitive for the determination of the CO_2 saturation state. To inspect

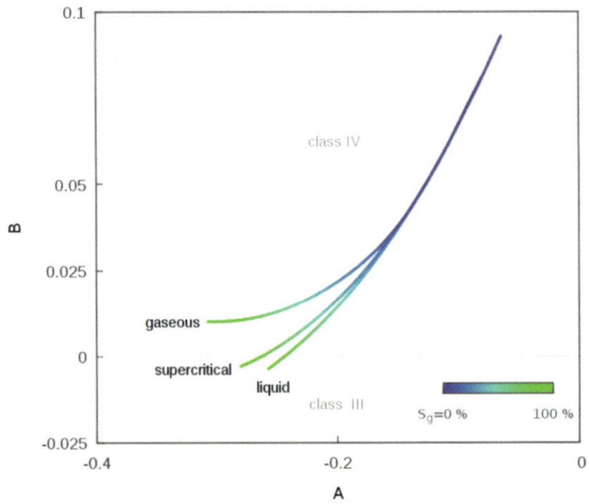

Figure 4. Intercept (A) vs. gradient (B) crossplot for variable saturation and physical states.

the behavior of the reflectivity in terms of the AVA classification, in Figure 4 we present typical intercept-gradient crossplots (B vs. A) to observe their trends for variable saturations and physical states. According to our model, while for gaseous CO_2 the AVA class is IV in all the saturation range, under liquid and supercritical conditions a change to class III can be expected for very high saturations. We also observe that since from intermediate to high gas saturations, the $B - A$ trajectories of each CO_2 state are different, a time-lapse crossplot analysis may bring useful information about the physical state of the underground CO_2. In order to analyze the influence of the state equations in our model estimates, in Figure 5 we display the coefficients A and B as a function of CO_2 saturation, for van der Waals [35], Peng & Robinson [34] and Duan et al. [33] EoS for the three physical states in Table 1. For gaseous and supercritical states the coefficients are very similar in all the saturation range. The most important discrepancies are observed when CO_2 is at liquid state. From these experiments we conclude that for seismic modeling applications the choice of the EoS will not significantly affect the results.

The sensitivity and robustness of the AVA coefficients to the presence of low fractions of methane (CH_4), which is a frequent impurity in CO_2 injection sites, was considered in [15].

3. Finite thickness CO_2 accumulations: frequency dependent AVA

So far we have modeled the CO_2 reflectivity response in terms of the simple single layer representation. However, due to the spatial variations in petrophysical properties and natural permeability barriers, the CO_2 tends to form accumulations of finite thickness.

In exploration geophysics, it is generally accepted that when the thickness of a layer is smaller than the predominant seismic wavelengths, the layer is said to be *thin*. Also, a common measure for the vertical seismic resolution is given by a quarter of the dominant wavelength [53]. As explained in [6],[8] and [54], the CO_2 accumulations within the Utsira sandstone in the Sleipner field can be regarded as thin layers. The interference effects between the

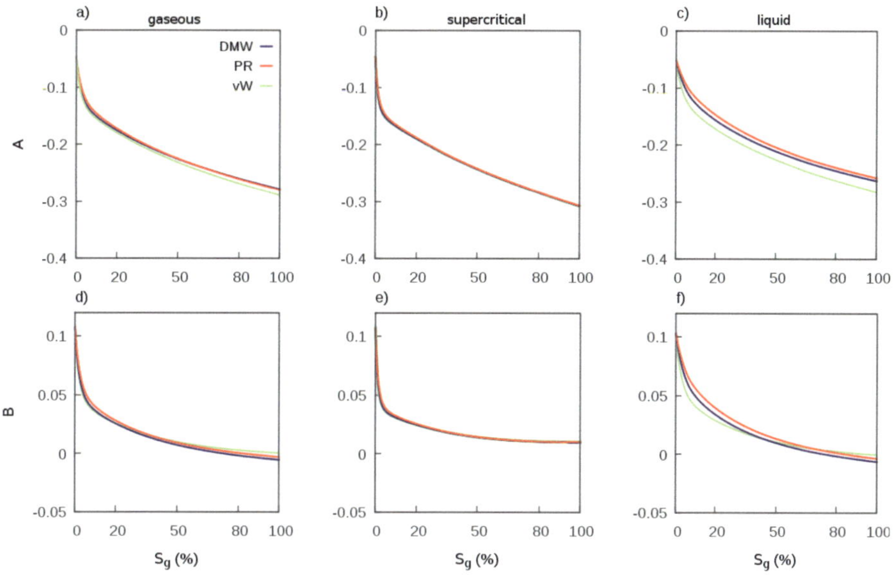

Figure 5. Intercept and gradient model estimates for variable saturation and physical states using different equations of state: DMW in blue, PR in red and vW in green lines.

multiple waves reflected at the different boundaries give rise to strong frequency dependence in the reflectivity of the medium, a phenomenon known as *reflectivity dispersion* [11],[55]. In this case the dispersive character has a geometric origin and is due to layer reverberation; however reflectivity dispersion can also take place at interfaces between highly dissipative media. The constructive interference between some frequency components of the up-going and down-going reflections at the physical boundaries gives rise to *tuning* effects. In the context of CO_2 sequestration these effects have been analyzed in several papers, such as [5], [8],[9], and [38].

As pointed out in [11] and [55], the modern use of time-frequency analysis and spectral decomposition tools [56] has shown that reflection events in practice are always frequency dependent. The spectral decomposition techniques allow obtaining and analyzing the seismic response of a medium at a given frequency. This makes it possible to investigate the reflected amplitude variation with angle at different frequencies (*spectral AVA*) and the amplitude variation with frequency (*AVF*) at a fixed angle.

At the time of this writing and to our knowledge, the AVF characteristics of CO_2 layers have not been modeled yet. With this motivation, in this section we investigate the existing correlations between reflectivity, frequency, saturation, physical state and thickness for a simple accumulation model. From these experiments our aim is to assess whether time lapse changes in the reflected amplitudes can be correlated with changes in such parameters, which are of utmost importance in CO_2 monitoring.

3.1. Description of the model

Here we present another simple but useful approach to model the reflectivity of the topmost CO_2 layer in the Sleipner field. For the computations we use the sandstone and caprock

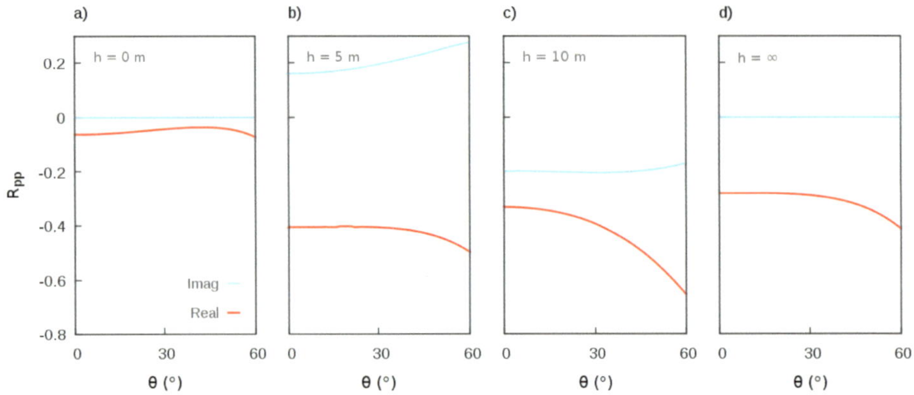

Figure 6. Generalized complex reflection coefficient R_{pp} vs. angle for f= 50 Hz, $S_g = 100\%$ and thicknesses $h = 0, 5, 10$ m and $h \to \infty$.

parameters described in Section 1.3. To simulate the supercritical conditions of the CO_2 injected into Utsira sandstone, carbon dioxide density and bulk moduli are calculated using the PR EoS at a temperature of 36°C and pressure of 10 MPa (Table 1).

According to the scheme used for the reflectivity method (described in Section 1.4), the upper halfspace plays the role of the caprock and the lower halfspace is the fully brine saturated Utsira sandstone. Below the caprock, the injected carbon dioxide is assumed to form a layer in the sandstone of thickness h with spatially constant CO_2 saturation S_g. The case of vertical variations in the CO_2 saturation or *saturation transitions* will be considered in Section 3.3. Thus for this model the generalized reflection coefficient is a complex, frequency-dependent function of the incidence angle, the CO_2 saturation and thickness h in the CO_2 bearing layer.

3.2. Behavior of reflectivity with saturation, frequency and thickness

In Figure 6 we display the generalized reflection coefficient R_{pp} vs. incidence angle for a frequency of $f = 50$ Hz, a saturation of $S_g = 100$ % and thicknesses of $h = 0, 5, 10$ m, considering also the limit $h \to \infty$. For this CO_2 saturation the P-wave velocity (see Figure 1) is $V_p \approx 1410$ m/s, so that the associated wavelength $\lambda = V_p/f$ is about 28 m and the vertical resolution limit, $\lambda/4$, at this frequency results in about 7 m. So, in these examples the 5 m layer is ultra-thin and the 10 m layer is thin.

When $h = 0$ and $h = \infty$, the model reduces to the classic two halfspace model, with a real reflection coefficient (shown in Figures 6 a) and b)), which can be analyzed in terms of the AVA classification. As can be seen in Figures 6 b) and c), the spectral AVA response for finite thickness layers differs considerably from that of the halfspaces, as expected. For $0 < h < \infty$ the AVA character is highly dependent on frequency, a dependence that can be clearly observed in Figure 7. In particular, the curves for the thicknesses $h = 5$ m and $h = 10$ m show a periodic behavior; meaning that the different frequency components of a seismic signal are reflected with different amplitudes. To understand this periodicity on a theoretical basis, it is useful to consider the analytical result for the reflectivity of an acoustic layer embedded between two halfspaces, treated in [52] and [53]. From that solution it can

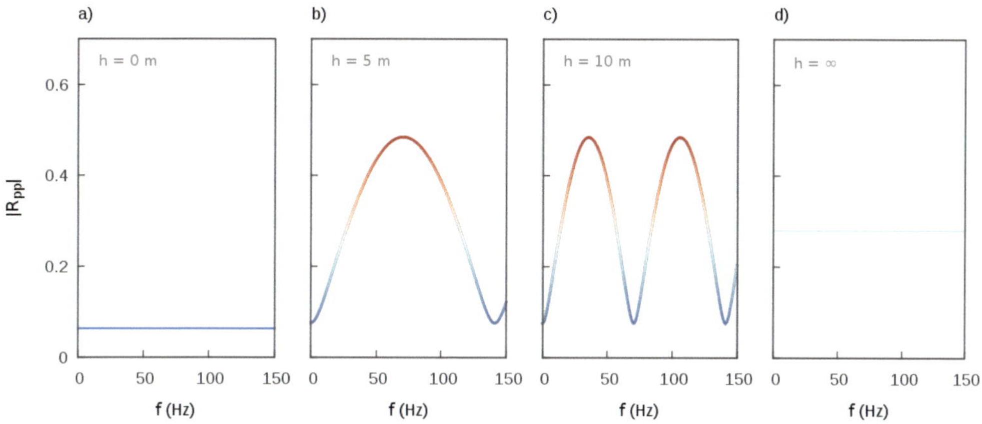

Figure 7. Modulus of the generalized reflection coefficient R_{pp} vs. frequency for normal incidence, $S_g = 100$ % and thicknesses $h = 0, 5, 10$ m and $h \to \infty$.

be shown that the function $R_{pp}(f, \theta)$ has multiple maxima or *peaks* at frequencies f^n_{max} and minima or *notches* located at frequencies f^n_{min}, given respectively by [15]

$$f^n_{max} = \left(\frac{1+2n}{2}\right)\frac{V_P}{2h\cos\theta_L}, \quad f^n_{min} = n\frac{V_P}{2h\cos\theta_L}, \quad \text{for} \quad n = 0, 1, 2, \cdots. \tag{7}$$

The occurrences of these extrema are directly proportional to the compressional velocity within the layer V_P (shown in Figure 1) and inversely proportional to its thickness h, where θ_L is the ray angle inside the layer. The frequencies at which the reflected amplitude has maxima are known as *tuning frequencies* [58] or *peak frequencies.* The first peak frequency f^0_{max} for normal incidence, hereafter denoted as f_P, results in

$$f_P = \frac{V_P}{4h}. \tag{8}$$

Equation (8) is a very useful relation for thickness determination by means of the f_P attribute, as shown in [38] in the context of CO_2 sequestration.

For an arbitrary number of layers with different physical properties and variable thicknesses the generalized reflection coefficient also displays a pattern of maxima and minima. However, for such cases there are no simple analytical expressions and consequently a numerical procedure such as the one presented in this work enables predicting the spectral characteristics of their reflectivity.

For the particular cases $h = 0$ and $h \to \infty$, plotted in Figures 7 a) and d), the reflectivity for normal incidence is real, constant and related to the impedance contrast between the caprock and the fully brine saturated sandstone and the fully CO_2 saturated layer. In Figures 7 b) and c), the location of the peak close to 70 Hz for $h = 5$ m and the peaks close to 35 Hz and 105 Hz for $h = 10$ m can be verified. Note also that for the continuous component $f = 0$ Hz, these AVF curves converge to the response in Figure 7 a), given that for such a long wavelength the layers $h = 5$ m and $h = 10$ m are invisible to the seismic waves.

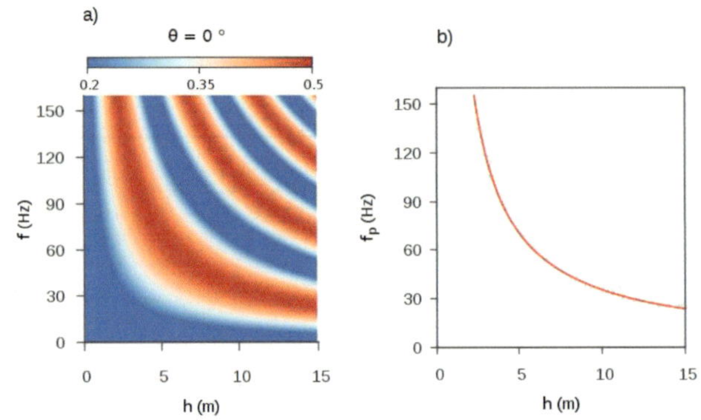

Figure 8. a) Modulus of R_{pp} vs. frequency and thickness for normal incidence and saturation S_g = 100%. b) Relation between the first peak frequency f_p and layer thickness h.

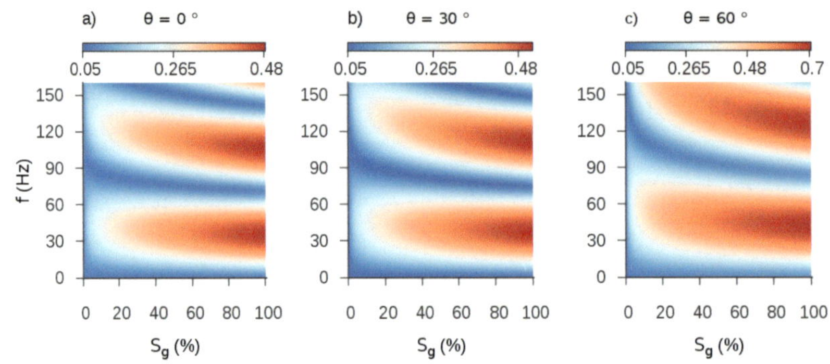

Figure 9. Modulus of R_{pp} vs. frequency and CO_2 saturation S_g = for thickness h = 10 m and three incidence angles $\theta = 0^\circ$, 30° and 60°.

The modulus of R_{pp} versus frequency and thickness for normal incidence is shown in Figure 8 a), in which the reciprocal relation between peak frequencies and thicknesses is clearly displayed. As seen, at each thickness we have a different peak frequency. To study this dependency in more detail, in Figure 8 b) we display the first peak frequency versus thickness, in which it is instructive to note that for a frequency of 35 Hz the corresponding thickness h is about 10 m. This is the case studied in [38], where spectral decomposition is used at Sleipner to map the topmost CO_2 accumulation. For very small thicknesses Figure 8 b) also indicates that the peak frequencies are well above the standard seismic frequency range, so high-resolution seismic acquisition is essential for these layers.

To analyze the sensitivity of the previous results to CO_2 saturation, next in Figure 9, we plot the modulus of R_{pp} versus frequency and saturation for thickness $h = 10$ m and incidence angles 0°, 30° and 60°. The peaks of the reflectivity are located in the high CO_2 saturation range, since in that case the impedance contrast between the caprock and the sandstone is maximum. The sensitivity to S_g depends on frequency showing bands of very weak reflectivity (in blue scale) and higher amplitudes (in red scale). The high reflectivity zones, which at the same time are sensitive to saturation, are located around the first and second peak frequencies for this example ($h = 10$ m), i.e. near 35 Hz and 105 Hz. These conclusions

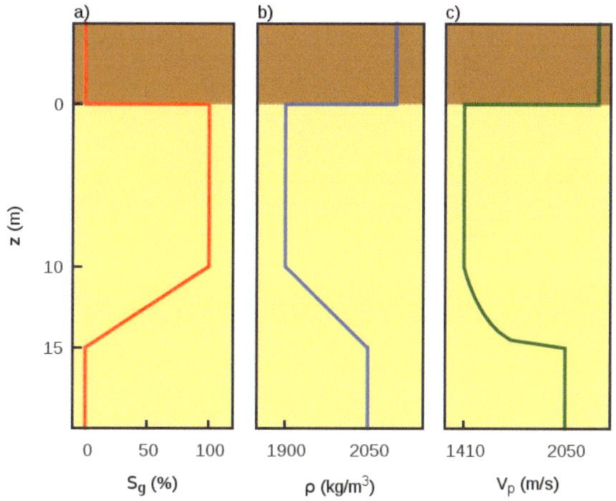

Figure 10. a) The carbon dioxide saturation profile and its effects on the Utsira sandstone P-wave velocity b) and bulk density c) for $h_0 = 10$ m and $h_t = 5$ m.

are valid for normal incidence $\theta = 0°$ and also for larger angles, although a slight shift of the peak frequencies to higher values is observed, as expected from equation (7).

3.3. The influence of saturation transitions

The seismic monitoring of carbon dioxide in geologic reservoirs is mostly focused on the characterization of accumulations of high saturation, due to their large seismic amplitudes. Nevertheless, low-saturation zones with dispersed CO_2, or *saturation transitions* may have an important role in the propagation of waves within the reservoir, giving rise to amplitude and phase changes of the seismic signals. Our aim is to analyze whether the presence of a saturation transition below the main accumulation substantially alters the previous modeling results. At the same time we investigate the influence of the vertical extent of the transition zone on the reflectivity of the reservoir.

To study these effects, we consider a modification of the model used in Section 3.2, by introducing a variable CO_2 saturation-depth profile. This model supposes that the migration of the injected CO_2 is controlled by buoyancy and that the reservoir permeability is homogeneous and isotropic, resulting in a laterally constant gas saturation field. The saturation-depth profile is inspired by the axisymmetric flow simulations presented in [9].

For the following examples we represent the saturation $S_g(z)$ with a piecewise linear function of the vertical depth z. The total thickness of the CO_2 bearing layer is given by

$$h = h_0 + h_t, \tag{9}$$

where h_0 is the thickness of a constant saturation zone below the caprock, with $S_g = 100$ % and h_t being the thickness of the saturation transition. The vertical variation in saturation $S_g(z)$ in turn produces a linear density profile $\rho(z)$ and a non-linear velocity with depth $V_P(z)$, as shown in Figure 10. The physical conditions of the CO_2 are those described in Section 3.1. To compute the generalized reflection coefficient using the reflectivity method,

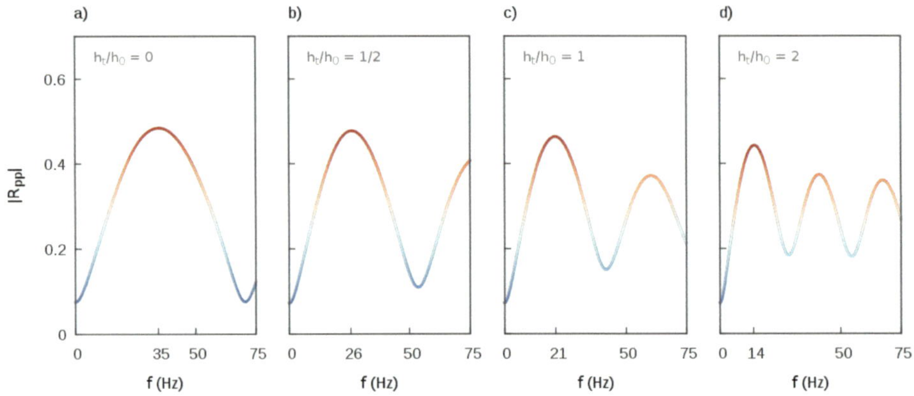

Figure 11. Normal incidence AVF for transition thicknesses $h_t = 0, 5, 10$ and 20 m.

the upper halfspace of the model is the caprock, followed by the constant saturation layer with $h_0 = 10$ m, the saturation transition with variable h_t and a fully brine saturated lower halfspace. The transition is discretized in a finite number of layers of 0.5 m with decreasing CO_2 saturation with depth.

In Figure 11 a)-d) we compare the normal incidence AVF response for transitions of different thicknesses given by $h_t/h_0 = 0, 0.5, 1$ and 2, that is $h_t = 0, 5, 10$ and 20 m. The overall bulk saturation within the transition, given by

$$\bar{S} = \frac{1}{h_t} \int_0^{h_t} S_g(z)\, dz, \tag{10}$$

is kept constant at $\bar{S} = 50$ % in the four cases. Due to the layered geometry and variable saturation of the model, the reflectivity vs. frequency shows a pattern of maxima and minima, with decreasing peak amplitudes with frequency. The presence of the saturation transition shifts the first peak to lower frequencies within the seismic band. In general for higher thicknesses h_t we observe lower values of the first peak frequency, which indicates that the inverse proportionality between these parameters still holds. Although the theoretical relation in equation (7) is not applicable to this multilayer geometry, the distinct periodicities can help to discriminate thin and thick transition zones.

The effect of the saturation transition can be noted in more detail in Figure 12, where we plot the first peak frequency f_p vs. thickness h for the case with and without saturation transition below the main accumulation. The discrepancies between the curves indicate that the CO_2 saturation transition introduces some uncertainty in the estimation of the accumulation thickness using the f_p attribute. In other words, if the existence of a transition is not taken into account, for a given value of f_p the thickness of the accumulation, as given by equation (8), is underestimated leading also to an incorrect estimation of its associated CO_2 volume.

A more extended and insightful study about the reflectivity and characterization of CO_2 saturation transitions was published by us in [13] and [14]. In those works, a comparative

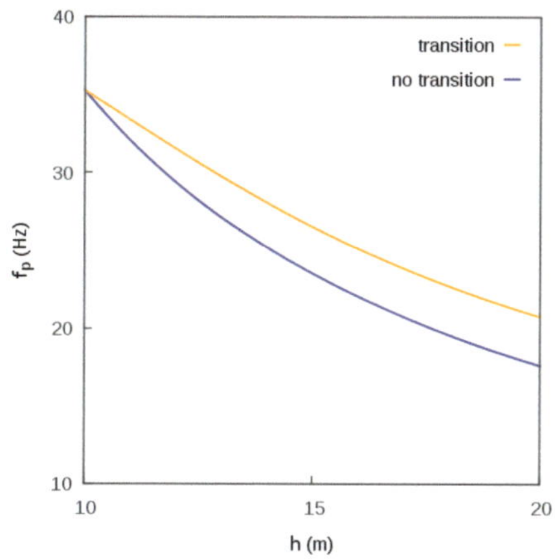

Figure 12. First peak frequency vs. thickness for the accumulation with and without saturation transition. The transition has $h_0 = 10$ m and h_t from 0 to 10 m.

analysis of the results obtained with the layered saturation model and those obtained using the classic Wolf ramp [10], [11] was also presented.

4. Conclusions

In this chapter we presented a review on the theoretical seismic reflectivity function for carbon dioxide accumulations. For monitoring purposes, emphasis was placed on the study of its correlation with parameters of interest, such as accumulation thickness and overall saturation. This study is intended to predict quantitatively the changes that can be expected in seismic time lapse data analysis. Although for the examples we calibrated our rock physics models with information of the well known Sleipner field, similar analysis can be implemented for any other geological site, including the case of CO_2–oil, or three phase fluid saturation with a proper adaptation of the physical models.

Given that the analysis of seismic AVA is a widely used tool in reservoir geophysics, using the classic Shuey's approximation, we modeled the behavior of intercept, gradient and curvature attributes under variable saturation and physical conditions. From our results we conclude that a monotonic increase in magnitude for the intercept can be expected in time lapse surveys, with significant variations with respect to the pre-injection state. This parameter shows clearly the decrease in the acoustic impedance in the sandstone for increasing CO_2 saturation.

The gradient is expected to decay strongly for low CO_2 saturations showing a different trend for saturations in the intermediate range, particularly for gaseous conditions. The curvature parameter is not expected to bring much more information about saturation, since in most cases its variations are smaller than those observed in the intercept and gradient. We have also shown that these AVA coefficients do not depend strongly on the choice of the equation of state to model the physical properties of CO_2 under *in-situ* reservoir conditions.

A loss of sensitivity on the AVA parameters, due to the stabilization of the seismic velocities, is characteristic of middle to high CO_2 saturations. This imposes a physical limitation on the monitoring of the CO_2 saturation. Our results suggest that a combined seismic time lapse analysis of the intercept and gradient AVA coefficients may be useful, not only to constrain the saturation state, but also to distinguish the physical state of the CO_2 accumulated below the caprock.

Taking into account that the vertical scale lengths in a layered medium provide physical constraints on the behavior of the propagating waves, we introduced this parameter in the computations of the reflectivity for a better characterization of this attribute. This is an issue of particular importance when the thicknesses of the CO_2 accumulations are small compared to the seismic wavelengths, a common situation in geologic repositories. The interference between multiple reflections gives rise to reflectivity dispersion effects and periodicities which can bring useful information. In this context, the use of modern spectral decomposition techniques plays an important role for the study of the amplitudes of reflected waves at different frequencies giving rise to amplitude versus frequency analysis, AVF. As shown in the examples, the periodicity of the reflectivity and the location of the first peak frequency can bring useful information about the thicknesses of the CO_2 bearing layers. In this sense, the utilization of the theoretical models presented here can be very useful for interpreting the observations and predicting peak frequencies. Our model estimates also show that the sensitivity of the reflectivity to saturation is frequency dependent, showing that low and high reflectivity bands can be expected. The high reflectivity zones show at the same time significant sensitivity to CO_2 saturation, a correlation that can be exploited for monitoring purposes.

We also analyzed the existing correlation between peak frequencies and thickness, to illustrate the utility of this spectral attribute in CO_2 accumulations. Our modeling results also indicate that the presence of saturation transitions below a main accumulation is a source of some inaccuracy in thickness and volume determination.

Regarding the potential utilization of these models and attributes (AVA coefficients, AVF, peak frequencies) using real seismic data to monitor changes in the CO_2 repository, some precautions must be taken to obtain meaningful estimations:

- it is very important to use high quality time-lapse seismic data, with good repeatability and frequency content as wide as possible;
- given that the amplitude spectrum of the seismic data is a combination of the wavelet spectrum and the reflectivity of the layers, it is very important for the processing sequence to include amplitude spectral balancing to remove wavelet overprint;
- correct identification and isolation of a window containing the seismic reflections associated with the top of CO_2 accumulations is necessary;
- the determination of peak frequencies and their time-lapse variations using an appropriate time-frequency decomposition on the selected seismic window is necessary.

In this way the theoretical results presented in this work can be tested for real case situations and may become useful tools for carbon dioxide monitoring problems.

Acknowledgements

This work was partially supported by CONICET, Argentina with grants PIP 00952, PIP 00777 and Universidad Nacional de La Plata, Argentina. Some of these results were obtained during the participation of the authors in CO2ReMoVe Project (http://www.co2remove.eu).

Author details

Claudia L. Ravazzoli* and Julián L. Gómez

*Address all correspondence to: claudia@fcaglp.unlp.edu.ar

CONICET and Faculty of Astronomical and Geophysical Sciences, National University of La Plata, Argentina

References

[1] Metz B., Davidson O., de Coninck H. C., Loos M. and Meyer L. A. (Eds.). IPCC Special Report on Carbon Dioxide Capture and Storage. Prepared by Working Group III of the Intergovernmental Panel on Climate Change. Cambridge University Press, 2005.

[2] Raistrick M. Carbon capture and storage projects to challenge governments, scientists and engineers. First Break, 2008; 26:35-368.

[3] Brown S., Bussod G. and Hagin P. AVO monitoring of CO_2 sequestration: A benchtop-modeling study. The Leading Edge 2007;26(12):1576-1583.

[4] Ravazzoli C. L. and Gómez J. L. AVA seismic reflectivity analysis in carbon dioxide accumulations: sensitivity to CO_2 phase and saturation. Journal of Applied Geophysics. 2011;73:93-100.

[5] Sturton S., Buddensiek M.L. and Dillen M. AVO analysis of thin layers - application to CO_2 storage at Sleipner. Proceedings of the 72th. Conference and Exhibition, EAGE, Barcelona, P303, 2010.

[6] Arts R.J., Eiken O., Chadwick R.A., Zweigel P., van der Meer L. and Zinszner B. Monitoring of CO_2 injected at Sleipner using time-lapse seismic data. Energy 2004;29:1383-1392.

[7] Arts R.J., Chadwick R.A., Eiken O., Trani M. and Dortland S. Synthetic versus real time-lapse data at the Sleipner CO_2 injection site. Society of Exploration Geophysicists San Antonio Meeting, Expanded Abstracts 2007; 2974-2978.

[8] Chadwick R.A., Arts R. and Eiken O. 4D Seismic quantification of a growing CO_2 plume at Sleipner, North Sea. In Doré A. and Vinning, B. Petroleum Geology: NW Europe and Global Perspectives. Proceedings of the 6th Petroleum Geology Conference. 1385-1399, 2005.

[9] Chadwick R.A., Noy D., Lindeberg E., Arts R.J., Eiken O. and Williams G. Calibrating reservoir performance with time-lapse seismic monitoring and flow simulations of the Sleipner CO_2 plume. Greenhouse Gas Technology Conference Proceedings, Trondheim. 1-6, 2006.

[10] Wolf A. The reflection of elastic waves from transition layers of variable velocity. Geophysics 1937; 2:357-363.

[11] Liner C.L, and Bodmann B. G.The Wolf ramp: reflection characteristics of a transition layer. Geophysics 2010;75(5):A31–A35.

[12] Gómez J. L., Ravazzoli C. L. and Carozzi F. E. Generalized reflectivity of CO_2 partially saturated layers. SEG Technical Program Expanded Abstracts 2010;488-492.

[13] Gómez J. L, Ravazzoli C. L. Reflection characteristics of linear carbon dioxide transition layers. Geophysics 2012;77(3):D75-D83.

[14] Gómez J. L. and Ravazzoli C. L. Modelling the reflectivity of a carbon dioxide transition zone. Proceedings of the Third EAGE CO_2 Geological Storage Workshop 2012;114-118.

[15] Gómez J. L. Seismic monitoring of CO_2: reflectivity modeling and attribute analysis. PhD Thesis (in Spanish). Facultad de Ciencias Astronómicas y Geofísicas, Universidad Nacional de La Plata; 2013.

[16] Dutta N. and Odé H. Seismic reflections from a gas-water contact. Geophysics 1983;48:148-162.

[17] Santos J.E., Corberó J., Ravazzoli C. L. and Hensley J. Reflection and transmission coefficients in fluid saturated porous media. Journal of the Acoustical Society of America 1992;91:1911-1923.

[18] Ravazzoli C. L. Analysis of reflection and transmission coefficients in three-phase sandstone reservoirs. Journal of Computational Acoustics 2001;9(4):1437-1454.

[19] Biot M.A. Theory of propagation of elastic waves in a fluid-saturated porous solid. I. Low frequency range. Journal of the Acoustical Society of America 1956;28:168-171.

[20] Biot M.A. Mechanics of deformation and acoustic propagation in porous media. Journal of Applied Physics 1962; 33: 1482-1498.

[21] Gassmann F. Uber die elastizitat poroser medien. Vierteljahrschrift der Naturforschenden Gessellshaft in Zurich 1951; 96:1-23.

[22] Domenico S.N. Effect of brine-gas mixture on velocity in an unconsolidated sand reservoir. Geophysics 1976; 41:882-894.

[23] Mavko G. and Mukerji T. Bounds on low-frequency seismic velocities in partially saturated rocks. Geophysics 1998; 63(3):918-924.

[24] Hill, R. Elastic properties of reinforced solids: some theoretical principles. J. Mech. Phys. of Solids 1963;11:357-372.

[25] Dutta N.C, and Odé H. Attenuation and dispersion of compressional waves in fluid-filled porous rocks with partial gas saturation (White model). Part I: Biot theory Geophysics 1979;44:1777-1788.

[26] Pride S.R., Berryman J.G., and Harris J.M. Seismic attenuation due to wave-induced flow. Journal of Geophysical Research 2004;109:B01201.

[27] Rubino J.G., Ravazzoli C.L., and Santos J.E. Equivalent viscoelastic solids for heterogeneous fluid-saturated porous rocks. Geophysics 2009;74(1):N1-N13.

[28] Carcione J.M., Picotti S., Gei D. and Rossi G. Physics and seismic modeling for monitoring CO_2 storage. Pure and Applied Geophysics 2006;163:175-207.

[29] Batzle M. and Wang Z. Seismic properties of pore fluids. Geophysics 1992;57:1396-1408.

[30] Bachu S. Screening and ranking of sedimentary basins for sequestration of CO_2 in geological media in response to climate changes. Environmental Geology 2003;44:277-289.

[31] Meadows M. Time-lapse seismic modeling and inversion of CO_2 saturation for storage and enhanced oil recovery. The Leading Edge 2008;27(4):506-516.

[32] Osborne M. J. and Swarbrick R. E. Mechanisms for generating overpressure in sedimentary basins: a reevaluation. Am. Assoc. Petrol. Geol. Bull.1997;81:1023-1041.

[33] Duan Z., Moller N. and Weare J. An equation for the CH_4-CO_2-H_2O system: I. Pure systems from 0 to 1000C and 0 to 8000 bar. Geochimica et Cosmochimica Acta 1992;56:2605-2617.

[34] Peng D. Y, and Robinson D.R. A new two-constant equation of state. Industrial and Engineering Chemistry Fundamentals 1976;15(1):59-64.

[35] van der Waals J. D. Over de continuiteit van den gas en vloeistoftoestand. Dissertation, Leiden 1873.

[36] Span R. and Wagner W. A new equation of state for carbon dioxide covering the fluid region from the triple-point temperature to 100 K at pressures up to 800 MPa. Phys. Chem. Ref. Data 1996;25:1509-1596.

[37] Picotti S., Carcione J.M., Gei D., Rossi G. and Santos J.E. Seismic modeling to monitor CO_2 geological storage: The Atzbach-Schwanenstadt gas field. Journal of Geophysical Research: Solid Earth 2012;117(B6):2156-2202.

[38] Williams G. and Chadwick A. Quantitative seismic analysis of a thin layer of CO_2 in the Sleipner injection plume. Geophysics 2012;77(6):R245-R256.

[39] Thomson W. Transmission of elastic waves through a stratified solid medium. Journal of Applied Physics 1950; 337-342.

[40] Haskell N. The dispersion of surface waves in multilayered media. Bulletin of the Seismological Society of America 1953;17-34.

[41] Kennett B. and Kerry N. Seismic waves in a stratified halfspace. Geophysical Journal of the Royal Astronomical Society 1979; 557-583.

[42] Kennett B. Seismic wave propagation in stratified media, ANU E-Press, http://epress.anu.edu.au, 2009.

[43] Jocker J., Smeulders D., Drijkoningen G., van der Lee C. and Kalfsbeek A. Matrix propagator method for layered porous media: Analytical expressions and stability criteria. Geophysics, 69(4):1071-1081, 2004.

[44] Mavko G., Mukerji T., and Dvorkin J. The rock physics handbook: Tools for seismic analysis of porous media. Cambridge University Press, 2nd. edition; 2011.

[45] Zoeppritz K. On the reflection and propagation of seismic waves. Göttinger Nachrichten 1919;1:66-84.

[46] Shuey R. A simplification of the Zoeppritz equations. Geophysics 1985;50(4):609-614.

[47] Castagna J. P. and Backus M.M. Offset-dependent reflectivity - theory and practice of AVO analysis. Investigations in Geophysics 8. Society of Exploration Geophysicists 1993.

[48] Castagna J. P., Swan H.W. and Foster D.J. Framework for AVO gradient and intercept interpretation. Geophysics 1998;63:948-956.

[49] Avseth P., Mukerji T. and Mavko G. Quantitative seismic interpretation: applying rock physics tools to reduce interpretation risk. Cambridge University Press; 2005.

[50] Ghaderi A. and Landro M. Estimation of thickness and velocity changes of injected carbon dioxide layers from prestack time-lapse seismic data. Geophysics 2009;74:O17-O28.

[51] Castagna J. P and Swan H. W. Principles of avo crossplotting. The Leading Edge 1997;17(2):120-127.

[52] Brekhovskik L. M. Waves in layered media, 2nd. Edition. Academic, New York; 1980.

[53] Liu, Y. and Schmitt D.R. Amplitude and AVO responses of a single thin bed. Geophysics 2003;68(4):1161-1168.

[54] Rubino J. G., Velis D. R. and Sacchi M. D. Numerical analysis of wave-induced fluid flow effects on seismic data: Application to monitoring of CO_2 storage at the Sleipner field. Journal of Geophysical Research 2011;116:B03306-84.

[55] Liner C. Elements of Seismic Dispersion: A Somewhat Practical Guide to Frequency-dependent Phenomena. SEG Distinguished Intructor Short Course, No.15 Society of Exploration Geophysicists; 2012.

[56] Partyka G., Gridley J. and Lopez J. Intepretational applications of spectral decomposition in reservoir characterization. The Leading Edge 1999;18:353-360.

[57] Lay T., and Wallace T.C. Modern Global Seismology. Academic Press; 1995.

[58] Castagna J.P, Sun S. and Siegfried R. W. Instantaneous spectral analysis: Detection of low-frequency shadows associated with hydrocarbons. The Leading Edge 2003;22(2):120-127.

Chapter 13

Optimization of CO_2 Sequestration in Saline Aquifers

Ramesh K. Agarwal and Zheming Zhang

Additional information is available at the end of the chapter

http://dx.doi.org/10.5772/57066

1. Introduction

For optimization of geological carbon sequestration (GCS) in saline aquifers, a genetic algorithm (GA) based optimizer has been developed and combined with the DOE multi-phase flow and heat transfer numerical simulation code TOUGH2 [1],[2]. Designated GA-TOUGH2, this combined solver/optimizer has been validated by performing optimization studies on a number of model problems and comparing the results with brute-force optimization, which requires a large number of simulations. Using GA-TOUGH2, an innovative reservoir engineering technique known as water-alternating-gas (WAG) injection has been investigated in the context of GCS, Additionally, GA-TOUGH2 has been applied to determine the optimal WAG operation for enhanced CO_2 sequestration capacity. GA-TOUGH2 is also used to perform optimization designs of time-dependent injection rate for optimal injection pressure management, and optimization designs of injection-well distribution for minimum well interference. The results obtained from these optimization designs suggest that over 50% reduction of in situ CO_2 footprint, greatly enhanced CO_2 dissolution, and significantly improved well injectivity can be achieved by employing GA-TOUGH2. The technique has also been employed to determine the optimal well placement in a multi-well injection operation. GA-TOUGH2 appears to hold great promise for studying a host of other optimization problems related to GCS.

2. Genetic algorithms

Genetic algorithms (GA) belong to a class of optimization techniques that are inspired by biological evolution [3],[4]. The algorithm begins with a set (identified as "generations") of vectors (identified as "individuals"). The individuals from one generation are used to create a new generation of individuals, which is supposed to be better than the previous generation.

Individuals used to form the next generation (identified as "offspring") are selected according to their function value in satisfying a certain criteria (identified as the "fitness function"). This process is repeated, creating the best individuals for each successive generation according to certain pre-defined criteria. Finally a generation of individuals is obtained where all the individuals in that generation produce the optimal values of the fitness function within a small tolerance. The algorithm is then considered to have achieved convergence. Detailed explanations of the GA can be found in Goldberg's book [4] and in our previous work [5].

3. GA-TOUGH2 integrated computer program

To realize the capability of numerical simulation and optimization for saline aquifer geological carbon sequestration (SAGCS), the GA optimizer is implemented in the TOUGH2 solver to obtain an integrated simulation-optimization computer program. Additional modules for pre- and post-simulation processes are introduced to enable the data sharing between GA and TOUGH2. A schematic of the program architecture and data flow is presented in Figure 1. Readers are encouraged to refer to our previous work for more details on the integrated code designated GA-TOUGH2 [6].

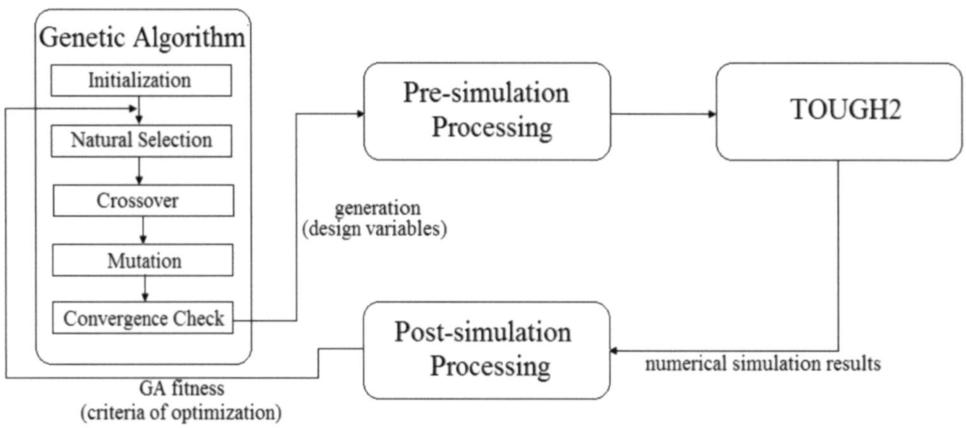

Figure 1. Graph of integrated GA-TOUGH2 code

4. Optimization of geological carbon sequestration in saline aquifers

The development of GA-TOUGH2 code and the successful simulation studies of GCS in large-scale saline formations have provided an understanding and the simulation tools for the study of optimization of some promising reservoir engineering techniques for more efficient and secure SAGCS practices. These optimization studies include (but are not limited to) the optimization of the constant-gas-injection rate for maximum CO_2 dissolution, the optimal design of water-alternating-gas (WAG) injection schemes (patterns) for maximum storage efficiency, the design of optimum injection scenarios for optimal pressure management, and the optimal placement of wells in a multi-well injection system. Some of these optimization studies are described below.

4.1. Optimization of reduction in CO_2 plume migration for the water-alternating-gas (WAG) injection scheme

A reservoir engineering technique known as water-alternating-gas (WAG) scheme is considered for SAGCS for improving the sequestration efficiency, although additional injection of water with CO_2 will inevitably increase the cost. GA-TOUGH2 is employed to determine the optimal WAG operation for maximum CO_2 sequestration efficiency while minimizing water usage.

The idea of WAG operation was originally introduced in the oil industry to improve the sweeping efficiency during gas flooding of the oil reservoir. A significant amount of remaining oil could be recovered by regularly injecting intermittent slugs of water and gas (usually CO_2). WAG operation has been widely applied in enhanced oil recovery since the late 1950s. Inspired by the practice in the oil industry, several investigators have surmised that intermittent injection of CO_2 and water could lead to better CO_2 storage efficiency by reducing the migration of CO_2 plume [5],[7], enhancing residual trapping [8],[9] and accelerating the CO_2 dissolution [10],[11]. Improved (reduced) CO_2-brine mobility ratio and accelerated CO_2 dissolution are the two most important characteristics that motivate the adoption of WAG operation in SAGCS. In multiphase flow, the non-wetting phase to wetting phase mobility ratio is defined as:

$$M = \frac{m_n}{m_w} = \frac{\mu_w \cdot k_{rn}}{\mu_n \cdot k_{rw}} \tag{1}$$

where μ_w is the wetting phase viscosity, k_{rw} is the wetting phase relative permeability, μ_n is the non-wetting phase viscosity, and k_{rn} is the non-wetting phase relative permeability. In the context of SAGCS, the pre-existing brine is considered as the wetting phase and injected supercritical CO_2 is considered as non-wetting phase. If the intermittent CO_2-water injection is treated as a quasi-mixture entering the aquifer, it will effectively bring down the mobility ratio compared to that of pure CO_2 injection. The effective mobility ratio is crucial for SAGCS efficiency for the following two reasons:

1. The mobility ratio determines whether the displacement of the reservoir fluid is stable. If $M < 1$, stable displacement occurs, i.e., the displacement of brine acts in a piston-like fashion; if $M > 1$, unstable displacement occurs, resulting in inefficient displacement of brine due to the formation of water/gas fingers.

2. The mobility ratio determines the speed of buoyancy-driven CO_2 migration. Investigation of the vertical migration of CO_2 plumes in porous media has shown that the front-end speed of a 1-D plume changes as the mobility ratio varies: the CO_2 plume front travels faster with higher mobility ratio and vice versa [5]. Since the buoyancy-driven upward

motion is the main cause of excessive lateral migration of in situ CO_2, it implies that in situ CO_2 will rise and spread more slowly by reducing the mobility ratio, resulting in a smaller environmental footprint.

Another key aspect of WAG operation is the enhanced CO_2 dissolution. In the literature, reservoir engineering techniques of injecting brine into the aquifer after the completion of CO_2 injection for achieving accelerated CO_2 dissolution have been studied by Leonenko and Keith [10]. Orr [8] and Bryant et al. [9] also claimed that CO_2-chasing water injection can expedite the process of residual trapping. Promising results have been obtained from both numerical simulations and feasibility analysis. The fundamental mechanism of accelerating CO_2 dissolution by water injection is enhanced convective mixing of CO_2 and brine/water. Since WAG operation consists of repeated cycles of CO_2-chasing water injection, it is expected that the CO_2 dissolution will be enhanced with the deployment of WAG. Considering these facts, optimal design of WAG operation for SAGCS is investigated below.

4.1.1. Optimization of WAG scheme using GA-TOUGH2

The WAG operation is studied for GCS in various saline aquifers (generic and identified large scale) and for different injection well orientations (vertical and horizontal). First, WAG operation for a generic saline aquifer with generic hydrogeological properties is investigated by considering both the vertical and horizontal injection wells. Vertical injection wells are the most common type of well with mature and economical well completion technology. Nevertheless, SAGCS with horizontal well injection is worth investigating since there are potential benefits of horizontal well injection, as has been noted by and Jikich and Sams [11] and Hassanzadeh et al. [12]. Next, WAG optimization is considered for identified large saline aquifers. The Frio formation and Utsira formation are considered in our study. For these formations, all simulation parameters are obtained from the history-matching simulations described in Chapter titled, "Numerical Simulation of CO2 Sequestration in Large Saline Aquifers".

One complete cycle of CO_2-water injection is identified as a WAG cycle. A complete WAG operation is composed of a series of such basic WAG cycles. For simplicity, it is assumed that WAG cycles are identical to each other. A schematic of the considered WAG operation is shown in Figure 2, with red blocks and blue blocks representing CO_2 injection and water injection respectively. The width of the blocks represents the duration of injection.

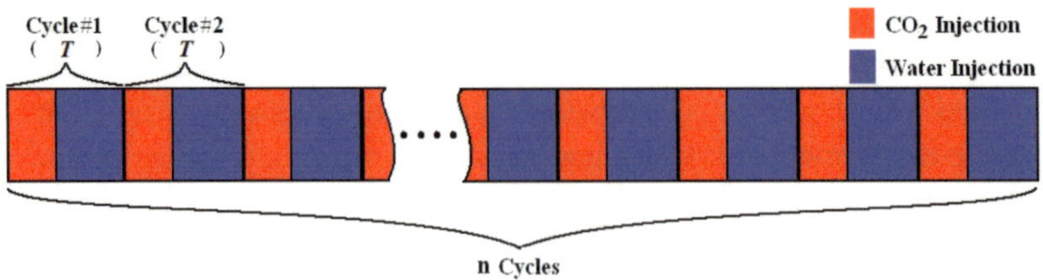

Figure 2. A typical schematic of the WAG operation

A set of four basic variables determines a unique cycle pattern, identified as: CO_2 injection rate I_{CO2}, water injection rate I_{water}, WAG ratio r_{WAG} (the ratio of injected CO_2 mass to injected water mass per cycle), and cycle duration time T. Assuming the duration of CO_2 injection in one WAG cycle as t_{CO2}, the WAG ratio is defined by the following equation (2).

$$r_{WAG} = \frac{m_{CO2}}{m_{water}} = \frac{I_{CO2}t_{CO2}}{I_{water}t_{water}} = \frac{I_{CO2}t_{CO2}}{I_{water}\left(T - t_{CO2}\right)} \tag{2}$$

By rearranging equation (2), the duration of CO_2 injection can be expressed by equation (3).

$$t_{CO2} = \frac{r_{WAG} \cdot I_{water}T}{r_{WAG} \cdot I_{water} + I_{CO2}} \tag{3}$$

Equation (3) suggests that a WAG operation can be uniquely defined if the four basic variables are given. Optimization of these four independent variables becomes a four-dimensional design problem, which can be computationally very expensive. To make the optimization more tractable, WAG cycle duration time T is determined prior to the simulation. It is rather tricky to determine the value of T beforehand. Nasir and Chong [13] have claimed that differences in WAG cycle duration time do not lead to significant differences in recovery efficiency for enhanced oil recovery. However, we have found in our research that WAG cycle duration time can significantly affect the performance of WAG operation under certain conditions. In our simulations/optimizations, we have set T at 30 days, which is an economically feasible and performance-acceptable choice. The effect of WAG cycle duration time on sequestration efficiency is discussed in a later section. With the above simplifications, the number of independent variables that uniquely determines a WAG operation is reduced by two. Since WAG cycle duration time T is pre-determined, any two variables from I_{CO2}, I_{water}, or r_{WAG} can be picked as the basic optimization variables for designing a WAG operation. There is no constraint or preference as to which of these two parameters should be chosen as the optimization design variables. Picking I_{CO2} and I_{water} as the two design variables, the remaining variable r_{WAG} is determined by equation (4).

$$\begin{aligned} r_{WAG} &= \frac{m_{CO2}}{m_{water}} = \frac{M_{CO2}/n}{I_{water}t_{water}} = \frac{M_{CO2}}{nI_{water}\left(T - M_{CO2}/nI_{CO2}\right)} \\ &= \frac{M_{CO2}}{I_{water}\left(nT - M_{CO2}/I_{CO2}\right)} = \frac{M_{CO2}I_{CO2}}{I_{water}\left(I_{CO2}nT - M_{CO2}\right)} \end{aligned} \tag{4}$$

where M_{CO2} is the total amount of CO_2 to be sequestered and n is the total number of WAG cycles.

A given amount of CO_2 to be sequestered is usually known as the sequestration target of a given SAGCS project. A medium-size coal-fired power plant typically generates approximately 1 million metric of CO_2 annually. For the purpose of our investigation, it is reasonable to assume 50% CCS efficiency, i.e., capture and sequestration of a half million metric tons of CO_2 for a proposed WAG operation in a hypothetical generic aquifer. For WAG operation in identified large aquifers, the target sequestration amount is set identical to that determined for the actual project. Equations (3) and (4) determine the unique WAG patterns. Simulations of non-optimized WAG operations are performed first to demonstrate the reduced CO_2 migration. Recalling that gaseous CO_2 reaches the caprock relatively quickly under buoyancy and then migrates underneath the caprock, it is the radial migration of gaseous CO_2 that causes enormous land use as well as the leakage risk. Therefore, the saturation of gaseous phase CO_2 (SG) directly underneath the caprock, originating from the injection well along the migration direction, should serve as an ideal indicator of storage efficiency. SG is the percentage of void space in the formation occupied by gaseous CO_2; thus it varies from 0 to 1. It becomes greater than zero when CO_2 displacement of brine occurs, and remains zero in CO_2 free zones. Therefore, the maximum migration of in situ CO_2 can be effectively determined by examining the SG profile underneath the caprock. Additionally, cross-sectional SG contours can also indicate the migration and dissolution of in situ CO_2.

4.1.2. WAG injection in a hypothetical generic saline formation with a vertical injection well

A hypothetical generic cylindrical domain with thickness of 100 m is considered as the target aquifer. The radius of the aquifer is set at 3000 m to minimize the influence of the boundary conditions. For generalization purposes, typical hydrogeological properties of deep saline aquifers are applied to the domain. CO_2 and water are injected at the center of the domain by an injection well fully perforating the aquifer. No water pumping is included in the simulation domain, with the assumption that water production is either far away from the storage site or comes from a nearby water reservoir. The WAG operation is assumed to consist of 20 WAG cycles, each lasting for 30 days. The injection operation therefore lasts for 600 days. CO_2 migration is examined 50 years after the inception of injection. Figure 3 shows the computational model and the mesh. Due to symmetry, only a radial slice of the aquifer is modeled. The computational mesh is highly refined near the injection well and near the caprock to accurately capture the migration of in situ CO_2 in those regions.

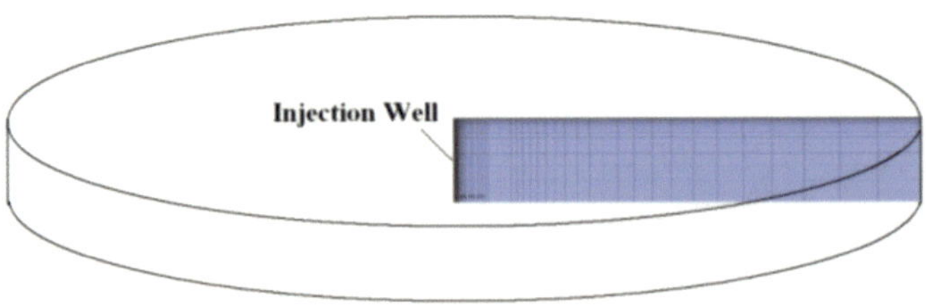

Figure 3. Generic domain for optimization of WAG operation with a vertical injection well

Table 1 summarizes the details of the model geometry, the hydrogeological properties and the simulation parameters.

Permeability	Isotropic, 100 mDarcy
Porosity	0.12
Residual Brine Saturation	0.2
Residual CO_2 Saturation	0.05
Relative Permeability	van Genuchten-Mualem
Capillary Pressure	van Genuchten-Mualem
Initial Conditions	$P = 12$ MPa, $T = 45$ °C
Initial CO_2 Mass Fraction	$X_{CO2} = 0$
Initial Salt Mass Fraction	$X_{sm} = 0.15$

Table 1. Hydrogeological properties and initial conditions for the cylindrical domain considered for the optimization study of WAG operation with vertical well injection

The fitness function of the optimization, i.e., the criteria for evaluating the performance of a certain WAG operation, is defined as the ratio of CO_2 migration reduction (with respect to that of CGI operation) to the total amount of water injection. It is mathematically represented by equation (5). This choice of fitness function arises from the consideration of the economic feasibility of implementing the WAG operation, since the transportation and pumping of water is likely to consume additional energy. It is obvious that a tradeoff exists between the water consumption and CO_2 migration reduction. Therefore, it is clear that the WAG operation leading to the maximum value of the fitness function would provide the optimal balance between the plume migration reduction and water requirement.

$$fitness = \frac{R_{CGI} - R_{WAG}}{m_{water}} \tag{5}$$

As mentioned earlier, I_{CO2} and I_{water} have been chosen as the two optimization design variables. The search space is [30 kg/s,100 kg/s] for both I_{CO2} and I_{water}, resulting in the search space for r_{WAG} as [0.19, 1.18]. A post-processing computational module has also been developed for determination of the migration reduction. The design variable values corresponding to optimal WAG operation and the optimal fitness function are determined as $I_{CO2,optimal}$ = 55.26 kg/s, $I_{water,optimal}$ = 39.19 kg/s, $r_{WAG,optimal}$ = 0.567, and $fitness_{optimal}$ = 0.0605 m/10^3 metric tons of water. Correspondingly, the durations of CO_2 and water injection in one WAG cycle can be calculated to be 8.6 and 21.4 days respectively. Thus, in each WAG cycle, CO_2 injection lasts for 8.6 days with an injection rate of 55.26 kg/s before it is cut off; then water injection begins with an injection rate of 39.19 kg/s until the 30-day cycle duration is completed. Identical WAG cycles repeat 20 times to complete the 600-day injection operation. Figure 4 shows the schematic of the optimal WAG operation.

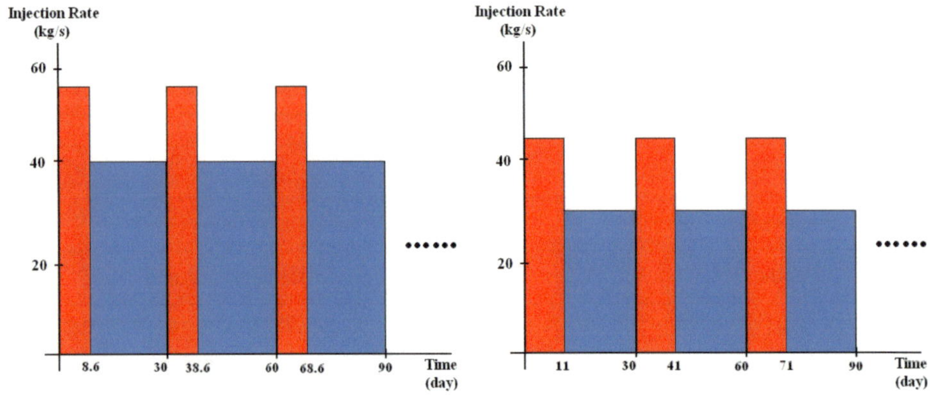

Figure 4. Graph of the optimized WAG operation with vertical well injection

Table 2 summarizes the technical benefits of applying the optimal WAG operation.

		Vertical Injection Well
CGI	CO_2 Radial Migration	1210 m
WAG	CO_2 Radial Reduction	87.66 m
	CO_2 Radial Reduction Ratio	7.24 %
	CO_2 Impact Area Reduction	642308 m^2
	CO_2 Impact Area Reduction Ratio	14%
	Total Water Injection Required	1448600 metric tons

Table 2. Summary of the benefits of implementing optimized WAG operation with vertical injection well

The corresponding CO_2 plume migration under the caprock is compared to the CGI operation in Figure 5. As summarized in Table 2 and shown in Figure 5, a 14% reduction in CO_2 impact area and significantly lowered CO_2 accumulation underneath the caprock can be achieved by replacing the conventional CGI operation with the optimal WAG injection. The cost of such a benefit is the pumping work required to inject 1448600 metric tons of water plus the extra CO_2 pumping work needed due to the increased injection pressure.

4.1.3. WAG injection in a hypothetical generic saline formation with a horizontal injection well

Jikich and Sams [11] and Hassanzadeh et al. [12] have suggested the potential benefits of utilizing horizontal injection wells for SAGCS. It has been claimed that vertical wells provide insufficient injectivity, while horizontal injectors can greatly improve injectivity and storage capacity. Hassanzadeh et al. have also suggested that horizontal injection well can lead to significantly higher CO_2 dissolution rate compared to the vertical injection well when water chasing injection is applied [12]. These suggested benefits of horizontal injection well have motivated us to apply and optimize WAG operation for horizontal well injection. Unlike the perfectly symmetric flow patterns with vertical injection well, a full 3D model is required when

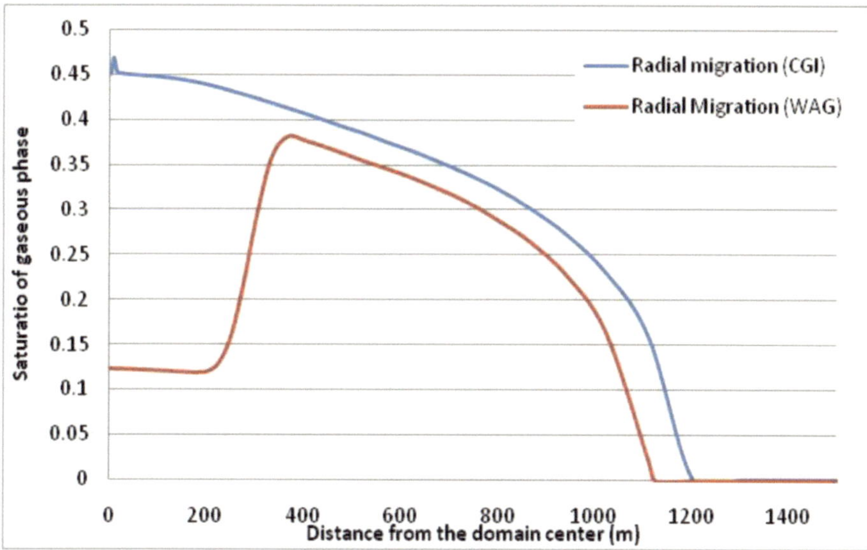

Figure 5. Radial gas saturation comparisons of optimized WAG operation and the non-optimized CGI operation for vertical injection well

horizontal injection well is considered since the flow patterns are no longer symmetric. As a result, the modeling and simulation of the SAGCS with horizontal injection well becomes computationally more intensive and requires higher computational cost. A hypothetical generic aquifer of dimensions 8000 m × 8000 m × 100 m is considered. It is assumed that an 800-m horizontal injection well sits in the middle of the aquifer. Due to symmetry, only a quarter of the domain is modeled, as shown in Figure 6. The modeled computational domain is therefore of the dimensions 4000 m × 4000 m × 100 m with a 400-m horizontal injector sitting in the middle of this domain. All the hydrogeological properties and simulation parameters are the same as used in case of vertical well injection (Table 1). The boundary conditions and the target injection amounts are adjusted for the quarter domain under consideration.

The introduction of horizontal injection well causes uneven CO_2 migration along the two principal axial directions. Since the top-planview of the CO_2 plume is expected to be elliptic instead of circular, it requires modification of the fitness function from the previous one used in the vertical well injection case. For simplicity, the average value of the migration distance along the two principal directions is employed to estimate the fitness function. Therefore, equation (5) is modified as:

$$fitness = \frac{R_{CGI} - R_{WAG}}{m_{water}} = \frac{\left(R_{CGI,x-direction} + R_{CGI,y-direction}\right)\big/2 - \left(R_{WAG,x-direction} + R_{WAG,y-direction}\right)\big/2}{m_{water}} \tag{6}$$

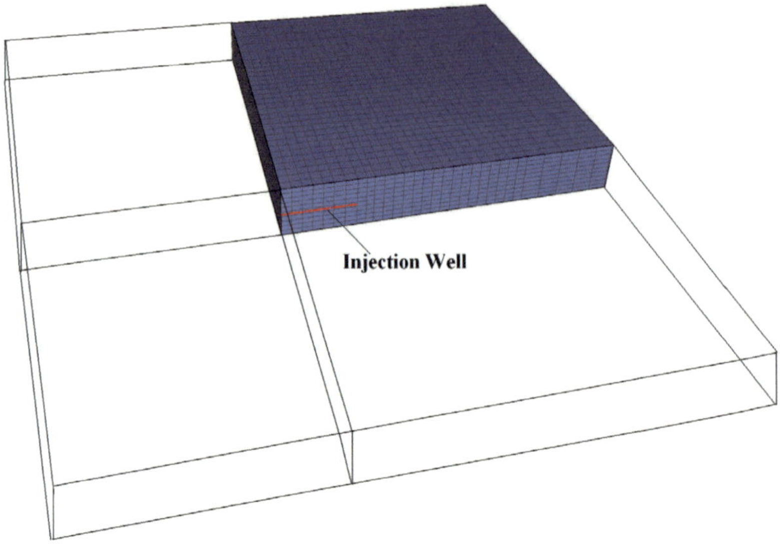

Figure 6. Quarter computational domain for WAG operation with horizontal injection well

where $R_{x\text{-}direction}$ and $R_{y\text{-}direction}$ represent the CO_2 migration distance along the direction of the injection well and in a direction perpendicular to it respectively. Again, I_{CO2} and I_{water} are chosen as the design variables. The search space of I_{CO2} and I_{water} is [20 kg/s, 80 kg/s] and the corresponding search space of r_{WAG} is [0.25, 1.95]. The values of design variables corresponding to optimal WAG operation and the optimal fitness function value are determined as $I_{CO2,optimal}$ = 44.87 kg/s, $I_{water,optimal}$ = 29.59 kg/s, $r_{WAG,optimal}$ = 0.823, and $fitness_{optimal}$ = 0.0718 m/10^3 metric tons of water. Correspondingly, the durations of CO_2 and water injection in one WAG cycle are calculated as 11 and 19 days respectively. Identical WAG cycles are repeated 20 times to complete the entire 600-day operation. Figure 7 shows the optimal WAG operation.

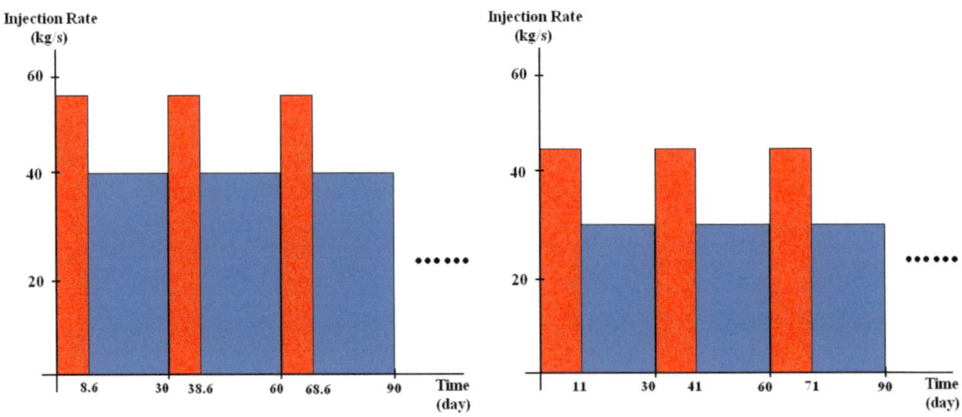

Figure 7. Graph of the optimal WAG operation with horizontal injection well

Table 3 summarizes the technical benefits of applying the optimal WAG operation with horizontal injection well. The corresponding CO_2 plume migration underneath the caprock is compared to that for the CGI operation in Figure 8.

		Horizontal Injection Well
CGI	CO_2 Radial Migration	*x*-direction: 1082.7 m, *y*-direction: 865.7 m
WAG	CO_2 Radial Reduction	*x*-direction: 116.1 m, *y*-direction: 26.9 m
	CO_2 Radial Reduction Ratio	*x*-direction: 10.7 m, *y*-direction: 3.1 m
	CO_2 Impact Area Reduction	397560 m^2
	CO_2 Impact Area Reduction Ratio	14%
	Total Water Injection Required	995635 metric tons

Table 3. Summary of the benefits for implementing optimized WAG operation with horizontal injection well

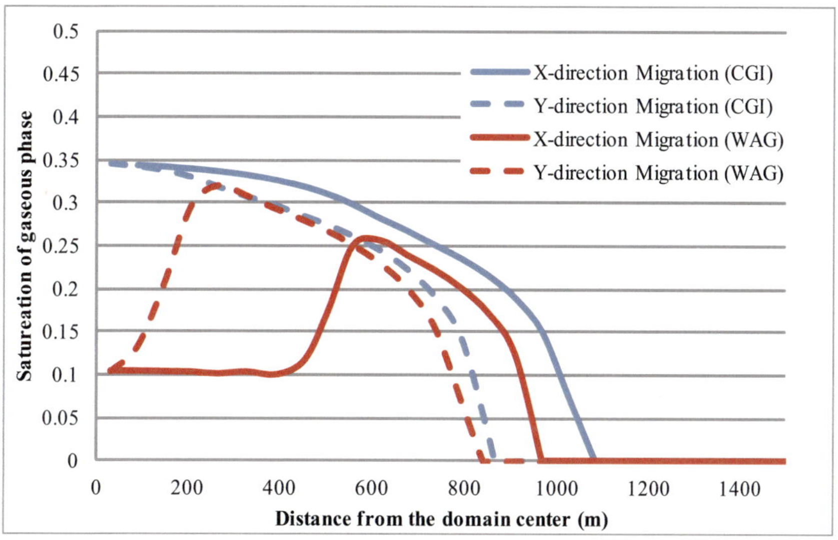

Figure 8. Radial gas saturation comparisons of optimized WAG operation and non-optimized CGI operation using a horizontal injection well

As summarized in Table 3 and shown in Figure 8, a 14% reduction in CO_2 impact area and significantly lowered CO_2 accumulation underneath the caprock can be achieved by replacing the conventional CGI operation with the optimal WAG injection. The cost of such benefits is the pumping work required to inject 995635 metric tons of water plus the extra CO_2 pumping work needed due to the increased injection pressure. These results are qualitatively similar to those obtained for the WAG operation with a vertical injection well. However, the water consumption of the WAG operation with a horizontal injection well is substantially less, implying significantly less energy penalty and improved technical and economic feasibility.

4.1.4. WAG operation with a vertical injection well over an anisotropic saline formation

Actual aquifers are generally heterogeneous in all aspects. It is generally agreed that heterogeneity can cause channeling and fingering of the CO_2 plume, thereby increasing the risk of leakage. It is also claimed that the heterogeneity can lead to locally enhanced trapping [8],[9]. Therefore, heterogeneity of aquifer properties should be taken into account if more realistic simulations with higher accuracy are desired. In section 4.2.3 and section 4.2.4, optimizations of WAG operation were performed for a hypothetical generic saline aquifer with generic hydrogeological properties. The results clearly suggested the potential benefits offered by the WAG technique. However, those simulations did not account for several types of uncertainties in the description of the reservoir conditions, among which heterogeneity is likely to have the greatest effect on the in situ migration of CO_2. In this section, we consider the optimization of WAG operation for an aquifer with anisotropy. Anisotropy of permeability, especially horizontal-to-vertical permeability anisotropy, is the most important property that can have a significant effect on vertical CO_2 migration. According to the laboratory studies of core samples, the horizontal permeability of a saline formation normally is 10~1000 times greater than the vertical permeability. On the other hand, geological stratification such as seen in the Mt. Simon formation also significantly reduces the effective vertical permeability by several orders of magnitude, resulting in drastic anisotropy in the effective permeability.

In this study, we consider the WAG operation with permeability anisotropy for vertical well injection. The model geometry, domain discretization, reservoir conditions, and all other hydrogeological properties are the same as used in the simulation described before in section 4.2.3. A horizontal-to-vertical permeability ratio of 10, i.e., $k_{horizontal}/k_{vertical}$=10, is considered. A quick estimation of the effective permeability of the Utsira formation indicates that setting the permeability ratio at 10 is actually a conservative value, as shown below.

Considering a stratified formation with all layers being horizontal, the directional flow through that formation will be as shown in Figure 9, in which each layer has its unique permeability as k_1, k_2, k_3, and thickness as h_1, h_2, h_3. The total thickness of the stratified formation is H. Flow transportation in the horizontal and vertical directions is considered separately.

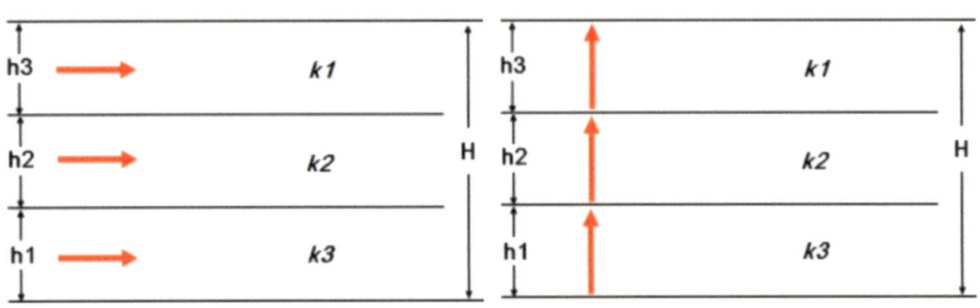

Figure 9. Graph of stratified formation with (a) horizontal flow, and (b) vertical flow

Using the mass conservation law and Darcy's law, equation (7) can be derived to evaluate the equivalent permeability along the two principal directions (horizontal and vertical) as follows.

$$k_{eq,horizontal} = \frac{k_1 h_1 + k_2 h_2 + k_3 h_3}{H} = \frac{\sum kh}{H} \qquad k_{eq,vertical} = \frac{H}{\frac{h_1}{k_1} + \frac{h_2}{k_2} + \frac{h_3}{k_3}} = \frac{H}{\sum \frac{k}{h}} \tag{7}$$

The Utsira formation can be described as a nine-layer structure with sandstone and shale alternatively overlapping each other, as suggested in Audigane et al.'s work [14]. It is assumed that each layer is isotropic in its hydrogeological properties. Detailed descriptions of this nine-layer structure can be found in Chapter titled, "Numerical Simulation of CO2 Sequestration in Large Saline Aquifers". Therefore, the equivalent permeability of the Utsira formation can be evaluated as:

$$\begin{aligned} k_{eq,horizontal} &= \frac{\sum_{i=1}^{9} k_i h_i}{\sum_{i=1}^{9} h_i} = \frac{3\times 70 + 4\times\left(0.01\times 5 + 3\times 25\right)}{70 + 4\times\left(5+25\right)} = 2.69 \text{ Darcy} \\ k_{eq,vertical} &= \frac{\sum_{i=1}^{9} h_i}{\sum_{i=1}^{9} \frac{h_i}{k_i}} = \frac{70 + 4\times\left(5+25\right)}{\frac{70}{3} + 4\times\left(\frac{5}{0.01} + \frac{25}{3}\right)} = 92.4 \text{ mDarcy} \end{aligned} \tag{8}$$

And the horizontal-to-vertical permeability ratio is obtained as:

$$\frac{k_{eq,horizontal}}{k_{eq,vertical}} = \frac{2.69 \text{ Darcy}}{92.4 \text{ mDarcy}} \approx 29 \tag{9}$$

The simple calculations above give a horizontal-to-vertical permeability anisotropy of 29 for the Utsira formation, i.e., the Utsira formation is 29 times more permeable horizontally than vertically. This demonstrates the existence of permeability anisotropy in actual aquifers, and also assures that assigning a horizontal-vertical permeability ratio of 10 is indeed a conservative choice in our investigation for the design of WAG scheme for an anisotropic aquifer.

Two modifications have been made from the original case of WAG operation study with a vertical injection well described in section 4.2.2. First, horizontal permeability of the formation is increased to 1 Darcy from the original value of 100 mDarcy. Vertical permeability is retained as 100 mDarcy to keep the permeability anisotropy of 10. Another modification is the perforation of the injection well. To take full advantage of the anisotropy, the injection perforation is reduced to one-third of its original length and is placed at the lower aquifer following Bryant's suggestion of "injection low let rise" [9]. Other than these two modifications, all other parameters of the model are retained. Same assumptions for WAG operation, i.e., 20 WAG cycles each lasting 30 days, and identical amount of CO_2 for sequestration, i.e., 0.822 million metric tons over the 600-day injection, are applied. Slightly different from previous studies of WAG operation over the isotropic formation, CO_2 migration beneath the caprock is examined immediately after the injection ceases. The values of the design variables corresponding to optimal WAG operation and the optimal fitness function value are found as $I_{CO2,optimal}$ = 36.13 kg/s, $I_{water,optimal}$ = 33.65 kg/s, $r_{WAG,optimal}$ = 0.847, and $fitness_{optimal}$ = 0.1438 m/10^3 metric tons of water. Correspondingly, the durations of CO_2 and water injection in one WAG cycle can be calculated to be 13 and 17 days respectively. Identical WAG cycles are repeated 20 times to complete the entire 600-day operation. Figure 10 shows the graph of the optimal WAG operation.

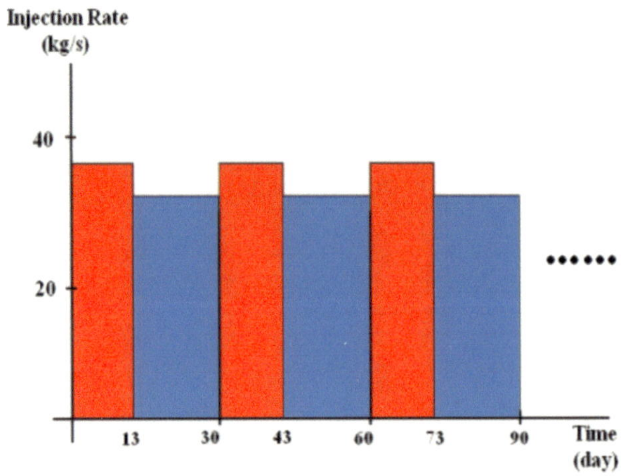

Figure 10. Graph of the optimal WAG operation for anisotropic saline formation

Table 4 summarizes the technical benefits due to adoption of the optimal WAG operation for the anisotropic saline formation.

CO_2 Radial Reduction	140 m
CO_2 Radial Reduction Ratio	32.56 %
CO_2 Impact Area Reduction	316673 m^2
CO_2 Impact Area Reduction Ratio	54.52 %
Total Water Injection Required	973574 metric tons

Table 4. Benefits of implementing the optimized WAG operation in an anisotropic saline aquifer

Figure 11 illustrates the SG curve underneath the caprock of the formation for the optimized WAG operation and that for the CGI operation with CO_2 injection rate of 15.85 kg/s.

As seen from Table 4 and Figure 11, consideration of permeability anisotropy has greatly improved the performance of WAG operation. Under a conservative horizontal-to-vertical permeability ratio of 10, the CO_2 footprint after a 600-day injection program is significantly

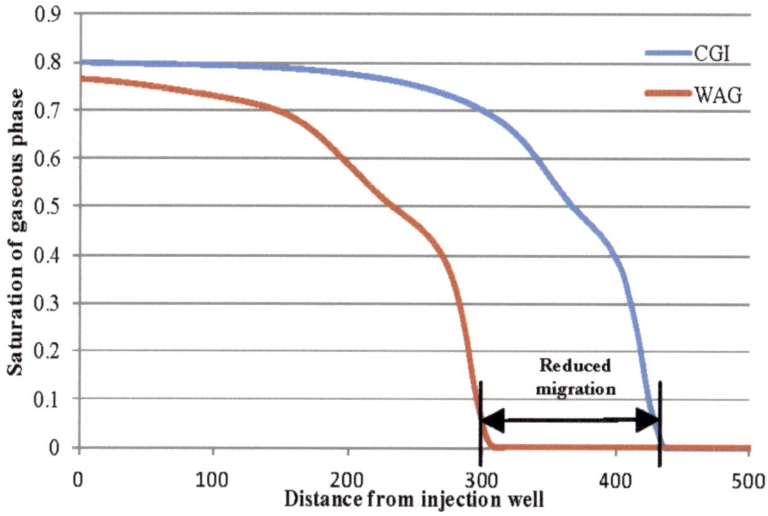

Figure 11. SG underneath the caprock showing migration reduction with optimized WAG operation in an anisotropic saline aquifer

reduced by 54.52%. Recalling the 14% reduction in CO_2 footprint for the isotropic formation, one can draw the conclusion that the anisotropy of formation permeability is an important parameter for high-performance WAG operation. It is also expected that even better performance of WAG operation would be achieved with a higher horizontal-to-vertical permeability ratio. Moreover, the CO_2 injection rate is significantly lower than that for the isotropic formation case. The duration of CO_2 injection in each WAG cycle is increased by about four days to maintain the overall injection amount. The lower injection rate and increased injection duration suggest improved injection conditions such as lower injection pressure.

To have a better illustration of the CO_2 migration reduction, simulations of three other non-optimized injection scenarios were conducted, namely constant-gas-injection with low injection rate (low-CGI), constant-gas-injection with high injection rate (high-CGI), and cyclic CO_2 injection. For the low-CGI case, CO_2 is injected with a constant mass flow rate of 15.85 kg/s for 600 days; for the high-CGI case, CO_2 is injected with a constant mass flow rate of 31.71 kg/s for 300 days; cyclic CO_2 injection is very similar to the optimal WAG injection except that water injection is removed from the operation. Therefore, all three additional cases have an identical amount of injected CO_2 but zero water injection. Comparison of the SG curves of the optimized WAG operation and the three non-optimized injection scenarios are shown in Figure 12 and Figure 13, and summarized in Table 5.

Figure 13 shows SG contours for the optimized WAG and three non-optimized injection scenarios after 600 days of injection at the radial cross-section of the formation.

Table 5 provides detailed comparisons between the optimized WAG operation and three non-optimized injection scenarios. The reduction of in situ CO_2 migration in optimized WAG is prominent.

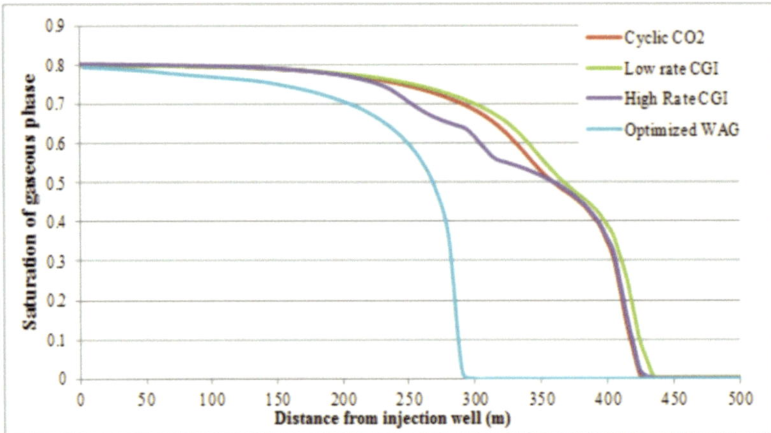

Figure 12. SG underneath the caprock; optimized WAG and non-optimized injection operations in an anisotropic saline aquifer

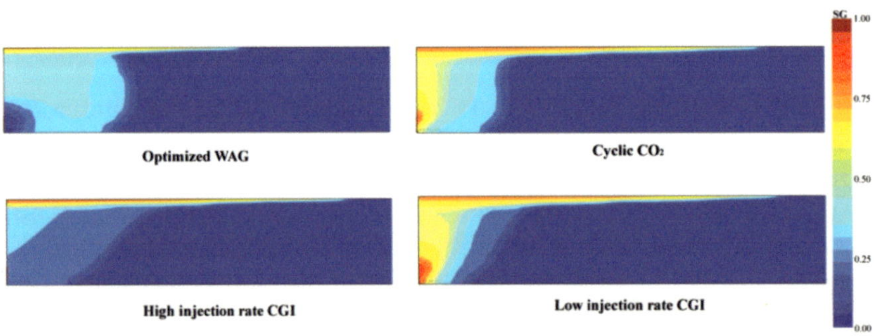

Figure 13. SG contours for optimized WAG and three non-optimized injection operations

Relative to Optimized WAG	Optimized WAG	Cyclic CO_2 Injection	High Rate CGI	Low Rate CGI
CO_2 Plume Migration	290 m	420 m	420 m	430 m
Additional CO_2 Migration	-	130 m	130 m	140 m
Increased Plume Radius	-	44.83 %	44.83 %	48.28 %
Increased Footprint Area	-	109.75 %	109.75 %	119.86 %

Table 5. CO_2 migration comparisons of optimized WAG with three other non-optimized injection scenarios

The results presented above clearly show the benefits of the WAG injection in reducing the in situ CO_2 migration. However, tradeoffs of such benefits need to be carefully considered for the safety and feasibility of SAGCS utilizing WAG operation. One of the most critical operational parameters of SAGCS is the pressure. The bottom line is that injection-induced pressure must not exceed the formation's fracture pressure under any circumstance. In practice, injection pressure is closely monitored and it is common to temporarily reduce the injection rate in order to reduce overly high injection pressure. Figure 14 shows the injection pressure (average value of the injection well) under the optimized WAG operation. According to our investigation, the optimized WAG operation causes the injection pressure to oscillate as the CO_2 injection and water injection alternate. Considering the peak pressure, an 8% increase of reservoir pressure from its hydrostatic condition can be noticed near the injection well under the optimized WAG operation. On the other hand, a maximum of 2% increase in reservoir pressure is induced in the three non-optimized injection scenarios. Therefore, one can draw the conclusion that harsher injection conditions are inevitable with WAG operations. However, the induced pressure elevation by WAG operation could be moderate enough not to pose significant concerns. It should also be pointed out that reservoir pressure response to the injection of CO_2 and water is very sensitive to the hydrogeological properties of the formation, such as porosity and permeability. Pressure analysis should be performed on a case-by-case basis for different saline formations to ensure the feasibility and safety of WAG operation.

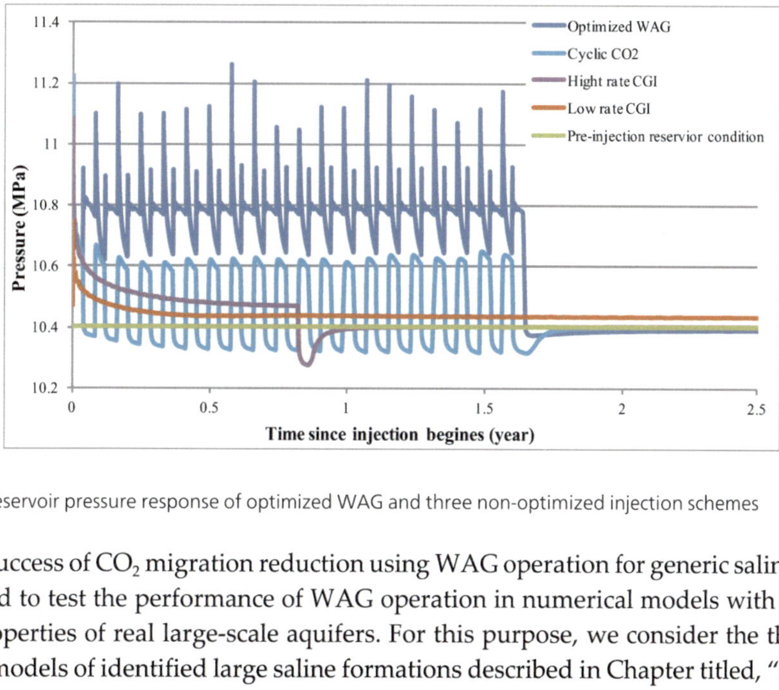

Figure 14. Reservoir pressure response of optimized WAG and three non-optimized injection schemes

With the success of CO_2 migration reduction using WAG operation for generic saline aquifers, we decided to test the performance of WAG operation in numerical models with hydrogeological properties of real large-scale aquifers. For this purpose, we consider the three representative models of identified large saline formations described in Chapter titled, "Numerical Simulation of CO2 Sequestration in Large Saline Aquifers" for numerical implementation and optimization of WAG operation. These three models are the one established for the Frio pilot project, the generalized cylindrical Utsira formation model, and a newly established cylindrical Utsira Layer #9 model. The Frio model represents the saline formation of relatively small thickness but with significant geological slope. The generalized Utsira model represents the saline formation with relatively large thickness and stratified hydrogeology. The cylindrical Utsira Layer #9 model is a good representative of saline formation with relatively small thickness and anisotropic hydrogeology.

4.1.5. WAG operation for generalized Utsira formation model

In contrast to the Frio formation, the generalized Utsira model for SAGCS has insignificant geological slope and relatively large thickness. Recalling the simulation studies of the Sleipner SAGCS project in Chapter titled, "Numerical Simulation of CO2 Sequestration in Large Saline Aquifers", the Utsira formation is a layered formation with about 200 m in thickness without evidence of any significant geological slope. Therefore, the cylindrical model of the layered Utsira formation presented in Chapter titled, "Numerical Simulation of CO2 Sequestration in Large Saline Aquifers" is used to study the WAG operations. All hydrogeological properties and numerical conditions used before as described in Chapter titled, "Numerical Simulation of CO2 Sequestration in Large Saline Aquifers" are retained, including the amount of CO_2 injection at a rate of 1 million metric tons per year. The conventional CGI operation used in the Chapter titled, "Numerical Simulation of CO2 Sequestration in Large Saline Aquifers" is replaced by the WAG operation. Reduction in radial CO_2 migration under the caprock is examined as the optimization criterion for five years of injection. The generalized Utsira formation model consists of nine alternating shale and sandstone layers, and the injection takes place at the middle of the bottom sandstone layer. The assumption of 30-day WAG cycle duration is retained, as employed in all the previous simulations discussed in sections 4.2.2 – 4.2.4. The WAG operation lasts for five years, during which 1 million metric tons of CO_2 is injected annually. The radial migration of CO_2 in the topmost sandstone layer is examined after two, three, and five years of injection. The values of the design variables corresponding to optimal WAG operation and the optimal fitness function value are found as $I_{CO2,optimal} = 95.72$ kg/s, $I_{water,optimal} = 75.32$ kg/s, $r_{WAG,optimal} = 0.64$, and $fitness_{optimal} = 0.0251$ m/10^3 metric tons of water. Correspondingly, the durations of CO_2 and water injection in one WAG cycle can be calculated as 10 and 20 days respectively. Identical WAG cycles are repeated 20 times to complete the entire five-year operation. Figure 15 shows the graph of the optimal WAG operation for the generalized Utsira formation model.

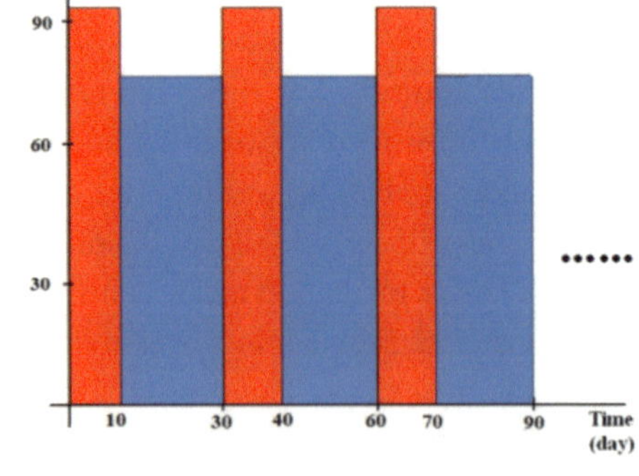

Figure 15. Graph of optimized WAG operation for Utsira formation

Figure 16 shows the CO_2 migration underneath the caprock at the second, third and fifth year for optimized WAG and conventional CGI operations. The reduction in radial CO_2 migration is prominent for the WAG operation.

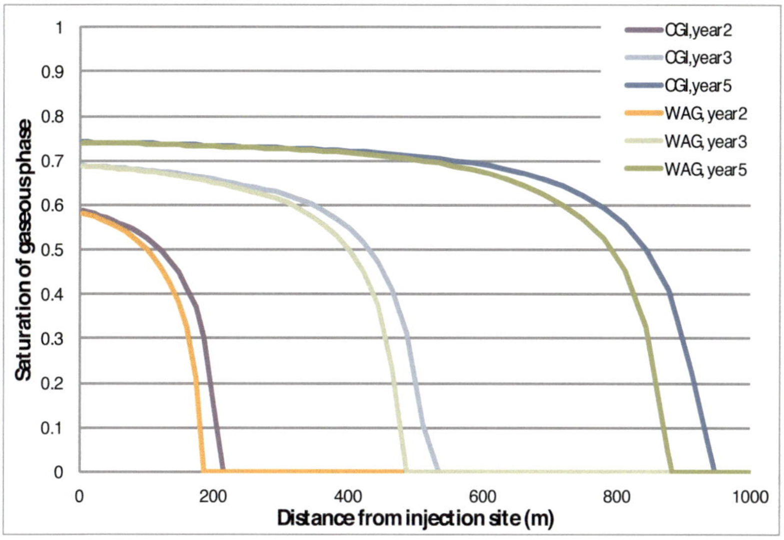

Figure 16. SG underneath the caprock showing plume reduction with optimized WAG operation for Utsira formation SAGCS

Table 6 summarizes the benefits of adopting the optimized WAG operation for the generalized Utsira formation model.

CGI	CO_2 Radial Migration	946.7 m
	Dissolution	16.89 %
WAG	CO_2 Radial Reduction	65.2 m
	CO_2 Radial Reduction Ratio	6.89 %
	CO_2 Impact Area Reduction	372,095 m^2
	CO_2 Impact Area Reduction Ratio	13.23 %
	Total Water Injection Required	1.5625 million metric tons annually
	Dissolution	23.43 %

Table 6. Benefits of implementing the optimized WAG operation for Utsira formation SAGCS

As seen from Figure 16 and Table 6, CO_2 migration under the caprock in the generalized Utsira formation has been significantly reduced by the WAG operation. Compared to the case of the Frio formation, a higher optimization fitness value for the Utsira formation model means more effective WAG operation. The time-elapsed CO_2 migration recorded in Figure 16 provides clear evidence that reduction in CO_2 migration can be observed as early as two years after injection. More importantly, it can also be seen that migration reduction in later years tends to be greater than that in earlier years, suggesting the development of greater reduction in CO_2 migration as injection proceeds. This is an encouraging result considering the decade-long lifespan of SAGCS projects. It is also useful to investigate how the layered structure of the formation affects the performance of the WAG operation. A closer look at the in situ CO_2 migration for conventional CGI operation (shown in Figure 17) and optimized WAG operation (shown in Figure 18) provides information on the effect of a layered structure on conventional CGI and optimal WG operation.

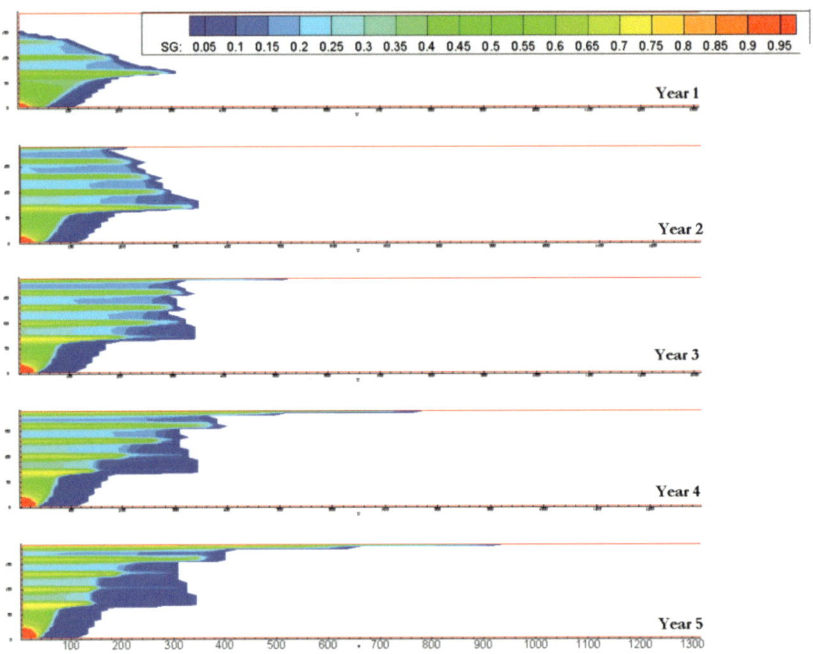

Figure 17. CO_2 plume migration during the first 5 years of CGI operation for the Utsira formation

The generalized Utsira formation model has an equivalent horizontal-to-vertical permeability ratio of 29 according to the calculation presented in Section 4.2.4. The simulations presented in Section 4.2.3 showed great improvement in WAG performance due to the anisotropic permeability. Similar conclusion can be also drawn from the simulation results presented in this section. Comparing Figure 17 and Figure 18 side by side, it can be seen that in situ CO_2 migration has been significantly reduced by the WAG operation, under which the displacement of brine in the lower sandstone layer becomes more stable. Storage efficiency increases under such scenario since more pore-space can now be occupied by supercritical CO_2. Additionally, storage safety is also improved due to the lowered concentration of supercritical CO_2.

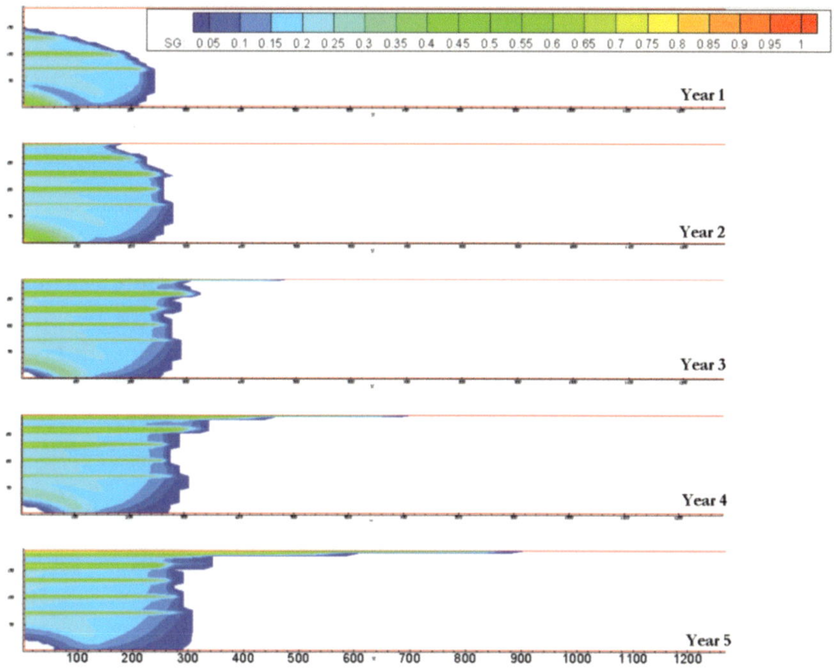

Figure 18. CO_2 plume migration during the first 5 years of optimized WAG operation for the Utsira formation

However, the fitness function value for the layered Utsira formation is not as satisfactory as that for the generic anisotropic aquifer (0.0251 versus 0.1438), although the former case has a higher permeability ratio. Two possible explanations are proposed for these results. First, the upward migration (as well as the resulting radial migration) of in situ CO_2 has already been significantly retarded by the secondary-sealing effect introduced by the layered structure of the formation. Recalling previous analysis, it can be seen that the migration reduction mechanism is similar for both WAG operation and the secondary sealing effect. Secondly, the CO_2 injection rate is set at 1 million metric tons annually for the layered Utsira formation, while it was 0.5 million metric tons annually for the generic anisotropic aquifer. This doubling of CO_2 injection rate for the Utsira formation effectively speeds up the upward migration of in situ CO_2. However, as has been previously discussed, CO_2-water interaction has to take place before the slug reaches the caprock to ensure the superior performance of WAG operation. It is therefore the enhanced upward migration of CO_2 together with the secondary sealing effect that make the WAG operation less satisfactory when applied to the generalized layered Utsira formation model compared to the WAG operation for the generic anisotropic aquifer.

4.1.6. WAG operation for Utsira Layer #9 model

The topmost sandstone layer (Layer #9) of the Utsira formation as presented in Chapter titled, "Numerical Simulation of CO2 Sequestration in Large Saline Aquifers" can serve as another excellent candidate to investigate WAG operation due to its well-understood hydrogeological properties. A cylindrical domain with the average thickness of the Utsira Layer #9 is modeled, which possesses identical characteristics of the detailed 3D Utsira Layer #9 model (from Chapter titled, "Numerical Simulation of CO2 Sequestration in Large Saline Aquifers") except for the absence of 3D topography. Although topographical details can be important in determining the accurate migration of in situ CO_2, such a simplification without compromising accuracy is necessary to analyze the effectiveness of WAG operation on CO_2 migration without incurring excessive computational cost.

The geometric and hydrogeological characteristics of the simplified Utsira Layer #9 model can be summarized as follows. We consider a cylindrical domain with thickness of 35 m with horizontal flat caprock. All hydrogeological properties are retained from the detailed 3D Utsira Layer #9 model described in the previous section, the most important being the horizontal-to-vertical permeability ratio of 10. CGI operation with nine-year average CO_2 injection rate of 2.7 kg/s is considered as the baseline case for comparison. The effect of WAG cycle durations on CO_2 migration is investigated for this relatively thin formation. The 30-day, 15-day, and 5-day WAG cycle durations are considered for the WAG optimization design. Our computations show that for the simplified Utsira Layer #9 model, only the WAG operation with 5-day cycle leads to noticeable migration reduction. Therefore, all results given below are for 5-day WAG cycle. The values of design variables corresponding to optimal WAG operation and the optimal fitness function value are found as $I_{CO2,optimal} = 11.56$ kg/s, $I_{water,optimal} = 7.62$ kg/s, $r_{WAG,optimal} = 0.646$, and $fitness_{optimal} = 0.506$ m/10^3 metric tons of water. Correspondingly, the durations of CO_2 and water injection in one WAG cycle can be calculated as 11 and 19 days, respectively. Identical WAG cycles are repeated 20 times to complete a two-year operation. Figure 19 shows the graph of optimized WAG operation for the simplified Utsira Layer #9 model with 5-day WAG cycle duration.

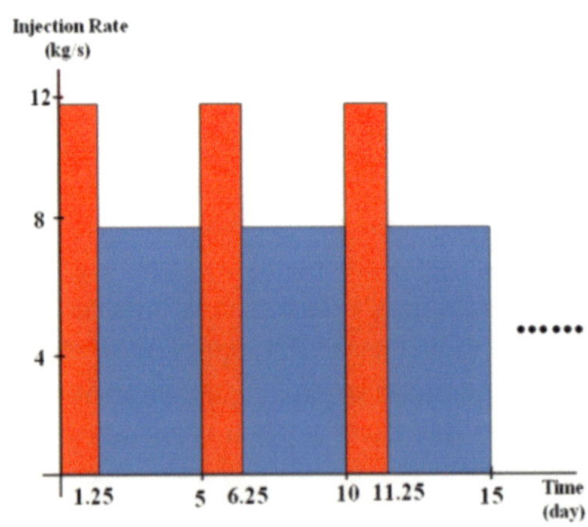

Figure 19. Graph of optimized WAG operation for Utsira Layer #9 model

Figure 20 shows the CO_2 migration underneath the caprock after two years of conventional CGI and optimized WAG operation. The reduction in radial CO_2 migration is significant for WAG operation.

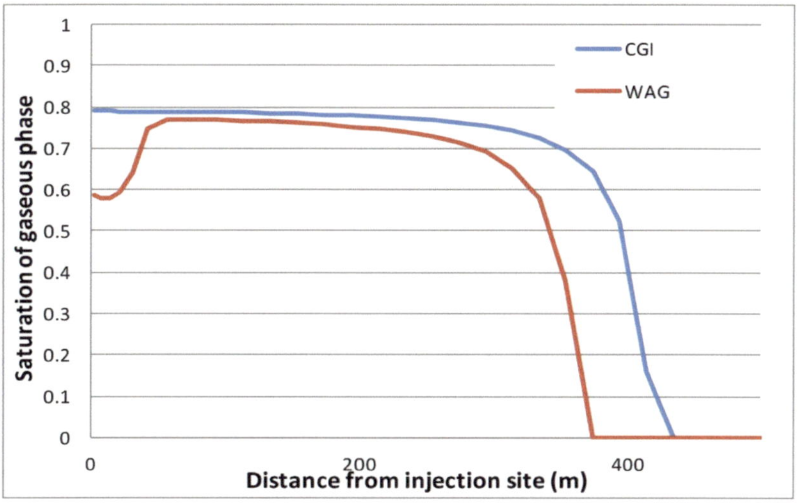

Figure 20. SG underneath the caprock showing plume reduction with optimized WAG injection for Utsira Layer #9 model

Table 7 summarizes the benefits of adopting the optimized WAG injection for the simplified Utsira Layer #9 model.

As seen from Figure 20 and Table 7, significant reduction in CO_2 migration has been achieved after only two years of WAG operation. Additionally, CO_2 dissolution is also significantly enhanced from about 9% to 22% of the total injected CO_2. More importantly, the results reveal the strong relationship between WAG cycle duration and the reservoir thickness regarding the performance of WAG operation. Surprisingly, our simulation results show that the 30-day cycle WAG operation actually "enhances" (not "reduces") the lateral migration of the CO_2 plume. Such a situation can be slightly mitigated when the 15-day WAG cycle duration is applied. However, no noticeable migration reduction is achieved. Considering all three cases of WAG operation for the identified formations, it appears that the aquifer thickness and WAG cycle duration are critical factors affecting the performance of a WAG operation. When the aquifer is thin, it takes less time for the CO_2 to reach the caprock. The assumption of treating alternative water and CO_2 slugs as a quasi-mixture is only valid when injected CO_2 interacts with the chasing water before it reaches the caprock. Failing to fulfill this requirement leads to poor WAG performance. It is the reservoir thickness and WAG cycle duration that determine the validity of quasi-mixture assumption for a given aquifer. Longer WAG cycle duration requires larger reservoir thickness and vice versa. Our simulations show that a minimum reservoir thickness may exist for a given WAG cycle duration under which the quasi-mixture assumption is valid and vice versa. This minimum thickness requirement may ultimately determine the technical feasibility of a WAG operation for an aquifer for achieving any reduction in CO_2 migration. Following this rationale, the success of WAG operation with 5-day cycle duration and its failure with the 15-day and 30-day cycle durations can be explained for Utsira Layer #9 model. This also implies that the CO_2 injected at the bottom of Layer #9

CGI	CO_2 Radial Migration	423 m
	Dissolution	8.97 %
WAG	CO_2 Radial Reduction	49 m
	CO_2 Radial Reduction Ratio	11.58 %
	CO_2 Impact Area Reduction	122689 m^2
	CO_2 Impact Area Reduction Ratio	21.83 %
	Total Water Injection Required	231916 metric tons
	Dissolution	23.02 %

Table 7. Benefits of optimized WAG operation for the Utsira Layer #9 model

reaches the caprock between 5 to 30 days (more likely in approximately 15 days since minor reduction in plume can be observed in this case) with the given reservoir hydrogeological properties and injection parameters.

With the simulation and optimization of WAG operation for the generalized Utsira formation and Utsira Layer #9 model, one can draw the conclusion that the WAG operation certainly holds technical promise to retard the spread of gaseous CO_2 in actual large-scale saline aquifers. It is also obvious from the results that the timeframe of in situ CO_2-water mixing versus the chosen WAG cycle duration are important considerations that must be carefully determined to assure improved reservoir performance from implementation of WAG operation. Various geological factors of the formation, such as geological slope and reservoir thickness, can contribute to insufficient mixing and thus compromise the performance of the WAG operation. Therefore, the operational parameters of WAG operation need to be designed on a case-by-case basis to achieve optimal performance.

4.1.7. Sensitivity of WAG operational parameters

In previous sections, it has been shown that the performance of WAG operation varies depending on the aquifer's geometric and hydrogeological parameters. It is therefore beneficial to look into the effect of various operational parameters on the performance of WAG operation. If the total amount of CO_2 for sequestration is given, any three out of the four operational parameters, namely the CO_2 injection rate (I_{CO2}), the water injection rate (I_{water}), the WAG ratio (r_{WAG}), and the WAG cycle duration (t_{WAG}), determine a unique WAG operation pattern. Using the case of generic anisotropic saline formation for SAGCS described in section 4.2.4, the effect of WAG operational parameters on the performance of WAG operation is investigated. The optimized case presented in section 4.2.4 is used as the baseline case. The CO_2 injection rate (I_{CO2}), water injection rate (I_{water}), and cycle duration (t_{WAG}) are chosen as the WAG operational parameters. Four additional cases are considered, the results of which are summarized in Table 8.

http://dx.doi.org/10.5772/57066

	I_{CO2} (kg/s)	I_{water} (kg/s)	Cycle Duration (day)	r_{WAG}
Baseline Case	36.13	33.35	30	0.847
Variation #1	50	33.35	30	0.697
Variation #2	36.13	50	30	0.565
Variation #3	36.13	33.35	15	0.847
Variation #4	36.13	33.35	50	0.847

Table 8. Various simulations used in the sensitivity study of WAG operation to its operational parameters

The following figures present radial cross-sectional views showing the migration of in situ CO_2 under the above four WAG cases of Table 8. Since all these cases are small variations from the optimal WAG baseline case, they all show significant reduction in CO_2 migration compared to the CGI case. However, the performance of WAG operation (fitness function), which is defined as plume reduction per unit amount of water injection, differs greatly from one case to another. Figure 21 shows the in situ CO_2 distribution in the reservoir for the optimized WAG operation and its variation #1 (with higher CO_2 injection rate compared to the baseline case).

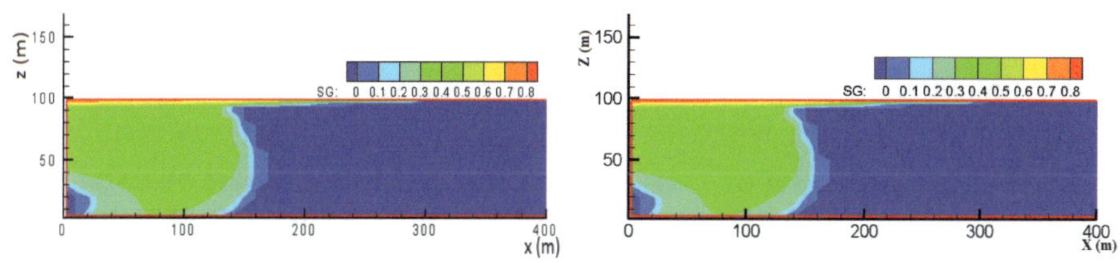

Figure 21. CO_2 distribution in the reservoir (left: optimized WAG; right: WAG with variation #1)

Figure 22 shows the in situ CO_2 distribution in the reservoir for the optimized WAG operation and its variation #2 (with higher water injection rate compared to the baseline case).

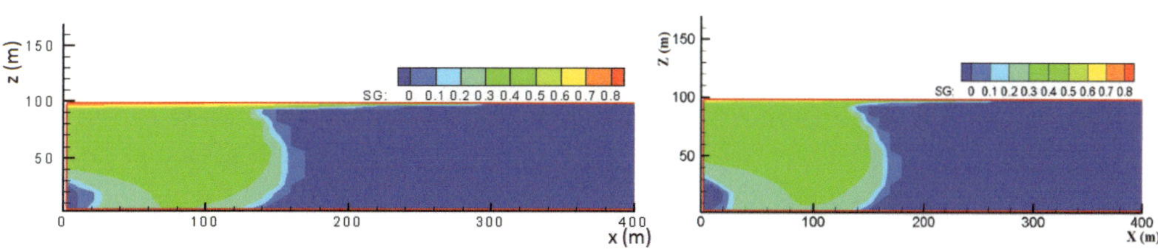

Figure 22. CO_2 distribution in the reservoir (left: optimized WAG; right: WAG with variation #2)

Figure 23 shows the in situ CO_2 distribution in the reservoir for variation #3 and variation #4 (with shorter and longer WAG cycle duration respectively compared to the baseline case).

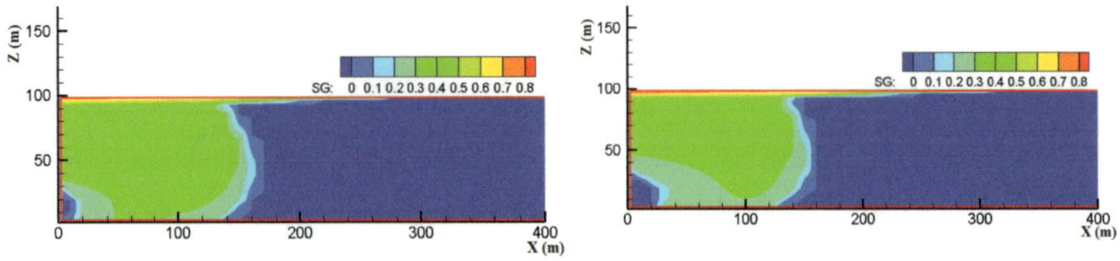

Figure 23. CO_2 distribution in the reservoir (left: WAG with variation #3; right: WAG with variation #4)

Since the lateral extent of the CO_2 plume is determined by the gaseous phase concentration beneath the caprock, its saturation is examined for the original optimized WAG operation and its four variations, as shown in Figure 24.

Table 9 shows the relative performance of the original optimized WAG operation and its four variations.

The results from the above sensitivity analysis are very informative. The following conclusions can be reached. First, none of the four additional cases with slight variations in WAG operational parameters led to higher fitness function value than the optimized baseline case. This result further validates the optimization capability of GA-TOUGH2. Second, CO_2 migration reduction is obtained in all the four cases. Variation case #2 and #4 even achieve greater reduction in migration compared to the baseline case. However, cases with greater migration reduction may not be desirable because the energy penalty (additional water requirement) is more severe for these cases. Third, the crucial role of WAG cycle duration on its performance is also evident. It is clear that the shorter WAG cycle duration is preferable for efficient WAG operation due to the resulting enhanced mixing of CO_2 and water. However, the frequent switching between CO_2 and water injection can be limited by the existing technology barriers.

5. Optimal pressure management

There are two reasons that make the injection pressure one of the most important operational parameters for the success of SAGCS. One is the well injectivity, which determines the total amount of CO_2 that can be injected in a given amount of time, and the other is the safety constraint on injection pressure, which should not exceed the formation's fracture pressure. In petroleum engineering, injectivity of an injection well is defined as the net fluid flow delivered per unit pressure differential between the mean injection pressure and the mean formation pressure. The definition of injectivity is given by equation (10) below.

$$injectivity = \frac{Q_{CO2}}{p_{\text{injection}} - p_{reservior}} \tag{10}$$

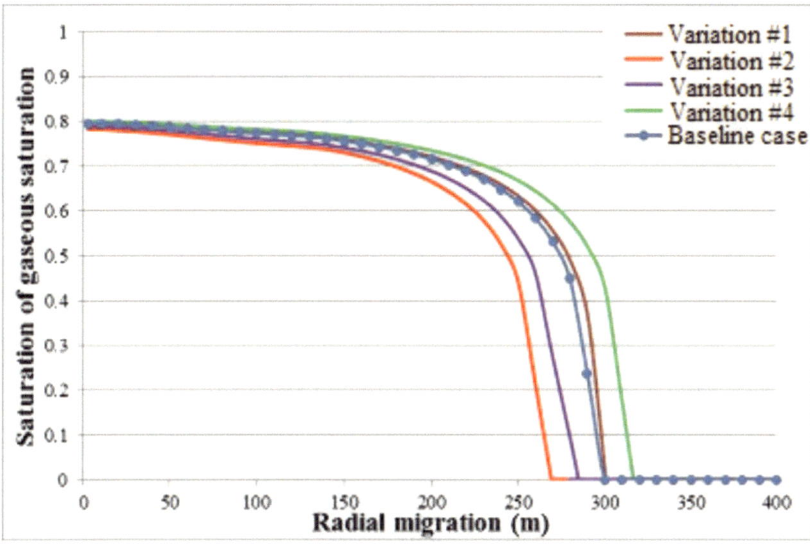

Figure 24. SG underneath the caprock, original optimized WAG operation and its four variations

	Baseline Case	Variation #1	Variation #2	Variation #3	Variation #4
Total CO_2 Injection (kg)	821917	821917	821917	821917	821917
Total Water Injection (kg)	973574	1179222	1454722	973574	973574
WAG ratio	0.847	0.697	0.565	0.847	0.847
Maximum Migration (m)	290	301	270	280	317
Migration Reduction (m)	140	129	160	150	113
Fitness (m/10^3 metric tons water)	0.1438	0.1096	0.1103	0.1545	0.1166

Table 9. Performance of the original optimized WAG operation and its variations

where Q_{CO2} is the injection mass rate and $p_{injection}$ is the injection pressure. The injectivity measure indicates the ability of an injection well to deliver supercritical CO_2 into the aquifer. The injection pressure response for a given SAGCS operation can be analyzed as follows. Applying Darcy's law to the region adjacent to the injection well, the achievable CO_2 injection mass rate Q_{CO2} is proportional to the product of the relative permeability $k_{r,g}$ of CO_2 and the pressure gradient near the injection well Δp. For two-phase flow of supercritical CO_2 and brine, $k_{r,g}$ is inversely proportional to the saturation of brine S_b. At an early stage of CO_2 injection, the pore space near the injection well is primarily occupied by the brine, which means high S_b in the adjacent region of the injection well. As a direct consequence, $k_{r,g}$ is relatively low, which results in considerable difficulty to displace brine by injecting CO_2. A direct indicator of this difficulty is the significant elevation of injection pressure, or in other words, very low injectivity. However, CO_2 injectivity does not remain unchanged. As injection continues, more brine is displaced from the pore space adjacent to the injection well, which effectively lowers S_b.

Simultaneously, $k_{r,g}$ increases. The increased $k_{r,g}$ at intermediate and later stages of CO_2 injection results in improvement of CO_2 injectivity. Therefore, if the injection rate is assumed constant, one can draw the conclusion that at the beginning of injection, high injection pressure is required to overcome the low effective permeability of CO_2. However, as more brine is displaced, injection pressure gradually drops because the permeability of CO_2 increases. Figure 25 schematically shows the effect of injection rate on injection pressure.

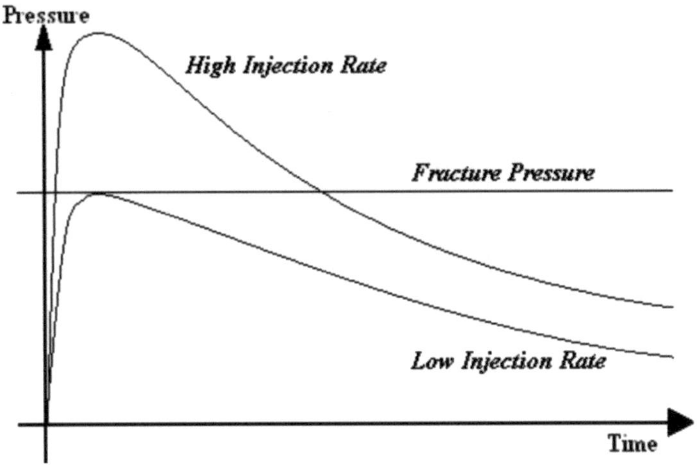

Figure 25. Graph of injection pressure response with time under various CO_2 injection rates

Intuitively, high injection rate is always preferred, since it can lead to more mass injection within a given time. However, a higher injection rate requires greater injection pressure. Regardless of the pumping capacity of the available injection equipment to provide the needed injection pressure, a critical constraint on allowable injection pressure exists. Like all mechanical structures, geological formations can also bear only a certain level of maximum stress while maintaining their integrity. They fracture when excessive stress is applied. Fractures in a formation can serve as pathways for the in situ mobile CO_2 to migrate to shallower aquifers and even all the way to the surface. The leakage of CO_2 through geological fractures is potentially threatening to the ecosystem near the storage site, so needless to say it will also significantly compromise its storage efficiency. Therefore, every attempt should be made to ensure the integrity of the formation, i.e., under no circumstance should the injection pressure exceed the formation's fracture pressure. Since the fracture pressure is an intrinsic property of the formation, it is likely to remain constant during the injection phase of SAGCS, shown by the horizontal line in Figure 25. Considering the injection pressure response under CGI operation and the fracture pressure guideline, Figure 25 reveals a crucial issue that must be addressed. If CO_2 is pumped into the aquifer with a relatively high injection rate (following the "High Injection Rate" scenario in Figure 25), the excessive pressure elevation at the early stage of injection can easily jeopardize the integrity of the formation; on the other hand, if CO_2 is pumped with a relatively low injection rate to ensure the formation's integrity, the injection will become inefficient at the intermediate and late stage since more CO_2 injection could have been achieved by a moderate increase in the injection pressure at these stages. Therefore, the overall injectivity can be improved while sustaining the sequestration security, if the injection rate can be adjusted with respect to time such that the injection pressure levels off as it approaches the fracture pressure and is maintained at that level during the later injection stage. Such a scenario is identified as the constant pressure injection (CPI) since the injection pressure is more or less maintained at a constant level. The concept of CPI fits perfectly into the category of the development of a "smart" injection well for SAGCS.

5.1. Method for designing constant pressure injection (CPI)

The optimization problem setup for CPI is rather straightforward. Before the optimization, a threshold pressure (the pressure limit chosen based on the formation's fracture pressure and other engineering concerns and regulations) is chosen as the optimization constraint. Since it is assumed that the injection rate is the only quantity to be adjusted for CPI, it becomes the design variable. The optimization is then carried out to minimize the fitness function defined by equation (9).

$$\text{fitness function} = \text{modified injectivity} = \frac{\left| p_{\text{threshold}} - p_{injection}(Q_{CO2}) \right|}{Q_{CO2}} \tag{11}$$

When the fitness function in equation (9) approaches zero, CPI operation is obtained and the corresponding injection scenario can then be determined. The optimization design of CPI operation is carried out using GA-TOUGH2. The optimization is essentially a solution-searching problem utilizing the GA optimization technique.

Unlike the optimization of the WAG operation, a new challenge emerges, namely how to describe the CO_2 injection rate as a time-dependent continuous function with limited discrete data. The concept of Bézier curve is introduced to address this problem. A Bézier curve is a parametric curve frequently used in computer graphics and related fields [15],[16]. It is defined by a set of control points, and uses them as coefficients of a certain polynomial to describe continuous curves. The control points of a Bézier curve can be denoted as P_0 through P_n, with (n-1) being the order of the Bézier curve. The order determines the complexity of the Bézier curve. A Bézier curve provides a simple means of creating arbitrary complex curves. A generalized mathematical expression of an nth order Bézier curve is given as:

$$B(t) = \sum_{i=0}^{n} \binom{n}{i} (1-t)^{n-i} t^i P_i \tag{12}$$

where (n,i) is the binomial coefficient, P_i is the ith control point defined prior to the curve's generation, and t is a variable defined on [0, 1]. Defining four control points as P_1, P_2, P_3, and P_4, an example of a cubic Bézier curve is shown in Figure 26.

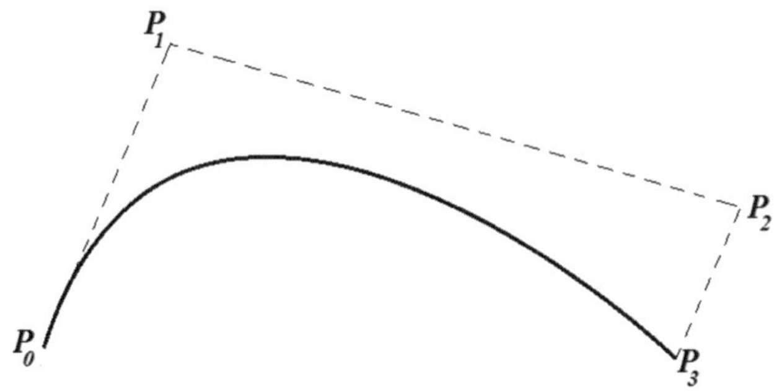

Figure 26. Graph of a cubic (3rd order) Bézier curve

In our work, each CO_2 injection scenario is described by a cubic Bezier curve. The CO_2 injection is a time dependent function of mass flow rate. Discretization of the injection with respect to time is needed to make the problem tractable for numerical simulation. With the discretization, CO_2 injection can be represented by step-functions for each time interval, and ultimately approximates the smooth injection as the time interval becomes small enough. The injection rate for each discrete time step is described at the midpoint of the interval, known as the sample point. Coordinates of other control points are arbitrarily generated for each GA individual. Any arbitrary CO_2 injection scenario can be generated by letting the parameter t increase from 0 to 1.

The design of CPI operation employs the identical hypothetical generic saline formation modeled for the optimization of WAG injection with a horizontal injector, as shown previously in Figure 20. All hydrogeological properties and numerical parameters remain unchanged. A threshold pressure of 180 bar is set for the maximum allowable injection pressure, with the assumption of a 50% increase from the initial pressure (120 bar). As mentioned earlier, the choice of threshold pressure is based on considerations of various aspects, such as the fracture pressure, injection regulation, safety factor, and risk analysis. The injection rate is allowed to vary between 0 kg/s and 150 kg/s, and the injection lasts for five years. The injection pressure response and the corresponding time-dependent injection rate of the optimized CPI operation are given in Figure 27. Two CGI cases, one with high injection rate (44 kg/s) and one with low rate (24 kg/s), are also included in these figures for comparison.

Several conclusions can be drawn by carefully examining the results of Figure 27. First, the injection pressure (green curve in Figure 27) is well behaved under the constraint of the threshold pressure. It increases rapidly at the early stage of the injection (on the order of days), and levels off as it approaches 180 bar. This is exactly the desired behavior of injection pressure response. Starting from 28 kg/s, the injection rate keeps increasing with the stabilized injection pressure. This means that the well injectivity gradually improves as CO_2 injection continues. Improved injectivity indicates more injected CO_2 after five years of operation. A direct indicator of the success of the designed CPI operation is the five-year average injection rate of 38 kg/s (compared to 34 kg/s for CGI operation). Second, both CGI operations first give an increase and then a decrease in injection pressure response, validating our previous conclu-

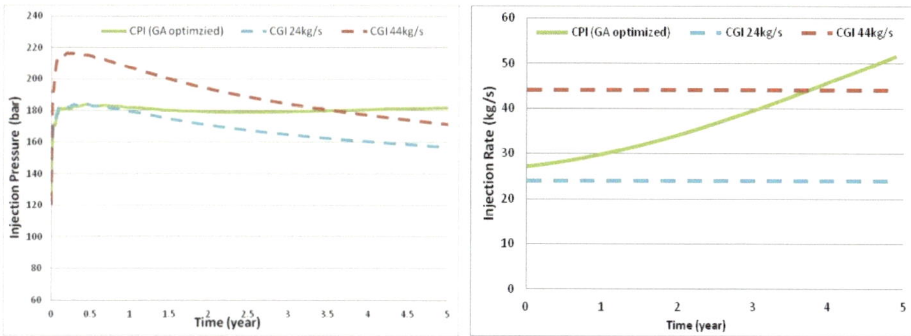

Figure 27. Injection pressure and injection rate of the optimized CPI operation, and its comparison with low-rate CGI and high-rate CGI

sion. Similar behavior of loss in injectivity has also been suggested by Burton et al. [17]. It can be seen that the injection pressure reaches about 220 bar with the high-rate CGI operation (44 kg/s), which is a 40 bar overshoot above the threshold pressure. Additionally, such pressure overshoot lasts for over 3.5 years before the injection pressure falls below 180 bar. Such large and prolonged pressure overshoot poses a significant risk to the formation's integrity. On the other hand, it is also seen that the injection pressure response with the low-rate CGI operation (24 kg/s) falls far below the threshold pressure after it peaks at the early stage. Although the integrity of the formation is not threatened, the injectivity has been severely compromised under such a low injection rate. Thus, only the CPI operation gives the optimal injection pressure management, which realizes the best injectivity while ensuring the injection safety by keeping the pressure always below the fracture pressure of the formation. Again, GA-TOUGH2 has successfully designed the CPI operation for a given pressure constraint.

6. Performance optimization of a multi-well system

It is likely that only a system of multiple injection wells would deliver enough injectivity for industrial level SAGCS. In the presence of multiple wells, the low compressibility of brine can potentially result in strong interference in the pressure generated by each well. This then brings up the question of how much the pressure interference is generated in a multi-well injection system, and how the injection wells should be placed to minimize interference. Two types of interference have been identified in a multi-well injection system, namely the CO_2 front interference and the pressure front interference, as described by Eccles et al. [18]. Neglecting complex in situ interactions such as phase shifting and mineralization, the interface between injection wells can be roughly estimated by superposition of the quantities from each single-well injection. If the aquifer is assumed to be relatively isotropic in its hydrogeological properties, Darvish et al. have shown that placing the wells on corners of regular polygons is preferred for uniform interference among wells [19]. Following such angular distribution of injection wells, the distance between the wells becomes the design variable for optimization. A four-well injection system is first considered to investigate the interference of plume migration and pressure disturbance between the wells. Afterwards, a two-well injection system is considered to study the relationship between well spacing and injectivity.

6.1. Four-well injection system

A hypothetical saline aquifer with dimensions of 4000 m × 4000 m × 70 m is modeled for this study. Generic hydrogeological properties and reservoir conditions similar to those used in the WAG operation study are assigned. The computational mesh is refined near the injection wells for accurate capture of the interference. Four cases with different inter-well distance are considered, in which injection wells are 600 m, 800 m, 1200 m, and 1600 m diagonally apart. The computational domains for the four cases are shown in Figure 28.

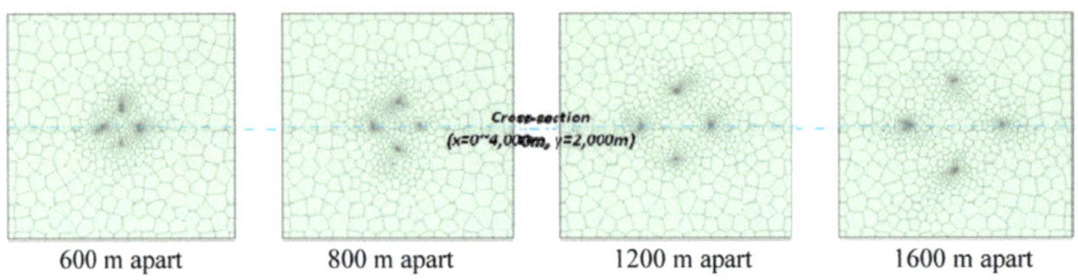

Figure 28. Computational domain of four-well injection systems with various inter-well distances

CO_2 is injected at a constant rate of 5 kg/s in each well. Mean injection pressure and gas saturation underneath the caprock after five years of injection are examined along the cross section indicated in Figure 29. The pressure response and CO_2 saturation curves are shown and compared in Figure 29 and Figure 30 respectively.

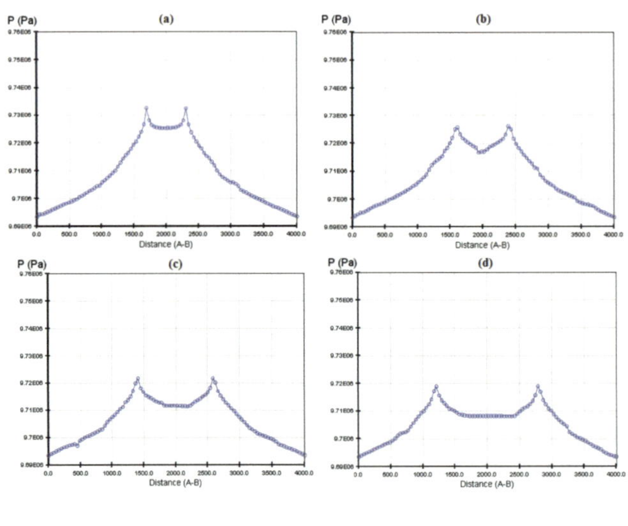

Figure 29. Pressure profile at the cross-section: wells (a) 600 m apart, (b) 800 m apart, (c) 1200 m apart, and (d) 1600 m apart

http://dx.doi.org/10.5772/57066

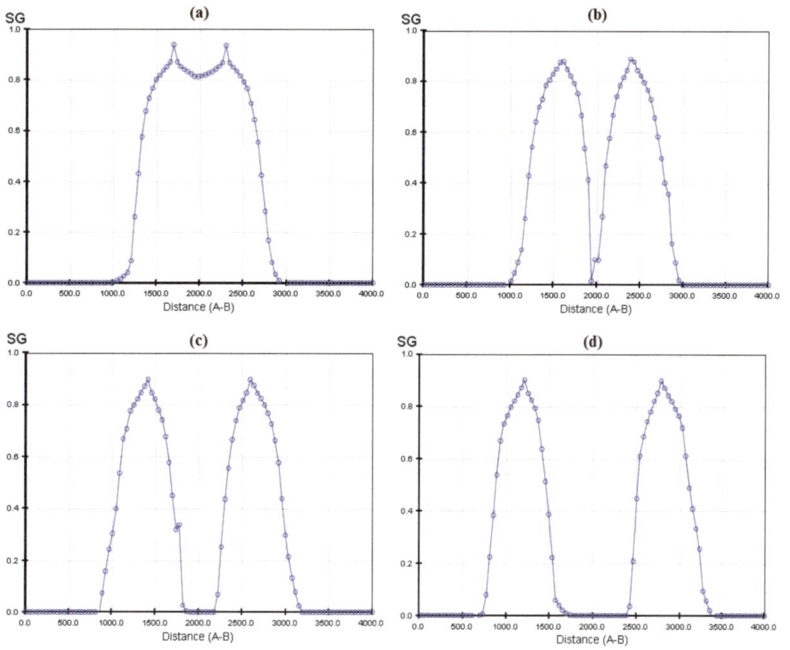

Figure 30. Gas saturation underneath the caprock at the cross-section: wells (a) 600 m apart, (b) 800 m apart, (c) 1200 m apart, and (d) 1600 m apart

The following conclusions can be reached based on the results shown in Figure 29 and Figure 30. First, well injectivity can be greatly improved by utilization of a multi-well injection system. The four-well system considered delivers CO_2 at total rate of 20 kg/s with no greater than 0.4% increase in reservoir pressure. This is orders of magnitude smaller than that of a single injection well for the same injection rate. This shows the technical benefit of utilizing a multi-well injection system for industrial level SAGCS. Second, well interference is prominent due to the presence of multiple injection wells. In Figure 29, the injection induced pressure elevation is 0.38% of the reservoir mean pressure for the case of 600-m inter-well spacing, while it drops to only 0.26% for the case of 1600-m inter-well spacing. That is to say, pressure interface is about 32% stronger when injection wells are 600 m apart compared to when they are 1600 m apart. The interference of the CO_2 plume is also seen in Figure 30. Third, comparison of Figure 29 and Figure 30 shows that the pressure interference is predominantly responsible for the compromised injectivity in a multi-well injection system. Plume interference can be easily avoided by moderately increasing the well spacing. For instance, plume interference is prominent for 600-m inter-well spacing, but almost disappears for 800-m inter-well spacing (as shown in Figure 30). On the other hand, the pressure interference remains persistent, which requires at least 1600-m inter-well spacing to become insignificant. However, large spacing between injection wells will result in greater land use. Therefore, optimal placement of wells is desirable to achieve an acceptable pressure and capacity interference as well as land use.

6.2. Two-well injection system

Because pressure interference is largely responsible for compromised injectivity of a multi-well injection system, in this section we examine the relationship between inter-well spacing and well injectivity. A half domain with dimensions 50000 m × 25000 m × 100 m is modeled. The computational domain is horizontally discretized by a uniform quadrilateral mesh with resolution of 200 m × 200 m and 500 m × 500 m. Two injection wells are assigned symmetrically at the center of the domain. The distance between these two injection wells is allowed to change freely and is considered as a design variable for the GA optimizer. Like the previous investigations of the four-well injection system, the injection operation is assumed to last for five years. Three cases with different injection rates and model parameters are considered, summarized in Table 10.

	Case #1	**Case #2**	**Case #3**
Single Well Injection Rate	2 kg/s	16 kg/s	16 kg/s
Hydrogeological Properties	Generic formation	Generic formation	Generalized Utsira formation
Mesh Resolution	200 m x 200 m	500 m x 500 m	500 m x 500 m

Table 10. Optimization cases for two-well injection system.

Denoting the injection pressure of the two-well injection system as $P_{two\text{-}well}$, and the injection pressure of the single-well injection system as $P_{single\text{-}well}$, the pressure difference ΔP between $P_{two\text{-}well}$ and $P_{single\text{-}well}$ is chosen as the fitness function. GA-TOUGH2 is employed to determine the minimal inter-well distance for a designated value of ΔP. For each case in Table 10, three optimization criteria are considered, namely ΔP being no greater than 0.1%, 0.5%, and 2% of $P_{single\text{-}well}$. The value of ΔP is examined and optimization is performed at the end of the five-year injection. In addition, well injectivity loss due to the pressure interference is also evaluated for the optimal well spacing given by GA-TOUGH2. Recalling the definition of well injectivity, equation (11), the injectivity loss of the two-well injection system can be evaluated as:

$$\begin{aligned} \text{injectivity loss} &= \text{injectivity}_{\text{two-well system}} - \text{injectivity}_{\text{single-well system}} \\ &= \left(\frac{Q_{CO2}}{P_{\text{two-well}} - P_{\text{reservior}}} - \frac{Q_{CO2}}{P_{\text{single-well}} - P_{\text{reservior}}} \right) \Bigg/ \frac{Q_{CO2}}{P_{\sin glewell} - P_{reservior}} \\ &= \frac{P_{\text{two-well}} - P_{\text{single-well}}}{P_{\text{single-well}} - P_{\text{reservior}}} \end{aligned} \quad (13)$$

where Q_{CO2}, $P_{two\text{-}well}$, and $P_{single\text{-}well}$ are as defined earlier. The optimization results are summarized in Table 11.

Several conclusions can be reached after careful examination of the optimization results in Table 11. First, it should be noted that the total amount of injected CO_2 is doubled from the

	Case #1		Case #2		Case #3	
Optimization Criteria	Inter-well Distance	Injectivity Loss	Inter-well Distance	Injectivity Loss	Inter-well Distance	Injectivity Loss
$\Delta P < 0.1$ % *Psingle-well*	5.4 km	-2.84%	32 km	-0.94%	14 km	-2.53%
$\Delta P < 0.5$ % *Psingle-well*	1.8 km	-12.76%	19 km	-4.54%	below mesh resolution	
$\Delta P < 2$ % *Psingle-well*	below mesh resolution		6.5 km	-28.57%	below mesh resolution	

Table 11. Optimal inter-well spacing and injectivity tradeoff for three cases under three optimization criteria.

single-well injection case for all optimization cases due to the presence of the second injection well. Second, the results show that for Case #1, pressure interference can be substantially reduced for ($\Delta P < 0.1\%$ of $P_{single\text{-}well}$) by placing injection wells 5400 m apart, for ($\Delta P < 0.5$ % of $P_{single\ well}$) by placing injection wells 1800 m apart, and for ($\Delta P < 2$ % of $P_{single\ well}$) by placing injection wells less than 200 m apart. The relative ease of mitigating the pressure interface for Case #1 can be explained by its low injection rate of 2 kg/s per well. However, a low injection rate leads to low injection pressure, which in turn makes the injectivity more sensitive to the injection pressure. 2.84% and 12.76% injectivity losses are found for Case #2 and Case #1 respectively. Case #2 is similar to Case #1 except for the significantly increased injection rate of 16 kg/s per well. Accordingly, the inter-well distance increases to meet the optimization criteria. It is estimated that at least 32 km, 19 km, and 6500 m distance between the wells is needed to realize the three levels of avoidance in pressure interference respectively. An encouraging result is that the injectivity loss for Case #2 is significantly smaller than that for Case #1. Such reduction in injectivity loss also implies the dominant role of injection rate when evaluating the injectivity of a multi-well injection system. Therefore, simply increasing the injection rate can be a direct and effective means to mitigate the injectivity loss due to pressure interference. However, it is worth noting that even for the reasonable avoidance of pressure interference in Case #2, the wells need to be placed 6500 m apart, which is still a significant distance considering land use. The exacerbated injectivity loss of 28.57% may also pose concerns on injection well performance. In Case #3, the hydrogeological properties of the Utsira sandstone formation are assigned to the modeled domain to obtain some real-life sense of the performance of a multi-well injection system. Due to the improved reservoir conditions, i.e., higher porosity and permeability, the inter-well distance to achieve the significant avoidance of pressure interference decreases greatly from 32 km in Case #2 to 14 km in Case #3. Moreover, the required inter-well distances even fall under the mesh resolution, i.e., less than 500 m, for the other less rigorous criteria in pressure interference. This suggests the great potential of implementing a multi-well injection system for the SAGCS of the Utsira formation without concern over large pressure interference. These results also show the promise of implementing a multi-well injection system for other large-scale saline formations.

7. Concluding remarks

In this chapter, the development of genetic-algorithm based optimizations has been described, integrated into the DOE multiphase flow and heat transfer numerical simulation code TOUGH2. The new code has been designated GA-TOUGH2. GA-TOUGH2 has the ability to determine optimal reservoir engineering techniques for improved CO_2 storage efficiency in saline aquifer carbon sequestration. Using GA-TOUGH2, the feasibility and technical benefits of adopting water-alternating-gas (WAG) injection technique has been investigated for SAGCS. In addition, problems such as optimal injection pressure management in SAGCS and injection well placement in a multi-well injection system have been investigated for the purpose of achieving higher storage efficiency and safer sequestration. Encouraging results have been obtained from all these optimization studies. Validated GA-TOUGH2 thus offers an innovative platform which holds great promise in studying a host of optimization/design problems for geological carbon sequestration.

As recommendations for the future work, more complex optimization studies could be performed to address a broader set of optimization problems, such as non-uniform WAG injection and maximization of capillary trapping. Some analytical solutions could be derived for more fundamental understanding of the WAG injection. Multi-objective GA optimization should also be introduced to obtained a higher level of optimization capability with consideration of multiple fitness functions. Additional real-life SAGCS projects such as the ADM project and FutureGen 2.0 should be continually studied over the years as more detailed field data become available. Optimization studies for these large-scale SAGCS projects should be performed for greater storage efficiency and reduced plume migration. GA-TOUGH2 should also be considered for the study of other aspects of GCS, such as enhanced oil or gas recovery in combination with carbon sequestration.

Acknowledgements

Financial support for this work was provided by the Consortium for Clean Coal Utilization (CCCU) at Washington University in St. Louis, MO, USA.

Author details

Ramesh K. Agarwal and Zheming Zhang

Department of Mechanical Engineering and Materials Science, Washington University in St. Louis, St. Louis, USA

References

[1] Pruess K. TOUGH2: A General Numerical Simulator for Multiphase Fluid and Heat Flow, Lawrence Berkeley Laboratory Report LBL-29400, California: Berkeley; 1999.

[2] Pruess K., Oldenburg C., and Moridis G. TOUGH2 User's Guide, Version 2.0 (revised), Lawrence Berkeley Laboratory Report LBL-43134, California: Berkeley; 2011.

[3] Genetic algorithm, Wikipedia website, http://en.wikipedia.org/wiki/Genetic_algorithm (accessed 24 May 2013).

[4] Goldberg D.E. Genetic Algorithms in Search, Optimization & Machine Learning. Addison-Wesley; 1989.

[5] Zhang Z. and Agarwal R. Numerical Simulation and Optimization of CO_2 Sequestration in Saline Aquifers. Computers & Fluids 2012; 16 (4) 891-899.

[6] Zhang Z. Numerical Simulation and Optimization of CO_2 Sequestration in Saline Aquifers. Ph.D dissertation. Washington University in St. Louis, USA; 2013.

[7] Tchelepi H., Durlofsky L., and Aziz K. A Numerical Simulation Framework for the Design, Management and Optimization of CO_2 Sequestration in Subsurface Formations. Global Climate and Energy Project (GCEP) Report, Stanford, 2009.

[8] Orr L. Carbon Capture and Sequestration: Where do We Stand? Presentation at NAE/AAES Convocation, Washington DC, 19 April, 2010.

[9] Bryant S.L., Lakshminarasimhan S., and Pope G.A. Buoyancy-dominated Multi-Phase Flow and Its Effect on Geological Sequestration of CO_2. Society of Petroleum Engineers Journal 2008; 447-454.

[10] Leonenko Y. and Keith D.W. Reservoir Engineering to Accelerate the Dissolution of CO_2 Stored in Aquifers. Environmental Science & Technology 2008; 42 2742-2747.

[11] Jikich S.A., Sams W.N., Bromhal G., Pope G., Gupta N., and Smith D.H. Carbon Dioxide Injectivity in Brine Reservoirs Using Horizontal Wells. In: 2nd Annual Conference on Carbon Sequestration, May 2003, Pittsburgh, PA, USA.

[12] Hassanzadeh H., Pooladi-Darvish M., and Keith D.W. Accelerating CO_2 Dissolution in Saline Aquifers for Geological Storage - Mechanistic and Sensitivity Studies. Energy and Fuels 2009; 23 3328-3336.

[13] Nasir F.M. and Chong Y.Y. The Effect of Different Carbon Dioxide Injection Modes on Oil Recovery. International Journal of Engineering & Technology 2009; 9 66-72.

[14] Audigane P., Gaus I., Czernichowski-Lauriol I., Pruess K., and Xu T., Two-Dimensional Reactive Transport Modeling of CO_2 Injection in a Saline Aquifer at the Sleipner Site. American Journal of Science 2007; 307 974-1008.

[15] Farin G. Curves and Surfaces for Computer-Aided Geometric Design, Fourth Edition. Academic Press, Waltham; 1996.

[16] Bézier curve, Wikipedia website, http://en.wikipedia.org/wiki/B%C3%A9zier_curve (accessed 24 May 2013).

[17] Burton M., Kumar N., and Bryant S.L. CO_2 Injectivity into Brine Aquifers: Why Relative Permeability Matters as Much as Absolute Permeability. Energy Procedia 2009; 1 (1) 3091-3098.

[18] Eccles J., Chandel M., and Pratson L. Large Scale Carbon Storage Deployment: Effects of Well Spacing on Geosequestration Site Costs and Capacity Estimates. In: 10th Annual Conference on Carbon Capture & Sequestration, May 2011,Pittsburgh, PA, USA.

[19] Pooladi-Darvish M., Moghdam S., and Xu D. Multiwell Injectivity for Storage of CO_2 in Aquifers. Energy Procedia 2011; 4 4252-4259.

Chapter 14

Alkaline and Alkaline-Earth Ceramic Oxides for CO_2 Capture, Separation and Subsequent Catalytic Chemical Conversion

Margarita J. Ramírez-Moreno,
Issis C. Romero-Ibarra, José Ortiz-Landeros and
Heriberto Pfeiffer

Additional information is available at the end of the chapter

http://dx.doi.org/10.5772/57444

1. Introduction

The amounts of anthropogenic carbon dioxide (CO_2) in the atmosphere have been raised dramatically, mainly due to the combustion of different carbonaceous materials used in energy production, transport and other important industries such as cement production, iron and steelmaking. To solve or at least mitigate this environmental problem, several alternatives have been proposed. A promising alternative for reducing the CO_2 emissions is the separation and/or capture and concentration of the gas and its subsequent chemical transformation. In that sense, a variety of materials have been tested containing alkaline and/or alkaline-earth oxide ceramics and have been found to be good options.

The aforementioned ceramics are able to selectively trap CO_2 under different conditions of temperature, pressure, humidity and gas mixture composition. The influence of those factors on the CO_2 capture (physically or chemically) seems to promote different sorption mechanisms, which depend on the material's chemical composition and the sorption conditions used. Actually, this capture performance suggests the feasibility of these kinds of solid for being used with different capture technologies and processes, such as: pressure swing adsorption (PSA), vacuum swing adsorption (VSP), temperature swing adsorption (TSA) and water gas shift reaction (WGSR). Therefore, the fundamental study regarding this matter can help to elucidate the whole phenomena in order to enhance the sorbents' properties.

2. CO_2 capture by different alkaline and alkaline-earth ceramics

Among the alkaline and/or alkaline-earth oxides, various lithium, sodium, potassium, calcium and magnesium ceramics have been proposed for CO_2 capture through adsorption and chemisorption processes [1-20]. These materials can be classified into two large groups: dense and porous ceramics. Dense ceramics mainly trap CO_2 chemically: the CO_2 is chemisorbed. Among these ceramics, CaO is the most studied one. It presents very interesting sorption capacities at high temperatures (T ≥ 600 °C). In addition to this material, alkaline ceramic oxides have been considered as possible captors, mostly lithium and sodium based ceramics (Li_5AlO_4 and Na_2ZrO_3, for example). In these cases, one of the most interesting properties is related to the wide temperature range in which some of these ceramics trap CO_2 (between 150 and 800 °C), as well as their high CO_2 capture capacity.

In these ceramics, the CO_2 capture occurs chemically, through a chemisorption process. At a micrometric scale, a general reaction mechanism has been proposed, where the following steps have been established: Initially, CO_2 reacts at the surface of the particles, producing the respective alkaline or alkaline-earth carbonate and in some cases different secondary phases. Some examples are:

$$Li_5AlO_4 + 2\,CO_2 \longrightarrow 2Li_2CO_3 + LiAlO_2 \quad (1)$$

$$Na_2ZrO_3 + CO_2 \longrightarrow Na_2CO_3 + ZrO_2 \quad (2)$$

$$CaO + CO_2 \longrightarrow CaCO_3 \quad (3)$$

The above reactions show that surface products can be composed of carbonates, but as well they can contain metal oxides or other alkaline/alkaline-earth ceramics. The presence of these secondary phases can modify (improve or reduce) the diffusion processes described below [1].

Once the external carbonate shell is formed, different diffusion mechanisms have to be activated in order to continue the CO_2 chemisorption, through the particle bulk. Some of the diffusion processes correspond to the CO_2 diffusion through the mesoporous external carbonate shell, and some others such as the intercrystalline and grain boundary diffusion processes [1, 18, 21].

Figure 1 shows the theoretical CO_2 chemisorption capacities (mmol of CO_2 per gram of ceramic) for the most studied alkaline and alkaline-earth ceramics. As it can be seen, metal oxides (Li_2O, MgO and CaO) are among the materials with the best CO_2 capture capacities. Nevertheless, Li_2O and MgO have not been really considered as possible options due to reactivity and kinetics factors, respectively. On the contrary, CaO is one of the most promising alkaline-earth based materials, with possible real industrial applications. Other interesting materials are ceramics with lithium or sodium phases, which present better thermal stabilities and volume variations than CaO. In addition, the sodium phases may present another advantage if the CO_2 capture is produced in the presence of steam. Under these conditions the sodium phases may produce sodium bicarbonate ($NaHCO_3$) as the carbonated phase, which is twice the amount of CO_2 could be trapped in comparison to the Na_2CO_3 product.

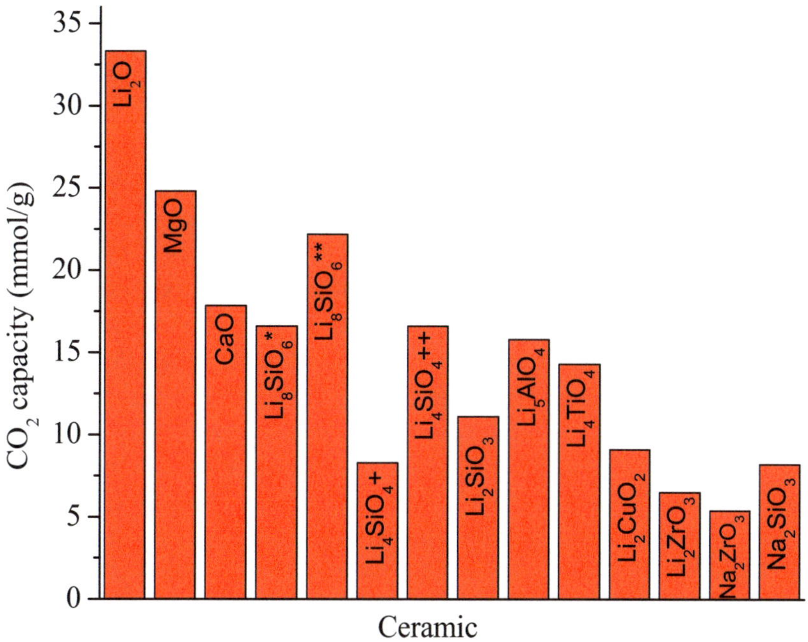

Figure 1. Theoretical CO_2 capture capacities for different alkaline and alkaline-earth ceramics. In the Li_8SiO_6 (labeled as *) and Li_4SiO_4 (labeled as +), the maximum capacity can depend on the CO_2 moles captured in each different phase formed ($Li_8SiO_6 + CO_2 \rightarrow Li_4SiO_4 + CO_2 \rightarrow Li_2SiO_3 + Li_2CO_3$).

Other ceramics containing alkaline-earth metals are the layered double hydroxides (LDH) or hydrotalcite-like compounds (HTLc). LDHs, also called anionic clays due to their layered structure and structural resemblance to a kind of naturally-occurring clay mineral. These materials are a family of anionic clays that have received much attention in the past decades because of their numerous applications in many different fields, such as antacids, PVC additives, flame retardants and more recently for drug delivery systems and as solid sorbents of gaseous pollutants [22-24]. The LDH structure is based on positively charged brucite like [$Mg(OH)_2$] layers that consist of divalent cations surrounded octahedrally by hydroxide ions. These octahedral units form infinite layers by edge sharing [25]. Due to the fact that certain fraction of the divalent cations can be substituted by trivalent cations at the centers of octahedral sites, an excess of positive charge is promoted. The excess of positive charge in the main layers of LDHs is compensated by the intercalation of anions in the hydrated interlayer space, to form the three-dimensional structure. These materials have relatively weak bonds between the interlayer and the sheet, so they exhibit excellent ability to capture organic or inorganic anions. The materials are easy to synthesize by several methods such as co-precipitation, rehydration-reconstruction, ion exchange, hydrothermal, urea hydrolysis and sol gel, although not always as a pure phase [26].

The LDH materials are represented by the general formula: $[M^{II}_{1-x}M^{III}_{x}(OH)_2]^{x+}[A^{m-}]_{x/m} \bullet nH_2O$ where M^{II} and M^{III} are divalent (Mg^{2+}, Ni^{2+}, Zn^{2+}, Cu^{2+}, etc.) and trivalent cations (Al^{3+}, Fe^{3+}, Cr^{3+}, etc.), respectively, and A^{m-} is a charge compensating anion such as CO_3^{2-}, SO_4^{2-}, NO_3^-, Cl^-, OH^-, where x is equal to the molar ratio of $[M^{III}/(M^{II} + M^{III})]$. Its value is commonly between 0.2 and 0.33, i.e., the M^{II}/M^{III} molar ratio is in the range of 4 - 2 [25], but this is not a limitation ratio and it depends on the M^{II} and M^{III} composition [27-29].

Among various CO_2 mesoporous adsorbents, LDH-base materials have been identified as suitable materials for CO_2 sorption at moderate temperatures (T ≤ 400 °C) [30-46] due to their properties such as large surface area, high anion exchange capacity (2-3 meq/g) and good thermal stability [37-40]. The LDH materials themselves do not possess any basic sites. For that reason, it is preferred to use their derived mixed oxides, formed by the thermal decomposition of LDH, which do exhibit interesting basic properties. Thermal decomposition of the material occurs in three stages, first at temperatures lower than 200 °C, at which the dehydration of superficial and interlayer water molecules takes place on the material. Then the second decomposition stage takes place in the range of 300-400 °C, at which the structure collapses due to a partial dehydroxylation process, typically associated with both the decomposition of Al-OH and the Mg-OH hydroxides. During dehydroxylation, changes occur in the structure. A portion of the trivalent cations of the brucite like layers migrates to the interlaminar region, allowing the preservation of the laminar characteristics of the material [41]. Finally, the total decomposition of the material occurs at temperatures higher than 400 °C, when the decarbonation process is completed [42].

Once the temperature reaches about 400 °C, LDH forms a three-dimensional network of compact oxygen with a disordered distribution of cations in the interstices, where the cations M^{+3} are tetrahedrally coordinated (interlayer region) and M^{+2} are octahedrally coordinated. The compressive-expansion stresses associated with the formation of the amorphous three-dimensional networks and their connection to the octahedral layer increases the surface area and pore volume, which can help improve the storage capacity properties, for example for gas sorption related applications, besides decreasing the ability of the Mg^{+2} cation to favor physisorption instead of chemisorption [30, 42]. For instance, the thermal evolution of the Mg/Al-CO_3 LDH structure is considered to be crucial in determining the CO_2 adsorption capacity, so there are several studies about this issue [42-44].

Reddy et al. [43] studied the effect of the calcination temperature on the adsorptive capacity of the Mg/Al-CO_3 LDH. They found out that the best properties were obtained at calcination temperature of 400 °C, which they attributed to the obtaining of a combination of surface area and the availability of the active basic sites. Actually, at this temperature the material is still amorphous, which allows having a relatively high surface area. Therefore, there is a high number of exposed basic sites, allowing the reversible CO_2 adsorption according to the following reaction:

$$\text{Mg-O} + CO_2 \rightarrow \text{Mg-O} \cdots CO_2\ (\text{ad}) \quad (4)$$

However, if the LDH is calcined under 500 °C, the material is able to transform back to the original LDH structure when it is exposed to a carbonate solution or another anionic containing solution. Finally, if the sample is heated to temperatures above 500 °C, the structural changes become irreversible because of the spinel phase formation [37].

As mentioned, the mixed oxides derived from the LDH calcination possess some interesting characteristics such as high specific surface area, excess of positive charge that needs to be compensated, basic sites and thermal stability at elevated temperatures (200 – 400 °C). Besides these aspects, it is important to consider the advantage of acid-base interactions on the CO_2 sorption applications, where acidic CO_2 molecules interact with the basic sites on the derived oxide. These characteristics make the LDH-materials acceptable CO_2 captors [43, 45]. However, the CO_2 adsorption capacity of this material is low compared with other ceramic sorbents; reaching mean values smaller than 0.1 mmol/g [46]. Nevertheless, many studies suggest that the adsorption capacity of LDH materials can be improved by modifying a factor set such as: composition, improvement of the material's basicity and contaminant gas stream composition [30-32, 36, 41-45, 47-59].

As previously mentioned, Reddy et al. [43] studied the influence of the calcination temperature of LDHs on their CO_2 capture properties. The Mg_3/Al_1-CO_3 material was calcined at different temperatures ranging from 200 to 600 °C. The results showed that when the calcination temperatures are under 400 °C, LDH is considered to be dehydrated and materials still keep the layered structure intact, wherein the CO_3^{2-} ions are occupying the basic sites. The obtained samples calcined at 400 °C have the maximum BET surface area of 167 m^2/g compared with samples calcined at lower temperatures. Moreover, during the calcination of the LDH at higher temperatures (T > 500 °C), most of the CO_3^{2-} decompose to release some basic sites for CO_2 adsorption. However, the final amount of basic sites decreases with the subsequent crystallization of the MgO and spinel ($MgAl_2O_4$). Hence, LDH materials obtained at 400 °C have the highest surface area and the maximum quantities of active basic sites exposed. Because of these characteristics, they achieved a total sorption capacity of 0.5 mmol/g [43]. The same researchers observed that 88% of the captured gas can be desorbed and during the material regeneration 98% of the original weight is gained. This is another important property of LDH materials in high temperature CO_2 separation applications as described later..

As mentioned, the thermal evolution of the layered structure has a great influence on the CO_2 capture. The loss of superficial interlayer water occurs at 200 °C. Then at temperatures between 300 and 400 °C the layer decomposition begins, resulting in an amorphous 3D network with the highest surface area [30], so the adsorption temperature improves the CO_2 capture in the order of 400 > 300 > 20 >200 °C [41-42, 47, 52]

Several researchers have investigated a set of different factors to improve the CO_2 sorption capacity. Yong et al. [47, 48] studied the factors which influence the CO_2 capture in LDH materials, such as aluminum content, water content and heat treatment temperature. Regarding the M/Al-CO_3 LDHs (M = Mg, Ni, Co, Cu or Zn), the best CO_2 sorption capacity was obtained for the Mg/Al materials degassed at 400 °C and with adsorption conditions of 25 °C. In general, the sorption capacity follows the trend Ni > Mg > Co > Cu = Zn. However, when the degassed temperature is increased, the trend is modified to Mg (400 °C) > Co (300°C) > Ni (350°C) > Cu (300°C) >Zn (200°C). These results show that Mg/Al-CO_3 is the best composition at the degassing temperature of 400 °C [47]. At this temperature, the material consists of an amorphous phase with optimal properties for use as CO_2 captor [42]. Also, the influence of Al $^{+3}$ has been studied as a trivalent cation at 25 and 300 °C adsorption temperatures, by Yong [41] and Yamamoto [49] respectively. Both samples were degassed at 300 °C and the results showed that the CO_2 capture is influenced by the adsorption temperature. At a temperature of 25 °C, the maximum adsorption was 0.41 mmol/g with an Mg/Al ratio equal to 1.5 (x = 0.375) [41] and at 300 °C the amount of CO_2 adsorbed was 1.5 mmol/g for a cation ratio of 1.66 (x = 0.4) [49]. The differences between the two capacities can be attributed to the Al content differences. The Al incorporation in the structure has two functions: 1) to increase the charge density on the brucite-like sheet; and 2) to reduce the interlaminar distance and the number of sites with high resistance to CO_2 adsorption [48].

On the other hand, Qian et al. [50] studied the effect of the charge compensation anions (A^- = CO_3^{-2}, NO_3^{-1}, Cl^-, SO_4^{-2} and HCO_3^{-1}) on the structural properties and CO_2 adsorption capacity of Mg/Al-A^- (molar ratio equal to 3). Despite all of the prepared LDH materials showed the typical XRD patterns of LDH materials, slight structural and microstructural differences were observed. In fact, the interlayer distance changed by varying the interlayer anions due to their difference in sizes and carried charges. These differences affect the morphology and the BET surface area of both fresh and heat-treated LDH materials. Additionally, thermal treatments were performed in order to optimize the adsorption capacity of these materials. The optimal temperature treatment was established for each Mg/Al-A^- based on the surface area of each calcined LDH. Then the CO_2 adsorption capacities of calcined LDH were tested at 200 °C. Mg_3/Al_1–CO_3 showed the highest CO_2 adsorption capacity (0.53 mmol/g). This value was much higher than those obtained for calcined Mg_3/Al_1-NO_3 > Mg_3/Al_1-HCO_3, Mg_3Al_1-Cl, and Mg_3/Al_1-SO_4 (≈ 0.1 mmol/g). The results indicated that BET surface area of calcined LDHs seems be the main parameter that determines the CO_2 adsorption capacity because the Mg-O active basic site [43, 45].

It has been demonstrated that the quasi-amorphous phase obtained by the thermal treatment of LDH at the lowest possible temperature has the highest CO_2 capture capacity. This finding is in line with the fact that high calcination temperature can decrease the number of active Mg–O sites due to the formation of crystal MgO [51].

Yong [41] and Yamamoto [49] investigated the influence of the several types of anions. The results suggested that the amounts CO_2 capture decrease as a function of the anion size, which promotes a larger interlayer spacing and the higher charge: $Fe(CN)_6^{4-}$(1.5 mmol/g) > CO_3^{2-} (0.5 mmol/g) > NO_3^- (0.4 mmol/g) > OH^- (0.4-0.25 mmol/g). The reason is that $Fe(CN)_6^{4-}$ and CO_3^{2-}, because they have more void space in the interlayer due size, and are able to accommodate higher CO_2 quantities. Calcined layered double hydroxide derivatives have shown great potential for high temperature CO_2 separation from flue gases. However, the presence of SOx and H_2O from flue gases can strongly affect CO_2 adsorption capacity and regeneration of hydrotalcite-like compounds. Flue gases emitted from power stations contain considerable amounts of water in the form of steam. The percentage of water found in the flue gas emitted from different sources varies between 7 and 22%, with the emissions from brown coal combustion having the highest water content [45]. For many other gas adsorption sorbents, steam generally has a negative effect on the adsorption performance because of competition for basic sites between CO_2 and H_2O. However, the presence of water or steam seems to be favorable for the adsorption capacity onto LDH [31,43,53,54]. This fact is the result of the increasing potential energy that is able to further activate basic sites, possibly by maintaining the hydroxyl concentration of the surface material and/or preventing site poisoning through carbonate or coke deposition [31]. An example of the above was reported by Yong et al. [47], who found that water or steam can increase the adsorption capacity of CO_2 by about 25%, from 0.4 mmol/g to 0.5 mmol/g.

Ding et al. [31] studied CO_2 adsorption at higher temperatures (480 °C) under conditions for steam reforming of methane. They found an adsorption capacity of 0.58 mmol/g, which was independent of water vapor content in the feed. In turn, Reddy et al. [45] investigated calcined LDHs' sorption performance influenced by CO_2 wet-gas streams. LDH samples were calcined at 400 °C [43] before measuring CO_2 sorption at 200 °C. The gas streams used were CO_2, CO_2 + H_2O, flue gas (14% CO_2, 4% O_2 and 82 % N_2) +12% H_2O.

For a pure CO_2 dry sorption, the maximum weight gain was 2.72% (~0.61 mmol/g) after 60 min, whereas the wet adsorption increased the weight of the calcined LDHs to 4.81%, showing an additional 2.09%, where He and He + H_2O were used to remove the H_2O water capture. The results showed that the helium has virtually no significant sorption affinity for the material, whereas the water-sorption profile of it clearly indicates a water weight gain of 1.67%, i.e., the gain was 0.1mmol/g due to steam presence, showing that water has a positive effect, shifting the CO_2 sorption by 0.42% as compared to dry CO_2 sorption. Also, these results revealed that in all cases about 70% sorption occurs during the first 5 min and reaches equilibrium after around 30 min.

To determine the influence of CO_2, Reddy et al. [43] tested a sample in both, wet and dry CO_2 stream conditions. The experiments showed that the same quantity of CO_2 can be trapped for the solid sorbent after two hours. The presence of water in the stream only affects the kinetics of the process. This result is in agreement with that reported by Ding et al. [31]. On the other hand, the results of the material tested suggest that the fact the CO_2 capture from flue gas was higher than in a pure stream of CO_2 might have been because the polluted gas was diluted in the stream. The presence of the water does not enhance de CO_2 capture; the maximum CO_2 adsorbed was 0.9 mmol/g. The differences between Reddy et al. results and the previously mentioned studies can be caused by the use of uncalcined LDHs, which already contain an -OH network.

To apply these materials in industrial processes, it is important to know the times during which each sorbent material can be used. Tests of the cyclability in LDH materials disclose that as function of the temperature the CO_2 capture time can vary. This can be attributed to CO_2 chemisorbed during each cycle [54] and/or to the formation of spinel-based aluminas, such as γ-Al_2O_3 (at temperatures higher than 400 °C). Hibino et al. [52] found that the carbonate content, acting as charge-compensating anion, continuously decreases in subsequent calcination – rehydration cycles. Reddy et al. tested LDH materials during six CO_2 adsorption (200 °C)-desorption (300 °C) cycles. The average amount gained was 0.58 mmol/g, whereas 75% of this value is desorbed, reaching desorption equilibrium after the third cycle. This can be attributed to the stabilization of the material phase and basic sites during the temperature swing.

Hufton et al. [54] studied a LDH material during several cycles in dry and wet CO_2 flows. As previously discussed, the presence of steam in the flow gas improves the CO_2 adsorption. However, after 10 adsorption cycles, the capture decreased 45%. The same behavior was observed in the dry gas flow. However, the final capture was similar to the wet gas stream, in agreement with Reddy et al. [43].

Recent studies have demonstrated that K-impregnated LDH or K-impregnated mixed oxides have a better CO_2 capture capacity due to the addition of K alkaline-earth element that improves the chemical affinity between the acidic CO_2 and alkaline surface of the sorbent material [32, 36, 55-56]. Additionally, it has been proposed that K-impregnation reduces the CO_2 diffusion resistance in the material. [57]. Hufton et al. [58] showed that the K-impregnation increases the CO_2 capture, but there is an optimal quantity of K to reach the maximum capture. Qiang et al. [50] tested an Mg_3/Al_1-CO_3 (pH = 10) impregnated with 20 wt.% K_2CO_3. The CO_2 adsorption capacity was increased between 0.81 and 0.85 mmol/g in the temperature range of 300 - 350 °C. This adsorption capacity is adequate for application in water gas shift reactions (WGS).

Lee et al. [59] tested the behavior of three commercial LDHs impregnated with K (K_2CO_3/LDH ratio between 0 and 1). Three Mg/Al-CO_3 LDH with different contents of magnesium were used. Results indicated that the sorption capacity of the LDH is improved by about 10 times with the optimal K_2CO_3 additions. Additionally, it was observed that impregnation is not the only factor that influences the adsorption but the composition too. The best value was obtained when the content of divalent cation was reduced and therefore, the material had a composition with the maximum trivalent cation content. The CO_2 adsorption capacity improved from 0.1mmol/g to 0.95mmol/g with K_2CO_3/LDH weight ratio equal to 0.35 at 400 °C. After determining the optimal alkaline source/LDH ratio, a set of samples was evaluated as a function of the temperature and the results showed a maximum of 1.35 mmol/g, at 50 °C. In the impregnated materials, CO_2 chemisorption can occur and the sorbed CO_2 can be further stored as metal carbonate forms.

Other alkaline elements can be used to improve the sorption capacity of materials. Martunus et al. [46] studied the impregnation of LDH with Na and K. The LDH samples were thermally treated at 450 °C for 5 min then calcined samples were re-crystallized in K_2CO_3 and Na_2CO_3 (1 M) solutions. The re-crystallized materials were tested as CO_2 captors and the capture was maximum with LDH-Na (0.688 mmol/g) > LDH- K (0.575 mmol/g) at 350 °C after five cycles. Finally, the re-crystallized material with the highest capture was calcined at 650 °C for 4 h and re-crystallized with a solution containing the appropriate quantities of K and Na to achieve alkaline metal loading up to 20%. When the sample was Impregnated with additional K and Na at 18.4% and 1.6%, respectively, the adsorption capacity rose from 0.688 to 1.21 mmol/g. This capacity increase was achievable despite the relatively low BET surface area, equal to 124 m^2/g.

Other alkaline elements such as cesium have been studied as reinforcement. Oliveira et al. [55] tested commercial Mg_1/Al_1-CO_3 and Mg_6Al_1-CO_3 impregnated with K and Cs carbonates. The materials were evaluated in the presence of steam (26% v/v water content) gas at different temperatures (306, 403 and 510 °C) at 0.4 bar of CO_2 partial pressure (total pressure 2 bar). The LDH with the highest sorption capacity was Mg_1/Al_1-CO_3–K with 0.76 mmol/g at 403 °C. Among the Cs impregnated samples, the Mg_6Al_1-CO_3-Cs presented the highest capacity with 0.41 mmol/g, while the commercial LDH samples presented CO_2 sorption capacities around 0.1 mmol/g.

The results suggest the existence of a sorption mechanism combining physical adsorption and chemical reaction. First the maximum physical adsorption is reached, then the chemisorption begins, but there is an optimal temperature. If the temperature is too low, the chemisorption does not happen, but with higher temperatures the loss of porosity impedes the contact of CO_2 molecules with active basic sites promoted by the alkaline element addition.

These results suggest there is an optimum amount of K_2CO_3 to impregnate the LDH that achieves a balance between the increase in the basicity of the sorbent material and its reduction of surface area, associated with CO_2 capture capacity. The influence of potassium is currently not clear and the relevant research is still ongoing. Finally, CO_2 adsorption capacity on the synthesized 20 wt.%$K_2CO_3/Mg_3/Al_1$–CO_3 (pH = 10) probably could be further increased in the presence of steam.

3. Ceramic oxide membranes as an alternative for CO_2 separation

Membrane-based processes, related to gas separation and purification, have achieved an important level of development for a variety of industrial applications [60]. Therefore, the use of separation membranes is one of the promising technologies for reducing the emissions of greenhouse gases such as CO_2. The term membrane is defined as a permselective barrier between two phases, the feed or upstream and permeate or downstream side [61]. This permselective barrier has the property to control the rate of transport of different species from the upstream to the downstream side, causing the concentration or purification of one of the species present in the feed gas mixture.

Membrane-based processes offer the advantage of large scale application to separate CO_2 from a gas mixture. Figure 2 schematizes the process where concentrated CO_2 is selectively separated from flue gas that is mainly composed of nitrogen and carbon dioxide along with other gases such as water vapor, SO_x, NO_x and methane. Subsequent to the membrane process, concentrated CO_2 obtained at the permeate side can be disposed or used as raw material for the synthesis of several chemicals such as fuel and value-added products [62].

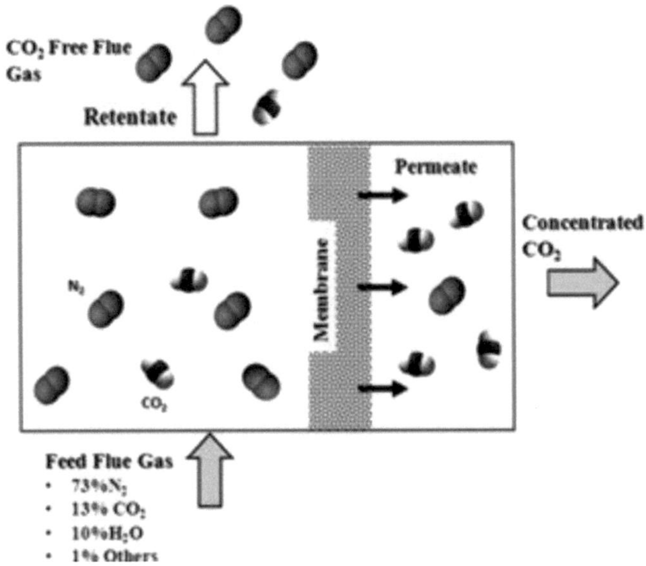

Figure 2. Membrane-based processes for the carbon dioxide separation from flue gases. The concentrated CO_2 is obtained in the permeate side.

Of course, the rate of transport or permeation properties of a particular gas through a given membrane depend on the nature of the permeant gas, as well as the physical and chemical properties of the membrane.

Inorganic membranes are more thermally and chemically stable and have better mechanical properties than organic polymer membranes; ceramic membranes offer both the advantage of large scale application and potential for pre- and post-combustion CO_2 separation applications, where membranes systems would be operating at elevated temperatures of 300-1000 °C [63].

Inorganic ceramic membranes can be classified as porous and nonporous or dense. These differ from each other not only in their structures but also in the mechanism of permeation. In porous membranes, the transport of species is explained with the pore-flow model, in which permeants are transported by pressure-driven convective flow through the pore network. Separation occurs because one of the permeants is excluded (molecular filtration or sieving) from the pores in the membrane and remains in the retentate while the other permeants move towards the downstream side. On the other hand, in nonporous membranes, separation occurs by solution-diffusion, in which permeants dissolve in the membrane material and then diffuse through the bulk membrane by a concentration gradient [60].

3.1. Porous membranes based on alkaline and alkaline-earth ceramic oxides for CO_2 separation

Among the porous systems for CO_2 separation, both microporous (carbon, silica and zeolite membranes) and modified mesoporous membranes have been reported [63-64].

Zeolites are hydrous crystalline aluminosilicates that exhibit an intracrystalline microporous structure as a result of the particular three-dimensional arrangement of their TO_4 tetrahedral units (T=Si or Al) [65]. Zeolite membranes are commonly prepared as thin films grown on porous alumina supports via hydrothermal synthesis and dry gel conversion methods [66]. Zeolite membranes of different structures have been developed to separate CO_2 from other gases via molecular sieving [67-69]. For example, membranes prepared with the 12-member ring faujasite (FAU)-type zeolite show high separation factors of 20-100 for binary gas mixtures of CO_2/N_2 [69]. In the same sense, T zeolite membranes exhibited very high selectivity, of about 400, for CO_2/CH_4 and 104 for CO_2/N_2. The high selectivity of CO_2/CH_4 exhibited by T zeolites is due to the small pore size of about 0.41 nm, which is similar in size to the CH_4 molecule but larger than CO_2 [69]. Table 1 shows the kinetic diameter of various molecules that are present in CO_2 containing gas mixtures such as flue and natural gas [70].

Molecule	**Kinetic diameter (Å)**
H_2O	2.65
H_2	2.69
CO_2	3.3
O_2	3.46
N_2	3.64
CH_4	3.80

Table 1. Kinetic diameter of various molecules based on the Lennard Jhones relationship.

Deca-dodecasil 3R (DDR) (0.36 nm x 0.44 nm), and pseudo-zeolite materials like silicoaluminophosphate (SAPO)-34 (0.38 nm) also show high CO_2/CH_4 selectivities due to narrow molecular sieving, which controls molecular transport into this material [69, 71-73]. For example, Tomita et al. [74] obtained a CO_2/CH_4 separation factor of 220 and CO_2 permeance values of 7×10^{-8} mol m^{-2} s^{-1} Pa^{-1} at 28 °C on a DDR membrane [75].

As discussed, one of the most important factors controlling permeation through microporous membranes is the restriction imposed by the molecular size of the permeant. However, the transport mechanism in microporous systems is more complex than just size exclusion and the permeation and selectivity properties are also affected by competitive adsorption among perment species that produce differences in mobility [76].

Thus, the diffusion mechanism for gas permeation through microporous membranes can be characterized by two modes: one controlled by adsorption and a second one where diffusion dominates [63]. In the case of adsorption-controlled mode with permeating gases having strong affinity with the membrane, a gas permeation flux equation is obtained by assuming steady-state single gas permeation, a constant diffusivity and a single gas adsorption described by a Langmuir-type adsorption isotherm, as in Eq. (5).

$$J = \phi q_s \frac{D_c}{L}\left(\frac{1 + bP_f}{1 + bP_p}\right) \; or \; J = \phi q_s \frac{D_c}{L}\left(\frac{1 - \theta_p}{1 + \theta_f}\right) \tag{5}$$

where J is the permeation flux, ϕ is a geometric correction factor that involves both membrane porosity and tortuosity, D_c is the corrected diffusivity of the permeating species, L is the membrane thickness, P_f and P_p represent the feed and permeate pressure respectively and θ_f and θ_p represent the relative occupancies.

Furthermore, if the adsorption isotherm of the permeating gas is linear (1 >> bP), then flux permeation is described by Eq. (6).

$$F = \phi q_s \frac{D_c}{L}\left(\frac{D_c}{L}\right)K \tag{6}$$

where F is the permeance and $K = q_s b$ is the adsorption equilibrium constant. Therefore, from Eq. (5) it can be concluded that permeance is determined by both diffusivity (Dc) and adsorption (K). Based on the above, an interesting option to enhance membrane properties is to intercalate zeolite membranes with alkaline and alkaline-earth cations. Zeolite intercalation can enhance the separation between CO_2 and other molecules such as N_2 by promoting preferential CO_2 adsorption [63, 77]. It is well known that zeolites show affinity for polar molecules, like CO_2, due to the strong interaction of their quadrupole moment with the electric field of the zeolite framework. In this sense, the adsorption properties of zeolites can be enhanced by the inclusion of exchangeable cations within the cavities of zeolites where the adsorbent-adsorbate interactions are influenced by the basicity and electric field of the adsorbent cavities [78-80]. Lara-Medina et al. [77] carried out separation studies of CO_2 and N_2 with a silicalite-1 zeolite membrane prepared via hydrothermal synthesis and subsequently modified by using lithium solutions in order to promote preferential CO_2 adsorption and diffusion. CO_2/N_2 separation factor increases from 1.46 up to 6 at 25 psi and 400 ∘C after lithium modification. An et al. [79] studied a series of membranes prepared starting from natural Clinoptilolite zeolite rocks. Disk membranes were obtained by cutting and polishing of the original minerals, which were subsequently chemically treated with aqueous solutions containing Li, Na, Sr or Ba ions. Ionic exchanged membranes showed better permeation properties due to the presence of the extra framework cations.

Although zeolite membranes offer certain advantages in comparison with polymer membranes, such as chemical stability, the main issues are related to the selectivity decrease as a function of the permeation temperature. This is explained in terms of the contribution of the adsorption to the separation, which decreases sharply as temperature increases. At high temperature, physical adsorption becomes negligible and permeation is mainly controlled by diffusion [63, 76]. Additionally, due to the fact that CO_2 and N_2 molecules have similar sizes (Table 1), the difference in diffusivity is not a strong controlling factor in determining selectivity.

Modified γ-Al_2O_3 mesoporous membranes have been also reported as a means for CO_2 separation [64]. Transport mechanisms in porous membranes have the contribution of different regimes. An overview of the different mechanisms is given in Table 2.

Transport Type	Pore diameter	Characteristics
Viscous flow	>20 µm	Non selective. Affects the total flow resistance of the membrane system.
Molecular diffusion	>10 µm	
Knudsen diffusion	2 – 100 nm	Occurs when the mean free path of the molecule is much larger than pore radius of the membrane. Shows selectivity based on molecular weights.
Surface diffusion		Shows selectivity due to interaction of molecules with membrane walls.
Capillary condensation		
Micropore diffusion (Configurational diffusion)	< 1.5 nm	

Table 2. Transport mechanisms in porous membranes.

Depending on the particular system, permeability of a membrane can involve several transport mechanisms that take place simultaneously. Considering no membrane defects and pore sizes in the range of 2.5-5 nm, γ-Al_2O_3 based membranes theoretically have two transport regimes: Knudsen diffusion and surface diffusion. Eq. (7) describes the permeability of a membrane by taking into consideration the Knudsen and surface diffusion.

$$F=\left(\frac{2\varepsilon\mu r}{3RTL}\right)\left(\frac{8RT}{\pi M}\right)^{0.5}+\frac{2\varepsilon\mu D_s}{r A_0 N_{av}}\frac{d x_s}{dP} \tag{7}$$

where r is the mean pore radius, µ is a shape factor, R is the universal gas constant, T is the temperature, P is the mean pressure, M is molar mass of the gas, A_o is the surface area occupied by a molecule, D_s is the surface diffusion coefficient, N_{av} is Avogadro's constant and X_s is the percentage of occupied surface compared with a monolayer [81].

For the cases when Knudsen diffusion dominates, selectivity can be correlated to the molecular weights of the permeating gases by the so called Graham's law of diffusion, which establishes that the transport rate of any gas is inversely proportional to the square root of its molecular weight. The CO_2/N_2 separation factor considering pure Knudsen diffusion is given by Eq. (8) and has a value of just 0.8. Therefore, Eq. (8) clearly shows that separation via Knudsen is limited for systems where species are of similar molecular weight.

$$\alpha\left(\frac{CO_2}{N_2}\right)=\sqrt{\frac{M_{CO_2}}{M_{N_2}}} \tag{8}$$

Based on the aforesaid, CO_2/N_2 separation factor can be better enhanced by promoting the surface diffusion mechanism (second term on the right hand side of Eq. (7)). Surface diffusion involves the adsorption of gas molecules on the surface of the pore and subsequent diffusion of the adsorbed species along the surface by a concentration gradient. Then separation properties of a membrane can be improved by generating such an interaction between one component of the feed gas mixture with the membrane; one option being via a chemical modification.

Cho et al [81] prepared a series of thin (2-5 μm thickness) γ-Al_2O_3 and CaO- or SiO_2-modified γ-Al_2O_3 membranes for CO_2 separation at temperatures between 25 and 400 °C. Impregnation of membranes with SiO_2 or alkaline CaO was done in order to improve the CO_2/N_2 selectivity by promoting adsorption between CO_2 gas molecules and the membrane pore wall. Although this kind of chemical modification of the membrane surface and the pore walls is able to activate the surface diffusion mechanism, an interesting behavior was observed. The CO_2/N_2 separation factor increased from 1.0 to 1.38 at 25 °C after modification of the γ-Al_2O_3 with SiO_2. On the other hand, CaO impregnated membranes showed a separation factor of 0.98, which is even lower than that of the unmodified γ-Al_2O_3. The same behavior has been reported by Uhlhorn et al. [82-83]. They reported MgO modified γ-Al_2O_3 membranes which did not show significant enhancement in the permeation and selectivity properties as a result of the modification process. This fact was explained in terms of the surface diffusion mechanisms. As discussed, it is expected that physicochemical modifications of the membrane can enhance preferential adsorption of the gas species in the feed. Impregnations with alkaline oxide such as calcium oxide or magnesia on the γ-alumina surface give more strong base sites than those promoted by silica. Therefore, it promotes a strong bonding of CO_2 on the alumina surface, causing CO_2 molecules to lose mobility, resulting in a smaller contribution of surface diffusion to the total transport mechanism.

There is another kind of membrane where alkaline and alkaline-earth ceramic oxides have been used for the fabrication of CO_2 permselective membranes. In these cases ceramic materials were chosen because of their well-known properties of physisorption of CO_2 at low and intermediated temperatures.

Kusakabe et al. [84] prepared both pure and modified $BaTiO_3$ CO_2 permselective membranes via the alkoxide based sol-gel method; impregnation and calcination at 600 °C. In order to establish the effects of CO_2 partial pressure, temperature and influence of the secondary oxide presence (CuO, MgO or La_2O) on the CO_2 adsorption properties of the membranes, pure and modified barium titanate powders were first evaluated by thermogravimetry and chromatography techniques. Dynamic CO_2 absorption was evaluated by applying the impulse response method, wherein the $BaTiO_3$ powder was packed in a separation column. The results suggested that the CO_2 molecules adsorbed on the $BaTiO_3$ powder are mobile at temperatures about 500 °C. Therefore, this membrane exhibits CO_2 permeation due to surface diffusion mechanism. Even though the prepared membranes showed selectivity, the Knudsen diffusion still has an important contribution to the gas transport due to the presence of membrane defects. The maximum separation factor of CO_2/N_2 through the membranes was estimated as 1.2. Therefore, further improvement of the permeation properties of this kind of membrane requires obtaining pinhole-free membranes.

Based on the same criteria, Nomura et al. [85] prepared Li_4SiO_4-based thin membranes on porous alumina supports. Membranes were obtained by the thermal treatment of different silica containing porous materials (Silicalite-1 and mesoporous silica) impregnated with lithium compounds. The authors called this method solid conversion. The use of different silica porous sources was proposed in order to enhance the reaction rate of Si and Li on the porous support at relatively low temperature, avoiding the reaction between the Li and alumina support itself. In the case of Silicalite-1 (MFI zeolite), a zeolite thin film was first prepared on the support by following the dry gel conversion technique. Then, the prepared Silicalite-1 layer was impregnated via dipping into a slurry containing lithium and silica fumed reactants (Li:Si = 4:1) and subsequently into a Li_2CO_3-K_2CO_3 slurry. The membrane was finally calcined at 600 °C for 2 h. It is believed that carbonate melts to fill the cracks and the pinholes of the Li_4SiO_4 formed membrane. A similar procedure of coating and calcination was carried out to prepare high quality membranes starting from mesoporous silica sources with pore sizes of 1.8-12.8 nm. Precursors react to form a Li_4SiO_4 membrane of 2-5 μm thickness that exhibits an N_2 permeance of 1.8×10^{-9} mol m^{-2} s^{-1} Pa^{-1} at 400 °C. This suggests there are no big defects after impregnation of the membrane with the binary mixture of Li_2CO_3-K_2CO_3carbonate. Due to the fact that the membrane operates in a rich CO_2 atmosphere, carbonates do not decompose even at temperatures of 600 °C. The maximum CO_2/N_2 permeance ratio was 0.85. The separation factor was higher than that for the Knudsen diffusion. Therefore, it can be conclude that Li_4SiO_4 layer was selective to CO_2 over N_2 at high temperature of 600 °C.

Nomura [86] reported a two -stage approach for the preparation of Li_4SiO_4-CO_2 selective membranes that involves the fabrication of a supported Li_4SiO_4 membrane and its subsequent modification by using a chemical vapor deposition (CVD) method. First, for the preparation of a thin Li_4SiO_4 membrane the so called solid conversion method described before was used, which is based on the reaction between a porous silica source and a lithium containing solution coated on a porous alumina membrane support. Although the formed membranes showed certain selectivity due to the preferential adsorption of CO_2 over N_2, the presence of pinholes and cracks caused low separation factors. Therefore, the membrane defects were fixed by using the counter diffusion CVD method to form a silica coating that fills the gaps between the lithium orthosilicate particles that make up the membrane. N_2 permeance was reduced about three orders of magnitude after CVD modification. Nitrogen permeance before and after the CVD treatment was 3.4×10^{-6} mol m^{-2} s^{-1} Pa^{-1} and 1.2×10^{-9} mol m^{-2} s^{-1} Pa^{-1} respectively. In the same sense, the CO_2/N_2 permeance rate increased from 0.7 to 1.2 at 600 °C. Some issues related with this system are the chemical and structural stability of the membranes observed during the permeation tests at elevated temperature. The membranes were broken when permeation tests were carried out at temperatures higher than 700 °C, with the consequent decrease in the CO_2/N_2 selectivity. The aforesaid is the result of the CO_2 chemisorption on the membrane. Lithium orthosilicate reacts with CO_2 to form lithium carbonate and lithium metasilicate (Li_2SiO_3) as products, as indicated by Eq. (9).

$$Li_4SiO_4 + CO_2 \longleftrightarrow Li_2CO_3 + Li_2SiO_3 \qquad (9)$$

Thermodynamically, this reaction is prone to occur at temperatures between room temperature and about 700 °C. However, experimentally it has been observed that reaction kinetics sharply increase above 550 °C. At these temperatures, the formation of carbonates involves an important change in volume that ends in the membrane's rupture.

Therefore, one of the issues related to the development of this kind of inorganic membrane is the thermochemical stability. Due to reactivity of alkaline and alkaline-earth ceramic oxides with CO_2 to form carbonates, not only preferential adsorption of CO_2 molecules over N_2 occurs, but CO_2 chemisorption and reaction.. Therefore, it is mandatory to establish the operational temperature within a range where CO_2 selective adsorption on the membrane layer promotes the separation process without reaction.

3.2. Nonporous membranes based on alkaline and alkaline-earth ceramic oxides for CO_2 separation

Some researchers have proposed the use of alkaline and alkaline-earth ceramic oxides to prepare membranes that are able to separate CO_2 at high temperatures via a different transport mechanism than those observed on porous membranes. Li_2ZrO_3 and Li_4SiO_4 based membranes are examples of the aforesaid. Permselectivity of CO_2 through these membranes takes place not only due to the selective CO_2 adsorption properties of ceramic phases but also via a mechanism of gas separation that involves the transport of CO_3^{2-} and O^{2-} ionic species through the electrolytes (carbonate-metal oxide) phases formed by the reaction of the membrane with the CO_2 [87-89].

Kawamura et al. [87] fabricated and characterized a membrane for CO_2 separation at high temperatures. The membrane was made of lithium zirconate (Li_2ZrO_3), an alkaline ceramic oxide that reacts with CO_2 to produced Li_2CO_3 and ZrO_2. These two reaction products are electrolyte materials produced *in-situ* when the membrane is exposed to the rich carbon dioxide atmosphere. The electrolytes formed thus are capable to transport both CO_2 and O_2 across the membrane via a dual ion conduction mechanism. The prepared membrane exhibited a separation factor of 4.9 between CO_2 and CH_4 gas molecules at a temperature of 600 °C. The obtained separation factor is higher than the Knudsen diffusion limit, 0.6. Therefore, the results clearly suggest the potential use of this kind of membrane system for CO_2 separation such as the case of CO_2 removal from natural gas.

Yamaguchi et al. [88] investigated the concept of the dual-ion conduction facilitated mechanism previously observed for the case of Li_2ZrO_3 membranes by focusing their efforts on the preparation of a CO_2 permselective membrane based on lithium orthosilicate (Li_4SiO_4). The supported membrane was prepared via a dip coating technique by using Li_4SiO_4 suspensions. The coating process was repeated several times before impregnation of the membrane with a Li_2CO_3/K_2CO_3 carbonate mixture and final sintering at 750 °C. In this membrane system, Li_4SiO_4 reacts *in-situ* with CO_2 to form Li_2CO_3 and Li_2SiO_3.

Gas separation studies were performed by using CO_2/N_2 mixtures as feed gas. The observed CO_2 permeance values were about 1×10^{-8} mol $m^{-2}s^{-1}Pa^{-1}$ in the temperature range of 525-625 °C. The CO_2/N_2 separation factor was estimated between four and six. Figure 3 shows a scheme of the dual-ion conduction mechanism explained as follows. In the feed side, carbon dioxide dissolves in the material and diffuses as carbonate ions through the molten carbonate electrolyte due to a concentration gradient. Then, in the downstream side of the membrane, the formation of gaseous CO_2 implies the formation of oxygen ions which must diffuse back to the feed side across the membrane and apparently through the formed Li_2SiO_3 skeleton to obtain the charge balance.

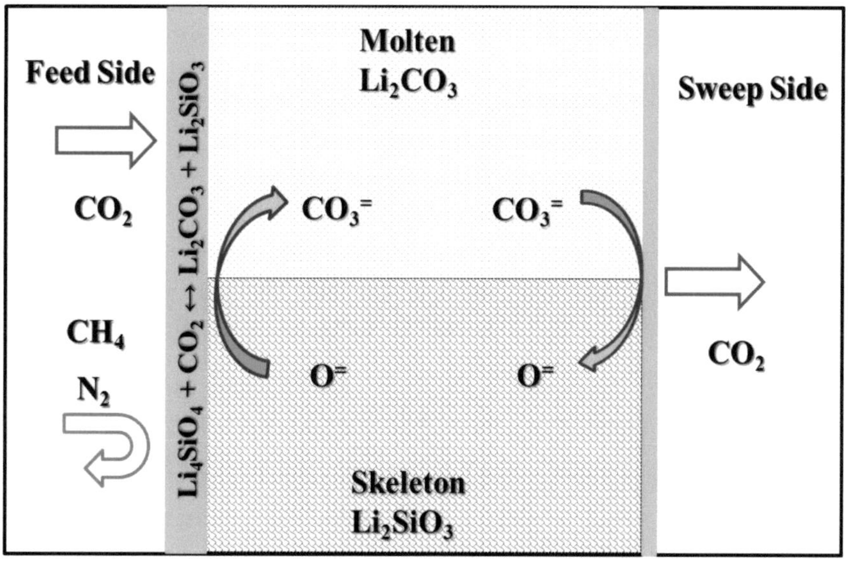

Figure 3. Schematic representation of a membrane system for the CO_2 separation via a dual-ion conduction mechanism.

The proposed transport mechanisms supports the higher selectivity values observed in the permeation test for both systems, Li_2ZrO_3 and Li_4SiO_4. Figure 4 shows the separation factor values (CO_2/N_2) obtained for different ceramic membranes described in the present report. The pure Knudsen value is written as baseline and separation factor of nonporous Li_4SiO_4 for comparison purposes. However, it is important to mention that the oxygen ion diffusion process is not totally clear. Indeed, there is no experimental study regarding the oxygen ionic conductivity properties of Li_2SiO_3 phase. On the other hand, pure ZrO_2 exhibits poor bulk oxygen ion conductivity. In fact, good conduction properties are observed only in acceptor-doped ZrO_2 based materials with oxygen vacancies being the predominant charge carriers [90]. Therefore, oxygen ion conduction through the membrane must be related to different transport paths, such as grain boundaries and interfacial regions formed between the ceramic and molten carbonate on the membrane.

More recently, the promising concept of ceramic oxide-carbonate dual-phase membranes has been proposed for carbon dioxide selective separation at intermediate and high temperatures (450-900 °C) [91-97].

This concept involves the fabrication of nonporous membranes capable of selectively separating CO_2 via its transport, as carbonate ions. Dual phase membranes are made of an oxygen ion

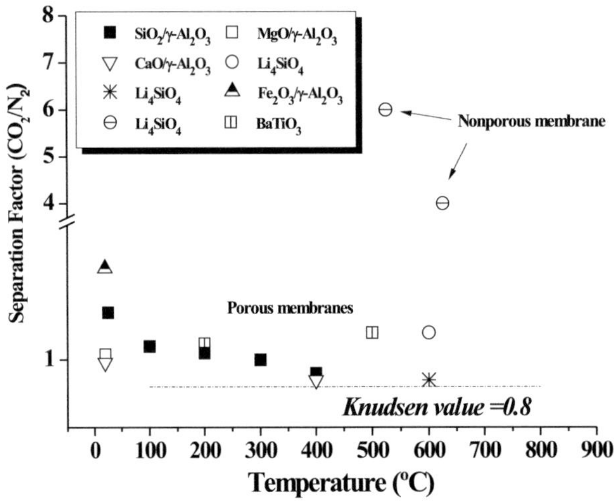

Figure 4. CO_2/N_2 separation factor of different ceramic oxide membranes.

conductive porous ceramic phase that hosts a molten carbonate phase. Rui et al. [98] proposed the CO_2 separation by the electrochemical conversion of CO_2 molecules to carbonate ions (CO_3^{2-}), which are subsequently transported across the membrane. Carbonate ionic species (CO_3^{2-}) are formed by the surface reaction between CO_2 and oxygen that comes from the ceramic oxide phase (feed side, Eq.(10)) and then transport of CO_3^{2-} takes place through the molten carbonate.

$$CO_2 + O_O^{x} \longleftrightarrow CO_3^{2-} + V_O^{\cdot\cdot} \tag{10}$$

Once carbonate ions have reached the permeate side, molecular CO_2 is released to the gas phase, delivering O_O^x species back to the ceramic oxide solid phase. This process takes place due to a chemical gradient of CO_2 in the system (Figure 5). Here, it is important to emphasize that dual-phase membranes are nonporous and therefore exhibit high separation selectivity as a result of the transport mechanism. Figure 5 also shows the SEM image of the cross section of a ceramic oxide-carbonate membrane prepared by pressing $La_{0.6}Sr_{0.4}Co_{0.8}Fe_{0.2}O_{3-\delta}$ powders and subsequent infiltration of the obtained porous ceramic (bright phase) with carbonate (dark phase).

Table 3 summarizes the different studies reported and certain advances that have been achieved so far regarding the dual-phase membrane concept. This table also includes the Li_2ZrO_3 and Li_4SiO_4 nonporous membranes previously described. Although the original reports do not clearly explain the operational mechanism [26-27], the dual-phase membrane concept gives a much better idea of the possible phenomenology involved [30,33,36].

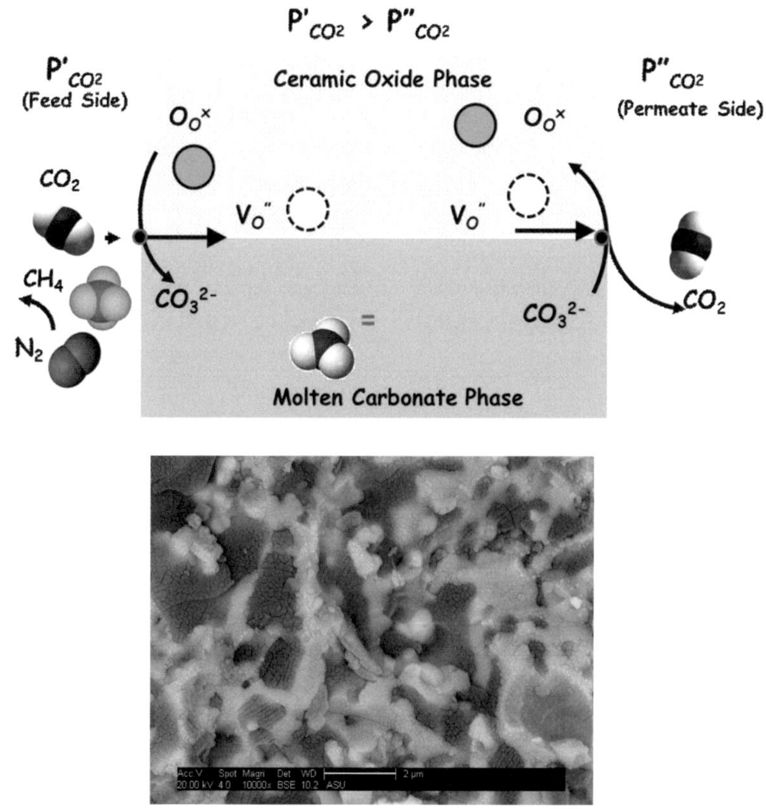

Figure 5. Schematic representation of a membrane system for the CO_2 separation and SEM image of a ceramic oxide-carbonate dual-phase membrane

3.3. Applications of CO_2 permselective ceramic oxide membranes for the design of membrane reactors.

As mentioned, CO_2 can be used as raw material for the synthesis of several chemicals [99]. Moreover, if CO_2 is concentrated or separated by a membrane system exhibiting high CO_2 permeation and permselectivity, this open up the possibility to develop a continuous process of membrane reaction to simultaneously capture and chemically convert CO_2. For example, if the membrane is able to separate CO_2 at intermediate and even high temperatures, it can be used for the design of a membrane reactor for the production and purification of hydrogen and syngas. Syngas is a gaseous fuel with a main chemical composition of CO, H_2, CO_2, and CH_4. Syngas can be used as feedstock for the synthesis of several other clean fuels such as H_2, methanol, ethanol, diesel and other hydrocarbons synthesized via the Fischer-Tropsch process [100-104].

Among the different processes for the synthesis of syngas and hydrogen, CO_2 methane reforming Eq. (11) and the water-gas shift reaction (WGS) Eq. (12) are the most promising options.

Ceramic Oxide phase	Molten Carbonate phase	Membrane features	Preparation method	Permeance ($mol.s.^{-1}m.^{-2}Pa.^{-1}$)	Separation Factor (CO_2/N_2)	Ref.
Li_2ZrO_3	Li_2CO_3	Thick membrane	*In situ* by exposing Li_2ZrO_3 to CO_2 atmosphere	1 x10^{-8}	4.9 (CO_2/CH_4 at 600°C)	[87]
Li_4SiO_4/Li_2SiO_3	K_2- Li_2CO_3	Thin supported membrane	Impregnation of carbonate	2x10^{-8}	5.5 (at 525°C)	[88]
$La_{0.6}Sr_{0.4}Co_{0.8}Fe_{0.2}O_{3-\delta}$	Li-Na-K_2CO_3	Thick membrane (0.35-1.5 mm)	Pressing and direct infiltration	4.77 x 10^{-8}	225 (at 900°C)	[91]
8 mol% Yttria doped zirconia (YZS) 10 mol% Gadolinia doped ceria (GDC)	Li-Na-K_2CO_3 Li-Na_2CO_3	Thin freestanding membranes (200-400 μm)	Tape casting and in situ infiltration	2.0 x10^{-8} (YSZ) 3.0 x10^{-8} (GDC)	> 2 (at 800 °C)	[92]
$Ce_{0.8}Sm_{0.2}O_{1.9}$	Na_2- Li_2CO_3	Thick membrane (1.2 mm)	Pressing of SDC-NiO powders where NiO is a sacrificial template	~1.2 x 10^{-6}	155-255 (at 700°C)	[93]
$Bi_{1.5}Y_{0.3}Sm_{0.2}O_3$	Li-Na-K_2CO_3	Thin supported membrane (50 μm)	Dip coating of modified thick support and infiltration	1.1 x 10^{-8}	2 (at 650°C)	[94]
8 mol% Yttria doped zirconia (YZS)	Li-Na-K_2CO_3	Thin supported membrane (10 μm)	Dip coating of YSZ on nonwettable thick support and infiltration	~ 7.8x10^{-8}	---	[95]
$Ce_{0.8}Sm_{0.2}O_{1.9}$	Li-Na-K_2CO_3	Thin tubular membrane	Centrifugal casting and direct infiltration			[96]
$La_{0.6}Sr_{0.4}Co_{0.8}Fe_{0.2}O_{3-\delta}$	Li-Na-K_2CO_3	Thick disk-shaped membrane	Pressing and direct infiltration			[97]

Table 3. Reported studies on dual-phase and related membranes for CO_2 separation.

$$CH_4 + CO_2 \longleftrightarrow 2CO + 2H_2 \tag{11}$$

$$CO + H_2O \longleftrightarrow CO_2 + H_2 \tag{12}$$

Figure 6 schematizes the membrane reactor concept considering the two reactions described above. Figure 6A shows a membrane reactor for dry reforming of methane to produce syngas at temperatures between 700 and 800 °C. Figure 6B illustrates the use of ceramic oxide membranes for hydrogen purification by separating the CO_2 from water-gas shift products at about 550 °C. Additionally, Figure 6B shows the possibility of using a ceramic sorbent to chemically trap the permeate CO_2 and therefore enhance the CO_2 permeation process by reducing the concentration of CO_2 in the permeate side.

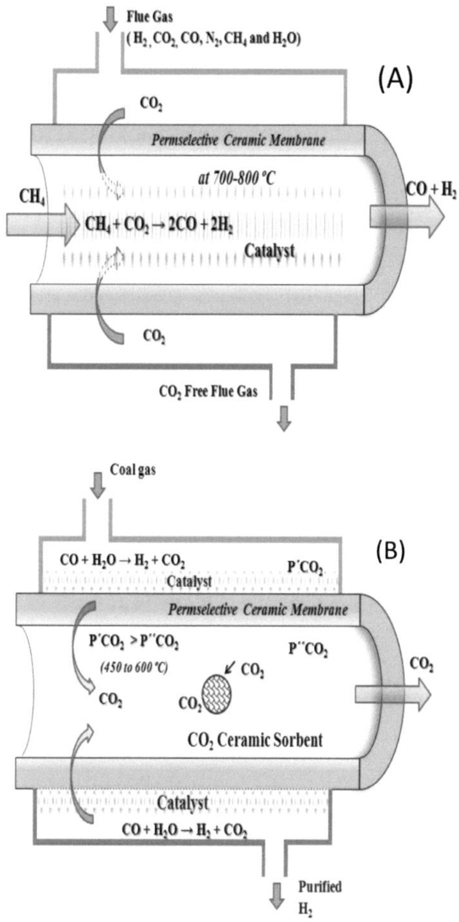

Figure 6. Schematic representation of the membrane reactor concept using a CO_2 permselective ceramic membrane: (a) CO_2 dry methane reforming and (b) water-gas shift reaction with hydrogen purification wherein CO_2 capture promotes the separation process.

4. Chemical transformation of CO_2 catalyzed by ceramic materials: the use of new alternatives.

One of the most widely used chemical absorption techniques for carbon capture and storage/sequestration (CCS) is CO_2 adsorption by ceramic materials. Once CO_2 has been captured-fixed, it can be converted into value-added products such as precursors in chemical transformation reactions. CO_2 is extensively used for enhanced oil recovery, as a monomer feedstock for urea and polymer synthesis, in the food and beverage industry as a propellant, and in production of chemicals. Therefore, the capture-fixation of CO_2 would make a system suitable for accomplishing chemical transformation of CO_2. The utilization of carbon dioxide is also very attractive because it is environmentally benign [105-115]. CO_2 conversion to fuel and value-added products is an ideal route for CO_2 utilization due to the simultaneous disposal of CO_2 and the benefit that many products can be used as alternate transportation fuels [116]. CO_2 chemical transformation methods include (i) reverse water-gas shift, (ii) hydrogenation to hydrocarbons, alcohols, dimethyl ether and formic acid, (iii) reaction with hydrocarbons to syngas, (iv) photo- and electrochemical/catalytic conversion, and (v) thermo-chemical conversion [100-122].

CO_2 can be catalyzed to valuable organic or inorganic compounds, where some basic catalytic materials (containing alkaline or alkaline-earth elements) are used. The activation of CO_2 by alkali metals has received considerable attention in various surface science studies, which have demonstrated the formation of intermediate CO_2, dissociation of CO_2 and formation of oxalate and carbonate alkali compounds [118-121]. Carbon dioxide has been identified as one such potential vector molecule (through reduction to syngas, methanol, methane, formic acid, formaldehyde, dimethylether (DME) and short-chain olefins) [117-118, 120-122]. CO_2 is a kinetically and thermodynamically stable molecule, so CO_2 conversion reactions are endothermic and need efficient catalysts to obtain high yield. CO_2 conversion to carbon monoxide (CO) looks like the simplest route for CO_2 reduction [121]. CO is a feedstock or intermediate product for the production of methanol and hydrocarbon fuels via Fischer-Tropsch synthesis of CH_4/CO_2 reforming to form syngas (CO/H_2) [122]. CO_2 reforming with CH_4 is an example of CO_2 being used as a soft oxidant, where the dioxide is dissociated into CO and surface oxygen, and oxygen abstracts hydrogen from methane to form water via the water-gas shift reaction (WGS) (Eqs. 11 and 12) [100-103, 121]. The catalytic chemistry of the reverse water-gas shift reaction and the following transformation to methanol/DME (or hydrocarbons via Fischer-Tropsch synthesis), and the subsequent production of gasoline (methanol-to-gasoline or diesel via hydrocracking of the alkanes produced in the Fisher-Tropsch process) are well established [102, 117-122]. On the other hand, methanol can be produced directly from carbon dioxide sources by catalytic hydrogenation and photo-assisted electrochemical reduction. A wide variety of CO_2 photo-reduction methods have been performed to oxygenate products, including formic acid (HCOOH) and formaldehyde (HCHO). HCOOH and HCHO are the simplest oxygenates produced from the reduction of CO_2 with H_2O (or proton solvents) [121]. Furthermore, CO_2 can be utilized as a monomeric building block to synthesize various value-added oxygen-rich compounds and polymers under mild conditions. As an example, chemical conversion of CO_2 through C–N bond formation can produce value-added chemicals such as oxazolidinones, quinazolines, carbamates, isocyanates and polyurethanes [105]. These commodity chemicals have been synthesized from green methods and have important applications in the pharmaceutical and plastic industries. The chemisorption of CO_2 based on C–N bond formation could be one of the most efficient strategies, utilizing liquid absorbents such as conventional aqueous amine solutions, chilled ammonia, amino-functionalized ionic liquids, and solid absorbents including amino-functionalized silica, carbon, polymers and resins. The processes by which chemicals for CO_2 capture are manufactured should also be considered in terms of their energy requirements, efficiencies, waste products, and CO_2 emissions [105, 123]. In that sense, dimethyl carbonate (DMC) is a promising target molecule derived from CO_2 catalyzed by inorganic dehydrating agents such as molecular sieves [107]. Dimethyl carbonate has received much attention as a safe, non-corrosive, and environmentally friendly building block for the production of polycarbonates and other chemicals, an additive to fuel oil owing to its high octane number and an electrolyte in lithium batteries due to its high dielectric constant. It can be synthesized through a two-step transesterification process utilizing CO_2 as raw material [105, 107].

As a complementary technology to carbon sequestration and storage (CSS), the chemical recycling of carbon dioxide to fuels is an interesting opportunity. Chemical compounds such as alkane products ($C_nH_{(2n+2)}$) are un-branched hydrocarbons suitable for diesel fuel and jet fuel [121]. In this regard, biofuels or biodiesel, catalyzed using ceramic materials, can provide a significant contribution in energy independence and mitigation of climate change [109-127]. Today the main renewable biofuels are bioethanol and biodiesel. Biodiesel is a liquid fuel consisting of mono alkyl esters (methyl or ethyl) of long-chain fatty acids derived from vegetable oils, animal fats or micro and macro algal oils [127]. Biodiesel is a sustainable, renewable, non-toxic, biodegradable diesel fuel substitute that can be employed in current diesel engines without major modification, offering an interesting alternative to petroleum-based diesel [106, 111-115, 124-128]. Besides this, it is free from sulfur and aromatic components, making it cleaner burning than petroleum diesel. Biodiesel has a high flash point, better viscosity and caloric power similar to fossil fuels. It can be mixed with petroleum fossil fuel at any weight ratio or percentage, and it can be used without blending with fossil fuel (B100) as a successful fuel [127, 128]. It has similar properties (physical and chemical) to petroleum diesel fuel. Recently, transesterification (also called alcoholysis) has been reported as the most common way to produce biodiesel with lipid feedstock (such as vegetable oil or algal oil) and alcohol (usually methanol or ethanol), in presence of an acid or base catalyst. Transesterification is the best method for producing higher-quality biodiesel and glycerol [108, 110-115, 124-132]. The reaction is facilitated with a suitable catalyst [129-131]. The catalyst presence is necessary to increase both, the reaction rate and the transesterification reaction conversion yield. The catalysts are classified as homogeneous or heterogeneous. Homogeneous catalysts act in the same liquid phase as the reaction mixture. Conversely, if the catalyst remains in a different phase, the process is called heterogeneous catalytic transesterification [113, 127-131]. Heterogeneous catalysts are mostly applied in transesterification reaction due to many advantages such as easy catalyst separation and reusability, improved selectivity, fewer process stages, no water formation or saponification reaction, including in green technology, and cost effectiveness [127, 132]. The heterogeneous catalysts increase the mass transfer rate during the transesterification reaction [127, 131]. Various ceramic materials have been investigated for the production of biodiesel [106, 109-115, 124-179]. Some of these solid catalysts include alkali and alkaline-earth metal carbonates and oxides such as magnesium oxide (MgO), calcium oxide (CaO), barium oxide (BaO), strontium oxide (SrO) [124-131, 133-143]; lithium base ceramics (Li_4SiO_4 and Li_2SiO_3 [144-146]); sodium silicate (Na_2SiO_3 [147]);

transition metal oxides and derivatives (titanium oxide, zinc oxide, mixed oxides catalysts [148-149]); ion exchange resin type acid heterogeneous catalysts [150]; MCM-metal impregnated materials [114]; layered double hydroxides (hydrotalcite-like hydroxides) [151-154]; hydrocalumite-like compounds [110,155]; supported bases [156-163]; and zeolites [164-165].

Among the alkaline earth metal oxides, CaO is a promising basic heterogeneous catalyst for synthesizing biodiesel at mild temperatures (below the boiling point of methanol, MeOH) and at atmospheric pressure due to its plentiful availability and low cost, but it is rapidly hydrated and carbonated upon contact with room temperature air. CaO is the most widely used catalyst for transesterification and produces a high yield of 98% of fatty acid methyl esters (FAME) during the first cycle of reaction [130]. Granados et al. [142] used activated CaO as a solid base catalyst in the transesterification of sunflower oil to investigate the role of water and carbon dioxide on the deterioration of the catalytic performance upon contact with air for different periods. The study showed that CaO was rapidly hydrated and carbonated in air. Consequently, the reusability of the catalyst for subsequent steps is a big question mark. Di Serio et al. [170] reported a 92% biodiesel yield with MgO catalyst, using 12:1 methanol to oil molar ratio with 5.0wt% of the catalyst at methanol supercritical condition for 1 h. Wen et al. [171] carried out transesterification from waste cooking oil with methanol at 170 °C for 6 h with 10wt % of MgO/TiO_2 and 50:1 M ratio of MeOH and oil. Guo et al. [172] studied the methyl ester yield produced via transesterification of soybean oil using sodium silicate as a catalyst. Sodium silicate was an effective catalyst for the microwave-irradiated production of biodiesel and hydrothermal production of hydrogen from by-product glycerol combined with Ni catalyst. The optimum reaction conditions obtained were 7.5:1 M ratio of alcohol/oil, 3wt% catalyst amount, 1 h reaction time and 60 °C reaction temperature. The FAME yield was ~100%. On the other hand, microwave-assisted transesterification of vegetable oil with sodium silicate is an effective and economical method for the rapid production of biodiesel. The reused catalyst after transesterification process for four cycles was recovered. Overall, sodium silicate was fully used in biodiesel production and glycerol gasification, and this co-production process provided a novel green method for biodiesel production and glycerol utilization [172].

Several techniques have been investigated for the transesterification reaction using heterogeneous catalysts for biodiesel production, as follows: transesterification via radio frequency microwaves, alcohol reflex temperature, alcohol supercritical temperature and ultrasonication [127, 173-177]. Recently, the use of ultrasonic irradiation has gained interest in biodiesel production [173-177]. Ultrasonic energy can emulsify the reactants to reduce the catalyst requirement, methanol-oil ratio, reaction time and reaction temperature and also provides the mechanical energy for mixing and the required activation energy for initiating the transesterification reaction [173-176]. The ultrasound phenomenon has its own physical and chemical effects on the liquid-liquid heterogeneous reaction system through cavitation bubbles, according to the following principles [175]: (1) the chemical effect, in which radicals such as H^+ and OH^- are produced during a transient implosive collapse of bubbles (in a liquid irradiated with ultrasound), which accelerates chemical reaction in the bulk medium; and (2) the physical effect of emulsification, in which the microturbulence generated due to radial motion of bubbles leads to intimate mixing (homogenizing the mixture) of the immiscible reactants.

Accordingly, the interfacial region between the oil and alcohol increases sharply, resulting in faster reaction kinetics and higher conversion of oil and biodiesel yield [127]. In 2000, the ultrasonication reactor was first introduced by Hielscher Ultrasonic GmbH for biodiesel production. Nishimura et al. [175] studied the transesterification of vegetable oil using low-frequency ultrasound (28-40kHz). An excellent yield (~98%) was obtained at a 28 kHz ultrasound while a significant reduction of reaction time was obtained by using 40 kHz ultrasound. Salamatinia et al. [176] used ultrasonic assisted transesterification to improve the reaction rate. In this study, they used SrO and BaO as heterogeneous catalysts in the production of biodiesel from palm oil. The results showed that the basic properties of the catalyst were the main cause for their high activity. The low-frequency ultrasonic assisted transesterification process had no significant mechanical effects on SrO, but BaO catalyst study confirmed that the ultrasound treatment significantly improved the process by reducing the reaction time to less than 50 min at a catalyst loading of 2.8wt% to achieve biodiesel yield higher than 95%. Another study of alkali earth metals was carried out by Mootabadi et al. [177]. They reported the effect of ultrasonic waves at 20 kHz and 200W on the regenerated catalyst and compared mechanical stirring and ultrasonic irradiation. They investigated the optimum conditions, using palm oil for biodiesel production with catalysts such as CaO, SrO and BaO. They concluded that catalyst leaching was the main cause for the catalyst inactivity in the case of the re-used catalyst. BaO catalyst was found to be stable during the leaching. At the optimized condition, 95.2% yield was achieved with 60 min of reaction time for both BaO and SrO catalysts. For CaO catalyst, 77.3% yield was achieved with the same conditions. The use of ultrasound showed great enhancement of the reaction parameters in terms of the obtained yield and reaction time. The obtained yields were 30 to 40% higher in comparison to the corresponding results obtained using a conventional stirring reactor system without ultrasonication. Deng et al. [178] prepared nano-sized mixed Mg/Al oxides. Due to their strong basicity, the nanoparticles were further used as catalyst for biodiesel production from jatropha oil. Experiments were conducted with the solid basic catalyst in an ultrasonic transesterification reaction. Under the optimum conditions, biodiesel yield was 95.2%. After removing the glycerol on the catalyst surface, the nano-sized mixed Mg/Al oxides were reused eight times. The authors concluded that calcination of hydrotalcite nanocatalyst under ultrasonic radiation is an effective method for the production of biodiesel from jatropha oil. The activity of base solid catalysts is associated to their basic strength, such that the most basic catalyst showed the highest conversion. In another work, Deng et al. [179] reported optimum conditions for biodiesel production in the presence of base solid catalysts. They studied BaO and Ca–Mg–Al hydrotalcite (the most effective). The 95% biodiesel yield from jatropha oils and Ca–Mg–Al hydrotalcite was established with 30 min of reaction time. Ca–Mg–Al hydrotalcite could be reused twelve times after washing of the adsorbed glycerol from the catalyst surface with ethanol. Other types of heterogeneous catalysts under ultrasonic irradiation were used for transesterification by Georgogianni et al. [114]. They studied a wide range of catalysts including Mg-MCM-41, Mg–Al hydrotalcite and K^+-impregnated zirconium oxide. They mixed frying oils, methanol and the catalyst in a batch reactor with mechanical stirring for 24 h and with ultrasonication for 5 h. The results suggested that the basic strength was the cause of the good activity of the catalysts. Mg–Al hydrotalcite achieved the highest reaction conversion of 87% at a reaction temperature of 60 °C. Overall, ultrasonic irradiation significantly enhanced the reaction rate, causing a reduction in reaction time, and the biodiesel yield increased [114]. Consequently, a better understanding of the use of ultrasonic sound waves to accelerate the transesterification process could lead to substantial future improvement of both batch and continuous production systems, to obtain a more sustainable biodiesel production process [127].

Author details

Margarita J. Ramírez-Moreno[1,2], Issis C. Romero-Ibarra[1], José Ortiz-Landeros[2] and Heriberto Pfeiffer[1*]

*Address all correspondence to: pfeiffer@iim.unam.mx

1 Instituto de Investigaciones en Materiales, Universidad Nacional Autónoma de México, Circuito exterior s/n, Ciudad Universitaria, Del. Coyoacán, México DF, Mexico

2 Departamento de Ingeniería en Metalurgia y Materiales, Escuela Superior de Ingeniería Química e Industrias Extractivas, IPN, UPALM, México DF, Mexico

References

[1] Ortiz-Landeros J.; A'valos-Rendo'n T. L.; Go'mez-Yáñez C.; Pfeiffer H. *Analysis and Perspectives Concerning CO_2 Chemisorption on Lithium Ceramics Using Thermal Analysis.* J. Therm. Anal. Calorim. 2012, 108, 647–655.

[2] A'valos-Rendo'n T.; Casa-Madrid J.; Pfeiffer H. *Thermochemical Capture of Carbon Dioxide on Lithium Aluminates ($LiAlO_2$ and Li_5AlO_4): A New Option for the CO_2 Absorption.* J. Phys. Chem. A 2009, 113, 6919–6923.

[3] Mejia-Trejo V. L.; Fregoso-Israel E.; Pfeiffer H. *Textural, Structural, and CO_2 Chemisorption Effects Produced on the Lithium Orthosilicate by Its Doping with Sodium ($Li_{4-x}Na_xSiO_4$).* Chem. Mater. 2008, 20, 7171–7176.

[4] Mosqueda H. A.; Vazquez C.; Bosch P.; Pfeiffer H. *Chemical Sorption of Carbon Dioxide (CO_2) on Lithium Oxide (Li_2O).* Chem. Mater. 2006, 18, 2307–2310.

[5] Shan S. Y.; Jia Q. M.; Jiang L. H.; Li Q. C.; Wang Y. M.; Peng J. H. *Novel Li_4SiO_4-Based Sorbents from Diatomite for High Temperature CO_2 Capture.* Ceram. Int. 2013, 39, 5437–5441.

[6] Olivares-Marín M.; Castro-Díaz M.; Drage T. C.; Maroto-Valer M. M. *Use of Small-Amplitude Oscillatory Shear Rheometry to Study the Flow Properties of Pure and Potassi-*

um-Doped Li_2ZrO_3 Sorbents During the Sorption of CO_2 at High Temperatures. Sep. Purif. Technol. 2010, 73, 415–420.

[7] Pacciani R.; Torres J.; Solsona P.; Coe C.; Quinn R.; Hufton J.; Golden T.; Vega L. F. *Influence of the Concentration of CO_2 and SO_2 on the Absorption of CO_2 by a Lithium Orthosilicate-Based Absorbent*. Environ. Sci. Technol. 2011, 45, 7083–7088.

[8] Xiao Q.; Tang X.; Liu Y.; Zhong Y.; Zhu W. *Citrate Route to Prepare K-Doped Li_2ZrO_3 Sorbents with Excellent CO_2 Capture Properties*. Chem. Eng. J. 2011, 174, 231–235.

[9] Xiao Q.; Liu Y.; Zhong Y.; Zhu W. *A Citrate Sol-Gel Method to Synthesize Li_2ZrO_3 Nanocrystals with Improved CO_2 Capture Properties*. J. Mater. Chem. 2011, 21, 3838–3842.

[10] Rodríguez-Mosqueda R.; Pfeiffer H. *Thermokinetic Analysis of the CO_2 Chemisorption on Li_4SiO_4 by Using Different Gas Flow Rates and Particle Sizes*. J. Phys. Chem. A 2010, 114, 4535–4541.

[11] Ortiz-Landeros J.; Gomez-Yañez C.; Palacios-Romero L. M.; Lima E.; Pfeiffer H. *Structural and Thermochemical Chemisorption of CO_2 on $Li_{4+x}(Si_{1-x}Al_x)O_4$ and $Li_{4-x}(Si_{1-x}V_x)O_4$ Solid Solutions*. J. Phys. Chem. A 2012, 116, 3163–3171.

[12] Alcerreca-Corte I.; Fregoso-Israel E.; Pfeiffer H. *CO_2 Absorption on Na_2ZrO_3: A Kinetic Analysis of the Chemisorption and Diffusion Processes*. J. Phys. Chem. C 2008, 112, 6520–6525.

[13] Pfeiffer H.; Vazquez C.; Lara V. H.; Bosch P. *Thermal Behavior and CO_2 Absorption of $Li_{2-x}Na_xZrO_3$ Solid Solutions*. Chem. Mater. 2007, 19, 922–926.

[14] Zhao T.; Ochoa-Fernández E.; Rønning M.; Chen D. *Preparation and High-Temperature CO_2 Capture Properties of Nanocrystalline Na_2ZrO_3*. Chem. Mater. 2007, 19, 3294–3301.

[15] Seggiani M.; Puccini M.; Vitolo S. *Alkali Promoted Lithium Orthosilicate for CO_2 Capture at High Temperature and Low Concentration.* Int. J. Greenhouse Gas Control 2013, 17, 25–31.

[16] Khokhani M.; Khomane R. B.; Kulkarni B. D. *Sodium-Doped Lithium Zirconate Nano-Squares: Synthesis, Characterization and Applications for CO_2 Sequestration*. J. Sol-Gel Sci. Technol. 2012, 61, 316–320.

[17] Veliz-Enriquez M. Y.; Gonzalez G.; Pfeiffer H. *Synthesis and CO_2 Capture Evaluation of $Li_{2-x}K_xZrO_3$ Solid Solutions and Crystal Structure of a New Lithium-Potassium Zirconate Phase*. J. Solid State Chem. 2007, 180, 2485–2492.

[18] Martínez-dlCruz L.; Pfeiffer H. *Microstructural Thermal Evolution of the Na_2CO_3 Phase Produced During a Na_2ZrO_3–CO_2 Chemisorption Process*. J. Phys. Chem. 2012, 116, 9675–9680.

[19] Santillan-Reyes G. G.; Pfeiffer H. *Analysis of the CO_2 Capture in Sodium Zirconate (Na_2ZrO_3). Effect of the Water Vapor Addition*. Int. J. Greenhouse Gas Control 2011, 5, 1624–1629.

[20] Iwana A.; Stephenson H.; Ketchie C.; Lapkin A. *High Temperature Sequestration of CO_2 Using Lithium Zirconates.* Chem. Eng. J. 2009, 146, 249–258.

[21] Tabarés F. L. Editor, *Lithium: Technology, Performance and Safety,* Nova Publishers, (2013). Chapter 6, *Lithium Ceramics as an Alternative for the CO_2 Capture. Analysis of Different Physicochemical Factors Controlling this Process,* pp 171-192.

[22] Bish D.L., *Anion-Exchange in Takovite: Applications to Other Hydroxide Minerals,* Bone Miner., 1980, 103, 170-175.

[23] Duan X.; Evans D. G., *Layered Double Hydroxides. Structure and Bonding,* Eds. Springer-Verlag: Berlin Heidelberg. Germany, 2006; vol. 119.

[24] Wang M. Z.; Hu Q. D. L.; Li Y.; Li S.; Zhang X.; Xi M.; Yang X., *Intercalation of Ga^{3+}-Salicylidene-Amino Acid Schiff Base Complexes into Layered Double Hydroxides: Synthesis, Characterization, Acid Resistant Property, in Vitro Release Kinetics and Antimicrobial Activity,* Appl. Clay Sci. 2013, 83&84, 182-190.

[25] Catti M.; Ferraris G.; Hull S.; Pavese A. *Static Compression and H Disorder in $Mg(OH)_2$ (Brucite) to 11 GPa: a Powder Neutron Diffraction Study.* Phys. Chem. Miner. 1995, 22, 200-206.

[26] He J.; Wei M.; Li B.; Kang Y.; Evans D. G.; Duan X., *Preparation of Layered Double Hydroxides,* Structure and Bonding, 2005, 119, 89-119.

[27] Gutmann N.; Müller B. *Insertion of the Dinuclear Dihydroxo-Bridged Cr(III) Aquo Complex into the Layered Double Hydroxides of Hydrotalcite-Type.* J. Solid State Chem. 1996, 122, 214-220.

[28] Fogg A.M.; Williams G.R.; Chester R.; O'Hare D. *A Novel Family of Layered Double Hydroxides—$[MgAl_4(OH)_{12}]$ $(NO_3)_2$ xH_2O (M = Co, Ni, Cu, Zn).* J. Mater. Chem., 2004, 14, 2369-2371.

[29] de Roy A.; Forano C.; Besse J.P. *Layered Double Hydroxides: Synthesis and Post-Synthesis Modification.* Review, Chapter 1, 2002.

[30] Choi S.; Drese J.H.; Jones C.W. *Adsorbent Materials for Carbon Dioxide Capture from Large Anthropogenic Point Sources.* ChemSusChem. 2009, 2, 796-854.

[31] Ding Y.; Alpay, E. *Equilibria and Kinetics of CO_2 Adsorption on Hydrotalcite Adsorbent.* Chem. Eng. Sci. 2000, 55, 3461-3474.

[32] Reijers H.T.J.; Valster-Schiermeier S.E.A.; Cobden P.D.; van der Brink R.W. *Hydrotalcite as CO_2 Sorbent for Sorption-Enhanced Steam Reforming of Methane.* Ind. Eng. Chem. Res. 2006, 45, 2522-2530.

[33] Wang X.P.; Yu J.J.; Cheng J.; Hao Z.P.; Xu Z.P. *High Temperature Adsorption of Carbon Dioxide on Mixed Oxides Derived from Hydrotalcite-Like Compounds.* Environ. Sci. Technol. 2008, 42, 614-618.

[34] Reynolds S.P.; Ebner A.D.; Ritter J.A. *Carbon Dioxide Capture from Flue Gas by Pressure Swing Adsorption at High Temperature Using a K-Promoted HTLC: Effects of Mass Transfer on the Process Performance*. Environ. Prog. 2006, 25, 334-342.

[35] Reynolds, S.P.; Ebner, A.D.; Ritter, J.A. *Stripping PSA Cycles for CO_2 Recovery from Flue Gas at High Temperature Using a Hydrotalcite Like Adsorbent*. Ind. Eng. Chem. Res. 2006, 45, 4278-4294.

[36] Ebner, A.D.; Reynolds, S.P.; Ritter, J.A. *Nonequilibrium Kinetic Model that Describes the Reversible Adsorption and Desorption Behavior of CO_2 in a K-Promoted Hydrotalcite-Like Compound*. Ind. Eng. Chem. Res. 2007, 46, 1737-1744.

[37] Cavani F.; Trifirb F.; Vaccari A. *Hydrotalcite-Type Anionic Clays: Preparation, Properties and Applications*, Catal. Today, 1991, 11, 173–301.

[38] Vaccari A. *Preparation and Catalytic Properties of Cationic and Anionic Clays*, Catal. Today, 1998, 41, 53–71.

[39] Das N.N.; Konar J.; Mohanta M.K.; Srivastava S.C. *Adsorption of Cr(VI) and Se(IV) from their Aqueous Solutions onto Zr^{4+}-Substituted ZnAl/MgAl-Layered Double Hydroxides: Effect of Zr^{4+} Substitution in the Layer*, J. Colloid Interf. Sci., 2004, 270, 1–8.

[40] Goh K.H.; Lim T.T; Dong Z. *Application of Layered Double Hydroxides for Removal of Oxyanions: A Review*, Water Research, 2008, 42, 1343–1368.

[41] Yong Z.; Mata, V.; Rodriguez, A.E. *Adsorption of Carbon Dioxide onto Hydrotalcite-Like Compounds (HTLCs) at High Temperatures*. Ind. Eng. Chem. Res. 2001, 40, 204-209.

[42] Bellotto M.; Rebours B.; Clause O.; Lynch J.; Bazin D.; Elkaîm E. *Hydrotalcite Decomposition Mechanism: A Clue to the Structure and Reactivity of Spinel-Like Mixed Oxides*, J. Phys. Chem. 1996, 100, 8535-8542.

[43] Ram Reddy M. K., Xu Z. P., Lu G. Q. (Max); Diniz da Costa J. C. *Layered Double Hydroxides for CO_2 Capture: Structure Evolution And Regeneration*. Ind. Eng. Chem. Res. 2006, 45, 7504-7509.

[44] Hufton J.; Mayorga S.; Gaffeney T.; Nataraj S.; Sircar S. *Sorption Enhanced Reaction Process (SERP),* Proceedings of the 1997 U.S., DOE Hydrogen Program Review, 1997, 1, 179-194.

[45] Ram Reddy M.K.; Xu Z.P.; Lu G.Q. (Max); Diniz da Costa J.C. *Influence of Water on High-Temperature CO_2 Capture Using Layered Double Hydroxide Derivatives*. Ind. Eng. Chem. Res. 2008, 47, 2630-2635.

[46] Martunus; Othman M.R; Fernando W.J.N. *Elevated Temperature Carbon Dioxide Capture Via Reinforced Metal Hydrotalcite*. Micropor. Mesopor. Mater. 2011, 138, 110–117.

[47] Yong Z.; Rodrigues A.E. *Hydrotalcite-Like Compounds as Adsorbents for Carbon Dioxide*. Energy Convers. & Manage. 2002, 43, 1865-1876.

[48] Newman S. P.; Jones W. *Supramolecular Organization and Materials Design*, Cambridge University Press, England, 2001.

[49] Yamamoto T.; Kodama T.; Hasegawa N.; Tsuji M.; Tamura Y. *Synthesis of Hydrotalcite with High Layer Charge for CO_2 Adsorbent*. Energy Convers Mgmt, 1995, 36, 637-640.

[50] Wang Q.; Wu Z.; Tay H. H.; Chen L.; Liu Y.; Chang J.; Zhong Z.; Luo J.; Borgna A. *High Temperature Adsorption of CO_2 on Mg-Al Hydrotalcite: Effect of the Charge Compensating Anions and the Synthesis pH*. Catal. Today, 2011, 164, 198-203

[51] Wang Q.; Tay H.H.; Ng D.J.W.; Chen L.; Liu Y.; Chang J.; Zhong Z.; Luo J.; Borgna A. *The Effect of Trivalent Cations on the Performance of Mg-M-CO_3 Layered Double Hydroxides for High-Temperature CO_2 Capture*. ChemSusChem. 2010, 3, 965-973.

[52] Hibino T.; Yamashita Y.; Kosuge K.; Tsunashima A. *Decarbonation Behavior of Mg-Al-CO_3 Hydrotalcite-Like Compounds During Heat Treatment,*. Clays Clay Minerals. 1995, 43, 427 – 432.

[53] Qian W.; Luo J.; Zhong Z.; Borgna A. *CO_2 Capture by Solid Adsorbents and their Applications: Current Status and New Trends*. Energy Environ. Sci., 2011, 4, 42-55.

[54] Hufton J. R.; Mayorga S.; Sircar S. *Sorption-Enhanced Reaction Process for Hydrogen Production*. AIChE J. 1999, 45, 248.

[55] Oliveira E.L.G.; Grande C.A.; Rodrigues A.E.; *CO_2 Sorption on Hydrotalcite and Alkali-Modified (K and Cs) Hydrotalcites at High Temperatures*. Sep. Purif. Technol. 2008, 62, 137-147.

[56] Yang J. I.; Kim J. N. *Hydrotalcites for Adsorption of CO_2 at High Temperature*. Korean J. Chem. Eng., 2006, 23, 77-80.

[57] Ida J.I.; Lin S. *Mechanism of High-Temperature CO_2 Sorption on Lithium Zirconate*. Environ. Sci. Technol., 2003, 37, 1999-2004.

[58] Hufton J.; Mayorga S.; Nataraj S.; Sircar S.; Rao M. *Sorption-Enhanced Reaction Process (SERP)*, Proceedings of the 1998, USDOE Hydrogen Program Review, 1998, 2, 693-705.

[59] Lee Jung M.; Min Yoon J.; Lee Ki B.; Jeon Sang G.; Na Jeong G.; Ryu Ho J. *Enhancement of CO_2 Sorption Uptake on Hydrotalcite by Impregnation with K_2CO_3*. Langmuir, 2010, 26, 18788–18797.

[60] Baker R. W. *Membrane Technology and Applications*, 2nd Editions, John Wiley and Sons (2004).

[61] Murder M., *Basic Principles of Membrane Technology*, Kluwer Academic Publishers (1991).

[62] Aresta M.; Dibenedetto A. *The Contribution of the Utilization Option to Reducing the CO_2 Atmospheric Loading: Research Needed to Overcome Existing Barriers for a Full Exploitation of the Potential of The CO_2 Use*, Cat. Today. 2004, 98, 455–462.

[63] Anderson M.; Wang H.; Lin Y. S. *Inorganic Membranes for Carbon Dioxide and Nitrogen Separation*, Rev. Chem. Eng., 2012, 28, 101-121.

[64] Yang H.; Xu Z.; Fan M.; Gupta R.; Slimane R. B.; Bland A. E.; Wright I. *Progress in Carbon Dioxide Separation and Capture: A Review*, J. Environ. Sci., 2008, 20, 14-27.

[65] Niwa M.; Katada N.; Okumura K. *Introduction to Zeolite Science and Catalysis, Characterization and Design of Zeolite Catalysts*, Springer Series in Materials Science, Vol. 141 (2010).

[66] Iwamoto Y.; Kawamoto H. *Science and Technology Trends: Quarterly Report*, 2009, 32, 42-59.

[67] Algieri C.; Barbieri G.; Drioli E.; *Zeolite Membranes for Gas Separations, in Membrane Engineering for the Treatment of Gases*. Royal Society of Chemistry Vol. 2 (2011).

[68] Fedosov D. A.; Smirnov A. V.; Knyazeva E. E.; Ivanova I. I. *Zeolite membranes: Synthesis, properties, and application*, Petroleum Chem., 2011, 51, 657-667.

[69] Caro J.; Noack M. *Zeolite membranes – Recent developments and progress.* Micropor. Mesopor. Mater. 2008, 115, 215–233.

[70] Jia M. D.; Peinemann K. V.; Behling R. D. *Ceramic Zeolite Composite Membranes. Preparation, Characterization and Gas Permeation*. J. Memb. Sci. 1993, 82, 15-26.

[71] Cui Y.; Kita H.; Okamoto K. I. *Preparation and Gas Separation Performance of Zeolite T Membrane.* J. Mater. Chem. 2004, 14, 924.

[72] Poshusta J.; Tuan V.; Pape E.; Noble R.; Falconer J. *Separation of Light Gas Mixtures Using SAPO-34 Membranes.* AIChE J. 2000, 46, 779-789.

[73] Li S.; Falconer J.; Noble R. *SAPO-34 Membranes for CO_2/CH_4 Separation.* J. Memb. Sci., 2004, 241, 121–135.

[74] Tomita T.; Nakayama K.; Sakai H. *Gas Separation Characteristics of DDR Type Zeolite Membrane.* Micropor. Mesopor. Mater. 2004, 68, 71.

[75] Caro J.; Noack M. *Zeolite Membranes: Recent Developments and Progress.* Micropor. Mesopor. Mater. 2008, 115, 215-233.

[76] Burggraaf A. J. *Fundamentals of Inorganic Membrane Science and Technology, Membrane Science and Technology*, Series, 4, Elsevier Science. Netherlands (1996).

[77] Lara-Medina J. J.; Torres-Rodriguez M.; Gutierrez-Arzaluz M.; Mugica-Alvarez V. *Separation of CO_2 and N_2 with a Lithium-Modified Silicalite-1 Zeolite Membrane.* Inter. J. Greenhouse Gas Control, 2012, 10, 494-500.

[78] Dyer A. *Introduction to Zeolite Science and Practice.* 3rd Revised Edition. J. Cejka, H. van Bekkum, A. Corma and F. Schiith (Editors) Elsevier B.V. (2007).

[79] An W.; Swenson P.; Gupta A.; Wu L.; Kuznicki T. M.; Kuznicki S. M. *Improvement of H_2/CO_2 Selectivity of the Natural Clinoptilolite Membranes by Cation Exchange Modification.* J. Memb. Sci., 2013, 433, 25–31.

[80] White J. C.; Dutta P. K.; Shqau K.; Verweij H. *Synthesis of Ultrathin Zeolite Y Membranes and their Application for Separation of Carbon Dioxide and Nitrogen Gases.* Langmuir 2010, 26, 12, 10287–10293.

[81] Cho Y.K., Han K., Lee K.H. *Separation of CO_2 by Modified γ-Al_2O_3 Membranes at High Temperature.* J. Membrane Sci. 1995, 104, 219-230.

[82] Keizer K.; Uhlhorn R.J.R.; van Vuren R.J.; Burggraaf A.J. *Gas Separation Mechanisms in Microporous Modified γ-$A1_2O_3$ Membranes.* J. Membrane Sci., 1988, 39, 285-300.

[83] Uhlhorn R.J.R.; Keizer K., Burggraaf A.J. *Gas and Surface Diffusion in Modified 33-Alumina Systems.* J. Membrane Sci., 1989, 46, 225-241.

[84] Kusakabe K., Ichiki K., Morooka S. *Separation of CO_2 with $BaTiO_3$ Membrane Prepared by the Sol—Gel Method.* J. Membrane Sci. 1944, 95, 171-177.

[85] Nomura M., Sakanishi T., Nishi Y., Utsumi K., Nakamura R.. *Preparation of CO_2 Permselective Li_4SiO_4 Membranes by Using Mesoporous Silica as a Silica Source.* Energy Procedia 2013, 37, 1004-1011.

[86] Nomura M.; Nishi Y.; Sakanishi T.; Utsumi K.; Nakamura R. *Preparation of Thin Li_4SiO_4 Membranes by Using a CVD Method,* Energy Procedia 2013, 37, 1012-1019.

[87] Kawamura H.; Yamaguchi T.; Nair B. N.; Nakagawa K.; Nakao S.I. *Dual-Ion Conducting Lithium Zirconate-Based Membranes for High Temperature CO_2 Separation.* J. Chem. Eng. Jpn. 2005, 38, (5) 322-328.

[88] Yamaguchi T.; Niitsume T.; Nair B. N.; Nakagawa K. *Lithium Silicate Based Membranes for High Temperature CO_2 Separation.* J. of Membrane Sci., 2007, 294, 16-21.

[89] Nair B.N.; Burwood R.P.; Goh V.J.; Nakagawa K.; Yamaguchi T. *Lithium Based Ceramic Materials and Membranes for High Temperature CO_2 Separation.* Progress in Materials Sci. 2009, 54, 511–541.

[90] Skinner S.J.; Kilner J.A. *Oxygen Ion Conductors.* Materials Today, 2003, 6, 30-37.

[91] Anderson M.; Lin Y.S. *Carbonate–Ceramic Dual-Phase Membrane for Carbon Dioxide Separation.* J. Membrane Sci. 2010, 357, 22.

[92] Wade J. L.; Lee C.; West A. C.; Lackner K. S. *Composite Electrolyte Membranes for High Temperature CO_2 Separation.* J. Membrane Sci. 2011, 369, 20.

[93] Zhang L.; Xu N.; Li X.; Wang S.; Huang K.; Harris W. H.; Wilson K.; Chiu S. *High CO_2 Permeation Flux Enabled by Highly Interconnected Three-Dimensional Ionic Channels in Selective CO_2 Separation Membranes*. Energy Environ. Sci. 2012, 5, 8310.

[94] Rui Z.; Anderson M.; Li Y.; Lin Y.S. *Ionic Conducting Ceramic and Carbonate Dual Phase Membranes for Carbon Dioxide Separation,* J. Membrane Sci. 2012, 417-418, 174.

[95] Lu B.; Lin Y.S. *Synthesis and Characterization of Thin Ceramic-Carbonate Dual-Phase Membranes for Carbon Dioxide Separation.* J. Membrane Sci. 2013, 444, 402–411.

[96] Dong X.; Ortiz-Landeros J.; Lin Y. S. *An Assymetric Thin Tubular Dual Phase Membrane.* Chem. Commun. 2013, 49, 9654.

[97] Ortiz-Landeros J.; Norton T.T.; Lin Y.S. *Effects of Support Pore Structure on Carbon Dioxide Permeation of Ceramic-Carbonate Dual-Phase Membranes.* Chem. Eng. Sci. 2013, 104, 891-898.

[98] Rui Z.; Anderson M.; Lin Y.S.; Li Y.; *Modeling and Analysis of Carbon Dioxide Permeation through Ceramic-Carbonate Dual-Phase Membranes*. J. Membrane Sci. 2009, 345, 110.

[99] Yu K.M.K.; Curcic I.; Gabriel J.; Tsang S.C.E. *Recent Advances in CO_2 Capture and Utilization*, ChemSusChem. 2008, 1, 893–899.

[100] Wender I. *Reactions of Synthesis Gas*, Fuel Process. Technol. 1996, 48, 3, 189-297.

[101] Rostrup-Nielsen J. R. *Syngas in Perspective*. Catal. Today 2002, 71, 3-4, 243-247.

[102] Dry M.E. *The Fischer–Tropsch Process: 1950–2000.* Catal. Today 2002, 71, 227-241.

[103] Yu K.M.K.; Curcic I.; Gabriel J.; Tsang S.C.E. *Recent Advances in CO_2 Capture and Utilization.* ChemSusChem. 2008, 1, 893 – 899.

[104] Gnanapragasam N.; Reddy B.; Rosen M. *Reducing CO_2 Emissions for an IGCC Power Generation System: Effect of Variations in Gasifier and System Operating Conditions.* Energ. Convers. Manage. 2009, 50, 1915-1923.

[105] Yang Z.Z.; He L.N.; Gao J.; Liu A.H.; Yu B. *Carbon Dioxide Utilization with C–N Bond Formation: Carbon Dioxide Capture and Subsequent Conversion.* Energy Environ. Sci. 2012, 5, 6602-6639.

[106] Chattopadhyay S.; Sen R. *Fuel Properties, Engine Performance and Environmental Benefits of Biodiesel Produced by a Green Process.* Appl. Energ. 2013, 105, 319–326.

[107] Choi J.C.; He L.N.; Yasuda H.; Sakakura T. *Selective and High Yield Synthesis of Dimethyl Carbonate Directly from Carbon Dioxide and Methanol.* Green Chem. 2002, 4, 230–234.

[108] Quispe C.A.; Coronado J.R.C; CarvalhoJr J.A. *Glycerol: Production, Consumption, Prices, Characterization and New Trends in Combustion.* Renew. Sust. Energ. Rev. 2013, 27, 475–493.

[109] Talebian-Kiakalaieh A.; Saidina Amin N. A; Mazaheri H. *A Review on Novel Processes of Biodiesel Production from Waste Cooking Oil.* Appl. Energ. 2013, 104, 683–710.

[110] Kuwahara Y.; Tsuji K.; Ohmichi T.; Kamegawa T.; Moria K.; Yamashita H. *Transesterifications Using a Hydrocalumite Synthesized from Waste Slag: An Economical and Ecological Route for Biofuel Production.* Catal. Sci. Tech. 2012, 2, 1842–1851.

[111] Atadashi I.M.; Aroua M.K.; Abdul Aziz A. *Biodiesel Separation and Purification: A Review.* Renew. Energ. 2011, 36, 437-443.

[112] Lam M.K.; Lee K.T. *Mixed Methanol–Ethanol Technology to Produce Greener Biodiesel from Waste Cooking Oil: A Breakthrough for $SO_4^{2-}/SnO_2–SiO_2$ catalyst.* Fuel Process. Technol. 2011, 92, 1639–1645.

[113] Lam M.; Lee K.T.; Rahman Mohamed A. *Homogeneous, Heterogeneous and Enzymatic Catalysis for Transesterification of High Free Fatty Acid Oil (Waste Cooking Oil) to Biodiesel: A Review.* Biotechnol. Adv. 2010, 28, 500–518.

[114] Georgogianni K.G.; Katsoulidis A.K.; Pomonis P.J.; Manos G.; Kontominas M.G. *Transesterification of Rapeseed Oil for the Production of Biodiesel Using Homogeneous and Heterogeneous Catalysis.* Fuel Process. Technol. 2009, 90, 1016–1022.

[115] Zhang Y.; Dube M.A.; McLean D.D.; Kates M. *Biodiesel Production from Waste Cooking Oil: 1. Process Design and Technological Assessment.* Bioresource Technol. 2003, 89, 1-16.

[116] Long Y.D.; Fang Z.; Su T.C.; Yang Q. *Co-production of Biodiesel and Hydrogen from Rapeseed and Jatropha Oils with Sodium Silicate and Ni Catalysts.* Appl. Energ. 2013, http://dx.doi.org/10.1016/j.apenergy.2012.12.076.

[117] Centi G.; Quadrelli E.A.; Perathoner S. *Catalysis for CO_2 Conversion: A Key Technology for Rapid Introduction of Renewable Energy in the Value Chain of Chemical Industries.* Energy Environ. Sci. 2013, 6, 1711-1731.

[118] Centi G.; Perathoner S. *Opportunities and Prospects in the Chemical Recycling of Carbon Dioxide to Fuels.* Catal. Today 2009, 148, 191-205.

[119] Hoffmann F.M.; Yang Y.; Paul J.; White M.G; Hrbek J. *Hydrogenation of Carbon Dioxide by Water: Alkali-Promoted Synthesis of Formate.* J. Phys. Chem. Lett. 2010, 1, 2130–2134.

[120] H. Yin, X. Mao, D. Tang, W. Xiao, L. Xing, H. Zhu, D. Wang, D.R. Sadoway. *Capture and Electrochemical Conversion of CO_2 to Value-Added Carbon and Oxygen by Molten Salt Electrolysis.* Energy Environ. Sci., 2013, 6, 1538-1545.

[121] Hu B.; Guild C.; Suib S.L. *Thermal, Electrochemical, and Photochemical Conversion of CO_2 to Fuels and Value-Added Products.* Journal of CO_2 Utilization. 2013, 1, 18–27.

[122] Kumar B.; Smieja J.M.; Kubiak C.P. *Photo-reduction of CO_2 on p-type Silicon Using Re(Bipy-But)$(CO)_3$Cl: Photovoltages Exceeding 600 mv for the Selective Reduction of CO_2 to CO.* J. Phys. Chem. C 2010, 114, 14220–14223.

[123] Bara J. *Review: The Chemistry of Amine Manufacture. What Chemicals Will We Need to Capture CO_2?* Greenhouse Gas Sci. Technol. 2012, 2, 162–171.

[124] Helwani Z.; Othman M.R.; Aziz N.; Fernando W.J.N.; Kim J. *Technology for Production of Biodiesel Focusing on Green Catalytic Techniques: A Review.* Fuel Process. Technol. 2009, 90, 1502–1515.

[125] Endalew K.; Kiros Y., Zanzi R. *Inorganic Heterogeneous Catalysts for Biodiesel Production from Vegetable Oils.* Biomass Bioenerg. 2011, 35, 3787-3809.

[126] Luque R.; Lovett J.C.; Datta B.; Clancy J.; Campeloa J.M.; Romero A.A. *Biodiesel as Feasible Petrol Fuel Replacement: A Multidisciplinary Overview.* Energy Environ. Sci. 2010, 3, 1706–1721.

[127] Ramachandran K.; Suganya T.; Gandhi N.N.; Renganathan S. *Recent Developments for Biodiesel Production by Ultrasonic Assist Transesterification Using Different Heterogeneous Catalyst: A Review.* Renew. Sust. Energ. Rev. 2013, 22, 410–418.

[128] Vyas A.P.; Verma J.L.; Subrahmanyam N. *A Review on FAME Production Processes.* Fuel 2010, 8, 1–9.

[129] Di Serio M., Ledda M., Cozzolino M., Minutillo G., Tesser R., Santacesaria E. *Transesterification of Soybean Oil to Biodiesel by Using Heterogeneous Basic Catalysts.* Ind. Eng. Chem. Res. 2006, 45, 3009-3014.

[130] Veljkovic V. B.; Stamenkovic O. S.; Todorovic Z. B.; Lazic M. L.; Skala D. U. *Kinetics of Sunflower Oil Methanolysis Catalyzed by Calcium Oxide.* Fuel 2009, 88, 554–1562.

[131] Singh Chouhan A.P.; Sarma A.K. *Modern Heterogeneous Catalysts for Biodiesel Production: A Comprehensive Review.* Renew. Sust. Energ. Rev. 2011, 15, 4378– 4399.

[132] Lee J.S.; Saka S. *Biodiesel Production by Heterogeneous Catalysts and Super-Critical Technologies: Review.* Bioresource Technol. 2010, 101, 7191–7200.

[133] Salamatinia B.; Abdullah A.Z.; Bhatia S. *Quality Evaluation of Biodiesel Produced through Ultrasound-Assisted Heterogeneous Catalytic System.* Fuel Process. Technol. 2012, 97, 1-8.

[134] Berrios M.; Martín M.A.; Chica A.F.; Martín A. *Purification of Biodiesel from Used Cooking Oils.* Appl. Energ. 2011, 88, 3625–3631.

[135] Zabeti M.; Daud W.M.A.W.; Aroua M.K. *Activity of Solid Catalysts for Biodiesel Production: A Review.* Fuel Process. Technol. 2009, 90, 770–777.

[136] Sharma Y. C.; Singh B.; Korstad J. *Latest Developments on Application of Heterogenous Basic Catalysts for an Efficient and Eco Friendly Synthesis of Biodiesel: A Review.* Fuel 2011, 90, 1309–1324.

[137] Wen Z.; Yu X.; Tu S.T.; Yan J.; Dahlquist E. *Synthesis of Biodiesel from Vegetable Oil with Methanol Catalyzed by Li-Doped Magnesium Oxide Catalysts.* Appl. Energ. 2010, 87, 743–748.

[138] Dossin T.F.; Reyniers M.F.; Berger R.J.; Marin G.B. *Simulation of Heterogeneously MgO-Catalyzed Transesterification for Fine-Chemical and Biodiesel Industrial Production.* Appl. Catal. B-Environ. 2006, 67, 136–148.

[139] Liu X.; He H.; Wang Y.; Zhu S.; Piao X. *Transesterification of Soybean Oil to Biodiesel Using CaO as a Solid Base Catalyst.* Fuel 2008, 87, 216–21.

[140] Liu X.; He H.; Wang Y.; Zhu S. *Transesterification of Soybean Oil to Biodiesel Using SrO as a Solid Base Catalyst.* Catal. Commun. 2007, 8, 1107–1111.

[141] Soares Días A.P.; Bernardo J.; Felizardo P.; Neiva Correia M.J. *Biodiesel Production by Soybean Oil Methanolysis over SrO/MgO Catalysts. The Relevance of the Catalyst Granulometry.* Fuel Process. Technol. 2012, 102, 146–155.

[142] Granados M.L.; Poves M.D.; Alonso D.; Mariscal R.; Galisteo F.C.; Moreno-Tost R. *Biodiesel from Sunflower Oil by Using Activated Calcium Oxide.* Appl. Catal. B-Environ. 2007, 73, 317–26.

[143] Watkins R.S.; Lee A.F.; Wilson K. *Li–CaO Catalysed Tri-Glyceride Transesterification for Biodiesel Applications.* Green Chem. 2004, 6, 335–340.

[144] Wang J.X.; Chen K.T.; Huang J.S.; Chen C.C. *Application of Li_2SiO_3 as a Heterogeneous Catalyst in the Production of Biodiesel from Soybean Oil.* Chinese Chem. Lett. 2011, 22, 1363–1366.

[145] Wang J.X.; Chen K.T.; Wu J.S.; Wang P.H.; Huang S.T.; Chen C.C. *Production of Biodiesel through Transesterification of Soybean Oil Using Lithium Orthosilicate Solid Catalyst.* Fuel Process. Technol. 2012, 104, 167–173.

[146] Chen K.T.; Wang J.X.; Dai Y.M.; Wang P.H.; Liou C.Y.; Nien C.W.; Wu J.S.; Chen C.C. *Rice Husk Ash as a Catalyst Precursor for Biodiesel Production.* J. Taiwan Inst. Chem. Eng. 2013, 44, 622-629.

[147] Guo P.; Zheng C.; Zheng M.; Huang F.; Li W.; Huang Q. *Solid Base Catalysts for Production of Fatty Acid Methyl Esters.* Renew. Energ. 2013, 53, 377-383.

[148] Singh A.K.; Fernando S.D. *Transesterification of Soyabean Oil Using Heterogenous Catalysts.* Energ. Fuels 2008, 22, 2067-2069.

[149] Omar W.N.N.W.; Amin N.A.S. *Biodiesel Production from Waste Cooking Oil over Alkaline Modified Zirconia Catalyst.* Fuel Process. Technol. 2011, 92, 2397-2405.

[150] Molaei Dehkordi A.; Ghasemi M. *Transesterification of Waste Cooking Oil to Biodiesel Using Ca and Zr Mixed Oxides as Heterogeneous Base Catalysts.* Fuel Process. Technol. 2012, 97, 45–51.

[151] Shibasaki-Kitakawa N., Honda H., Kuribayashi H., Toda T., Fukumura T., Yonemoto T. *Biodiesel Production Using Anionic Ion-Exchange Resin as Heterogeneous Catalyst.* Bioresource Technol. 2007, 98, 416–421.

[152] Shumaker J. L.; Crofcheck C.; Tackett S.A.; Santillan-Jimenez E.; Morgan T.; Ji Y.; Mark Crocker; Toops T.J. *Biodiesel Synthesis Using Calcined Layered Double Hydroxide Catalysts.* Appl. Catal. B-Environ. 2008, 82, 120–130.

[153] Corma A., Hamid S.B.A., Iborra S., Velty A. *Lewis and Bronsted Basic Active Sites on Solid Catalysts and their Role in the Synthesis of Monoglycerides.* J. Catal. 2005, 234, 340-347.

[154] Sankaranarayanan S.; Churchil Antonyraj A.; Kannan S. *Transesterification of Edible, Non-Edible and used Cooking Oils for Biodiesel Production Using Calcined Layered Double Hydroxides as Reusable Base Catalysts.* Bioresource Technol. 2012, 109, 57–62.

[155] Navajas A.; Campo I.; Arzamendi G.; Hernandez W.Y.; Bobadilla L.F.; Centeno M.A.; Odriozola J.A.; Gandia L.M. *Synthesis of Biodiesel from the Methanolysis of Sunflower Oil Using PURAL® Mg–Al Hydrotalcites as Catalyst Precursors.* Appl. Catal. B-Environ 2010, 100, 299–309.

[156] Gomes J.F.P.; Puna J.F.B.; Gonçalves L.M.; Bordado J.C.M. *Study on the use of MgAl Hydrotalcites as Solid Heterogeneous Catalysts for Biodiesel Production.* Energy 2011, 36, 6770-6778.

[157] Sánchez-Cantú M.; Pérez-Díaz L.M.; Tepale-Ochoa N.; González-Coronel V.J.; Ramos-Cassellis M.E.; Machorro-Aguirre D.; Valente J.S. *Green Synthesis of Hydrocalumite-Type Compounds and their Evaluation in the Transesterification of Castor Bean Oil and Methanol.* Fuel 2013, 111, 23-31.

[158] Sun H.; Ding Y.; Duan J.; Zhang Q.; Wang Z.; Lou H.; Zheng X. *Transesterification of Sunflower Oil to Biodiesel on ZrO_2 Supported La_2O_3 Catalyst.* Bioresource Technol. 2010, 101, 953–958.

[159] Kim H.J.; Kang B.S.; Kim M.J.; Park Y.M.; Kim D.K.; Lee J.S.; Lee K.Y. *Transesterification of Vegetable Oil to Biodiesel Using Heterogeneous Base Catalyst.* Catal. Today 2004, 93–95, 315–320.

[160] Ebiura T.; Echizen T.; Ishikawa A.; Murai K.; Baba T. *Selective Transesterification of Triolein with Methanol to Methyl Oleate and Glycerol Using Alumina Loaded with Alkali Metal Salt as a Solid-Base Catalyst.* Appl. Catal. A-Gen. 2005, 283, 111–116.

[161] Xie W.; Li H. *Alumina-Supported Potassium Iodide as a Heterogeneous Catalyst for Biodiesel Production from Soybean Oil.* J. Mol. Catal. A-Chem. 2006, 255, 1–9.

[162] Lukic I.; Krstic J.; Jovanovic D.; Skala D. *Alumina/Silica Supported K_2CO_3 as a Catalyst for Biodiesel Synthesis from Sunflower Oil.* Bioresource Technol. 2009, 100, 4690–4696.

[163] Evangelista J.P.C.; Chellappa T.; Coriolano A.C.F.; Fernandes Jr. V.J.; Souza L.D.; Araujo A.S.; *Synthesis of Alumina Impregnated with Potassium Iodide Catalyst for Biodiesel Production from Rice Bran Oil.* Fuel Process. Technol. 2012, 104, 90–95.

[164] Arzamendi G.; Campo I.; Arguiñarena E.; Sánchez M.; Montes M.; Gandía L.M. *Synthesis of Biodiesel with Heterogeneous NaOH/Alumina Catalysts: Comparison with Homogeneous NaOH.* Chem. Eng. J. 2007, 134, 123–130.

[165] Baroutian S.; Aroua M.K.; Raman A.A.; Sulaiman N.M.N. *Methanol Recovery During Transesterification of Palm Oil in a TiO_2/Al_2O_3 Membrane Reactor: Experimental Study and Neutral Network Modeling*. Sep. Purif. Technol. 2010, 76, 58–63.

[166] Wu H.; Zhang J.; Wei Q.; Zheng J.; Zhang J. *Transesterification of Soybean Oil to Biodiesel Using Zeolite Supported CaO as Strong Base Catalysts.* Fuel Process. Technol. 2013, 109, 13-18.

[167] Babajide O.; Musyoka N.; Petrik L.; Ameer F. *Novel Zeolite Na-X Synthesized from Fly Ash as a Heterogeneous Catalyst in Biodiesel Production.* Catal. Today 2012, 190, 54-60.

[168] Alves C.T.; Oliveira A.; Carneiro S.A.V; Silva A.G.; Andrade H.M.C.; Vieira de Melo S.A.B.; Torres E.A. *Transesterification of Waste Frying Oil Using a Zinc Aluminate Catalyst.* Fuel Process. Technol. 2013, 106, 102–107.

[169] Borges M.E.; Díaz L. *Recent Developments on Heterogeneous Catalysts for Biodiesel Production by Oil Esterification and Transesterification Reactions: A Review.* Renew. Sust. Energ. Rev. 2012, 6, 2839–2849.

[170] Di Serio M.; Tesser R.; Pengmei L.; Santacesaria E. *Heterogeneous Catalysts for Biodiesel Production.* Energ. Fuels 2008, 22, 207–17.

[171] Wen Z., Yu X., Tu S.T., Yan J., Dahlquist E. *Biodiesel Production from Waste Cooking Oil Catalyzed by TiO_2–MgO Mixed Oxides.* Bioresource Technol. 2010, 101, 9570–9576.

[172] Guo F.; Peng Z.G.; Dai J.Y.; Xiu Z.L. *Calcined Sodium Silicate as Solid Base Catalyst for Biodiesel Production.* Fuel Process. Technol. 2010, 991, 322–328.

[173] Singh A.K.; Fernando S.D.; Hernandez R. *Base Catalyzed Fast Transesterification of Soybean Oil Using Ultrasonication.* Energ. Fuel 2007, 21, 1161–1164.

[174] Kalva A.; Sivasankar T.; Moholkar V.S. *Physical Mechanism of Ultrasound-assisted Synthesis of Biodiesel.* Ind. Eng. Chem. Res. 2008, 48, 534–544.

[175] Nishimura CSMV; Maeda Y.R. *Conversion of Vegetable Oil to Biodiesel Using Ultrasonic Irradiation.* Chem. Lett. 2003, 32, 716–717.

[176] Salamatinia B., Mootabadi H., Bhatia S., Abdullah A.Z. *Optimization of Ultrasonic-Assisted Heterogeneous Biodiesel Production from Palm Oil: A Response Surface Methodology Approach.* Fuel Process. Technol. 2010, 91, 441–448.

[177] Mootabadi H.; Salamatinia B.; Bhatia S.; Abdullah A.Z. *Ultrasonic-assisted Biodiesel Production Process from Palm Oil Using Alkaline Earth Metal Oxides as the Heterogeneous Catalysts*. Fuel 2010, 89, 1818–1825.

[178] Deng X.; Fang Z.; Liu Y.; Yu C. *Production of Biodiesel from Jatropha Oil Catalyzed by Nanosized Solid Basic Catalyst.* Energy 2011, 36, 777–784.

[179] Deng X.; Fang Z.; Hu Y.; Zeng H.; Liao K.; Yu C.L. *Preparation of Biodiesel on Nano Ca–Mg–Al Solid Base Catalyst under Ultrasonic Radiation in Microaqueous Media*. Petrochemical Technology 2009, 38, 1071–1075.

Chapter 15

Predicting the Phase Equilibria of Carbon Dioxide Containing Mixtures Involved in CCS Processes Using the PPR78 Model

Romain Privat and Jean-Noël Jaubert

Additional information is available at the end of the chapter

http://dx.doi.org/10.5772/57058

1. Introduction

Carbon dioxide is an extremely important product of the chemical, pharmaceutical and petrochemical industries. Its main applications are production of coal liquids, petroleum processes such as enhanced oil recovery and separation and supercritical fluid extraction. CO_2 is however a greenhouse gas that affects the Earth's temperature and many efforts are devoted to the reduction of CO_2 emissions. The design and operation of many processes dealing with the CO_2 capture and storage (CCS) greatly depend on knowledge about pressure-volume-temperature-composition (PVTxy) and mixing properties (h^M, $c_P{}^M$) of mixtures involved in these processes. A typical CO_2 capture and storage (CCS) chain consists of four main steps: CO_2 capture, conditioning (dehydration, non-condensable gas separation and/or liquefaction, and compression/pumping), transport and storage.

Carbon dioxide captured from an energy-conversion process always contains impurities that impact the design and operation of CCS systems. The type and amount of the impurities contained into the CO_2 depend on the fuels used and the type of capture technology. A non-exhaustive list of these impurities is: CH_4, H_2S, N_2, O_2, CO, H_2, COS, Ar, SO_x, NO_x, NH_3, SO_2, amines and H_2O.

Because of the various natures of the impurities encountered in CCS processes and because these processes cover a large range of operating conditions (from atmospheric pressure to supercritical states), it is often necessary to fully guesstimate the phase equilibrium of mixtures CO_2 + impurities in both the sub-critical and critical regions by using an appropriate thermodynamic model (this is detailed in section 2.1). In order to meet this objective, Jaubert and Privat developed a group-contribution method (GCM) allowing estimation of the temperature-dependent binary interaction parameters $k_{ij}(T)$ for the widely used Peng-Robinson equation of state (EoS). A key point in this approach is that the k_{ij} between two components i and j is a function of temperature (T) and of the pure component critical temperatures (T_{ci} and T_{cj}), critical pressures (P_{ci}, P_{cj}) and acentric factors (ω_i, ω_j). This means that no additional properties besides those required by the EoS itself (T_c, P_c, ω) are needed. The proposed model, named PPR78 for Predictive-Peng-Robinson 1978, can predict any phase equilibrium of mixtures containing CO_2 and light alkanes, N_2, H_2, H_2S, H_2O.

In this chapter, the ability of the PPR78 model for prediction of CO_2 + impurity binary mixtures is discussed and analyzed in terms of phase-equilibrium properties and enthalpies of mixing. The PPR78 model is also applied to the prediction of multicomponent CO_2-containing mixtures.

In the first part, the theoretical foundations of the PPR78 equation of state are carefully described. In the second part, the capabilities of the PPR78 model to predict both the global phase behavior (i.e., the phase behavior in both the sub and supercritical area) and the enthalpy data of various binary systems involved in CCS processes are graphically shown and quantitatively assessed.

It is often written in the literature that the greater the number of compounds in a mixture, the more a thermodynamic model may fail in representing its behavior. In that regard, the last part of the chapter deals with multicomponent systems (containing more than two pure species) encountered in CCS processes.

2. Presentation of the PPR78 thermodynamic model

2.1. On the need for an accurate thermodynamic model

Today, the synthesis design and optimization of CCS processes is carried out with the help of process simulators such as PRO/II, ProSim, HYSIS, ASPEN, UniSim, etc. It is however well known that the accuracy of the simulated results mainly depends on the quality of the chosen thermodynamic model. In most cases, the phase behavior of multicomponent systems, for which nearly no data are available, has to be known. The phase behavior can obviously be measured, but measurements are very time consuming (the VLE measurement of a 10-component system at atmospheric pressure in 10 mole% steps - a total of 92, 378 data points - would require 37 years!).

When dealing with fluids involved in CCS processes, many difficulties appear. Indeed, such mixtures contain a large number of various compounds, the proper representation of which involves accurately quantifying the interactions between each pair of molecules. This quantification obviously becomes increasingly difficult, if not impossible, as the number of molecules grows. To avoid such a time-consuming effort, an alternative solution lies in using a thermodynamic model able to estimate the interactions from mere knowledge of the structure of the molecules contained in the blend.

2.2. The Equation of State (EoS)

A few decades ago, Peng and Robinson [1] published their well-known equation of state, called in this chapter PR76. Some years later, the same authors published an improved version of this equation [2], which yields more accurate vapor pressure predictions for heavy hydrocarbons than those obtained by using PR76. This improved equation is called PR78 in this paper. For a pure component, the PR78 EoS is:

$$P(T,v)=\frac{RT}{v-b_i}-\frac{a_i(T)}{v(v+b_i)+b_i(v-b_i)} \tag{1}$$

with:

$$\begin{cases} R=8.314472\ J\cdot mol^{-1}\cdot K^{-1} \\ X=\dfrac{-1+\sqrt[3]{6\sqrt{2}+8}-\sqrt[3]{6\sqrt{2}-8}}{3}\approx 0.253076587 \\ b_i=\Omega_b\dfrac{RT_{c,i}}{P_{c,i}}\ \text{with}\ \Omega_b=\dfrac{X}{X+3}\approx 0.0777960739 \\ a_i(T)=\Omega_a\dfrac{R^2T_{c,i}^2}{P_{c,i}}\left[1+m_i\left(1-\sqrt{\dfrac{T}{T_{c,i}}}\right)\right]^2\ \textit{with}\ \Omega_a=\dfrac{8(5X+1)}{49-37X}\approx 0.457235529 \\ \text{if}\ \omega_i\le 0.491\quad m_i=0.37464+1.54226\omega_i-0.26992\omega_i^2 \\ \text{if}\ \omega_i>0.491\quad m_i=0.379642+1.48503\omega_i-0.164423\omega_i^2+0.016666\omega_i^3 \end{cases} \tag{2}$$

where P is the pressure, R the gas constant, T the temperature, a and b are EoS parameters, v the molar volume, T_c the critical temperature, P_c the critical pressure and ω the acentric factor. In this chapter, the PR78 EoS is used. To apply such an EoS for mixtures, *mixing rules* are necessary to calculate the values of a and b of the mixtures.

A widely employed way to extend the cubic EoS to mixtures, the mole fractions of which are x_i, is via the so-called Van der Waals one-fluid mixing rules [quadratic composition dependency for both parameters] and the classic combining rules, i.e., the geometric mean rule for the cross-energy and the arithmetic mean rule for the cross co-volume parameter:

$$\begin{cases} a=\sum_{i=1}^{N}\sum_{j=1}^{N}x_ix_ja_{ij} \\ b=\sum_{i=1}^{N}\sum_{j=1}^{N}x_ix_jb_{ij} \\ a_{ij}=\sqrt{a_ia_j}(1-k_{ij}) \\ b_{ij}=\frac{1}{2}(b_i+b_j)(1-l_{ij}) \end{cases} \tag{3}$$

where a_i and b_i denote the EoS parameters for pure component i, T is the temperature, and N is the number of components in the mixture.

Through this, two new parameters, the so-called *binary interaction parameters* (k_{ij} and l_{ij}), appear in the combining rules. One of them, namely k_{ij}, is by far the most important one. Indeed, a non-null l_{ij} is only necessary for complex polar systems and special cases. This is the reason why, when dealing with petroleum fluids, phase equilibrium calculations are generally performed with $l_{ij}=0$ and the mixing rule for the co-volume parameter simplifies to:

$$b=\sum_{i=1}^{p} x_i b_i \tag{4}$$

Note that the chosen mixing rules are used by most petroleum companies and above all are available in any computational package.

In our approach and in order to define a predictive model, the binary interaction parameters appearing in the mixing rules are calculated by the group-contribution method (GCM), which means that a k_{ij} calculation requires each molecule of a given mixture to be divided into elementary groups of atoms.

$$k_{ij}(T)=\frac{-\frac{1}{2}\left[\sum_{k=1}^{N_g}\sum_{l=1}^{N_g}(\alpha_{ik}-\alpha_{jk})(\alpha_{il}-\alpha_{jl})A_{kl}\cdot\left(\frac{298.15}{T/K}\right)^{\left(\frac{B_{kl}}{A_{kl}}-1\right)}\right]-\left(\frac{\sqrt{a_i(T)}}{b_i}-\frac{\sqrt{a_j(T)}}{b_j}\right)^2}{2\frac{\sqrt{a_i(T)\cdot a_j(T)}}{b_i\cdot b_j}} \tag{5}$$

In Equation (5), N_g is the number of different groups defined by the method. In the current state, the PPR78 model can cover a large spectrum of petroleum fluids, from natural gases to crude oils [3-18]. 20 groups exist and make it possible to accurately predict the phase behavior of mixtures containing alkanes, alkenes, aromatic compounds, cycloalkanes, permanent gases (CO_2, H_2, N_2, H_2S), water and mercaptans.

In this setup, α_{ik} is the fraction of molecule i occupied by group k (occurrence of group k in molecule i divided by the total number of groups present in molecule i); and $A_{kl}=A_{lk}$ and $B_{kl}=B_{lk}$ (where k and l are two different groups) are constant parameters determined in our previous studies ($A_{kk}=B_{kk}=0$).

In order to deal with mixtures containing CO_2, light alkanes, N_2, H_2, H_2S and H_2O, a set of N_g = 7 groups is necessary:

1. the methane group (CH_4)
2. the ethane group (C_2H_6)

3. the carbon dioxide group (CO_2)

4. the nitrogen group (N_2)

5. the hydrogen sulfide group (H_2S)

6. the hydrogen group (H_2)

7. the water group (H_2O)

Since the carbon dioxide and the impurities are all individually considered as single groups, Equation (5) can be simplified as follows:

$$k_{12}(T) = \frac{A_{12} \cdot \left(\frac{298.15}{T/K}\right)^{\left(\frac{B_{12}}{A_{12}}-1\right)} - \left(\frac{\sqrt{a_1(T)}}{b_1} - \frac{\sqrt{a_2(T)}}{b_2}\right)^2}{2\frac{\sqrt{a_1(T)\cdot a_2(T)}}{b_1 \cdot b_2}} \tag{6}$$

where k_{12} is the binary interaction parameter between molecules 1 and 2.

For these 7 groups, a total of 42 parameters (expressed in MPa) were determined. They are summarized in Table 1 [3, 6-9, 13, 15].

	CH_4	C_2H_6	CO_2	N_2	H_2S	H_2O	H_2
CH_4	0	-	-	-	-	-	-
C_2H_6	A=13.04 B=6.863	0	-	-	-	-	-
CO_2	A=137.3 B=194.2	A=135.5 B=239.5	0	-	-	-	-
N_2	A=37.90 B=37.20	A=61.59 B=84.92	A=98.42 B=221.4	0	-	-	-
H_2S	A=181.2 B=288.9	A=157.2 B=217.1	A=134.9 B=201.4	A=319.5 B=550.1	0	-	-
H_2O	A=2265 B=4722	A=2333 B=5147	A=559.3 B=277.9	A=2574 B=5490	A=603.9 B=599.1	0	-
H_2	A=156.1 B=92.99	A=137.6 B=150.0	A=265.9 B=268.3	A=65.20 B=70.10	A=145.8 B=823.5	A=830.8 B=−137.9	0

Table 1. Group interaction parameters of the PPR78 model: $(A_{kl}=A_{lk})$ / MPa and $(B_{kl}=B_{lk})$ / MPa.

Note that this formulation is very useful for someone having commercial software working with the PR EoS. As shown in one of our previous studies [4], working with hydrocarbon binary mixtures, Equations 4 and 5 are able to predict the different k_{ij} trends commonly encountered. Indeed, k_{ij} may increase, decrease or pass through a minimum with respect to temperature and chain length.

2.3. On the importance of enthalpy-of-mixing (h^M) data

Engineers use principles drawn from thermodynamics to analyze and design industrial processes. The application of the first principle (also named energy rate balance) to an open multi-component system at steady state is written as:

$$\dot{W} + \dot{Q} = \sum \dot{n}_{\text{out}} h_{\text{out}} - \sum \dot{n}_{\text{in}} h_{\text{in}} \tag{7}$$

where $\dot{W}$ and $\dot{Q}$ are the net rates of energy transfer by work and by heat, respectively; $\dot{n}$ is the molar flowrate and h denotes the molar enthalpy of a stream. Subscripts *in* and *out* mean inlet and outlet streams, respectively. Note that kinetic-energy and potential-energy terms were not considered in Equation (7). According to classic thermodynamics, the molar enthalpy of a p-component homogeneous system at a given temperature T, pressure P and composition z (mole fraction vector) is:

$$\underbrace{h(T, P, z)}_{\substack{\text{molar enthalpy}\\ \text{of an inlet or}\\ \text{an outlet stream}}} = \sum_{i=1}^{p} z_i h_{pure\ i}(T, P) + h^M(T, P, z) \tag{8}$$

where $h_{pure\ i}(T, P)$ is the molar enthalpy of pure component i at the same temperature and pressure as the mixture, estimated from the PR78 EoS (i.e., the equation of state applied to pure components). $h^M(T, P, z)$ is the molar enthalpy change on isothermal and isobaric mixing that will be estimated from the PPR78 model (i.e., the equation of state for mixtures).

From Equations (7) and (8), it thus clearly appears that good estimations of h^M data are essential to perform reliable energy balance for CCS processes.

3. Predicting the phase equilibrium behavior of systems containing CO_2 + impurities

The results obtained by using the PPR78 EoS to reproduce the phase behavior of such systems are graphically shown in this section. In order to numerically evaluate the performances of the model, absolute and relative deviations between experimental data points and their prediction are systematically provided. We now define the different quantities we will refer to hereafter:

- the absolute mean deviation on the liquid phase composition:

http://dx.doi.org/10.5772/57058

$$\overline{\Delta x_1} = \frac{\sum_{i=1}^{n_{bubble}} \left(\left| x_{1,\exp} - x_{1,cal} \right| \right)_i}{n_{bubble}} \tag{9}$$

- the relative mean deviation on the liquid phase composition:

$$\overline{\Delta x_1 \%} = \frac{100}{n_{bubble}} \sum_{i=1}^{n_{bubble}} 0.5 \left(\frac{|\Delta x|}{x_{1,\exp}} + \frac{|\Delta x|}{x_{2,\exp}} \right)_i \tag{10}$$

with

$$|\Delta x| = \left| x_{1,\exp} - x_{1,cal} \right| = \left| x_{2,\exp} - x_{2,cal} \right| \tag{11}$$

- the absolute mean deviation on the gas phase composition:

$$\overline{\Delta y_1} = \frac{\sum_{i=1}^{n_{dew}} \left(\left| y_{1,\exp} - y_{1,cal} \right| \right)_i}{n_{dew}} \tag{12}$$

- the relative mean deviation on the gas phase composition:

$$\overline{\Delta y_1 \%} = \frac{100}{n_{dew}} \sum_{i=1}^{n_{dew}} 0.5 \left(\frac{|\Delta y|}{y_{1,\exp}} + \frac{|\Delta y|}{y_{2,\exp}} \right)_i \tag{13}$$

with

$$|\Delta y| = \left| y_{1,\exp} - y_{1,cal} \right| = \left| y_{2,\exp} - y_{2,cal} \right| \tag{14}$$

n_{bubble} and n_{dew} are the number of bubble points and dew points, while x_1 is the mole fraction in the liquid phase of the most volatile component and x_2 is the mole fraction of the heaviest component. Similarly, y_1 is the mole fraction in the gas phase of the most volatile component and y_2 is the mole fraction of the heaviest component.

3.1. Phase diagrams of CO_2 + light alkane systems

Mixtures of CO_2 + methane have been measured extensively and a huge amount of reliable experimental phase equilibrium and critical data are available in the open literature. To our best knowledge, 424 bubble points, 418 dew points and 17 critical points have been published in the literature. An comprehensive list of references can be found in [6].

Figure 1a shows the isothermal phase diagrams for the system methane(1) + CO_2(2) at six different temperatures. For this system, from low to high temperature the binary interaction parameter varies from 0.093 to 0.112.

Figure 1b is a projection of the **critical locus** (i.e., the locus of the vapor-liquid mixture's critical points) of this binary system on the (P_c, T_c) plane. Accurate results are obtained at any temperature although the critical pressure is slightly overestimated at intermediate temperatures.

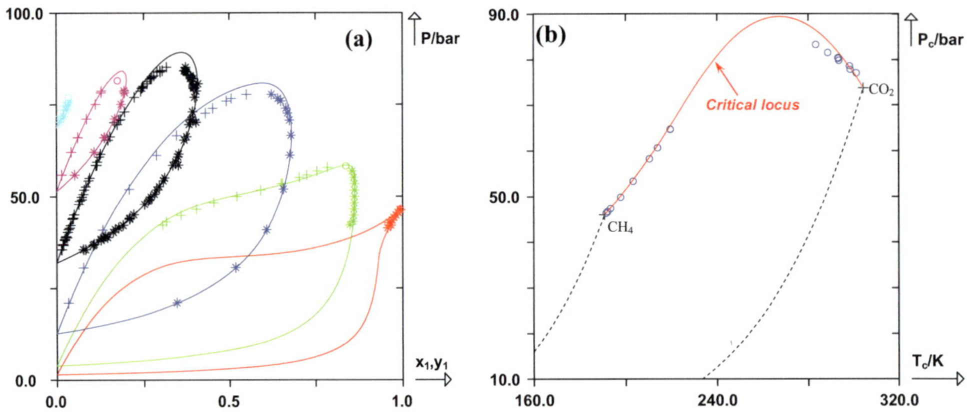

Figure 1. Prediction of isothermal dew and bubble curves and prediction of the critical locus for the binary system (methane + CO_2) using the PPR78 model. +: experimental bubble points, ∗: experimental dew points, ○: experimental critical points. Solid line: predicted curves with the PPR78 model. Dashed line: vaporization curves of pure compounds. 1a: system methane(1)/CO_2(2) at 6 different temperatures: T_1 = 191.20 K (k_{ij} = 0.0931), T_2 = 210.15 K (k_{ij} = 0.0956), T_3 = 240.00 K (k_{ij} = 0.100), T_4 = 270.00 K (k_{ij} = 0.106), T_5 = 288.50 K (k_{ij} = 0.109), T_6 = 301.00 K (k_{ij} = 0.112). 1b: critical locus of the methane/CO_2 system.

Mixtures of CO_2 + ethane have also been measured extensively. The open literature even offers a few more data than for the CO_2 + methane system: 483 bubble points, 438 dew points and 22 critical points. For more details about the references, see reference [6].

Figure 2a, 2b and 2c show the isothermal phase diagrams for the system CO_2(1) + ethane(2) at twelve different temperatures. For this system, a homogeneous positive azeotrope always exists. The PPR78 model is able to perfectly predict the phase behavior of this system including the temperature minimum on the critical locus (see Figure 2d).

The deviations observed in both these systems (i.e., CO_2 + methane and CO_2 + ethane) are:

http://dx.doi.org/10.5772/57058

	$\overline{\Delta x_1}$(absolute)	$\overline{\Delta x_1}$%(relative)	$\overline{\Delta y_1}$(absolute)	$\overline{\Delta y_1}$%(relative)	n_{bubble}	n_{dew}
$CH_4(1) + CO_2(2)$	0.012	4.87	0.008	3.13	424	418
$C_2H_6(1) + CO_2(2)$	0.020	6.58	0.015	4.16	483	438

Table 2. Absolute and relative mean deviations observed in the CO_2 + methane and CO_2 + ethane systems.

3.2. Phase diagrams of $CO_2 + N_2$ system

Mixtures of nitrogen + carbon dioxide have been measured extensively and there is a vast amount of reliable experimental phase equilibrium and critical data (no fewer than 635 VLE data points were found for this system, including the critical points). An accurate list of references can be found in references [7, 8]. This system is so asymmetrical in terms of size and interactions that its critical locus is no longer continuous (see Figure 3e). Such behavior is called Type III according to the classification of Van Konynenburg and Scott [16]. It can be seen that the PPR78 model is able to predict the experimental data with relatively good accuracy. A few examples are shown Figure 3.

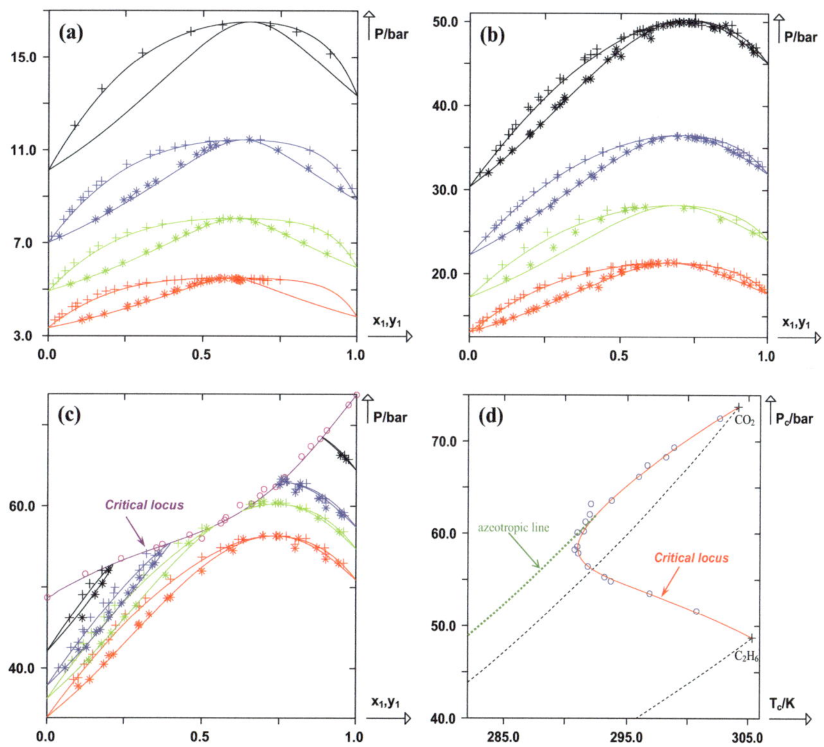

Figure 2. See caption of Figure 1. Dotted line: locus of azeotropic points. 2a: system $CO_2(1)$/ethane(2) at 4 different temperatures: $T_1 = 210.00$ K ($k_{ij} = 0.133$), $T_2 = 220.00$ K ($k_{ij} = 0.131$), $T_3 = 230.00$ K ($k_{ij} = 0.130$), $T_4 = 241.45$ K $k_{ij} = 0.129$). 2b: system $CO_2(1)$/ethane(2) at 4 different temperatures: $T_1 = 250.00$ K ($k_{ij} = 0.128$), $T_2 = 260.00$ K ($k_{ij} = 0.127$), $T_3 = 270.00$ K ($k_{ij} = 0.127$), $T_4 = 283.15$ K ($k_{ij} = 0.126$). 2c: system $CO_2(1)$/ethane(2) at 4 different temperatures: $T_1 = 288.15$ K ($k_{ij} = 0.126$), $T_2 = 291.15$ K ($k_{ij} = 0.126$), $T_3 = 293.15$ K ($k_{ij} = 0.126$), $T_4 = 298.15$ K ($k_{ij} = 0.126$). 2d: critical locus of the CO_2/ethane system.

Note that in the vicinity of the critical temperature of pure carbon dioxide, the critical locus is perfectly predicted. Similarly to what was observed in the CO_2 + methane system, the overestimation of the mixture's critical pressure by the PPR78 model increases when the temperature decreases. In the present case, the binary interaction parameter k_{ij} is a decreasing function of the temperature (see Figure 3f). k_{ij} is positive at temperatures lower than 208 K, and negative at temperatures higher than 208 K.

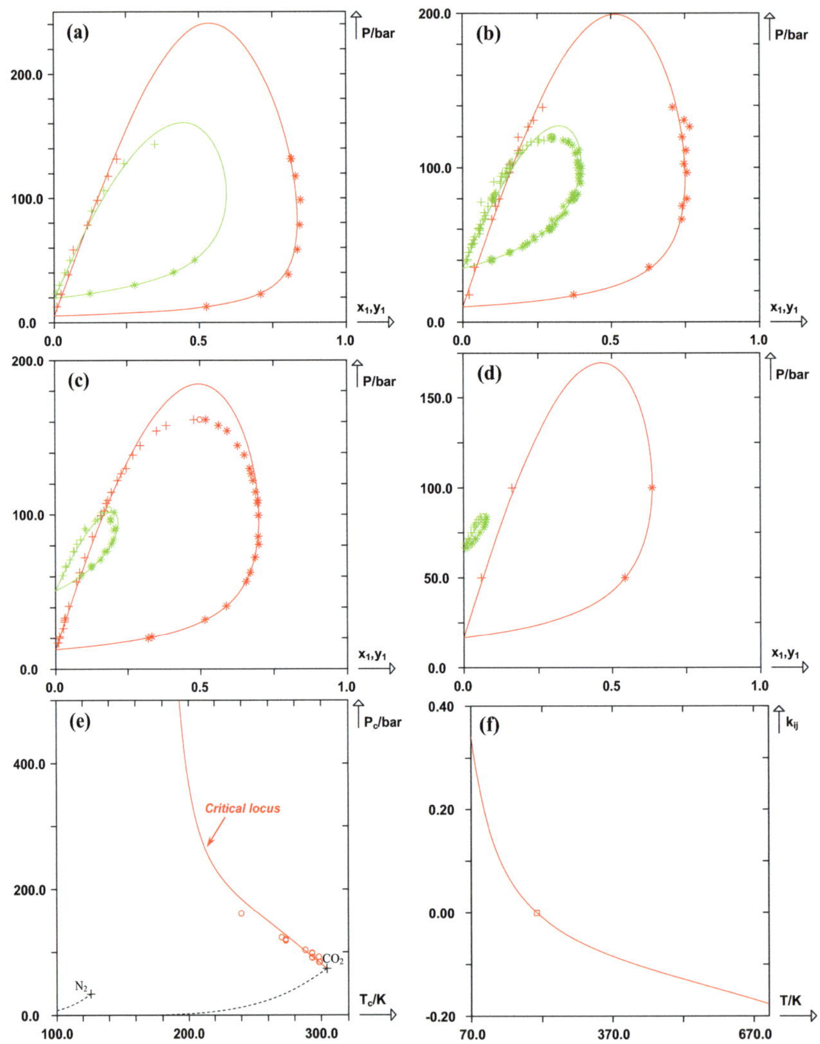

Figure 3. See caption of Figure 1. 8a: System $N_2(1) + CO_2(2)$ at two different temperatures: T_1 = 218.15 K (k_{ij} = -0.00856) and T_2 = 253.15 K (k_{ij} = -0.0325). 8b: System $N_2(1) + CO_2(2)$ at two different temperatures: T_1 = 232.85 K (k_{ij} = -0.0194) and T_2 = 273.20 K (k_{ij} = -0.0437). 8c: System $N_2(1) + CO_2(2)$ at two different temperatures: T_1 = 240.00 K (k_{ij} = -0.0242) and T_2 = 288.20 K (k_{ij} = -0.0514). 8d: System $N_2(1) + CO_2(2)$ at two different temperatures: T_1 = 248.15 K (k_{ij} = -0.0294) and T_2 = 298.80 K (k_{ij} = -0.0564). 8e: critical locus. 8f: k_{ij} vs. temperature (the square indicates the temperature at which k_{ij} = 0).

http://dx.doi.org/10.5772/57058

The deviations observed in the $N_2 + CO_2$ system are:

	$\overline{\Delta x_1}$ (absolute)	$\overline{\Delta x_1}\%$ (relative)	$\overline{\Delta y_1}$ (absolute)	$\overline{\Delta y_1}\%$ (relative)	n_{bubble}	n_{dew}
$N_2(1) + CO_2(2)$	0.008	6.03	0.008	2.32	320	301

Table 3. Absolute and relative mean deviations observed in the $N_2 + CO_2$ system.

3.3. Phase diagrams of $CO_2 + H_2S$ system

For this system, 177 bubble points, 176 dew points and 10 critical points were found in the open literature. These references are reported in [9]. Eight predicted isobaric dew and bubble curves are shown in Figure 4a and 4b. As can be seen, the PPR78 model is able to predict the behavior of this system with relatively good accuracy. However, as shown by the critical curve displayed in Figure 4c and in accordance with what we previously observed, the PPR78 model tends to overestimate the critical pressures although the absolute deviations remain quite small (< 5 bar).

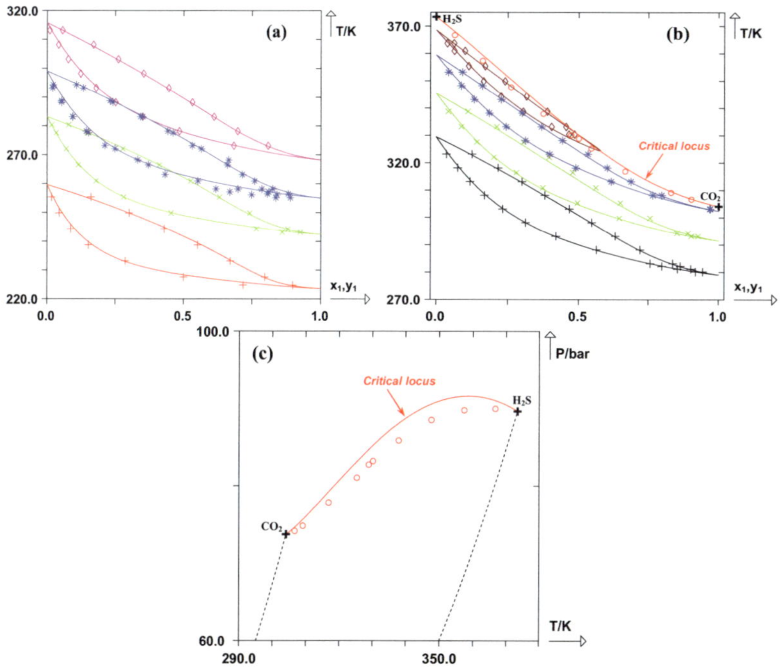

Figure 4. See caption of Figure 1. 4a: System $CO_2(1) + H_2S(2)$ at four different pressures: $P_1 = 6.90$ bar, $P_2 = 13.79$ bar, $P_3 = 20.68$ bar, $P_4 = 30.40$ bar. Experimental dew and bubble points at $P_1 = 6.90$ bar (+), $P_2 = 13.79$ bar (X), $P_3 = 20.68$ bar (∗), $P_4 = 30.40$ bar (◊). 4b: System $CO_2(1) + H_2S(2)$ at four different pressures: $P_1 = 40.53$ bar, $P_2 = 55.16$ bar, $P_3 = 70.93$ bar, $P_4 = 82.74$ bar. Experimental dew and bubble points at $P_1 = 40.53$ bar (+), $P_2 = 55.16$ bar (X), $P_3 = 70.93$ bar (∗), $P_4 = 82.74$ bar (◊). 4c: critical locus of the $CO_2(1) + H_2S(2)$ system.

The deviations observed in this system are:

	$\overline{\Delta x_1}$ (absolute)	$\overline{\Delta x_1}\%$ (relative)	$\overline{\Delta y_1}$ (absolute)	$\overline{\Delta y_1}\%$ (relative)	n_{bubble}	n_{dew}
$CO_2(1) + H_2S(2)$	0.009	5.13	0.008	3.06	177	176

Table 4. Absolute and relative mean deviations observed in the $CO_2 + H_2S$ system.

3.4. Phase diagrams of $CO_2 + H_2$ system

Mixtures of $H_2(1) + CO_2(2)$ present many reliable experimental data (302 bubble points, 300 dew points and 11 critical points). For further details, see reference [13]. As shown in Figure 5a, a type III phase behavior is observed for this system. In the vicinity of the critical point of the least volatile component (CO_2), the critical locus is perfectly predicted (for T > 250 K), as well as the corresponding isothermal P–xy phase diagrams (see Figures 5b and 5c). Once again, critical pressures are systematically overestimated, as highlighted by the isotherms in Figure 5c. Generally, liquid-liquid equilibrium (LLE) and vapor-liquid equilibrium (VLE) data associated with these two binary systems are fairly well predicted over wide ranges of temperature and pressure.

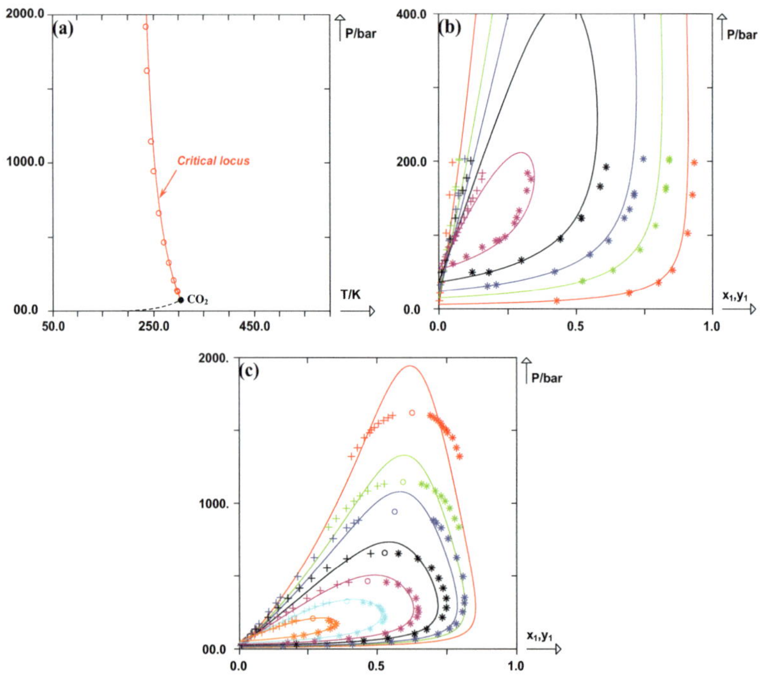

Figure 5. See caption of Figure 1. 5a: prediction of the critical locus of the hydrogen(1) + carbon dioxide(2) system. 5b: prediction of isothermal dew and bubble curves for the system hydrogen(1) + carbon dioxide(2) at five different temperatures: T_1 = 219.90 K (k_{ij} = –0.0211), T_2 = 244.90 K (k_{ij} = 0.0286), T_3 = 259.90 K (k_{ij} = 0.0577), T_4 = 274.90 K (k_{ij} = 0.0865), T_5 = 290.15 K (k_{ij} = 0.1153). 5c: Prediction of isothermal dew and bubble curves for the system hydrogen(1) + carbon dioxide(2) at seven different temperatures: T_1 = 237.00 K (k_{ij} = 0.0130), T_2 = 245.00 K (k_{ij} = 0.0288), T_3 = 250.00 K (k_{ij} = 0.0386), T_4 = 260.00 K (k_{ij} = 0.0579), T_5 = 270.00 K (k_{ij} = 0.0771), T_6 = 280.00 K (k_{ij} = 0.0962), T_7 = 290.00 K (k_{ij} = 0.1150).

http://dx.doi.org/10.5772/57058

The deviations observed in this system are:

	$\overline{\Delta x_1}$ (absolute)	$\overline{\Delta x_1}\%$ (relative)	$\overline{\Delta y_1}$ (absolute)	$\overline{\Delta y_1}\%$ (relative)	n_{bubble}	n_{dew}
$H_2(1) + CO_2(2)$	0.018	12.8	0.018	7.11	302	300

Table 5. Absolute and relative mean deviations observed in the $H_2 + CO_2$ system.

3.5. Phase diagrams of $CO_2 + H_2O$ system

3.5.1. On the difficulty to model aqueous systems

Binary aqueous systems containing hydrocarbons or permanent gases all exhibit vapor–liquid, liquid–liquid and liquid–liquid–vapor equilibria. Here we want to graphically illustrate the well–known dilemma that at a given temperature, the k_{ij} suitable to describe the solubility of CO_2 in water is different from the one required to correlate the solubility of water in CO_2. By way of example, the phase diagram of the CO_2 (1) + H_2O (2) system is plotted at T = 298.15 K in Figure 6.

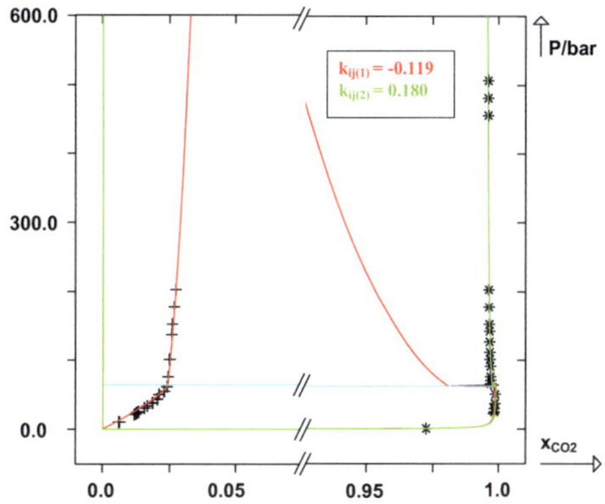

Figure 6. See caption of Figure 1. System $CO_2(1) + H_2O(2)$ at T = 298.15 K with two different k_{ij} values: $k_{ij(1)} = -0.119$ and $k_{ij(2)} = 0.180$.

At this temperature, both pure components are subcritical and a three–phase line (locus of liquid-liquid-vapor equilibrium) is found since the temperature of the three–phase critical endpoint, noted CEP (which is the terminating point of the three-phase line) is about 308 K. A negative k_{ij} value of –0.12 perfectly correlates the bubble curve, the dew curve and the solubility of CO_2 in water in the liquid–liquid region. On the other hand, such a k_{ij} value fails to correlate the solubility of water in liquid CO_2, which can only be captured with a positive k_{ij} value of +0.18. In order for the PPR78 model to find the best compromise between these two situations, we tried our best to maintain a similar number of data points to describe the solubility of water in the CO_2 and the solubility of CO_2 in H_2O respectively. As shown in Figure 7a, for such a system, the PPR78 model predicts $k_{12} = -0.098$.

3.5.2. What the PPR78 model can do for such complex systems

The behavior of the $CO_2 + H_2O$ system has been studied experimentally by different authors (as reported by Qian et al. [15]) in the high–temperature/high–pressure region with value of 539 K for the minimum temperature of the critical locus beginning at the water critical point. As shown in Figure 7 f, reliable predictions of the critical coordinates by the PPR78 EoS are obtained until 4000 bar. As previously discussed, at T = 298.15 K (see Figure 7a), a k_{ij} value of –0.098 has difficulties in capturing the solubility of water in CO_2. Taking into account the scatter of the experimental data, Figure 7b-d highlights that our model predicts the phase behavior of this system with reasonable accuracy.

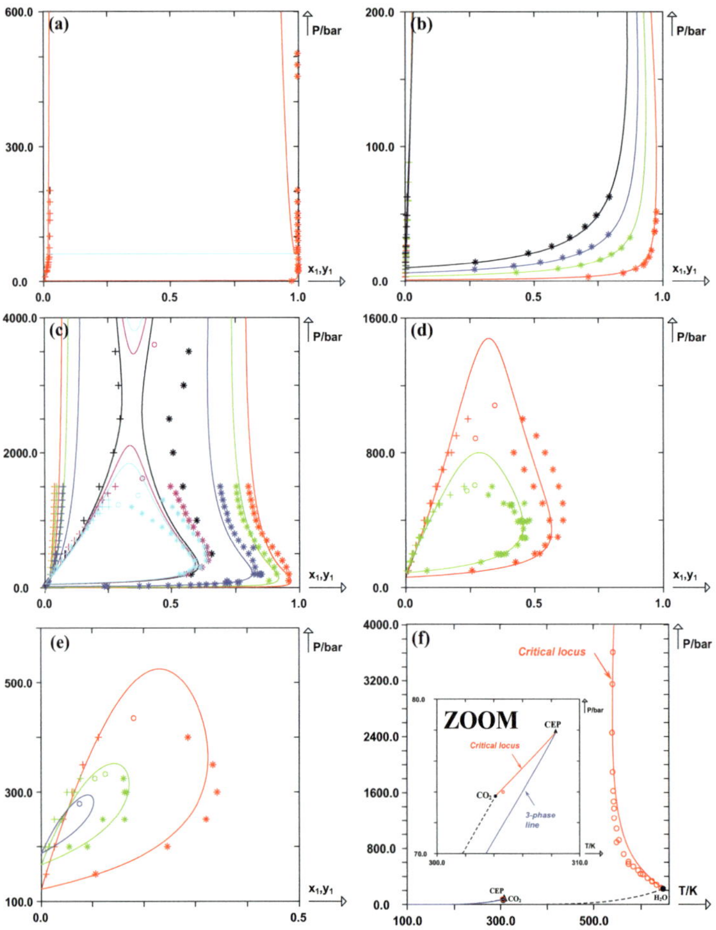

Figure 7. See caption of Figure 1. 7a: System $CO_2(1) + H_2O(2)$ at T = 298.15 K (k_{ij} = –0.098). 7b: System $CO_2(1) + H_2O(2)$ at four different temperatures: T_1 = 373.15 K (k_{ij} = –0.059), T_2 = 413.15 K (k_{ij} = –0.036), T_3 = 433.15 K (k_{ij} = –0.023), T_4 = 453.15 K (k_{ij} = –0.010). 7c: System $CO_2(1) + H_2O(2)$ at six different temperatures: T_1 = 383.15 K (k_{ij} = –0.054), T_2 = 423.15 K (k_{ij} = –0.030), T_3 = 473.15 K (k_{ij} = 0.004), T_4 = 538.15 K (k_{ij} = 0.054), T_5 = 541.15 K (k_{ij} = 0.056), T_6 = 543.15 K (k_{ij} = 0.058). 7d: System $CO_2(1) + H_2O(2)$ at two different temperatures: T_1 = 548.15 K (k_{ij} = 0.062), T_2 = 573.15 K (k_{ij} = 0.084). 7e: System $CO_2(1) + H_2O(2)$ at three different temperatures: T_1 = 598.15 K (k_{ij} = 0.108), T_2 = 623.15 K (k_{ij} = 0.133), T_3 = 633.15 K (k_{ij} = 0.144). 7f: Critical locus of the $CO_2(1) + H_2O(2)$ system.

The deviations observed in this system are:

	$\overline{\Delta x_1}$ (absolute)	$\overline{\Delta x_1}\%$ (relative)	$\overline{\Delta y_1}$ (absolute)	$\overline{\Delta y_1}\%$ (relative)	n_{bubble}	n_{dew}
$CO_2(1) + H_2O(2)$	0.006	12.9	0.027	15.8	1068	543

Table 6. Absolute and relative mean deviations observed in the $CO_2 + H_2O$ system.

4. Predicting enthalpies of mixing of CO_2 + impurities systems with the PPR78 model

When molecules are few-polar and few-associating (e.g., in mixtures of alkanes), pure-component terms provide an excellent estimation of the molar enthalpy of the mixture. Therefore, the enthalpy-of-mixing terms can be seen as a correction, just aimed at improving the first estimation given by the pure-component ground terms. In other words, with few-polar and few-associating mixtures, h^M terms are generally nearly negligible with respect to pure-component terms in the energy rate balance. Typically, h^M terms are very small in alkane mixtures (several J/mol or several tens of J/mol) but are not negligible in petroleum mixtures containing CO_2, H_2O, alcohols, etc. (likely to reach up to several tens of kJ/mol).

When the parameters involved in the k_{ij} correlations are not directly fitted on enthalpy-of-mixing data (this is, for instance, the case for the PPR78 model), the relative deviations between the calculated and experimental h^M data ($\overline{\Delta h^M \%}$) can be very important and reach values sometimes greater than 100%.

However, since h^M quantities are only used to evaluate the molar enthalpies, h_{in} and h_{out}, involved in the energy rate balance (see Equation (7)), the relative h^M deviations do not necessarily matter: only their impact on the accuracy of the energy balance should be actually addressed.

In order to assess how errors in h^M may affect the energy balance, we decided to adopt a criterion converting the difference $\Delta h^M = \left| h^M_{calc} - h^M_{\exp} \right|$ in terms of temperature effect ΔT_h through:

$$\begin{cases} \text{For one datapoint: } \Delta T_h = \dfrac{\Delta h^M}{c_P} \\ \text{For a series of data: } \overline{\Delta T_h} = \dfrac{\sum_{i=1}^{n_h} \left(\Delta T_h\right)_i}{n_h} \end{cases} \tag{15}$$

where c_P is the molar isobaric heat capacity of the mixture, and z_i, the mole fraction of component i.

For all the data, it was possible to find for the considered systems (references are reported in reference [17]), that the deviations are:

	$\Delta\overline{T}_h$ (K)	Number of data $n_{hM\ data}$
CO_2 + methane	0.63	636
CO_2 + ethane	1.35	408
CO_2 + N_2	1.01	693
CO_2 + H_2O	6.88	539

Table 7. Prediction of h^M data with the PPR78 model for CO_2-containing mixtures

Satisfactory results are thus obtained for the first three systems. This leads us to claim that accurate energy balances can be performed with the PPR78 model in processes involving mixtures of carbon dioxide, nitrogen and light alkanes as well.

The CO_2 + H_2O system shows a much more important deviation of around 7 K, which may certainly introduce inaccuracy in an energy balance. However, in mixtures mainly containing CO_2 and small proportions of water, errors in the estimation of h^M are much more reasonable. The performances of the PPR78 model in terms of h^M prediction are illustrated in Figure 8.

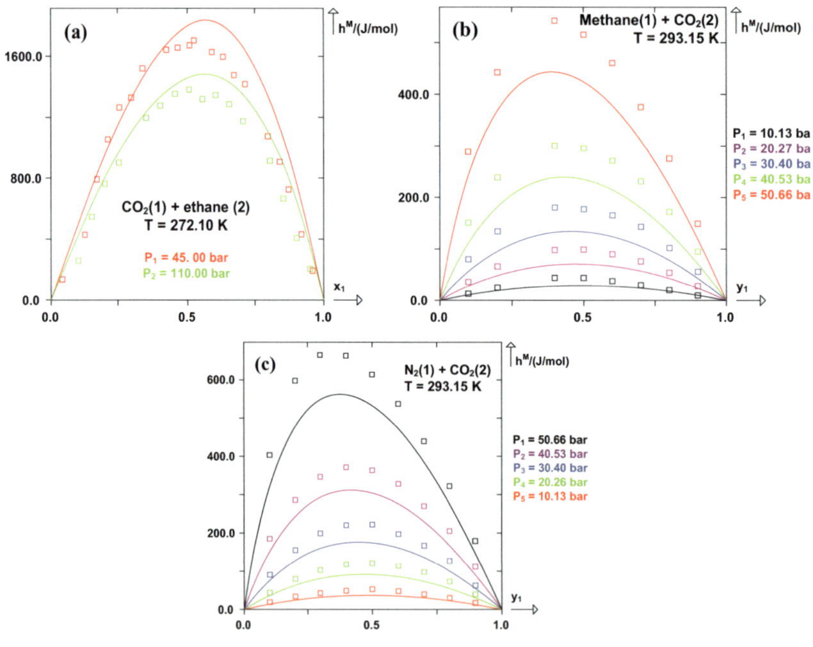

Figure 8. Prediction by the PPR78 EoS of mixing enthalpies of three CO_2-containing binary systems. Continuous curves: calculated with the PPR78 model. (□) experimental data. (a) h^M-curve of a liquid binary system. (b) and (c) h^M-curves of two gaseous binary systems.

http://dx.doi.org/10.5772/57058

5. Predicting the phase behavior of CO_2–containing multicomponent systems

Many phase-equilibrium data have been published on multicomponent CO_2-containing mixtures. These mixtures are generally synthetic natural gases containing not only the few molecules studied in this chapter but also light alkanes (generally from methane to hexane). The capability of the PPR78 model is now illustrated in such mixtures. All the references of the experimental datapoints can be found in [12].

a. Yarborough et al.'s fluids.

In 1970, working with 45 different natural gases containing between 2 and 7 components (N_2, CO_2, methane, ethane, propane, n-butane, n-pentane), Yarborough et al. measured 52 bubble and dew-point pressures (the compositions of Yarborough et al.'s fluids are given in [12]). The PPR78 model is able to predict these 52 pressures with an absolute average deviation of 2.0 bar (i.e., 3.2%) which is certainly close to the experimental uncertainty.

b. Jarne et al.'s fluids.

In 2004, Jarne et al. measured 110 upper and lower dew-point pressures for two natural gases containing nitrogen, carbon dioxide and alkanes up to n-C_6. The composition of the fluids and the accuracy of the PPR78 model can be seen in Figure 9. The average deviation of these 110 pressures is only 2.0 bar. This is an extremely good result because many data points are located in the vicinity of the cricondentherm, where the slope of the dew curve is very steep.

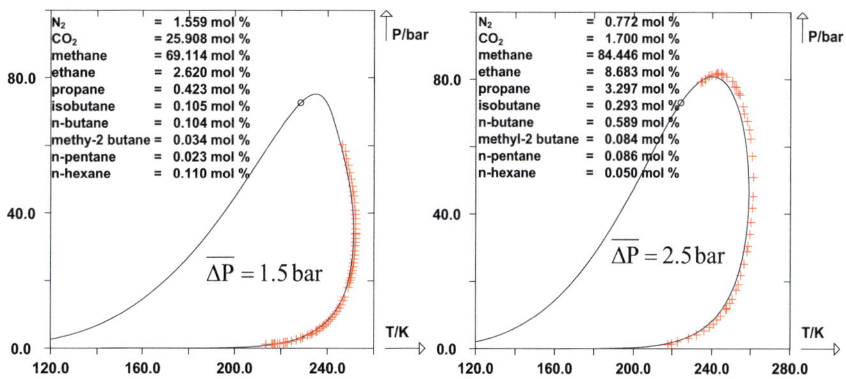

Figure 9. Solid line: (P, T) phase envelopes of Jarne et al.'s natural gases predicted with the PPR78 model. +: experimental upper and lower dew-point pressures. o: predicted critical point.

c. Zhou et al.'s fluid.

In 2006, Zhou et al. measured 6 dew point-pressures for a natural gas containing N_2, CO_2 and 7 alkanes. Figure 10 shows that with an average deviation lower than 1.2 bar (i.e., 1.5%), the PPR78 model is able to accurately predict these data.

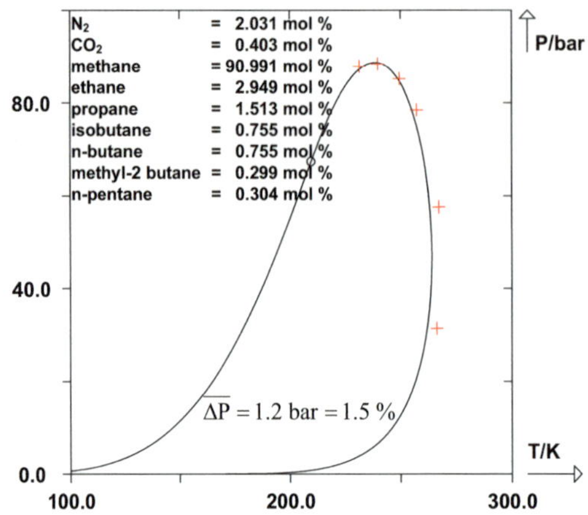

Figure 10. Solid line: (P, T) phase envelopes of Zhou et al.'s natural gas predicted with the PPR78 model. +: experimental upper and lower dew-point pressures. o: predicted critical point.

6. Conclusion

This chapter has demonstrated that the PPR78 model is capable of predicting the phase equilibrium behavior of mixtures containing carbon dioxide, light alkanes, nitrogen, hydrogen sulfide, hydrogen and water with good accuracy. This model can also be used to perform energy balance calculations and can be successfully extended to multicomponent mixtures.

Although this chapter shows phase-equilibrium predictions obtained with the PPR78 EoS over the whole composition range (i.e., from CO_2-rich to CO_2-poor mixtures), CCS processes essentially involve mixtures containing high proportion of CO_2 and small fractions of impurities.

Because PPR78's authors did not want to limit the range of applicability of their model, the A and B parameters involved in the k_{ij}-estimation formula (see Equation (5)) were fitted on all the data we could find in the open literature. Nevertheless, as shown in this chapter, good to excellent predictions can still be obtained for mixtures containing large amounts of CO_2.

We believe that the PPR78 model can be safely used to model CCS processes. Whereas interactions between CO_2 and impurities were thoroughly described in this chapter, interactions between impurities themselves were not (and are still present in multicomponent mixtures). A proof that a similar quality of predictions can actually be obtained is given in some of the papers mentioned in the bibliographic section [3, 6, 7-9, 13, 15, 17].

Finally, as a limitation on the use of the PPR78 model, some molecules such as NH_3, SO_2 or NOx cannot be presently described by the model. The PPR78 model is currently still under development by its authors and the missing groups should be added in the near future.

Author details

Romain Privat and Jean-Noël Jaubert

Ecole Nationale Supérieure des Industries Chimiques, Université de Lorraine, LRGP (Laboratoire Réactions et Génie des Procédés), Nancy, France

References

[1] Peng, D.Y. & Robinson, D.B. A New Two-Constant Equation of State. Ind. Eng. Chem. Fundam. 1976;15 59-64

[2] Peng, D.Y. & Robinson, D.B. The characterization of the heptanes and heavier fractions for the GPA Peng-Robinson programs, Gas processors association, Research report RR-28 1978.

[3] Jaubert, J.N. & Mutelet, F. VLE predictions with the Peng-Robinson equation of state and temperature dependent k_{ij} calculated through a group contribution method. Fluid Phase Equilib. 2004;224(2) 285-304

[4] Jaubert, J.N.; Vitu, S.; Mutelet, F. & Corriou, J.P. Extension of the PPR78 model to systems containing aromatic compounds. Fluid Phase Equilib. 2005;237(1-2) 193-211

[5] Vitu, S.; Jaubert, J.N. & Mutelet, F. Extension of the PPR78 model (Predictive 1978, Peng Robinson EoS with temperature dependent k_{ij} calculated through a group contribution method) to systems containing naphtenic compounds. Fluid Phase Equilib. 2006;243 9-28

[6] Vitu, S.; Privat, R.; Jaubert, J.N. & Mutelet, F. Predicting the phase equilibria of CO_2 + hydrocarbon systems with the PPR78 model (PR EoS and k_{ij} calculated through a group contribution method). J. Supercrit. Fluids. 2008;45(1) 1-26

[7] Privat, R.; Jaubert, J.N. & Mutelet, F. Addition of the Nitrogen group to the PPR78 model (Predictive 1978, Peng Robinson EoS with temperature dependent k_{ij} calculated through a group contribution method). Ind. Eng. Chem. Res. 2008;47(6) 2033-2048.

[8] Privat, R.; Jaubert, J.N. & Mutelet, F. Use of the PPR78 model to predict new equilibrium data of binary systems involving hydrocarbons and nitrogen. Comparison with other GCEOS. Ind. Eng. Chem. Res. 2008;47(19) 7483-7489

[9] Privat, R.; Mutelet, F. & Jaubert, J.N. Addition of the Hydrogen Sulfide group to the PPR78 model (Predictive 1978, Peng Robinson EoS with temperature dependent k_{ij} calculated through a group contribution method). Ind. Eng. Chem. Res. 2008;47(24) 10041-10052

[10] Privat, R.; Jaubert, J.N. & Mutelet, F. Addition of the sulfhydryl group (-SH) to the PPR78 model (Predictive 1978, Peng-Robinson EoS with temperature dependent k_{ij}

calculated through a group contribution method). J. Chem. Thermodyn. 2008;40(9) 1331-1341

[11] Privat, R. & Jaubert, J.N. Addition of the sulfhydryl group (-SH) to the PPR78 model : Estimation of missing group-interaction parameters for systems containing mercaptans and carbon dioxide or nitrogen or methane, from newly published data. Fluid Phase Equilibria. 2012;334 197-203

[12] Jaubert, J.N.; Privat, R. & Mutelet, F. Predicting the phase equilibria of synthetic petroleum fluids with the PPR78 approach. AIChE J. 2010;56(12) 3225-3235

[13] Qian, J.-W., Jaubert, J.-N., Privat, R. Phase equilibria in hydrogen-containing binary systems modeled with the Peng–Robinson equation of state and temperature-dependent binary interaction parameters calculated through a group-contribution method. J. of Supercritical Fluids 2013;75 58–71

[14] Qian, J.-W., Jaubert, J.-N., Privat, R. Prediction of the phase behavior of alkene-containing binary systems with the PPR78 model. Fluid Phase Equilibria 2013;54 12–235

[15] Qian, J.-W., Privat, R., Jaubert, J.-N. Predicting the phase equilibria, critical phenomena and mixing enthalpies of binary aqueous systems containing alkanes, cycloalkanes, aromatics, alkenes and gases (N2, CO2, H2S, H2) with the PPR78 equation of state. Ind. Eng. Chem. Res. 2013; 52 16457–16490.

[16] Privat, R. & Jaubert, J.N. Classification of global fluid-phase equilibrium behaviors in binary systems. Chemical Engineering Research and Design 2013; 91 1807-1839

[17] Qian, J.-W., Privat, R., Jaubert, J.-N, Duchet-Suchaux, P. Prediction of enthalpy and heat capacity changes on mixing by means of the PPR78 (Predictive Peng-Robinson, 1978) cubic equation of state. Energy and Fuels 2013; 27 7150–7178.

[18] Jaubert, J.-N. & Privat, R. Relationship between the binary interaction parameters (kij) of the Peng-Robinson and those of the Soave-Redlich-Kwong equations of state : Application to the definition of the PR2SRK model. Fluid Phase Equilibria 2010; 295(1) 26–37.